W0037384

Macromolecular Sequences in Systematic and Evolutionary Biology

MONOGRAPHS IN EVOLUTIONARY BIOLOGY

Series Editors:

MACROMOLECULAR SEQUENCES IN SYSTEMATIC AND EVOLUTIONARY BIOLOGY
Edited by Morris Goodman

Macromolecular Sequences in Systematic and Evolutionary Biology

Edited by
MORRIS GOODMAN
Wayne State University
Detroit, Michigan

PLENUM PRESS • NEW YORK AND LONDON

Library of Congress Cataloging in Publication Data

Main entry under title:

Macromolecular sequences in systematic and evolutionary biology.

(Monographs in evolutionary biology)
Includes bibliographical references and index.
1. Chemical evolution. 2. Biology—Classification. 3. Macromolecules. I. Goodman, Morris, 1925– II. Series.
QH371.M27 1982 575 82-15138
ISBN 978-1-4684-4285-4 ISBN 978-1-4684-4283-0 (eBook)
DOI 10.1007/978-1-4684-4283-0

Softcover reprint of the hardcover 1st edition 1982
A Division of Plenum Publishing Corporation
233 Spring Street, New York, N.Y. 10013

Contributors

Jaap J. Beintema Biochemisch Laboratorium, Nijenborgh 16, 9747 AG Groningen, The Netherlands

Michael Coates School of Australian Environmental Studies, Griffith University, Nathan, Queensland, Australia 4111

John Czelusniak Departments of Biology and Anatomy, Wayne State University School of Medicine, Detroit, Michigan 48202

Margaret O. Dayhoff National Biomedical Research Foundation, Georgetown University Medical Center, Washington, D.C. 20007

Wilfried W. De Jong Department of Biochemistry, University of Nijmegen, Geert Grooteplein N21, 6525 EZ Nijmegen, The Netherlands

Howard Dene Department of Anatomy, Wayne State University School of Medicine, Detroit, Michigan 48201

Morris Goodman Department of Anatomy, Wayne State University School of Medicine, Detroit, Michigan 48201

David Hewett-Emmett Department of Human Genetics, University of Michigan Medical School, Ann Arbor, Michigan 48109

Richard Holmquist Space Sciences Laboratory, University of California at Berkeley, Berkeley, California 94720

Lois T. Hunt National Biomedical Research Foundation, Georgetown University Medical Center, Washington, D.C. 20007

Thomas H. Jukes Space Sciences Laboratory, University of California at Berkeley, Berkeley, California 94720

Johannes A. Lenstra Biochemisch Laboratorium, Nijenborgh 16, 9747 AG Groningen, The Netherlands

Michael J. Novacek Department of Zoology, San Diego State University, San Diego, California 92182

Dennis Pearl Department of Statistics, University of California at Berkeley, Berkeley, California 94720

Alejo E. Romero-Herrera Department of Anatomy, Wayne State University School of Medicine, Detroit, Michigan 48201

Alan F. Scott The Johns Hopkins University School of Medicine, and Howard Hughes Medical Institute, Laboratory for Human Biochemical Genetics, Baltimore, Maryland 21205

Kirby D. Smith The Johns Hopkins University School of Medicine, and Howard Hughes Medical Institute, Laboratory for Human Biochemical Genetics, Baltimore, Maryland 21205

Simon Stone Department of Botany, University of Adelaide, Adelaide, South Australia 5001

Richard E. Tashian Department of Human Genetics, University of Michigan Medical School, Ann Arbor, Michigan 48109

Patrick J. Venta Department of Human Genetics, University of Michigan Medical School, Ann Arbor, Michigan 48109

Preface

George H. F. Nuttall pioneered the study of phylogeny through the genetically encoded sequence structures of proteins. His classic monograph, *Blood Immunity and Blood Relationship*, was published in 1904. The findings described in this monograph testified that immunologic comparisons of serum proteins could help reveal the phyletic relationships of primates and other animals. Although Nuttall had no way of knowing that a correspondence between the nucleotide sequences of genes and the amino acid sequences of proteins was the genetic basis for the immunologic specificities of animal sera, he clearly saw the implications of his findings. Thus he wrote in the introduction of his monograph, "The persistence of the chemical blood-relationship between the various groups of animals serves to carry us back into geological times, and I believe we have but begun the work along these lines, and that it will lead to valuable results in the study of various problems of evolution." Nuttall's prophecy is being fulfilled. Through the first two-thirds of the 20th century immunology led the way in the molecular analysis of the phyletic relationships of animal taxa above the species level. Amino acid sequencing of proteins began in earnest during the 1960s. It overtook immunology during the 1970s and provided more exact molecular data for investigating the history of life and the forces of chance and selection which drive evolution. The dramatic development of recombinant DNA and nucleotide sequencing technologies are now bringing us to a new watershed in comparative molecular studies. A far more detailed reconstruction of the history of genes and of the phylogeny of organisms containing the genes than hitherto attempted is becoming possible. Genealogical reconstructions carried out by the parsimony method on considerable amino acid sequence data, as well as on the initial body of globin gene nucleotide sequence data, are already providing valuable results for systematic and evolutionary biology. These results are described in the present volume.

This volume is based on a symposium which explored how sequence data from proteins and DNA are used in studying phylogeny and evolution. This symposium was held at the Third International Congress of Systematic and Evolutionary Biology during the summer of 1980 in Vancouver, British Columbia. In addition to the chapters based on papers presented at this symposium, several further contributions are included, one to provide background information on placental mammalian evolution and two for wider coverage of new directions made possible by recombinant DNA and nucleotide sequencing technologies. The various orders of placental mammals in the Eutheria are presently better represented by amino acid sequence data than are any other major group of eukaryotes. Thus these data, which come from amino acid sequencing mainly of ribonucleases, lens α-crystallins, myoglobins and α- and β-hemoglobins, cytochromes *c*, fibrinopeptides A and B, and carbonic anhydrases, are being used to investigate eutherian systematics. In turn, evidence from the fossil record on times of divergence between eutherian lineages is testing the clock model of molecular evolution and provides data needed for investigating tempo and mode in molecular evolution. Our vision, of course, should not be limited to the Eutheria. Sequence data on certain proteins and genes can encompass the whole panorama of living species and thus provide the means for investigating the history of life from its earliest stages on earth to the present. That the study of phylogeny and evolution through protein and DNA sequences is very much in its infancy, despite a vigorous beginning, should readily be apparent to the readers of this volume, as it is to the authors. The best years lie ahead.

Morris Goodman
Wayne State University

Contents

Part I. Eutherian Phylogeny and Protein Evolution

Chapter 1
Information for Molecular Studies from Anatomical and Fossil Evidence on Higher Eutherian Phylogeny
Michael J. Novacek

1. Introduction 3
2. A Historical Sketch 5
3. Eutherian Monophyly 10
4. Higher Level Eutherian Phylogeny 12
   4.1. Edentata, Pholidota 16
   4.2. Carnivora, Creodonta 19
   4.3. Rodentia, Lagomorpha, Macroscelidea 20
   4.4. "Archontans" and Insectivora 21
   4.5. The Ungulate Radiation: Perissodactyla, Artiodactyla, Cetacea, Proboscidea, Tubulidentata, Hyracoidea, Sirenia 25
5. Divergence Times for Eutherian Orders 27
6. Eutherian Phylogeny, Fossil History, and Evolutionary Rates 30
   Appendix 34
   References 35

Chapter 2

Evolution of Mammalian Pancreatic Ribonucleases

Jaap J. Beintema and Johannes A. Lenstra

1. Introduction 43
2. Trees 44
   2.1. Most Parsimonious Tree 48
   2.2. Biologic Tree 50
3. Evolutionary Rates 52
4. Three-Dimensional Structure of Ribonuclease 56
5. Enzymic Activity on Low-Molecular-Weight Nucleotide Substrates and on RNA 58
6. Enzymic Activity on Double-Stranded RNA 61
7. Interaction with Inhibitor Proteins 64
8. Glycosylation and the Function of Carbohydrate 64
   References 71

Chapter 3

Eye Lens Proteins and Vertebrate Phylogeny

Wilfried W. De Jong

1. Introduction 75
2. Comparative Anatomy and Evolution of the Lens 75
3. Comparative Studies of Lens Protcins 78
   3.1. The Crystallins 78
   3.2. Intraspecies Variation in Lens Proteins 82
   3.3 Crystallin Variation between Species 82
4. Structural Analysis of α-Crystallin 84
   4.1. Isolation of α-Crystallin 84
   4.2. Electrophoresis of the α-Crystallin Chains 87
   4.3. Sequence Analysis of αA Chains 88
   4.4. Phylogenetic Reconstructions 91
5. Phylogenetic Inferences from α-Crystallin A Sequences 93
   5.1. Mammalian Phylogeny 93
   5.2. Relationships among Vertebrate Classes and Subclasses 104
6. Molecular Aspects of α-Crystallin Evolution 106
   6.1. Parallel and Back Substitutions 108
   6.2. Changes in Charge Are Avoided 108
   6.3. "Covarions" 109
   6.4. Variable Rates of Change and a Directional Trend in Substitutions 109
   6.5. Unequal Distribution of Substitutions over the αA Chain 110
   References 112

Chapter 4

Amino Acid Sequence Evidence on the Phylogeny of Primates and Other Eutherians

Morris Goodman, Alejo E. Romero-Herrera, Howard Dene, John Czelusniak, and Richard E. Tashian

1. Introduction 115
2. Genealogic Evidence from Amino Acid Sequences 116
2.1. Tree Construction Strategy 116
2.2. Sequences Analyzed 122
2.3. Gnathostome Myoglobin Genealogy 130
2.4. Genealogic Evidence from Other Gene Phylogenies 136
2.5. Species Phylogeny from Combined Sequence Data 145
3. Utilizing the Clock Model of Protein Evolution 148
3.1. Rationale and Procedure 148
3.2. Results Obtained by the Clock Model 152
4. The Tempo and Mode of Protein Evolution 161
4.1. Pattern of Rate Variations 161
4.2. The Central Role of Natural Selection 167
5. Conclusions 183
Epilogue: New Cladistic Findings on Globin Phylogeny 185
References 187

Chapter 5

Evolution of Chromosomal Proteins

Lois T. Hunt and Margaret O. Dayhoff

1. Introduction 193
2. Nucleosome Structure 194
2.1. The Chromatin Fiber 194
2.2. The Nucleosome and Nucleosome Core 195
3. Computer Methods 199
3.1. Detection of Protein Relationships 199
3.2. Protein Alignments and Evolutionary Trees 200
3.3. Protein Families and Superfamilies 202
3.4. Mutation Acceptance Rates 203
4. Chromosomal Proteins 203
4.1. Prokaryote DNA-Binding Proteins 203
4.2. Viral Nucleic Acid-Binding Proteins 205

4.3. Protamines 206
4.4. Nonhistone Chromosomal Proteins 207
4.5. Histones H1 and H5 212
4.6. Histone H2A and Nuclear Protein A24 212
4.7. Histone H2B 217
4.8. Histone H3 219
4.9. Histone H4 221
4.10. Mutation Acceptance Rates 224
4.11. The Core Histone Superfamily 225
References 230

Part II. Modeling the Process of Sequence Divergence

Chapter 6

Simulation of the Evolution of Macromolecular Sequences by Random Fixation of Allowed Codons

Michael Coates and Simon Stone

1. Introduction 243
2. The Simulation Model—RFAC 247
   2.1. The Input Data 247
   2.2. The Starting DNA Sequence 247
   2.3. Random Point Mutation 248
   2.4. Allowed Amino Acids 248
   2.5. Fixation 251
   2.6. Phylogenies 251
3. Comparison of RFAC with Other Models 254
4. Methods of Comparison of Macromolecular Sequence Divergence in Real and Simulated Evolution 256
5. Results 257
   5.1. Effects of Rate of Fixation and Amount of Allowed Variation on Macromolecular Sequence Divergence 257
   5.2. Fitting of Simulated to Real Results of Evolution 258
   5.3. Effect of Amount of Allowed Variation on "Efficiency" of Evolution 269
6. Conclusions and Discussion 273
References 277

Chapter 7

Nonuniform Molecular Divergence: The Quantitative Evolutionary Analysis of Genes and Messenger RNAs under Selective Structural Constraints

Richard Holmquist, Dennis Pearl, and Thomas H. Jukes

1. Introduction 281
2. Constrained Stochastic Theory 282
3. Magnitude of Selective Constraints 284
   3.1. Base Composition 284
   3.2. Base Replacement Probabilities 285
   3.3. Density of Fixed Mutations among Variable Codons 286
   3.4. Distribution of Fixed Mutations within Codons 288
4. Estimation of Parameters 291
   4.1. The Observational Principle Underlying Parameter Estimation in Gene Sequences 291
   4.2. Evolutionary Measures 294
5. A Comparison of Evolutionary Estimates Made from Protein and Nucleic Acid Sequence Data and from Nonrandom REH Theory 296
   5.1. A Closely Related Divergence: Mouse and Rabbit β-Hemoglobin 296
   5.2. A Distantly Related Divergence: Rabbit α- and β-Hemoglobin 298
6. Comparison of Calculation with Experiment 299
7. Generality of Results 302
8. Extension to Noncoding Regions 305
9. Importance of Accurate Genetic Distance Estimates to Systematics 306
10. Discussion 309
11. Conclusions 312
References 312

Part III. Prospects for Investigating Evolution through Genomic DNA

Chapter 8

Genomic DNA: New Approaches to Evolutionary Problems

Alan F. Scott and Kirby D. Smith

1. Introduction 319
2. Overview of Methods 320

3. Comparative Studies of Genomic DNA ... 327
3.1. Y-Chromosome DNA ... 327
3.2. Other Middle-Order Reiterated DNAs ... 329
3.3. Studies of the Ribosomal Genes ... 331
3.4. Studies of Single-Copy DNA ... 336
3.5. DNA Polymorphisms ... 344
3.6. Other DNAs ... 346
4. Conclusions ... 349
References ... 350

Chapter 9

Features of Gene Structure, Organization, and Expression That Are Providing Unique Insights into Molecular Evolution and Systematics

David Hewett-Emmett, Patrick J. Venta, and Richard E. Tashian

1. Introduction ... 357
2. Recombinant DNA Technology ... 358
2.1. Cloning ... 358
2.2. DNA and RNA Sequencing ... 362
3. Expression and Organization of Eukaryotic Genes ... 364
3.1. DNA Transcription and Processing of RNA ... 364
3.2. The Intervening Sequences of Genes ... 367
3.3. The Role of Intervening Sequences ... 369
3.4. Gene Duplication ... 373
3.5. Pseudogenes ... 377
4. Use of Nucleotide Sequences in Evolutionary Tree-Building ... 382
4.1. Why Nucleotide Sequences? ... 382
4.2. Tree-Building Strategy ... 384
4.3. A Hemoglobin Nucleotide Tree ... 384
5. Concluding Remarks ... 397
References ... 398

Index ... 407

PART I

Eutherian Phylogeny and Protein Evolution

CHAPTER 1

Information for Molecular Studies from Anatomical and Fossil Evidence on Higher Eutherian Phylogeny

MICHAEL J. NOVACEK

1. Introduction

The mammals seem ideally suited for taxonomy. As the eminent Theodore Gill wrote in 1873, "it takes no penetrating acumen to recognize man, the monkeys, the bats, the typical ruminants, and the typical cetaceans as distinct forms existent in nature" (Gill, 1873). This clarity of vision, however, does not extend to the question of relationships among the major mammalian groups. Although there is nearly unanimous agreement on the tripartite division of mammals into monotremes, marsupials, and eutherians, agreement generally stops at this point. The majority of living mammals, the eutherians or "placentals," are often presented as a series of some 15 ordinal-level taxa without reference to a hierarchic classification. Similarly, the phylogenetic history of eutherians is commonly depicted as a bushlike radiation sprouting from mysterious roots at the end of the Mesozoic.

To some extent, such portrayals are justified. The fossil record does suggest that the rate of diversification for eutherians was rapid in comparison with that for many other animal groups. Unfortunately, the early members of these diverse lineages are often poorly represented. Most

MICHAEL J. NOVACEK • Department of Zoology, San Diego State University, San Diego, California 92182.

Late Mesozoic and Early Tertiary eutherians are represented by only partial dentitions. Some taxa we suspect as early members of different mammalian orders can be distinguished only by a few bumps and ridges on tooth crowns. We know virtually nothing of other parts of the anatomy in these taxa. These difficulties have doubtless influenced the course of mammalian taxonomy over the past few decades. Our progress in this field since the efforts of 19th and early 20th century systematists has been limited primarily to revisions of subordinal groups.

It is becoming more difficult, however, to decry lack of understanding of higher eutherian phylogeny and still ignore the problem. Some recent proposals for a pattern of higher resolution [notably, that of McKenna (1975)] reflect a trend toward the adoption of cladistic approaches by an increasing number of systematists. Such approaches emphasize the need for an explicit statement of relationship at any taxonomic level—the conversion of phylogenetic "bushes" to "trees," if you will—and a more rigorous analysis of the evidence for such statements. Moreover, the activity during the last decade in the field of molecular systematics has naturally been aimed at higher level problems. Except in the case of primates, where sampling has been fairly extensive, molecular systematists are of necessity concerned with questions of interordinal relationships; their data for most mammalian orders are limited to one or a few representative taxa. Interaction between molecular workers and those concerned with the more traditional evidence provided by paleontology and comparative anatomy depends on a common attention to higher level problems.

Despite these exhortations, considerations here of higher eutherian phylogeny fall far short of providing rigorously tested, explicit statements of relationships. In fact, many proposals in this chapter show less resolution than molecular-based phylogenies presented in other chapters of this volume and elsewhere. It is worthwhile, nonetheless, to offer a realistic overview of the evidence, with the following questions in mind: What degree of resolution is actually permitted by the evidence? Where do discrepancies arise among various kinds of molecular, morphologic, and paleontologic evidence? Where does the prognosis for phylogenetic resolution seem brightest or poorest? Finally, how does our understanding of higher level eutherian relationships influence our interpretations of rates of evolution and related phenomena? The "guts" of this discussion—the anatomical and paleontologic details so painstakingly collected and analyzed over the years—do not receive sufficient coverage here. It is hoped that the works cited in this chapter offer a reasonable, albeit small, sampling of the vast literature concerned with such evidence.

2. A Historical Sketch

The history of mammalian classification is clearly in need of review. The most comprehensive work on the subject is still Gregory's monograph, "The Orders of Mammals" (Gregory, 1910). The brief summary of early work presented below relies heavily on Gregory's treatment. The standard reference for mammalian systematics, which includes a summary and update of Gregory's coverage, is Simpson's classification (Simpson, 1945). Tracing this taxonomic history through more recent decades is difficult; the literature on the subject is widely scattered. Useful discussions are provided in McDowell (1958), Butler (1956, 1972), McKenna (1969, 1975), Van Valen (1966, 1967), and Szalay (1977).

Delineating the distinct features of mammals was a concern of Aristotle and other Greek philosophers, as well as of a handful of Medieval and Renaissance scholars. The development of a rational and comprehensive characterization of the group is, however, generally attributed to John Ray's work in the late 17th century. Ray's (1693) *Synopsis Methodica Animalium Quadrapedum et Serpentini Generis* included, among other notable achievements, an explicit definition of warm-blooded, viviparous, hairy quadrupeds as a distinct class of animals. Ray employed this definition by formally recognizing that whales were mammals and not fish, an allocation anticipated in the writings of Aristotle. Ray also established criteria for the recognition of various mammalian orders. Many of these criteria, such as variations in features of limbs, number of digits, number and arrangement of anterior teeth, and presence of claws or hooves, are regarded as perfectly valid by present-day mammalogists.

Ray's opus did not go uncriticized. Linnaeus (1735) and others found fault with Ray for basing his classification on only a few aspects of the skeletal system. The critics were not immune to committing such flaws in setting up their own classifications. In fact, the fusion of a broadly based comparative anatomy with mammalian systematics was a slow process, which is not reviewed here, but is given excellent coverage by Gregory (1910). Such a movement perhaps began with the work of the Perrault group in the early 1700s and was finally realized with the striking syntheses of Blainville (1816, 1836, 1839–1864). Blainville based many of his systematic arrangements on skull structure and reproductive anatomy, as well as on traditionally emphasized features of teeth and digits. His insights, which included the threefold division of mammals into monotremes, marsupials, and eutherians, were not immediately embraced. Gregory (1910, p. 75) remarked that Blainville's inspiration had the misfortune of being "essentially esoteric, recondite, and repugnant to long accepted opinions and usages and, moreover, as a product of 'philosophical zoology,' it encountered the powerful and very effective opposition of Cuvier." Despite these setbacks, Blainville's efforts were given due recognition in later classifications, and his basic conclusions represent the least controversial aspect of our view of higher mammalian phylogeny.

By the end of the 19th century, a battle plan for the higher level classification of the Eutheria had been drawn up by several authorities. Following Huxley's (1880) proclamations, it was decided that the Insectivora represented the ancestral trunk of the eutherian tree, and that untangling the relationships of this problematic order might be the key to a general solution. Weber's (1904) review of insectivoran relationships was an influential work in this phase of study. His classification reflected both first-hand comparative work and a synthetic review of earlier studies by Dobson (1882, 1883), Edwards (1868), Haeckel (1866), Leche (1885), and Gill (1870, 1872). Weber clarified the distinctions between Haeckel's Menotyphla (elephant shrews and tupaiids) and Lipotyphla (moles, shrews, hedgehogs, chrysochlorids, tenrecids, and solenodontids). This division proved popular and enduring; the validity of the lipotyphlan grouping has withstood the scrutiny of many modern taxonomists, while the menotyphlan concept was not seriously questioned until the mid-20th century (Simpson, 1945; Butler, 1956, 1972; McDowell, 1958).

A notable contemporary of Weber's was Herluf Winge, who published a series of studies on higher level mammalian systematics between 1887 and 1918 [translated into English by Winge (1941)]. Winge's studies offer a useful compendium of anatomical diversity in mammals. Unfortunately, many of his phylogentic hypotheses were constructed from lengthy and problematic arguments that emphasized the adaptive implications of characters rather than the distributions of characters he so ably summarized. Many of his proposals for interordinal relationships were rejected by later workers [see also remarks of Szalay (1977)].

Review of early 20th century work on mammalian systematics would not be complete without consideration of Gregory's (1910) monumental "Orders of Mammals." Gregory originally intended "only" to review the history of mammalian systematics prior to 1910, but he applied that knowledge and his expertise in comparative anatomy and paleontology to the construction of his own classification. With respect to eutherians, Gregory suggested several superordinal groups. He allocated the Insectivora and the Ferae (creodonts, carnivorans, and pinnipeds) to the Therictoidea. He recognized a monophyletic group, the Archonta, for the inclusion of Dermoptera, Primates, Chiroptera, and Menotyphla. Gregory also considered lagomorphs as members of the Rodentia, and he argued that the Rodentia and Edentata (including edentates in the present sense as well as aardvarks, pangolins, and fossil taeniodonts) were separately derived from an undiscovered ancestral stock. In venturing these hypotheses, Gregory placed emphasis on features of the skull, dentitions, and foot structure.

Gregory's important work does show inconsistencies. For example, the need for a distinction between primitive and derived characters in phylogenetic analysis was stressed early in the monograph. Yet this discrimination was not actually employed in many of Gregory's phylogenetic proposals. Gregory also strongly emphasized the need for concise, informative diagnoses at every taxonomic level, but he recognized some major categories, such as the Archonta, without morphologic definition. These shortcomings do not detract from the fact that Gregory's work remains a remarkably comprehensive treatment of mammalian phylogeny.

During the first decades of the 20th century, paleontologic research played a major role in the development of mammalian classification. Gregory (1910), for example, relied heavily on the comparative study of fossil teeth and skulls in fashioning his arguments for higher level relationships. Paleontologic input was even more evident in studies by Matthew (1909, 1937), who derived many of his theories on the basic relationships of primates, insectivorans, carnivorans, and early ungulates from his comparative studies of dental and tarsal features of Early Tertiary mammals from North America. George Gaylord Simpson, who wrote the standard reference work on mammalian classification (Simpson, 1945), invested much of his early career in studies of Mesozoic and Early Tertiary mammals.

Simpson's work proved strongly influential for other reasons as well. His status as a leading figure in theoretical and evolutionary biology lent authority to the outline of systematic principles included in his 1945 classification. Moreover, Simpson provided an exhaustive listing of mammalian genera as well as discussions of familial revisions. Earlier synthetic treatments were generally uneven in coverage at these lower taxonomic levels. Nevertheless, some of Simpson's more sweeping and original proposals for superordinal groupings have met with resistance. For example, Simpson's recognition of the cohorts Unguiculata and Ferungulata are not clearly supported by shared derived anatomical characters, and these categories have been dropped from more recent classifications (e.g., McKenna, 1975; Szalay, 1977).

In the years following the publication of Simpson's classification, discoveries of Mesozoic mammals [see review in Lillegraven *et al.* (1979)] contributed to an increased understanding of the reptile–mammal transition and the early diversification of mammals. Problems concerning the phylogenesis of early eutherians, however, continued to be obfuscating.

Some headway was made in revising the systematics of the putative ancestral stock of the Eutheria, the insectivorans. Important studies by Butler (1956), McDowell (1958), and McKenna (1960) outlined the paleontologic and anatomical evidence for a monophyletic order (essentially equivalent to Haeckel's Lipotyphla) and excluded poorly understood fossil taxa that were previously assigned by default to the Insectivora. Although these authors more effectively characterized the Insectivora, they did not clearly demonstrate the relationships of this order to other mammals. The studies did, however, help to dispell the long-held notion that "true" insectivorans were the ultimate ancestors of the majority of other eutherian groups.

By the end of the 1960s, an improved picture of early eutherian history was gained through dramatic new evidence of Late Cretaceous fossils (Lillegraven, 1969; Kielan-Jaworowska, 1969; Clemens, 1973). Some of this information led McKenna (1975) to question traditional interpretations of the evolution of the dental replacement pattern in therians. McKenna's reassessment of such patterns and his review of other anatomical features formed the basis of a new classification of the Mammalia. Many of his proposals were novel, not only for their departure from standard arrangements, but also for the explicit nature of the resulting classification. In order to develop a classification that would more accurately reflect cladistic hypotheses of relationship, McKenna established several new superordinal categories for mammals (e.g., magorders, mirorders, and grandorders), to which he applied a variety of new names. Among some of McKenna's more important conclusions with respect to eutherian relationships were his recognition of the Edentata as a primitive out-group of other Eutheria, and his recognition of Recent Lagomorpha, macroscelideans, and a variety of Late Cretaceous and Early Tertiary taxa previously recognized as "insectivores" as an early, divergent clade (Magnorder Ernotheria). He also abandoned usage of Simpson's Unguiculata and Ferungulata and resurrected Gregory's Archonta (excluding macroscelideans).

McKenna's (1975) classification has proven controversial. Hecht (1976) and Szalay (1977) have remarked on the lack of firm anatomical evidence necessary to justify such an explicit and cladistically oriented classification. Criticisms were also directed at McKenna's reinterpretation of dental replacement patterns in mammals (Szalay, 1977; Novacek, 1977a; Bown and Kraus, 1979), and some of his suggestions for superordinal groupings have been rejected by other workers (Novacek, 1977a; Szalay, 1977; Luckett, 1980). Nonetheless, this work offers a stimulating new look at higher mammalian relationships. Many critics seem overly distracted by the nomenclatural originality of McKenna's classification, whereas a more important aspect of this work—namely, the comparison of certain phylogenetic hypotheses with viable alternatives—has not received the attention it deserves, particularly within the paleontologic community.

Another recent review of higher level mammalian classification is Szalay's (1977) treatment, based largely on his comprehensive studies of the eutherian tarsus. Szalay was strongly opposed to developing a cladistic classification comparable to McKenna's, although he did suggest certain superordinal groupings that agreed with those recognized by the latter author. For the monophyly of the Archonta, Szalay emphasized the evidence provided by features of the astragalus and calcaneum, an argument that has met some opposition (Cartmill and MacPhee, 1980; Luckett, 1980; Novacek, 1980). Szalay also advocated the grouping of lagomorphs with elephant shrews, and he suggested possible close affinities between edentates and pholidotans. Some of Szalay's hypotheses are considered in more detail below.

The last few years have seen an increasing number of studies of higher eutherian phylogeny based on evidence not accessible to paleontologists. Among these are Luckett's (1977, 1980) detailed surveys of mammalian fetal membranes, Campbell's (1980) studies of the nervous system, and of course the numerous contributions devoted to comparative biochemistry and immunology presented in this volume and elsewhere. Despite such recent trends, it is clear that paleontology has played a major role in shaping our views of mammalian phylogeny. This influence is perhaps stronger than is the case for certain other vertebrate groups; C. Patterson (1977), for example, claimed that paleontologic investigation has not played such a dominant role in the development of teleostan phylogeny, even though the fossil record for this group is comparable to that of mammals in quantity and quality. What does seem significantly responsible for the integration of paleontology in the study of higher mammalian phylogeny is the fact that many notable students of fossil mammals were active and influential in the study of higher level patterns among Recent taxa. At times, the strong influence of this paleontologic perspective can have detrimental effects. For example, many mammalogists accept Matthew's (1909) reconstruction of the primitive carotid arterial system in eutherians, even though Matthew derived his reconstruction of soft anatomy indirectly from tracings or grooves in the bony ear region of fossil carnivorans. Only in recent years have these theories been placed under suspicion, due to much-needed embryologic and anatomical studies of living taxa (Presley, 1979; Conroy and Wible, 1978). Objections have also been raised concerning the great emphasis on dental comparisons in mammalian taxonomy, an emphasis clearly influenced by the common occurrence of teeth in the fossil record. Skeptics claim that heavy dependence on this limited sample of the whole animal has resulted in many erroneous systematic conclusions [see comments of McDowell (1958) and Szalay (1977)].

Given these criticisms, one is tempted to judge the value of the study of the fossil record in the analysis of higher eutherian phylogeny. It seems reasonable to conclude that this record will continue to provide new and important phylogenetic evidence if fossil taxa are carefully viewed in the context of comparative features of living mammals. The potential for new, phylogenetically important information from the fossil record is also better in cases where a greater range of anatomical features—as opposed to merely a greater number of "fossil tooth taxa"—is sampled and analyzed. Fitting examples of such contributions are the recent discoveries of skulls and skeletons of early eutherians from the Late Cretaceous mammals of Mongolia (see Kielan-Jaworowska, 1969, 1977). Not only do these fossils provide us with our earliest records of eutherians, but they also display an interesting mosaic of primitive and derived traits of different regions of the cranioskeleton. Knowledge of these mosaics is vital to judging alternative hypotheses for relationships. Of course, one cannot expect to find exquisitely preserved fossils around every butte, but it is clear that an enormous amount of skeletal material in various available collections has never been studied. In this respect, it is encouraging that some paleomammalogists (e.g., Szalay and Drawhorn, 1980) are conducting broad-scale investigations of fossil taxa for anatomical features poorly sampled in the past.

3. Eutherian Monophyly

Before proceeding with a discussion of higher level relationships within Eutheria, it is worthwhile to consider the basis for recognition of this diverse group. Little controversy concerning the validity of the Eutheria is apparent in the current literature, a situation that attests to the importance of Blainville's 19th century contributions. The most convincing support for eutherian monophyly is derived from the structures associated with fetal development. Unlike marsupials and monotremes, all living eutherians show the presence of a trophoblast, an epithelial covering of the blastocyst that functions importantly in gas exchange, hormone and prostaglandin production, and long-term protection of the fetus from maternal immune response. These functions, particularly the last, are no doubt related to the fact that intrauterine gestation is much longer, on the average, in eutherians than in marsupials. As Lillegraven (1975) concludes, the origin of the trophoblast in the earliest eutherians may represent one of the most important innovations in therian evolution. Another developmental trait unique to eutherians is the villous chorioallantoic placenta. Marsupials, by contrast, possess a choriovitelline placenta (one formed by an enlarged yolk sac instead of an allantois). The only exception to this pattern occurs in peramelid marsupials (the bandicoots). In these animals, the placenta is formed by the chorion and the allantois, but, unlike the condition in eutherians, the absorptive surface of the placenta has simple corrugations instead of villi. As in other marsupials, the gestation period of peramelids is short and the suckling period is long. There is little doubt that the chorioallantoic placenta represents a derived feature within Mammalia (Lillegraven, 1969, 1975; Luckett, 1977).

Several other ontogenetic traits as well as features of the reproductive tract, the cranioskeleton, the brain, and the dentition provide additional evidence for eutherian monophyly. These are listed in Table I and are discussed at some length in Lillegraven (1969, 1975, 1979), McKenna (1969, 1975), and Kielan–Jaworowska *et al.* (1979). Some of these characters are more clearly supportive than others. For example, the distinctions between marsupials and placentals with respect to brain development and molar morphology are often vaguely treated. Aspects of the reproductive tract in marsupials seem clearly primitive in comparison to the eutherian condition [e.g., the positional relationships of the ureters and the Mullerian ducts; see Lillegraven (1969)], while other features of the marsupial system (e.g., the intermittent appearance of the pseudovaginal canal) seem autapomorphous (i.e., uniquely derived) for the group. Hence, the simple generalization that eutherians show a more specialized reproductive system than do metatherians can be misleading.

Several characters cited as distinctive features of Eutheria (e.g., Kielan-Jaworowska *et al.*, 1979) are omitted from Table I because they either provide ambiguous evidence for monophyly or they merely represent primitive mammalian features retained by eutherians. That many eutherians have well-developed auditory bullae hardly seems significant. This structure can be constructed from a variety of elements; hence, an ossified auditory bulla was probably acquired independently in several eutherian lineages (Novacek, 1977b; MacPhee, 1979, 1981). Soricids and tubulidentates, as well as several fossil eutherians, show an abullate condition similar to that of monotremes, and the lack of an osseous bulla is probably a primitive mammalian condition (Novacek, 1977b). It is also clear that the multiple replacement of the antemolar teeth, sometimes cited as a distinguishing character of the Eutheria, is a conservative mammalian feature. In contrast, the limited replacement of premolars in marsupials (where only the last premolar of the upper and lower jaws, the P4/4, is replaced by a permanent tooth) is a derived feature, diagnostic for this infraorder. Other problematic, but frequently cited, trademarks of the

Eutheria are the presence of a baculum (a variable condition within the group), the occurrence of precocial young (in many eutherians, the young are born in an altricial state, though they are usually better developed than marsupial neonates), and the lack of a pouch or marsupium (the pouch may be a secondary feature developed independently in monotremes and marsupials). It should also be noted that although epipubic bones are often associated with the development of the marsupium, the presence of these elements is here regarded as a primitive mammalian character. Epipubic bones are present in both sexes of living marsupials (including species of pouchless females) in both sexes of the monotremes, and in some Mesozoic mammals (Kielan-Jaworowska, 1969, 1975b). Vaughn (1956) effectively argued that the epipubic bones were remnants of the puboischiadic plate in reptiles, and that this pair of elements was secondarily used for support of the pouch in some female marsupials. All Recent eutherians show the loss of the epipubic elements, although there is indirect evidence that these elements were present in certain Late Cretaceous eutherians (Kielan-Jaworowska, 1975b). Lillegraven (1969) related loss of the epipublics to prolonged gestation and the lack of a need for a marsupium in eutherian mammals.

4. Higher Level Eutherian Phylogeny

Figure 1 shows a hypothesis for the relationships of the major eutherian groups as suggested by the anatomical and paleontologic evidence reviewed in this section. Omitted from the figure are several ordinal-level taxa represented only by fossils. Many of these extinct groups are of course important to eutherian phylogenetic history, but their bearing on the interrelationships of the Recent orders is moot. However, certain fossil "orders" are considered that seem inseparable from the question of relationships of modern forms (e.g., the rodent–lagomorph association shown in Fig. 1). Finally, it is emphasized that hypotheses for the relationships of extant orders are heavily dependent on evidence from fossil members of these orders.

The degree of resolution depicted in Fig. 1 is clearly not ideal in a cladistic sense. Like others, I have tried, and failed, to pick through the reticulate phylogenetic possibilities and reach a general solution. Szalay (1977) came to a similar point in a comprehensive analysis of the Eutheria based on his survey of the tarsus and other cranioskeletal features, although he explicitly rejected the notion that one should attempt to resolve all phylogenies into cladistic dichotomies. In contrast to Szalay's perspective, McKenna (1975) stressed the need for maximum resolution, both

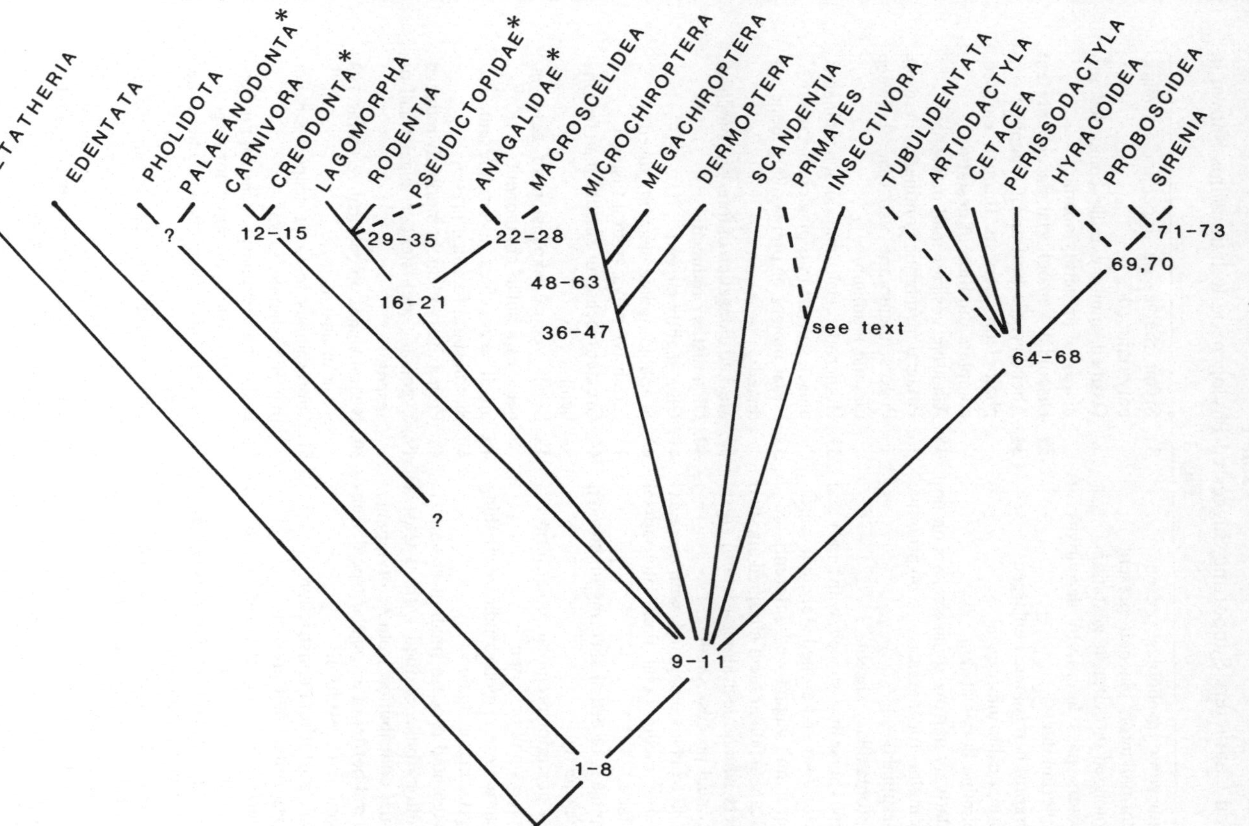

Figure 1. Suggested phylogenetic relationships for major groups of Recent and selected fossil (indicated by asterisk) eutherians. Numbers at nodes refer to shared derived characters listed in Table 1. Dashed lines indicate highly tentative or alternative relationships. See text for further explanation.

Table I
Derived Characters Supporting Suggested Phylogenetic Relationships Shown in Fig. 1[a]

1. Embryonic trophoblast present
2. Chorioallantoic placenta present
3. Prolonged intrauterine gestation
4. Ureters pass lateral to derivitives of Mullerian ducts
5. Cerebral hemispheres enlarged
6. Corpus callosum present
7. Epipubic bones absent
8. Relatively narrow stylar shelves on upper molars (in comparison with primitive marsupials)
9. Septomaxillary absent
10. Stapes horseshoe-shaped, with large basal perforation and arched stapedial crura
11. Uterus and vagina well-differentiated
12. Calcaneal fibular facet modified into two facets which meet along a lateral crest
13. Peroneal process small
14. Tibial trochlea elongated, with more extensive mediodorsal than laterodorsal surface
15. Emphasis of orthal shearing function in cheek teeth
16. Last posterior premolar molariform
17. Canine small or absent
18. Wear pattern of molars indicate crushing or grinding occlusion
19. Postglenoid process small or absent
20. Auditory bulla ossified with at least a partial contribution from ectotympanic
21. Lateral border of astragalar trochlea enlarged and crestiform
22. Weak postorbital constriction
23. Supraorbital shelf present
24. Anterior border of the coronoid process of lower jaw inclined
25. Jaw condyle positioned markedly higher than angular process
26. Distinct anterior palatal foramina
27. Recessus meatus of ectotympanic expanded into a tube
28. Curvature of astragalar trochlea sharply limited on posterior border
29. Number of lower incisors reduced and canine lost
30. One pair of incisors enlarged above and below
31. Stylar shelves on upper molars markedly reduced
32. Ossified auditory bulla composed primarily of an ectotympanic
33. Discoidal, hemochorial placenta present
34. Abembryonic portion of bilaminar omphalopleure persists through early or complete duration of gestation
35. Allantoic vesicle small or vestigial
36. Patagial membrane continuous between fingers; membrane also spans between tail and hindlimb
37. Humeropatagialis and coracocutaneous muscles insert into plagiopatagium
38. Nasal moiety of premaxillary bone reduced
39. Marked elongation of fore- and hindlimb
40. Ulna distally reduced
41. Ulna slightly curved
42. Scaphoid, lunate elements fused
43. Extensive pelvic–sacral fusion
44. Greater trochanter smaller than femoral head
45. Small ventral expansion of ectotympanic as a bullar component
46. Small basicochlear fissure present
47. Supraorbital foramen present
48. Premaxillary element markedly reduced
49. Zygomatic arch slender; jugal small or absent
50. Neural spines on cervical vertebrae 1–3 weak or absent
51. Supraspinous fossa of scapula smaller than infraspinous fossa
52. Distal end of humerus lacks supratrochlear depression and supinator ridge
53. Entepicondylar foramen of humerus absent
54. Ulna markedly reduced, incomplete distally
55. Olecranon process of ulna vestigial
56. Digits 2–5 of manus greatly elongated to support wing membrane
57. Manus rotated 90°
58. Hindlimbs rotated from normal quadrapedal orientation
59. Calcanealastragalar articulation not af-

Table I *(continued)*

fected by convex articular facet (articulation via large depression or bony ridges)

60. Occipitopollicalis muscle present on leading edge of the propatagium
61. Preplacenta broad, horseshoe-shaped with definitive placenta more localized
62. Definitive yolk sac "gland-like"
63. Prominent "interstitial membrane" occurs in chorioallantoic placenta
64. Primitive forms with bunodont cusp morphology on molars (crown patterns greatly modified in more "advanced" ungulates)
65. Primitive forms show consistent presence of hypocone on upper molars
66. Lower molars of primitive forms show swelling around the base of the metaconid and elongation of M_3 through expansion of the talonid
67. Terminal phalanges on all digits; elongated, but not fissured; tendency for development of hooves in many lineages
68. Tendency for mesaxonic arrangement of digits around digit III (modified in various "advanced" ungulates)
69. Serial arrangement of carpal bones
70. Loss of astragalar cuboid contact; astragalus articulates distally only with navicular
71. Bilophodont molars with tendency to form additional lobe on posterior part of cingulum
72. Extensive premaxillary–frontal contact; retraction of nasofacial region; geometry and position of nasal and maxillary elements modified
73. Squamosal contribution to the glenoid region and the zygoma robust and laterally expanded

[a]Numbers correspond to numbers in Fig. 1. For explanations of characters see text and Appendix.

in cladograms and in classifications. Despite this objective, McKenna (1975) failed to derive a notably more cladistic arrangement for the Eutheria than presented here. For instance, he recognized four superordinal categories (grandorders)—the Ferae, the Archonta, the Insectivora, and the Ungulata—but he was not able to develop a hierarchic sequence for this tetrachotomy. Many molecular-based trees are of course more explicit in their depiction of higher level relationships. This distinction seems partly due to the fact that molecular systematists commonly apply parsimony algorithms to their phylogenetic analyses, a procedure not readily adopted by many morphologists and paleontologists. Moreover, "molecular phylogenies" are often based on substitution patterns in only one or a few proteins. Although parallel and back substitutions of a single protein can occur rampantly in these trees, the molecular systematist is not yet faced with the bewildering mosaic of widely different features and systems that the comparative anatomist must consider.

Despite these problems, it seems realistic to evaluate a set of broader questions on eutherian phylogeny in light of available evidence from comparative anatomy and paleontology. These can be organized in the following fashion:

1. The relationships of the Edentata and the question of their alleged affinities with Pholidota and the extinct Palaeanodonta.
2. The phylogenetic position of the Carnivora and their extinct relatives, the Creodonta, within the Eutheria.
3. The evidence for and against a monophyletic group including Rodentia, Lagomorpha, a variety of Asian Tertiary forms, and possibly macroscelideans.
4. The question of validity for the Archonta, comprising Dermoptera, Chiroptera, Scandentia (tupaiids), and Primates, and the possible relationships of alleged archontans with the Insectivora.
5. The basis for recognition of the ungulate supergroup, including the great diversity of Early Tertiary "condylarths," extinct notoungulates, the Recent artiodactyls, perissodactyls, cetaceans, proboscideans, hyracoids, sirenians, and tubulidentates.

Evidence for the monophyly of each of the orders in the above categories will not be reviewed here, unless there has been some recent contention regarding the validity of a particular order. The characters listed in Table I are suggested as evidence for the superordinal clusters depicted in Fig. 1.

4.1. Edentata, Pholidota

The edentates, represented by the living armadillos, anteaters, and tree sloths, show a variety of bizarre specializations not found in other mammals, and the relationships of this group have long puzzled taxonomists. It has been popular to recognize edentates as a derivative of the Insectivora through some vague phylogenetic connection. Matthew (1918) identified as Early Tertiary North American group, the Palaeanodonta, as close relatives of edentates, and his conclusions were essentially endorsed by Simpson (1945). Emry (1970) has provided a detailed argument against this view, and has favored a closer relationship between palaeanodonts and pholidotans (see further remarks below). Several workers, including Simpson (1978), have come to accept Emry's arguments, at least in part, and the Edentata are now viewed by some as a group of South American origin with no clearly substantiated affinities to mammals from other continents (Hoffstetter, 1970; McKenna, 1975, 1980; Simpson, 1978). This distinction is important; no other South American eutherians apparently had such an isolated evolutionary history. Even the variety of extinct notoungulates, pyrotheres, and liptopterns known from the South American Tertiary have plausible ties to ancient ungulate groups that may have been distributed widely at the beginning of the Cenozoic [see review in McKenna (1980)].

Both Hoffstetter (1970) and McKenna (1975) have taken the long-term isolation of edentates in South America as corroboration for the view that this order represents the earliest divergent branch among the living Eutheria. The biogeographic evidence for this view is compelling, but, as McKenna (1975, 1980) has readily acknowledged, the anatomical evidence for the early split between edentates and other Eutheria is not firmly established. McKenna (1975) listed several primitive mammalian features: the presence of septomaxillary bones, the complete ossification of the ribs, low body temperature, poor thermoregulation, and the apparent lack of a well-differentiated uterus and vagina as characters retained in edentates but modified or lost in other eutherians. The physiologic features in this listing are, however, present in some other placentals (e.g., tenrecs) and there is convincing evidence that the heterothermy characteristic of tree sloths is a special adaptation correlated with the folivorous dietary habits of these forms (McNab, 1978). The anatomy of the edentate reproductive tract is intriguing and shows a departure from the plan observed in most other eutherians (Grassé, 1955), but polarity assessment of this feature based on detailed comparisons between edentates and other mammals has not been accomplished. Similarly, there is ambiguity with respect to the extensive ossification of the ribs in edentates. This feature does not really match the conditions seen in nontherians or reptiles; in edentates the distal portion of the ribs are not only ossified, but they articulate with the sternum in a uniquely complex fashion. This condition is plausibly related to the xenarthrous construction of the vertebrae, and, hence, completely ossified ribs may represent an autapomorphy for edentates.

Among the characters noted above, only the presence of the septomaxillary offers some support to the notion that edentates are a very primitive eutherian group. The septomaxillary is present only in dasypodids (armadillos), but its absence in other families is usually regarded as a secondary loss. This element occurs in monotremes and reptiles, but not in marsupials or nonedentate eutherians (Hoffstetter, 1958). Loss of the septomaxillary in all likelihood occurred independently in marsupials and nonedentate eutherians.

Another character that might well serve as evidence for the remote divergence of the Edentata is the condition of the stapes. In some edentates, this bone is either columnar with a small perforation at its base (e.g., *Tolypeutes*, a dasypodid), or it is more flaired toward its base and without a perforation (e.g., *Bradypus*, a tree sloth). In contrast, anteaters (Myrmecophagidae) have a stapes with a large basal opening flanked by slender stapedial crura, a condition commonly found in nonedentate eutherians (Fleischer, 1970). Myrmecophagid ontogeny shows, however, that the stapes develops from the imperforate condition seen in adults of other edentate families (Doran, 1878). Marsupials also present a range of conditions for the stapes, but some members of this order show either the imperforate form (e.g., *Notorcyctes*) or the slightly perforate form (*Petaurus, Thylogale*) that resemble closely the conditions in armadillos and sloths. It has been argued that an imperforate, column-like stapes is a primitive mammalian character (Gregory, 1910; Fleischer, 1970). This condition is also found in monotremes, a fact that substantiates Fleischer's interpretations, unless the monotreme ossicle chain evolved independently from that of therians (Crompton and Jenkins, 1979). Among Eutheria, a stapes of this general form occurs only in edentates, pholidotans, pinniped carnivorans, and cetaceans. Marine mammals display an ossicle chain of extremely aberrant construction, and the stapes in these taxa was probably secondarily modified from the perforate, horseshoe-shaped stapes characteristic of terrestrial carnivorans, ungulates, and most other eutherians. On the other hand, it seems reasonable to postulate that the monotreme-like stapes of some edentates and pholidotans indicates the retention of a primitive mammalian condition.

Thus, a few bits of evidence support the concept that edentates derive from the earliest split within Eutheria as represented by the Recent orders. For the present, it seems reasonable to hold to this hypothesis despite Simpson's (1978) discouraging words on the issue. As McKenna (1975) points out, the question of edentate relationships is begging for thorough consideration, possibly through studies of the reproductive tract, the ear, and comparative biochemistry. The earliest edentate fossils are *bona fide* armadillos [see review in McKenna (1980)], and they reveal little about the morphotypic plan for this order.

It is important in this context to consider the relationships of the Pholidota. As noted above, Emry (1970) documented a case for the close relationship between pholidotans and extinct palaeanodonts. Rose (1978) has effectively argued that palaeanodonts also bear affinity to the pantolestids, a primitive group that shows a combination of carnivoran and insectivoran characters in the dentition. Possibly the nearest living relatives of the Pholidota are indeed either insectivorans or carnivorans, although the matter seems far from settled. Szalay (1977), for one, has objected to Emry's association of pholidotans and palaeanodonts, arguing that the tarsal evidence suggests a closer relationship between palaeanodonts and edentates. He thus resurrects the idea that edentates have an identifiable sister group from the Early Tertiary North American record. These counterarguments certainly warrant further attention. Pholidotans are regarded here as Eutheria *incertae sedis*, though the stapes morphology suggests an earlier divergence than is the case for many other lineages within this infraorder. Advocating a close affinity between pholidotans and carnivorans or insectivorans requires either (a) that the stapes in pangolins is actually a derived structure, or (b) that the common ancestors of these groups had a primitive eutherian stapes resembling that in pholidotans. The first alternative is contradicted by the out-group comparisons made above. The second alternative is problematic because the stapes condition thought to be primitive for eutherians is not known to occur in any Recent or fossil member of the Carnivora or Insectivora.

4.2. Carnivora, Creodonta

Represented within the Carnivora are not only a diversity of specialized terrestrial groups, but also the independently derived groups of pinnipeds (Tedford, 1976). Nevertheless, the morphotypic condition for this order is very primitive. Such a conservative plan is shown, for example, in the basicranium (MacIntyre, 1972) and the tarsus (Szalay, 1977) of Paleocene and Eocene miacids. Evidence for the monophyly of the Carnivora is limited to derived features of the cheek teeth which relate to shearing occlusion. Details of the dentition and the tarsus suggest a close relationship between the Carnivora and the extinct Creodonta (Matthew, 1909, 1937; Szalay, 1977). Other than this association, the relationships of the Carnivora are by no means established. Lillegraven (1969) has noted some interesting dental similarities between early miacids and the Late Cretaceous palaeoryctids, but the affinities of the latter group seem at least as problematic as the affinities of the Carnivora, and the association has limited relevance to the question of relationships among Recent Eutheria.

One of the important conclusions of Simpson's (1945) classification was the suggested grouping of carnivorans with the great diversity of ungulate orders within the cohort Ferungulata. Simpson did not extensively review the anatomical evidence for the monophyly of the Ferungulata, but his arrangement was endorsed by certain other authors. MacIntyre (1972), for example, identified features of the petrosals from Late Cretaceous faunas of North America as assignable to a "ferungulate type," and he thus inferred that the taxa with petrosals of this morphology were early members of the cohort. Much of MacIntyre's (1972) discussion, however, merely suggests that early ungulates and early carnivorans resemble each other in sharing a large number of primitive eutherian characters. It is difficult to find any convincing anatomical evidence for the Ferungulata, and recent classifications by McKenna (1975) and Szalay (1977) distinguish carnivorans and ungulates as separate and not necessarily closely related clades.

4.3. Rodentia, Lagomorpha, Macroscelidea

The concept Glires, uniting rodents and lagomorphs (rabbits and pikas), can be traced back to Linnaeus (1735), who modified this grouping rather dramatically in subsequent publications by adding oppossums, rhinoceroses, and bats. The distinction of lagomorphs and the basic rodent "grades," the Sciuromorpha, Myomorpha, and Hystricomorpha, were first clarified by Gervais (1849), Brandt (1855), and Tullberg (1899). Gregory (1910), who simply included lagomorphs within the Rodentia, provided an extended list of both primitive and derived characters that he claimed as evidence for the monophyly of the order. Simpson (1945) emphasized the distinctions between the Lagomorpha and Rodentia, but he accepted, with reservations, the natural unity of the Glires. There has long been a diversity of opinion, however, concerning the validity of Glires. Gidley (1912) concluded that a few special similarities between rodents and rabbits did not necessarily imply close relationship. Many authors accepted Gidley's interpretation and all but ruled out the possibility of close ties between rodents and lagomorphs. Wood (1957) suggested that lagomorphs were derived from extinct periptychid condylarths, and thus shared a common ancestor with ungulate groups. Other suggested ancestors for lagomorphs were "zalambdodont insectivores" (Russell, 1959) or the curious Pseudictopidae from the Paleocene of Mongolia (Van Valen, 1964). Rodents, on the other hand, were associated with primates by McKenna (1961), Wood (1962), Van Valen (1966), and several other authors.

Skeptics notwithstanding, there is evidence that Rodentia and Lagomorpha share a closer common ancestry with each other than with any other Recent eutherian group. The key qualifier here is "Recent"; a diverse assortment of extinct pseudictopids, zalambdalestids, eurymilids, and anagalids from the Late Cretaceous and Early Tertiary of Asia might be closer relatives of either rodents or lagomorphs [see reviews in Van Valen (1964), Szalay and McKenna (1971), McKenna (1975), Szalay (1977), Novacek (1977a), Kielan-Jaworowska *et al.* (1979)]. Szalay (1977) has also proposed a close relationship for Rodentia and the Leptictimorpha, an archaic group of insectivore-like forms from North America. There is no convincing morphologic basis for this association, although the Leptictimorpha may be an *insertae sedis* member of a diverse clade that includes Rodentia, Lagomorpha, Macroscelidea, and several of the fossil taxa noted above (McKenna, 1975; Novacek, 1977a).

Shared derived characters suggesting a close relationship between lagomorphs and rodents are listed in Table I. Some of the cranioskeletal and dental characters are present in other early Asian eutherians, but this distribution does not damage seriously the Glires concept with respect to comparisons among Recent orders. As Szalay (1977) has remarked, there are no convincing alternatives. Luckett (1977; in preparation) has found a number of special features of fetal development that are consistent with the monophyly of the Glires (see also Table I). Cranioskeletal and dental evidence for this grouping has been considered by Van Valen (1964), Sych (1971), Evans (1942), McKenna (1975), Kielan-Jaworowska (1975a), Kielan-Jaworowska *et al.* (1979), Novacek (1977a), and Szalay (1977), though not all these authors advocate recognition of a close lagomorph–rodent association. Paleontologic input on the problem will come with further study of eurymilids, pseudictopids, and anagalids.

Macroscelideans, the curious elephant shrews of Africa, are here tentatively regarded as the nearest living out-group to rodents and lagomorphs. Szalay (1977) actually suggested a more recent common ancestry for macroscelideans and lagomorphs to the exclusion of Rodentia, based on shared derived features of the pes. Macroscelideans do not, however, show special features of fetal development common to rodents and lagomorphs, and, on the balance, the distribution of apomorphs favors their out-group position. This concept departs significantly from earlier groupings of macroscelideans and tupaiids as Menotyphla. There is now a general consensus that this menotyphlan concept is erroneous, and that macroscelideans should be recognized as a distinctly separate order of mammals without close ties with tupaiids of lipotyphlous insectivorans (Butler, 1956, 1972; McDowell, 1958; B. Patterson, 1965). McKenna (1975) is to be credited for emphasizing the possible relationships between macroscelideans and lagomorphs, though his arrangement differs in several respects from that advocated here. McKenna grouped elephant shrews, lagomorphs, leptictimorphs, anagalids, pseudictopids, zalambdalestids, and *Kennalestes* and *Asioryctes* (two "palaeoryctid" genera from the Late Cretaceous of Mongolia) in his magnorder Ernotheria. His primary evidence for this category was a theory of dental replacement in mammals that has been disputed by other authors [see comments in Hecht (1976), Szalay (1977), Bown and Kraus (1979)]. However, the association of lagomorphs and macroscelideans recognized here is partly based on other features noted by McKenna (1975).

4.4. "Archontans" and Insectivora

Although the variety of taxa recognized as archontans (tupaiids, primates, dermopterans, chiropterans) have been set apart from the lipotyphlous insectivorans (here equated with Insectivora), the evidence does not clearly suggest this segregation. As discussed below, there still remains the possibility that primates share a more recent common ancestry with some insectivoran clade [e.g., the Erinaceomorphs, as suggested by

Szalay (1975), Cartmill (1975), and in this chapter] than with tupaiids or other archontans. In contrast, the occasional suggestion of a close relationship between bats and insectivorans—usually framed in terms of a scenario depicting insectivores as the terrestrial ancestors of their volant descendants—is contradicted by a more plausible alternative; namely, the kinship of Chiroptera and Dermoptera. Since the time of Winge's (1917) publication, there has been a tendency to rule out the possibility of a bat–dermopteran association. Winge argued convincingly that the flying lemurs do not represent the ideal ancestral condition for bats. However, close relationship between these taxa does not require that dermopterans provide the perfect prototype. Leche (1885), in a detailed comparative study, noted special similarities in the musculature, patagium, and digestive tract that were common to bats and flying lemurs. Some of his comparisons were in error (Winge, 1917, 1941; Novacek and Luckett, unpublished study), but there is a sufficient collection of additional evidence for the close affinity of these groups. This evidence comprises features of the wing, forelimb skeleton and musculature, premaxillary region, pectoral girdle, carpus, basicranium, pelvic girdle, hindlimb, tarsus, and fetal membranes (Table I). These items of synapomorphy are discussed by Novacek and Luckett (unpublished study). Like other hypotheses for superordinal relationships, the bat–dermopteran grouping warrants detailed investigation. At present, there are no apparent cases that offer strong alternatives. The resemblance of bats to insectivorans resides mainly in dental features, such as the dilambdodont crown patterns, that are not morphotypic for either order (Butler, 1972; Novacek, 1976). Moreover, this alternative rests on a much more limited range of anatomical features than the thesis advocated here. There is also no clear reason for recognizing bats as closer relatives of tupaiids and primates than dermopterans. The association of bats with these other "archontans" is rarely based on an explicit list of characters (e.g., Gregory, 1910). Szalay's (1977) concept of the Archonta, a group to which he allocated chiropterans, was based on special similarities in foot structure, but did not include a discussion of the tarsus in bats. The foot elements of chiropterans are, in fact, bizarre in structure, and show no obvious similarity to those in tupaiids, primates, or dermopterans (Novacek, 1980).

The foregoing discussion assumes the monophyly of the Chiroptera; that is, the natural unity of the suborders Megachiroptera (often misleadingly dubbed "the Old World fruit bats") and the Microchiroptera. The fact that bats are the only true flying mammals and that both suborders show numerous shared derived features that are attributed to flying abilities has convinced most mammalogists of bat monophyly for over a century. Recently, however, Smith (1976, 1977) and Smith and Madkour

(1980) have challenged this concept, arguing that certain derived features of the penis, neurosensory system, and limb joints point to a diphyletic origin for bats. Smith and Madkour (1980) regard Megachiroptera, along with dermopterans, primates, and tupaiids, as members of the Archonta, whereas they exclude Microchiroptera from this association by associating this latter group with lipotyphlous insectivorans. Their hypothesis faces massive contradiction, not only from shared derived features of the flight apparatus, but from features of the nasofacial and basicranial regions, fetal membranes, nervous system, and muscular system (Table I, and Novacek and Luckett, unpublished study).

The highly publicized trio represented by primates, tupaiids, and lipotyphlous insectivores has perhaps been the subject of more heated controversy than any other problem set of eutherians. Traditionally, tupaiids and primates were regarded as the closely related descendants of some ancient insectivoran stock. Gregory's (1910) association of tupaiids with primates and other alleged archontans was based on the observation that tree shrews differed markedly in skull structure from insectivorans and "approached" in morphology skulls of the lemuroid type. Thus began a complicated history of taxonomy involving these groups. Carlsson (1922) mustered evidence for the actual allocation of the tupaiids to the primates, a hypothesis bolstered by a series of well-known publications by Le Gros Clark (1924, 1925, 1926, 1959, 1971). Many authors have subsequently expressed disenchantment with this allocation [for important historical reviews, see Campbell (1974) and Luckett (1980)]. Several recent contributions (Novacek, 1980; Luckett, 1980; Cartmill and MacPhee, 1980; Butler, 1972, 1980) present numerous lines of evidence discrepant with the tupaiid–primate grouping. These studies also follow Simpson (1945) in rejecting the Archonta. Other authors (e.g., McKenna, 1975; Szalay, 1977) acknowledge the difficulty in demonstrating close primate and tupaiid relationships but retain recognition of the Archonta (*sans* Macroscelidea). A few molecular-based phylogenetic trees (Dene *et al.*, 1980; Goodman, 1975; Sarich and Cronin, 1976) also support an archontan grouping, although bats are not included in some of these analyses. The most explicit case for the Archonta has been presented in recent studies of the tarsus by Szalay (1977) and Szalay and Drawhorn (1980). These authors recognized details of the proximal tarsus in various alleged archontans as components of a form–function complex for arboreal or volant locomotion. Different interpretations of the distribution of "archontan" tarsal features and their phylogenetic significance have been expressed (Novacek, 1980), to the effect that the proximal tarsus does not provide a clear set of synapomorphies for the Archonta. For the moment it is perhaps better to view the data as ambiguous, forcing one to the conclusion that tupaiids, primates, and insectivorans represent three distinct ordinal-level taxa of uncertain relationships among themselves and with other Eutheria.

Before completely resigning oneself to this conservative conclusion, it is worthwhile to entertain the possibility of an alternative phylogenetic relationship between primates and insectivorans. This connection seems particularly strong in the case of the erinaceomorph insectivores, a family that shares with primates the following similarities: (a) Both of these groups show an ossified petrosal element in the auditory bulla. A bulla largely composed of a petrosal is a unique primate feature, perhaps the most diagnostic character of the group. The petrosal element of the bulla in erinaceomorphs is relatively smaller, sharing this structure with an enlarged alisphenoid wing. However, any petrosal contribution to the bulla is a rare occurrence in eutherians. (b) The dentitions of the earliest known primates, such as *Purgatorius* (Clemens, 1974), show a striking resemblance to derived aspects of the erinaceomorph dentition (Novacek, 1977a). (c) The petromastoid region of the tympanic cavity is developed in a similar manner in primates and erinaceomorphs, and this condition departs from the plausible primitive state for eutherians (Novacek, 1982; Szalay, 1975). (d) Lemuroid primates and erinaceomorphs (as well as other insectivorans) show an orbital mosaic with a well-developed maxillary element that contacts the frontal anterodorsally but excludes the palatine from contact with the lacrimal. This condition is reasonably interpreted as derived for eutherians (Butler, 1956; McDowell, 1958; McKenna, 1975). As a potential synapomorphy for primates and erinaceomorphs, the orbital condition is ambiguous; other primates show a more primitive condition wherein the maxillary is virtually excluded from the orbital wall by the enlarged palatine. Cartmill (1975) has, however, suggested that this condition in non-lemuroids is secondary and that the arrangement common to lemuroids and insectivorans is morphotypic for primates.

Given the above qualifications, the evidence for primate–insectivoran relationships via the erinaceomorphs is tentative at best. The evidence does, however, at least provide *some* basis for this connection, whereas the evidence for primate–tupaiid affinities is even more open to complications (Cartmill and MacPhee, 1980; Luckett, 1980; Novacek, 1980). Acceptance of the erinaceomorph ties to primates might disrupt significantly the current classifications for these orders. If primates shared a special relationship with erinaceomorphs that excluded other insectivorans, the latter group would again be fragmented into monophyletic and paraphyletic subunits. The principal evidence for the monophyly of the Insectivora is, however, the condition of the orbital mosaic, a condition shared by all Recent insectivoran families as well as lemuroid primates.

If one postulates that primates had this condition as a primitive state, then primates become a subgroup of the Insectivora. This proposal might appal some systematists, but it would be necessary in order to derive a classification that demonstrated the close kinship of primates and erinaceomorphs and still retained a concept of monophyly for erinaceomorphs and other insectivorans.

4.5. The Ungulate Radiation: Perissodactyla, Artiodactyla, Cetacea, Proboscidea, Tubulidentata, Hyracoidea, Sirenia

There seems to be a general consensus that the vast radiation of groups referred to variously as "ungulates," "subungulates," "paenungulates," and "condylarths" have a single origin. It is cautioned that these terms are used to mean quite different taxonomic concepts by many authors; the nomenclatural problems involving these lineages are as complex as the phylogenetic tangles they betray (Matthew, 1937; Simpson, 1945; McKenna, 1975; Van Valen, 1971; Szalay, 1977). Nevertheless, anatomical information on fossil and Recent taxa suggests a relationship between several living ungulate groups and early Tertiary forms here recognized informally as condylarths. It should be emphasized that different condylarth lineages, such as arctocyonids, mesonychids, hyopsodontids, and periptychids, are not necessarily members of a group excluded from ungulates of a more modern aspect. Special relationships between certain condylarth families and certain Recent ungulate orders have been convincingly demonstrated (e.g., Van Valen, 1968). The most primitive ungulates, the arctocyonids, show the bunodont features of the dentition and details of the tarsus that are characteristic of many ungulate lineages and that represent a marked departure from the inferred primitive condition for eutherians. Some lineages—the Cetacea and Sirenia, for instance—secondarily lost or modified this basic plan. Other kinds of evidence for the monophyly of the ungulate orders have proven elusive.

Gregory's (1910) scheme for the ungulate cluster included perissodactyls, hyracoids, proboscideans, sirenians, tubulidentates (allocated with reservations), and extinct notoungulates, liptopterns, toxodonts, and several condylarth or "protoungulate" groups. Gregory excluded whales and artiodactyls from this broad group. Simpson (1945) followed Gregory in separating cetaceans from the ungulate clades, but he did associate artiodactyls with perissodactyls and other alleged ungulates in his cohort Ferungulata. There now seems a strong case for the close relationships of cetaceans with mesonychid condylarths (Van Valen, 1968), whereas, as noted above, the lumping of carnivorans with ungulates in the Ferungulata has not been widely affirmed. The affinities of artiodactyls with early condylarths are not at all clear. McKenna (1975) assigned this modern order, along with arctocyonids, tillodonts, tubulidentates, dinoceratans, and embrithopods, to his mirorder Eparctocyona, but this action was not accompanied by a discussion of supportive anatomical evidence. Artiodactyls may be more primitive than many other Recent ungulate lineages in retaining a clavicle and in lacking modifications of the femur and other aspects of the postcranial skeleton (Earl Manning, personal communication).

The myriad possibilities for relationships among the ungulate orders will not be reviewed here, with the exception of a brief consideration of the Paenungulata. Simpson (1945) established this superorder Proboscidea (comprising elephantoids and the extinct *Moeritherium*), Hyracoidea, Sirenia, Pantodonta, Dinocerata, Pyrotheria, Deinotheroidea, and Embrithopoda. In his recognition of the Tethytheria, McKenna (1975) retained only a close grouping of Proboscidea and Sirenia and he dispatched other alleged paenungulates to different ungulate categories. Some authors have, however, recently endorsed Simpson's paenungulate concept, although they remove *Moeritherium* from the Proboscidea and they allocate Tubulidentata to the Paenungulata, recognizing this taxon as a plesiomorphous sister group to Hyracoidea, Proboscidea, and Sirenia (Shoshani *et al.*, 1978, 1981; J. Shoshani, personal communication). These authors cite immunologic, protein sequence, dental, and cranioskeletal evidence for their hypothesis (Shoshani *et al.*, 1978; J. Shoshani, personal communication). In Chapter 3 of the present volume, De Jong reviews evidence from lens α-crystallin A amino acid sequences for grouping Tubulidentata, Hyrocoidea, Sirenia, and Proboscidea together in the superorder Paenungulata. Allocation of the tubulidentates to the Paenungulata does, however, seem problematic; few convincing characters support this link. The tubulidentates fit more poorly within the Ungulata than does any other alleged member of this group. The ear region of aardvarks, for instance, seems much more primitive than the plan shown by other putative ungulates (Klauuw, 1931; Novacek, 1977b). The membership of hyracoids within the Paenungulata cannot be confidently accepted until the shared derived dental similarities between hyracoids and perissodactyls can be ruled out altogether as evidence for phylogenetic relationship. The most convincing aspect of the paenungulate grouping still pertains to its core; namely the close association of proboscideans and sirenians. Gregory (1910, p. 409) reviewed many specialized anatomical features consistent with this hypothesis, features derived from comparative studies of the nasal region of the cranium, the pectoral and pelvic girdles, the teeth, the jaw joint, and other systems [also Shoshani *et al.* (1978); J. Shoshani, personal communication; and Table I herein]. Useful summaries of distinctive traits for the Proboscidea are provided by Osborn (1936, 1942) and Maglio (1973); similar compilations for the Sirenia are provided by Reinhart (1959) and Domning (1978).

5. Divergence Times for Eutherian Orders

The fossil record provides at least minimum estimates, as suggested by first fossil occurrences, for the ages of major eutherian taxa. In some cases (e.g., Primates), these first occurrences seem not appreciably younger than plausible divergence times. In other cases (e.g., Scandentia) the fossil record represents only a small segment of the history of the lineage. Both minimum estimates and postulated times of divergence are listed in Table II. The suggested range of error for minimum estimates is primarily attributable to problems encountered in assigning an age, in millions of years before present (my bp), to the assemblage containing the oldest fossil for a particular taxon. This error is compounded in the case of postulated times of divergence, because these approximations are based on fossil occurrences, phylogenetic hypotheses, and inferences that transform these interpretations into dates measured in millions of years.

In view of these uncertainties, the rationale behind some of the dates listed in Table II warrants explanation. The date for divergence of the Eutheria reflects the belief that the mammal assemblage from the Late Cretaceous of Mongolia described by Kielan-Jaworowska (1969, 1975a) and her associates represents an early radiation of primitive eutherians. Indeed, several of these taxa, including the shrew-like *Asioryctes*, show cranioskeletal characters that are more primitive than in all other known Eutheria (Kielan-Jaworowska, 1977; Novacek, 1980). At least some authors (Slaughter, 1971) believe that there is evidence for still earlier eutherians from the Trinity assemblage, a sample of isolated teeth and jaws from the Middle Cretaceous (approximately 105 my bp) of Texas. Kermack and Kielan-Jaworowska (1971) have argued that the Trinity mammals are not definitely assignable to either marsupials or eutherians, and they prefer calling these forms "Theria of eutherian–metatherian grade." This title has been endorsed by others (Clemens *et al.*, 1979), but there are some undeniable similarities between certain Trinity specimens and primitive eutherians. It is reasonable to suggest a date for the origin of Eutheria that is at least as old as the Trinity fauna, with the proviso that this date may be too old if discovery of more complete material from the middle Cretaceous does not reveal the existence of eutherians.

Similar inference is necessary in order to estimate splitting times for the majority of the eutherian orders. Edentates, for example, are known

Table II
Estimated Times of Origin for Selected Eutherian Taxa[a]

Taxon	Minimum estimate[b]	Postulated divergence time	Reference
Eutheria	Late Cretaceous 78 ± 2	100 ± 10	McKenna (1975), Clemens *et al.* (1979)
Edentata	?Late Paleocene 55 ± 4	80 ± 5	Simpson (1978), McKenna (1980)
Pholidota[c]	Late Paleocene 55 ± 4	?60 ± 5	Emry (1970), B. Patterson (1978)
Carnivora	Middle Paleocene 62 ± 2	70 ± 5	Van Valen (1966), Tedford (1976)
Lagomorpha	Late Paleocence 52 ± 2	62 ± 2	Wood (1957), Van Valen (1964), McKenna (1975)
Rodentia	Late Paleocene 52 ± 2	62 ± 2	Wood (1962), Dawson (1967)
Macroscelidea[d]	Late Oligocene 30 ± 3	?55 ± 5	B. Patterson (1965), Butler (1978)
Chiroptera	Early Eocene 50 ± 1	60 ± 5	Smith (1976), Van Valen (1979)
Dermoptera[e]	Middle Paleocene 57 ± 3	65 ± 5	Rose and Simons (1977)
Scandentia[f]	Miocene 10 ± 2	70 ± 5	Jacobs (1980)
Primates	Late Cretaceous 65 ± 2	70 ± 5	Van Valen and Sloan (1965), Clemens (1974)
Insectivora[g]	Late Cretaceous 65 ± 2	70 ± 5	Van Valen (1967), Butler (1972)
Tubulidentata	Miocene 20 ± 2	?55 ± 5	B. Patterson (1978)
Artiodactyla[h]	Early Eocene 53 ± 1	57 ± 5	McKenna (1969), Rose (1981)
Cetacea	Middle Eocene 45 ± 3	52 ± 3	Van Valen (1968), Barnes and Mitchell (1978)
Perissodactyla[h]	Early Eocene 53 ± 1	58 ± 4	McKenna (1969), Rose (1981)
Hyracoidea[i]	?Eocence ?45 ± 5	?58 ± 3	Meyer (1978), Sudre (1979)
Proboscidea	Early Oligence 30 ± 3	50 ± 5	Maglio (1973), Coppens *et al.* (1978)
Sirenia	Early–Middle Eocene 45 ± 5	50 ± 5	Domning (1978)

[a]All dates are in millions of years before the present.

[b]Dates for first occurrences are based on correlation charts in Berggren *et al.* (1978) and Szalay and Delson (1979).

[c]A Late Paleocene first occurrence for Pholidota supports Emry's (1970) allocation of Metacheiromyidae to this order, an action queried by B. Patterson (1978). The oldest undisputed pholidotans are Early Oligocene in age (Emry, 1970).

[d]Macroscelidea are members of a clade that diverged early in eutherian history, but they probably shared a more recent common ancestry with anagalids or other Early Tertiary

Table II *(continued)*

groups from Asia (see Fig. 1). Hence, the divergence date for elephant shrews may be more recent than the rodent–lagomorph split.

[e]The assignment of the extinct Plagiomenidae (Rose and Simons, 1977) to the Dermoptera is endorsed here, and accordingly the first occurrence of this order is listed as Middle Paleocene. Other than plagiomenids, dermopterans do not have a fossil record.

[f]The Siwalik sequence of Pakistan is the source for the earliest tree shrews. The lowest occurrence of the tupaiid specimens are from an assemblage considered Middle Miocene in age (Jacobs, 1980). Some Early Tertiary groups (adapisoricids, microsyopids, mixodectids) alleged to be fossil tupaiids are not recognized as such here.

[g]Insectivora as used here is essentially equivalent to Haeckel's Lipotyphla (Butler, 1972). This concept excludes a diversity of primitive eutherians from the Late Cretaceous and Paleocene that were formerly assigned to the Insectivora [see also comments in McKenna (1975), Novacek (1980)].

[h]The first appearances of the Artiodactyla and Perissodactyla in the fossil record are now put near the Paleocene–Eocene boundary [end of the Clarkforkian land-mammal age, beginning of the Wasatchian land-mammal age (Rose, 1981)].

[i]The exact age of the North African Eocene assemblage with Hyracoidea (Sudre, 1979) is in doubt. The earliest undisputed occurrence for this group is in the Lower Oligocene sequence from the Fayum Depression of Egypt (Meyer, 1978).

from the late Paleocene of South America (Ameghino, 1906; Simpson, 1978; McKenna, 1980), but the early divergence of the group from other Eutheria, as is depicted in Fig. 1, would suggest a notably more remote time of origin for this order. The affinities of pholidotans, tubulidentates, and a few others are so poorly understood that divergence times for these groups can be suggested only as very tentative approximations.

Several estimates of divergence times listed in Table II fall roughly into the same category, an expected result in light of the multichotomies shown in Fig. 1. Carnivora, Dermoptera, Primates, and Insectivora (*sensu stricto*) are all well represented by the middle Paleocene, but they are either absent or very sparsely represented in the latest Cretaceous. A range of divergence times slightly older than the Cretaceous–Paleocene boundary is suggested for some of these taxa, while other taxa allow for a range that actually brackets this boundary (see Table II). Hence, several eutherian orders may have arisen in the Late Mesozoic but may not have diversified enough to be effectively sampled before the early or middle Paleocene.

Various ungulate groups are suggested here as having divergence dates in the later Paleocene or earlier Eocene. Even though condylarths appear in the Late Cretaceous record, radiations of these primitive forms and their more "modern" ungulate sister taxa do not occur until the Paleocene. Unfortunately, the Early Tertiary fossil record of Africa, a continent which served as an important evolutionary theatre for many ungulate and "paenungulate" groups, is extremely spotty. Explanations for additional estimates of divergence times are given as footnotes to Table II. Relevant paleontologic references for a particular taxon are also given in the table.

6. Eutherian Phylogeny, Fossil History, and Evolutionary Rates

A cladistic perspective does not demand that all evolution consist of a series of dichotomous events, but it does assume that a dichotomous pattern is one of maximum resolution, and is therefore a primary objective of phylogenetic reconstruction. Some view this as an unreasonable expectation of systematic analysis, arguing that dichotomous patterns can misrepresent many phylogenetic histories. This objection has certainly been raised with respect to the problem of eutherian diversification. Thus, Simpson (1978, p. 326) writes: "Incidentally, I do not believe that the Xenarthra are the oldest branch from the ancestral Eutheria or that the orders of mammals arose in a series of successive dichotomies. I believe that most of the orders, particularly those known from the late Cretaceous and Paleocene, arose by what might be called polytomy, and that a Hennigian cladogram is not in this connection, or in most others, a likely representation of a true phylogeny, but is an artificial key."

Simpson's position would doubtless be shared by many paleomammalogists. However, it is far from clear what exactly constitutes evidence of polytomy. Multichotomous patterns like those shown in Fig. 1 might be a reflection of true phylogeny, but they might also be the expected result of a limited data set. A point emphasized throughout this chapter is that the morphologic data set for higher level eutherian phylogeny could stand some improvement. Hence, the simple acceptance of polytomy without attempts to resolve phylogenies into more discrete events does not seem very constructive. This argument applies regardless of the manner in which phylogenies are converted into classifications.

The impression that polytomy is a true picture of higher eutherian phylogeny is obviously influenced by the nature of the fossil record. At least seven of the 18 ordinal-level taxa listed in Table II have recorded first occurrences in the Late Cretaceous or Paleocene. However, the range of first occurrences of these orders spans a considerable length of time—about 15 million years—which seems a more than adequate period for some dichotomous branching. The range of first occurrences for all modern mammals is of course greater—about 38 million years—even if groups with strongly biased records are excluded from this estimate. Postulated divergence times shown in Table II represent a span of some 30 million years. By these accounts, the diversification of eutherian mammals was not an overnight phenomenon.

Even if one were to admit that the time frame for the radiation of eutherians was somewhat broader than popularly portrayed, one might still claim that this group demonstrates rates of diversification that are notably greater than in many other animals. This distinction was first effectively documented by Simpson (1944), who compared various measures of evolutionary rates in mammals and bivalves. Several recent studies are in basic support of Simpson's conclusions. Stanley (1979), for example, estimates that the early diversification of placental mammals requires net rates of increase of species R of 0.39 my^{-1} (Stanley, 1979, p. 139), in comparison to R values of 0.061 my^{-1} for the average bivalve family. These figures are calculated from the standard growth equation

$$N = N_0 \, e^{Rt}$$

where N is the number of species at the end of time t, N_0 is the original number of species (unity for monophyly), and e is the base of the natural logarithms.

The phylogenetic considerations presented here, as well as the postulated divergence times listed in Table II, somewhat temper conclusions regarding comparative evolutionary rates in clams and mammals. Stanley calculates R for eutherian mammals by assuming the existence of only three basic lineages in the Late Cretaceous (at about 70 my bp) and postulating the existence of about 1000 species (or about one-quarter of the living species) near the end of the Paleocene, some 15 million years later. Acceptance of the divergence dates in Table II would lead one to conclude that species diversity by the end of the Cretaceous was considerably greater than Stanley's calculation allows. In fact, a recent compilation by Kielan-Jaworowska *et al.* (1979) shows that at least 33 species of Late Cretaceous eutherians are now formally recognized. Nevertheless, Stanley appropriately notes that different values for N and N_0 may not greatly alter the results, because these variables are entered logarithmically into the calculation of R. What does alter the results significantly is the value of t, and, as noted above, this variable is subject to a considerable range of creative choice. For instance, if a range of 90–110 my bp for the origin time of eutherians is applied to the calculation of net rate of species increase, the corresponding R for the minimum date is 0.91 my^{-1} and R for the maximum date is 0.12 my^{-1}. These values are still greater than the average rates given for bivalve families, but considerably less than Stanley's value of 0.39 my^{-1} for the early Cenozoic diversification of eutherians.

The lack of precision concerning paleontologic estimates of divergence times can affect other kinds of evolutionary rates. A case in point is the calibration of the molecular clock, or the measure of metronomic or stochastic rates of protein evolution. Fitch (1977) concluded that the clock may not work well for individual proteins, but, if averaged over enough substitutions in enough protein genes and enough taxa, a rather good fit should result for divergence times based on nucleotide substitutions and dates from the fossil record. As evidence, Fitch (1977) plotted accumulated substitutions against paleontologic divergence dates using a set of 687 nucleotide substitutions for seven proteins of 17 mammalian taxa. This plot showed that most points (except those for primates) fall near a line drawn through the origin and a point representing the eutherian divergence date determined by accumulated substitutions and paleontologic estimates. The paleontologic divergence date for Eutheria used by Fitch was 120 my bp. However, application of the postulated divergence times given in Table II shifts the points away from the line (Fig. 2). This discrepancy does not mean that the original portrayal of the fit is necessarily wrong; merely that the "slop" in estimates for paleontologic divergence dates can significantly affect the calibration of the clock. In this case it seems that molecular studies require more precision than the fossil record can give.

To sum, both molecular and taxonomic assessments of evolutionary rates in eutherians are limited critically by the imprecision of estimates of divergence dates from fossil and phylogenetic data. What we can say is that there was a pulse of diversification—if a radiation occurring anywhere between 75 and 60 my bp can be called a "pulse"—that was part of the early history of eutherians. One of the more interesting items to come from recent paleontologic work is that the familiar Late Cretaceous–Paleocene radiation may not represent the first such event for eutherians. It is conceivable, as noted above, that an earlier radiation is represented by a morphologically diverse fauna from the Late Cretaceous of Central Asia. A more complete picture of mammalian evolution from such earlier times and a reemphasis of broad-based comparative anatomy for living and fossil eutherians should contribute to a resolved pattern of higher level phylogeny for this group. This pattern must be established in a more satisfactory way if the many inferences concerning eutherian rates of evolution are to rest on firmer ground.

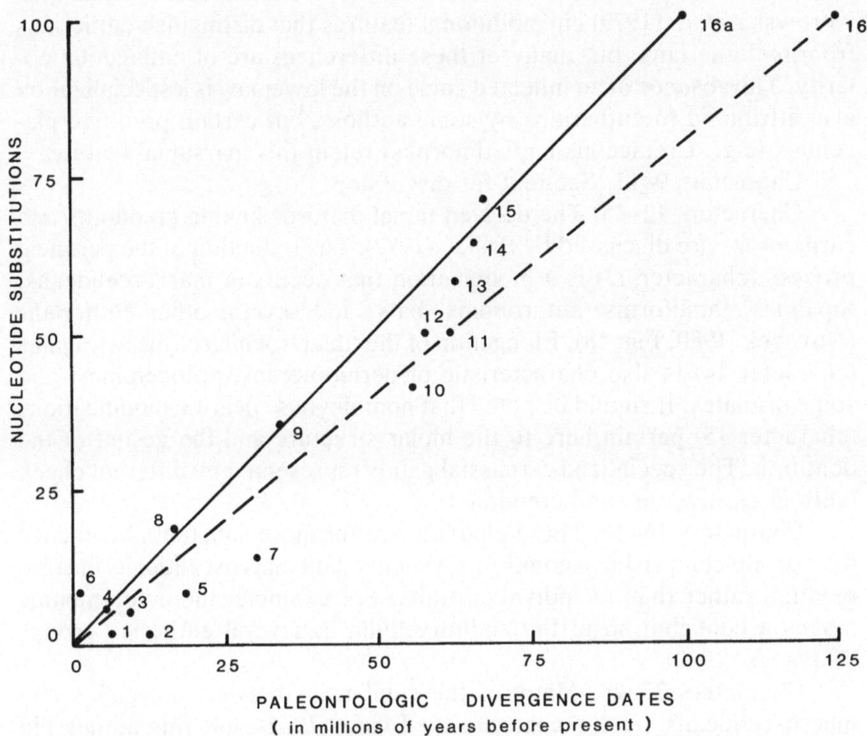

Figure 2. Relation of nucleotide substitutions from a maximum parsimony procedure and estimated dates of divergence based on the fossil record (abscissa). Nucleotide substitutions are from Fitch (1977, Figs. 8 and 9). Dashed line is the line between the origin and point 16, which represents the coordinates for accumulated substitutions and the marsupial–placental divergence date used by Fitch (1977, Fig. 9). Solid line is drawn through point 16a, which represents a fix based on a divergence date of 100 my bp, as suggested by Table II here. All other points represent divergences used by Fitch readjusted for estimates of divergence times used in this chapter. These points are: marsupial–placental divergence (16a); lagomorph, rodent, primate–carnivore, perissodactyl, artiodactyl (15); carnivore–perissodactyl, artiodactyl (14); lagomorph, rodent–primate (13); perissodactyl–artiodactyl (12); lagomorph–rodent (11); camelid, bovid–suid (10); camelid–bovid (9); sheep, goat–cow (8); hominoids–lower anthropoids (7); horse–donkey (6); man, chimp, gorilla–gibbon (5); sheep–goat (4); rhesus–"*Aethiops*" (3); man, chimp–gorilla (2); man–chimp (1). All divergence estimates have a subjective range of deviation between 1 and 8 million years. Point 16a is the midpoint of the range of 90–110 my bp suggested here as the divergence date for Eutheria.

Appendix. Notes on Characters Listed in Table I

Characters 1–8. See text and references cited therein. Kielan-Jaworowska *et al.* (1979) cite additional features that distinguish eutherians from metatherians, but many of these differences are of ambiguous polarity. The absence of an inflected angle on the lower jaw is a specialization also attributed to eutherians by some authors, but certain primitive placentals (e.g., Cretaceous leptictimorphs) retain this marsupial feature.

Characters 9–11. See text for discussion.

Characters 12–15. The derived tarsal features linking creodonts and carnivorans are discussed by Szalay (1977). The reduction of the peroneal process (character 13) is a modification that occurs in macroscelideans, tupaiines, tarsiiforms, anthropoids, bats, and several other eutherians (Novacek, 1980, Fig. 18). Elongation of the tibial trochlea of the astragalus (character 14) is also characteristic of dermopterans, ptilocercines, and some primates. It should be noted that homologous shearing modifications (character 15) pertain here to the molar structure and the gestalt of the dentition. The specialized carnassial pair is represented by different cheek teeth in carnivorans and creodonts.

Characters 16–21. These characters offer more substantial evidence for a group comprising lagomorphs, rodents, and macroscelideans in combination rather than as individual traits. For example, the ectotympanic makes a contribution to the auditory bulla in several eutherian groups (Novacek, 1980).

Characters 22–28. Many of the similarities between anagalids and macroscelideans were recognized by Evans (1942), but this author did not make distinctions between derived and primitive traits in most cases. More recent discussions of these features are provided by McKenna (1975), Szalay (1977), and Novacek (1980).

Characters 29–35. Dental characters cited here as evidence for the close relationship of lagomorphs and rodents have been recognized by several authors as cases of convergence (see text). The marked expansion of the ectotympanic bone in the bulla (character 32) is regarded here as a further modification of the trend established with character 20 (see above). The composite bulla of Macroscelidea, however, is a divergent and highly specialized condition. The derived fetal membrane traits (characters 33–35) must be interpreted with respect to development as well as structure; for example, hemochorial placenta develop from more incipient membranes in a variety of ways within Eutheria (Luckett, 1977).

Characters 36–63. Many characters uniting dermopterans and chiropterans are further modified in characters that support the monophyly of bats. For example, the patagium of dermopterans (character 36) became an effective airfoil for powered flight in chiropterans. It is emphasized that many of the characters cited as evidence for these relationships are not restricted to the flight apparatus (e.g., characters 38, 45–49, 59, 61–63). Polarities and distributions of characters 36–68 are discussed in Leche (1885), Winge (1917, 1941), Miller (1907), Jepsen (1970), Van Valen (1979), Novacek and Luckett (unpublished data), and others.

Characters 69–73. Several characters cited by Gregory (1910) in his detailed comparisons of hyraxes, proboscideans, and sirenians are not applied here, because they are either primitive, nonhomologous, or of ambiguous polarity. As Domning (1978) emphasizes, there is compelling paleontologic and biogeographic evidence for association of proboscideans, sirenians, and the extinct moeritheres. These groups probably diversified in the region of the Tethys Sea in the Early Tertiary.

Note added in proof: Storch (1981) has recently described *Eurotamandua*, an alleged myrmecophagid from the middle Eocene of West Germany. This discovery contradicts statements made above that edentates lack a documented early evolutionary history outside of South America. Storch's assignment of *Eurotamandua* to the Edentata is, however, open to question. His reconstruction of the xenarthrous condition of the vertebrae in *Eurotamandua* (Fig. 8) does not resemble the xenarthry characteristic of edentates. Skull features cited by Storch as evidence for a close relationships between the Eocene genus and modern anteaters are primarily primitive traits. Based on the published illustrations and text, it is still not clear why *Eurotamandua* is excluded from the Pholidota. For the present, the idea of long term isolation of edentates in South America is not seriously challenged.

References

Ameghino, F., 1906, Les Formations sédimentaires du Cretácé Supérieur et du Tertiaire de Patagonie avec un parallèle entre leurs faunes mammalogiques et celles de l'Ancien continent, *An. Mus. Nac. Buenos Aires* **15**(3):1–568.

Barnes, L. G., and Mitchell, E., 1978, Cetacea, in: *Evolution of African Mammals* (V. J. Maglio and H. B. S. Cooke, eds.), Harvard University Press, Cambridge, Massachusetts, pp. 582–602.

Berggren, W. A., McKenna, M. C., Hardenbol, J., and Obradovich, J. D., 1978, Revised Paleogene polarity time scale, *J. Geol.* **86**:67–81.

Blainville, H. M. D., De, 1816, Prodrome d'une nouvelle distribution systématique de règne animal, *Bull. Soc. Philom.* **1816**:105–124.

Blainville, H. M. D., De, 1836, Classification des mammifères, 1834, in: *Dictionnaire pittoresque d'Histoire naturelle* (F. L. P. Gervais, ed.), Volume 4, p. 619, M. M. Renard, Martinet et Cie, Paris.

Blainville, H. M. D., De, 1839–1864, *Ostéographie ou description iconographique comparée du squellette et du système dentaire des mammifères*, J. B. Baillière et Fils, Paris.

Bown, T. M., and Kraus, M. J., 1979, Origin of the tribosphenic molar and metatherian and eutherian dental formulae, in: *Mesozoic Mammals: The First Two-Thirds of Mammalian History* (J. A. Lillegraven, Z. Kielan-Jaworowska, and W. A. Clemens, eds.), University of California Press, Berkeley, California, pp. 172–182.

Brandt, J. F., 1855, Beiträge zur nähern Kenntniss der Säugethiere Russlands, *Mem. Acad. Imp. Sci. St. Pétersbourg, Sér. 6* **9:**1–365.

Butler, P. M., 1956, The skull of *Ictops* and a classification of the Insectivora, *Proc. Zool. Soc. Lond.* **126:**453–481.

Butler, P. M., 1972, The problem of insectivore classification, in: *Studies in Vertebrate Evolution* (K. A. Joysey and T. S. Kemp, eds.), Winchester Press, New York, pp. 253–265.

Butler, P. M., 1978, Insectivora and Chiroptera, in: *Evolution of African Mammals* (V. J. Maglio and H. B. S. Cooke, eds.), Harvard University Press, Cambridge, Massachusetts, pp. 56–68.

Butler, P. M., 1980, The tupaiid dentition, in: *Comparative Biology and Evolutionary Relationships of Tree Shrews* (W. P. Luckett, ed.), Plenum Press, New York, pp. 171–204.

Campbell, C. B. G., 1974, On the phyletic relationships of tree shrews, *Mammal Rev.* **4:**125–143.

Campbell, C. B. G., 1980, The nervous system of the Tupaiidae: its bearing on phyletic relationships, in: *Comparative Biology and Evolutionary Relationships of Tree Shrews* (W. P. Luckett, ed.), Plenum Press, New York, pp. 219–242.

Carlsson, A., 1922, Uber die Tupaiidae und ihre Beziehungen zu den Insectivora und den Prosimiae, *Acta Zool.* **3:**227–270.

Cartmill, M., 1975, Strepsirhine basicranial structures and affinities of Cheirogaleidae, in: *Phylogeny of the Primates* (W. P. Luckett and F. S. Szalay, eds.), Plenum Press, New York, pp. 313–354.

Cartmill, M., and MacPhee, R. D. E., 1980, Tupaiid affinities: The evidence of the carotid arteries and cranial skeleton, in: *Comparative Biology and Evolutionary Relationships of Tree Shrews* (W. P. Luckett, ed.), Plenum Press, New York, pp. 95–132.

Clemens, W. A., 1973, Fossil mammals of the type Lance Formation, Wyoming, Part III. Eutheria and Summary, *Univ. Calif. Publ. Geol. Sci.* **94:**i–vi, 1–102.

Clemens, W. A., 1974, *Purgatorius*, an early paromomyid primate (Mammalia), *Science* **184:**903–905.

Clemens, W. A., Lillegraven, J. A., Lindsay, E. H., and Simpson, G. G., 1979, Where, when and what—A survey of known Mesozoic mammal distribution, in: *Mesozoic Mammals: The First Two-Thirds of Mammalian History* (J. A. Lillegraven, Z. Kielan-Jaworowska, and W. A. Clemens, eds.), University of California Press, Berkeley, California, pp. 7–58.

Conroy, G. C., and Wible, J. R., 1978, Middle ear morphology in *Lemur variegatus*: Some implications for primate paleontology, *Folia Primatol.* **29:**81–85.

Coppens, Y., Maglio, V. J., Madden, C. T., and Beden, M., 1978, Proboscidea, in: *Evolution of African Mammals* (V. J. Maglio and H. B. S. Cooke, eds.), Harvard University Press, Cambridge, Massachusetts, pp. 336–367.

Crompton, A. W., and Jenkins, F. A., Jr., 1979, Origin of mammals, in: *Mesozoic Mammals: The First Two-Thirds of Mammalian History* (J. A. Lillegraven, Z. Kielan-Jaworowska, and W. A. Clemens, eds.), University of California Press, Berkeley, California, pp. 59–73.

Dawson, M. R., 1967, The fossil history of the families of recent mammals, in: *Recent Mammals of the World, A Synopsis of Families*, Ronald Press, New York, pp. 12–53.

Déne, H., Goodman, G., Prychodko, W., and Matsuda, G., 1980, Molecular evidence for the affinities of Tupaiidae, in: *Comparative Biology and Evolutionary Relationships of Tree Shrews* (W. P. Luckett, ed.), Plenum Press, New York, pp. 269–291.

Dobson, G. E., 1882, *A Monograph of the Insectivora, Systematic and Anatomical*, Part I, John Van Voorst, London, pp. 1–96.

Dobson, G. E., 1883, *A Monograph of the Insectivora, Systematic and Anatomical*, Part II, John Van Voorst, London, pp. 97–172.

Domning, D. P., 1978, Sirenia, in: *Evolution of African Mammals* (V. J. Maglio and H. B. S. Cooke, eds.), Harvard University Press, Cambridge, Massachusetts, pp. 573–581.

Doran, A. H. G., 1878, Morphology of the mammalian *Ossicula auditus*, *Trans. Linn. Soc. Lond. (Zool.)* **1**:371–497.

Edwards, H. M., 1868, *Recherches pour servir à l'histoire naturelle des mammifères comprenant des considérations sur la classification de ces animaux*, G. Masson, Paris.

Emry, R. J., 1970, A North American Oligocene pangolin and other additions to the Pholidota, *Bull. Am. Mus. Nat. Hist.* **142**(6):455–510.

Evans, F. G., 1942, The osteology and relationships of the elephant shrews (Macroscelididae), *Bull. Am. Mus. Nat. Hist.* **80**:85–125.

Fitch, W. M., 1977, The phyletic interpretation of macromolecular sequence information: Sample cases, in: *Major Patterns in Vertebrate Evolution* (M. K. Hecht, P. C. Goody, and B. B. Hecht, eds.), Plenum Press, New York, pp. 211–248.

Fleischer, G., 1970, Studien am Skelett des Gehörorgans der Säugetiere, einschliesslich des Menschen, *Saeugetierkd. Mitt.* **21**:131–239.

Gervais, F. L. P., 1849, Rongeurs, in: *Dictionnaire universal d'histoire naturelle*, Volume 11 (M. Ch. d'Orbigny and M. M. Renard, eds.), Martinet et Cie, Paris, pp. 198–204.

Gidley, J. W., 1912, The lagomorphs, An independent order, *Science N. S.* **36**(922):285–286.

Gill, T., 1870, On the relationships of the orders of mammals, in: *Proceedings of the American Association for the Advancement of Science, 19th Meeting*, pp. 267–270.

Gill, T., 1872, Arrangements of the families of mammals with analytical tables, *Smithson. Misc. Collect.* **11**(1):i–vi, 1–98.

Gill, T., 1873, On the status of Aristotle in systematic zoology, *Am. Nat.* **vii**:458–463.

Goodman, M., 1975, Protein sequence and immunological specificity: Their role in phylogenetic studies of Primates, in: *Phylogeny of the Primates* (W. P. Luckett and F. S. Szalay, eds.), Plenum Press, New York, pp. 219–248.

Grassé, P.-P., 1955, Ordre des Édentés, in: *Traite de Zoologie*, Volume 17 (P.-P. Grassé, ed.), Masson, Paris, pp. 1574–1653, 1705–1712.

Gregory, W. K., 1910, The orders of mammals, *Bull. Am. Mus. Nat. Hist.* **27**:1–524.

Haeckel, E., 1866, *Generelle Morphologie der Organismen*, Volume ii, Georg Reimer, Berlin.

Hecht, M. K., 1976, Phylogenetic inference and methodology as applied to the vertebrate record, in: *Evolutionary Biology*, Volume IX (M. K. Hecht, W. C. Steere, and B. Wallace, eds.), Plenum Press, New York, pp. 335–363.

Hoffstetter, R., 1958, Xenarthra, in: *Traite de Paléontologie*, Volume 6(2) (J. Piveteau, ed.), Masson, Paris, pp. 535–636.

Hoffstetter, R., 1970, Radiation initiale des mammifères placentaires et biogéographie, *C.R. Acad. Sci. Paris D* **270**:3027–3030.

Huxley, T. H., 1880, On the application of the laws of evolution to the arrangement of the Vertebrata and more particularly of the Mammalia, *Proc. Zool. Soc. Lond.* **1880**:649–662.

Jacobs, L. L., 1980, Siwalik fossil tree shrews, in: *Comparative Biology and Evolutionary*

Relationships of Tree Shrews (W. P. Luckett, ed.), Plenum Press, New York, pp. 205–216.

Jepsen, G. L., 1970, Bat origins and evolution, in: *Biology of the Bats*, Volume I (W. A. Wimsatt, ed.), Academic Press, New York, pp. 1–65.

Kermack, K. A., and Kielan-Jaworowska, Z., 1971, Therian and non-therian mammals, in: *Early Mammals* (D. M. Kermack and K. A. Kermack, eds.) [*Linn. Soc. Zool. J.* **50**(Suppl. 1)], Academic Press, New York, pp. 103–115.

Kielan-Jaworowska, Z., 1969, Preliminary data on the Upper Cretaceous eutherian mammals from Bayn Dzak, Gobi Desert, *Palaeontol. Pol.* **1969:**171–191.

Kielan-Jaworowska, Z., 1975a, Preliminary description of two new eutherian genera from the Late Cretaceous of Mongolia, *Palaeontol. Pol.* **33:**5–16.

Kielan-Jaworowska, 1975b, Possible occurrence of marsupial bones in Cretaceous eutherian mammals, *Nature* **255:**698–699.

Kielan-Jaworowska, Z., 1977, Evolution of the therian mammals in the Late Cretaceous of Asia. Part II. Postcranial skeleton in *Kennalestes* and *Asioryctes*, *Palaeontol. Pol.* **37:**65–84.

Kielan-Jaworowska, Z., Bown, T. M., and Lillegraven, J. A., 1979, Eutheria, in: *Mesozoic Mammals: The First Two-Thirds of Mammalian History* (J. A. Lillegraven, Z. Kielan-Jaworowska, and W. A. Clemens, eds.), University of California Press, Berkeley, California, pp. 221–258.

Klaauw, C. J., van der, 1931, The auditory bulla in fossil mammals, with a general introduction to this region of the skull, *Bull. Am. Mus. Nat. Hist.* **62:**1–352.

Leche, W., 1885, Uber die Säugerthiergattung *Galeopithecus*, *Svenska Vet. Akad. Handl.* **21:**1–92.

Le Gros Clark, W. E., 1924, The myology of the tree-shrew (*Tupaia minor*), *Proc. Zool. Soc. Lond.* **1924:**461–497.

Le Gros Clark, W. E., 1925, On the skull of *Tupaia*, *Proc. Zool. Soc. Lond.* **1925:**559-567.

Le Gros Clark, W. E., 1926, On the anatomy of the pen-tailed tree-shrew (*Ptilocercus lowii*), *Proc. Zool. Soc. Lond.* **1926:**1179–1309.

Le Gros Clark, W. E., 1959, *The Antecedents of Man*, Edinburgh University Press, Edinburgh.

Le Gros Clark, W. E., 1971, *The Antecedents of Man*, 3rd ed., Edinburgh University Press, Edinburgh.

Lillegraven, J. A., 1969, Latest Cretaceous mammals of upper part of Edmonton Formation of Alberta, Canada, and review of marsupial–placental dichotomy in mammalian evolution, *Univ. Kans. Paleontol. Contrib.* **50**(Vertebrata 12):1–122.

Lillegraven, J. A., 1975, Biological considerations of the marsupial–placental dichotomy, *Evolution* **29:**707–722.

Lillegraven, J. A., 1979, Reproduction in Mesozoic mammals, in: *Mesozoic Mammals: The First Two-Thirds of Mammalian History* (J. A. Lillegraven, Z. Kielan-Jaworowska, and W. A. Clemens, eds.), University of California Press, Berkeley, California, pp. 259–276.

Lillegraven, J. A., Kielan-Jaworowska, Z., and Clemens, W. A. (eds.), 1979, *Mesozoic Mammals: The First Two-Thirds of Mammalian History*, University of California Press, Berkeley, California.

Linnaeus, C., 1735, Systema naturae, sive Regna tria Naturae systematice proposita per Classes, Ordines, Genera et species, *Folia Lugduni Batavorum*.

Luckett, W. P., 1977, Ontogeny of amniote fetal membranes and their application to phylogeny, in: *Major Patterns in Vertebrate Phylogeny* (M. K. Hecht, P. C. Goody, and B. M. Hecht, eds.), Plenum Press, New York, pp. 439–516.

Luckett, W. P., 1980, The use of reproductive and developmental features in assessing tupaiid affinities, in: *Comparative Biology and Evolutionary Relationships of Tree Shrews* (W. P. Luckett, ed.), Plenum Press, New York, pp. 245–266.

MacIntyre, G. T., 1972, The trisulcate petrosal pattern in mammals, in: *Evolutionary Biology*, Volume 6 (M. K. Hecht, W. C. Steere, and B. Wallace, eds.), Plenum Press, New York, pp. 275–303.

MacPhee, R. D. E., 1979, Entotympanics, ontogeny, and primates, *Folia Primatol.* **31**:23–47.

MacPhee, R. D. E., 1981, Auditory regions of primates and eutherian insectivores, Morphology, ontogeny, and character analysis, *Contrib. Primatol.* **18**:i–xv, 1–282.

Maglio, V. J., 1973, Origin and evolution of the Elephantidae, *Trans. Am. Phil. Soc. N.S.* **63**(3):1–149.

Matthew, W. D., 1909, The Carnivora and Insectivora of the Bridger Basin, Middle Eocene, *Mem. Am. Mus. Nat. Hist.* **9**:291–567.

Matthew, W. D., 1918, A revision of the Lower Eocene Wasatch and Wind River faunas, Part 5—Insectivora (continued), Glires, Edentata, *Bull. Am. Mus. Nat. Hist.* **38**:565–657.

Matthew, W. D., 1937, Paleocene faunas of the San Juan basin, New Mexico, *Trans. Am. Phil. Soc. N.S.* **30**:i–viii, 1–510.

McDowell, S. B., 1958, The Greater Antillean insectivores, *Bull. Am. Mus. Nat. Hist.* **115**:113–214.

McKenna, M. C., 1960, Fossil Mammalia of the early Wasatchian Four Mile fauna, Eocene of northwest Colorado, *Univ. Calif. Publ. Geol. Sci.* **37**(1):1–130.

McKenna, M. C., 1961, A note on the origin of rodents, *Am. Mus. Novit.* **2037**:1–5.

McKenna, M. C., 1969, The origin and early differentiation of therian mammals, *Ann. N.Y. Acad. Sci.* **167**(1):217–240.

McKenna, M. C., 1975, Toward a phylogenetic classification of the Mammalia, in: *Phylogeny of the Primates* (W. P. Luckett and F. S. Szalay, eds.), Plenum Press, New York, pp. 21–46.

McKenna, M. C., 1980, Early history and biogeography of South America's extinct land mammals, in: *Evolutionary Biology of the New World Monkeys and Continental Drift* (R. L. Chiochon and A. B. Chiarelli, eds.), Plenum Press, New York, pp. 43–77.

McNab, B., 1978, Energetics of arboreal folivores: Physiological problems and ecological consequences of feeding on an ubiquitous food supply, in: *Ecology of Arboreal Folivores* (C. G. Montgomery, ed.), Smithsonian Institute Press, Washington, D.C., pp. 153–162.

Meyer, G., 1978, Hyracoidea, in: *Evolution of African Mammals* (V. J. Maglio and H. B. S. Cooke, eds.), Harvard University Press, Cambridge, Massachusetts, pp. 284–314.

Miller, G. S., 1907, The families and genera of bats, *Bull. U.S. Nat. Mus.* **57**:1–282.

Novacek, M. J., 1976, Insectivora and Proteutheria of the Later Eocene (Uintan) of San Diego County, California, *Nat. Hist. Mus. Los Angeles County Contrib. Sci.* **283**:1–52.

Novacek, M. J., 1977a, Evolution and relationships of the Leptictidae (Eutheria: Mammalia), Ph.D. dissertation, University of California, Berkeley, California.

Novacek, M. J., 1977b, Aspects of the problem of variation, origin and evolution of the eutherian auditory bulla, *Mammal Rev.* **7**:131–149.

Novacek, M. J., 1980, Cranioskeletal features in tupaiids and selected Eutheria as phylogenetic evidence, in: *Comparative Biology and Evolutionary Relationships of Tree Shrews* (W. P. Luckett, ed.), Plenum Press, New York, 35–93.

Novacek, M. J., 1982, *Diacodon alticuspis*, An erinaceomorph insectivore from the early Eocene of northern New Mexico, *Univ. Wyo. Contrib. Geol.* **20**:135–149.

Osborn, H. F., 1936, *Proboscidea* I, American Museum Press, New York.

Osborn, H. F., 1942, *Proboscidea* II, American Museum Press, New York.

Patterson, B., 1965, The fossil elephant shrews (family Macroscelididae), *Bull. Mus. Comp. Zool. Harv. Univ.* **133**:295–335.

Patterson, B., 1978, Pholidota and Tubilidentata, in: *Evolution of African Mammals* (V. J. Maglio and H. B. S. Cooke, eds.), Harvard University Press, Cambridge, Massachusetts, pp. 268–278.

Patterson, C., 1977, The contribution of paleontology to teleostan phylogeny, in: *Major Patterns in Vertebrate Evolution* (M. K. Hecht, P. C. Goody, and B. M. Hecht, eds.), Plenum Press, New York, pp. 579–643.

Presley, R., 1979, The primitive course of the internal carotid artery in mammals, *Acta Anat.* **103**:238–244.

Ray, J., 1693, *Synopsis Methodica Animalium Quadrupedum et Serpentini Generis*, London.

Reinhart, R. H., 1959, A review of the Sirenia and Desmostylia, *Univ. Calif. Publ. Geol. Sci.* **36**(1):1–146.

Rose, K. D., 1978, A new Paleocene epoicotheriid (Mammalia), with comments on the Palaeanodonta, *J. Paleontol.* **52**(3):658–674.

Rose, K. D., 1981, The Clarkforkian land-mammal age and mammalian faunal composition across the Paleocene–Eocene boundary, *Univ. Mich. Papers Paleontol.* **26**:i–x, 1–197.

Rose, K. D., and Simons, E. L., 1977, Dental function in the Plagiomenidae: Origin and relationships of the mammalian order Dermoptera, *Contrib. Mus. Paleontol. Univ. Mich.* **24**:221–236.

Russell, L. S., 1959, The dentition of rabbits and the origin of lagomorphs, *Bull. Nat. Mus. Canada* **166**:41–45.

Sarich, V. M., and Cronin, J. E., 1976, Molecular systematics of the primates, in: *Molecular Anthropology* (M. Goodman and R. E. Tashian, eds.), Plenum Press, New York, pp. 141–170.

Shoshani, J., Goodman, M., and Prychodko, W., 1978, Cladistic analysis of the Paenungulata by computer, *Am. Zool.* **18**:601.

Shoshani, J., Goodman, M., Barnhart, M., Prychodko, W., Vereshchagin, N. K., and Mikhelson, V. M., 1981, Blood cells and proteins in the Magadan mammoth calf: Immunodiffusion comparisons of *Mammuthus* to extant paenungulates and tissue ultrastructure, in: *The Magadan Mammoth* (N. K. Vereshchagin, ed.), Nauka, Leningrad.

Simpson, G. G., 1944, *Tempo and Mode in Evolution*, Columbia University Press, New York.

Simpson, G. G., 1945, The principles of classification and a classification of mammals, *Bull. Am. Mus. Nat. Hist.* **85**:1–350.

Simpson, G. G., 1978, Early mammals in South America: Fact, controversy, and mystery, *Proc. Am. Phil. Soc.* **122**(5):318–328.

Slaughter, B. H., 1971, Mid-Cretaceous (Albian) therians of the Butler Farm local fauna, Texas, in: *Early Mammals* (D. M. Kermack and K. A. Kermack, eds.) [*Linn. Soc. Zool. J.* **50**(Suppl. 1)], Academic Press, New York, pp. 131–143.

Smith, J. D., 1976, Chiropteran evolution, in: *Biology of the Bats of the New World Family Phyllostomatidae*, Part 1 (R. J. Baker, J. Knox Jones, Jr., and D. C. Carter, eds.), *Spec. Publ. Mus. Texas Tech. Univ.* **10**:48–69.

Smith, J. D., 1977, Comments on flight and the evolution of bats, in: *Major Patterns in Vertebrate Evolution* (M. K. Hecht, P. C. Goody, and B. M. Hecht, eds.), Plenum Press, New York, pp. 427–437.

Smith, J. D., and Madkour, G., 1980, Penial morphology and the question of chiropteran phylogeny, in: *Proceedings Fifth International Bat Research Conference* (D. E. Wilson and A. L. Gardner, eds.), Texas Tech Press, Lubbock, Texas, pp. 347–365.

Stanley, S. M., 1979, *Macroevolution: Pattern and Process*, Freeman, San Francisco.

Storch G., 1981, *Eurotamandua Joresi*, ein Myrmecophagide aus dem Eozän der "Grube Messel" bei Darmstadt (Mammalia, Xenarthra). *Senchenbergiana, lethaea* **61**:247–289.

Sudre, J., 1979, Nouveaux mammifères du Sahara occidental, *Palaeovertebrata* **9**(3):84–115.

Sych, L., 1971, Mixodontia, A new order of mammals from the Paleocene of Mongolia, *Palaeontol. Pol.* **1971**(25):147–158.

Szalay, F. S., 1975, The origin of primate higher categories: An assessment of basicranial evidence, in: *Phylogeny of the Primates* (W. P. Luckett and F. S. Szalay, eds.), Plenum Press, New York, pp. 91–125.

Szalay, F. S., 1977, Phylogenetic relationships and a classification of the eutherian Mammalia, in: *Major Patterns in Vertebrate Evolution* (M. K. Hecht, P. C. Goody, and B. M. Hecht, eds.), Plenum Press, New York, pp. 315–374.

Szalay, F. S., and Delson, E., 1979, *Evolutionary History of the Primates*, Academic Press, New York.

Szalay, F. S., and Drawhorn, G., 1980, Evolution and diversification of the Archonta in an arboreal milieu, in: *Comparative Biology and Evolutionary Relationships of Tree Shrews* (W. P. Luckett, ed.), Plenum Press, New York, pp. 133–169.

Szalay, F. S., and McKenna, M. C., 1971, Beginning of the age of mammals in Asia: The late Paleocene Gashato fauna, Mongolia, *Bull. Am. Mus. Nat. Hist.* **144**:269–318.

Tedford, R. H., 1976, Relationship of pinnipeds to other carnivores (Mammalia), *Syst. Zool.* **25**:363–374.

Tullberg, T., 1899, *Ueber das System der Nagethiere: eine phylogenetische Studie*, Akademischeu Buchdruckerei, Upsala.

Van Valen, L., 1964, A possible origin for rabbits, *Evolution* **18**(3):484–491.

Van Valen, L., 1966, Deltatheridia, A new order of mammals, *Bull. Am. Mus. Nat. Hist.* **132**:1–126.

Van Valen, L., 1967, New Paleocene insectivores and insectivore classification, *Bull. Am. Mus. Nat. Hist.* **135**:217–284.

Van Valen, L., 1968, Monophyly or diphyly and the origins of whales, *Evolution* **22**(1):37–41.

Van Valen, L., 1971, Adaptive zones and the orders of mammals, *Evolution* **25**:420–428.

Van Valen, L., 1979, The evolution of bats, *Evol. Theory* **4**:129–142.

Van Valen, L., and Sloan, R. E., 1965, The earliest primates, *Science* **150**:743–745.

Vaughn, P. P., 1956, The phylogenetic migrations of the ambiens muscle, *J. Elisha Mitchell Sci. Soc.* **72**:243–262.

Walton, D. W., and Walton, G. M., 1970, Postcranial osteology of bats, in: *About Bats* (B. H. Slaughter and D. W. Walton, eds.), Southern Methodist University Press, Dallas, Texas, pp. 93–126.

Weber, M., 1904, *Die Säugetiere: Einführing in die Anatomie und Systematik der recenten und fosilen Mammalia*, Jena.

Winge, H., 1917, Udsigt over Insektaedernes indbyrdes Slaegtskab, *Vidensk. Medd. Dan. Naturhist. Foren.* **68**:83–203.

Winge, H., 1941, *The Interrelationships of the Mammalian Genera*, Volume 1, *Monotremata, Marsupialia, Insectivora, Chiroptera, Edentata*, C.A. Reitzels Forlag, Copenhagen.

Wood, A. E., 1957, What, if anything, is a rabbit? *Evolution* **11**:417–425.

Wood, A. E., 1962, The Early Tertiary rodents of the family Paramyidae, *Trans. Am. Phil. Soc. N.S.* **52**(1):1–261.

CHAPTER 2

Evolution of Mammalian Pancreatic Ribonucleases

JAAP J. BEINTEMA and JOHANNES A. LENSTRA

1. Introduction

Pancreatic ribonucleases form a group of homologous proteins found in considerable quantities in the pancreas of a number of mammalian taxa and a few reptiles (Barnard, 1969; Beintema *et al.*, 1973). The ribonuclease activity varies greatly in different species. Large quantities are found in ruminants and species that have a ruminant-like digestion, and in a number of species with cecal digestion (Fig. 1). Barnard (1969) proposed that an elevated level of pancreatic ribonuclease is the response to the necessity of digesting large amounts of ribonucleic acid derived from the microflora of the stomach of ruminants. This explanation agrees with the observation of Dobson and Wilson (1980) that the level of stomach lysozyme is also elevated in several ruminants and species that have a ruminant-like digestion.

In this review we will discuss the molecular evolution of pancreatic ribonuclease with special emphasis on possible relationships between properties of the enzyme and the nature of the digestive systems of the species involved. As rumination and ruminant-like digestion have evolved several times independently during the evolution of mammals (Moir, 1968; Fig. 2), pancreatic ribonucleases may well show different solutions to the same evolutionary pressure.

JAAP J. BEINTEMA and JOHANNES A. LENSTRA • Biochemisch Laboratorium, Nijenborgh 16, 9747 AG Groningen, The Netherlands.

Order	Species	Diet	less than 5 μg/g	5-50 μg/g	50-500 μg/g	more than 500 μg/g
MONOTREMES	Echidna		●●			
MARSUPIALS	Oppossum			●		
	Phalanger				●	
	Kangaroos	XX			●●●●●●●●●	
	Rat kangaroo			●		
INSECTIVORES			●●●●			
BATS	Fruit bats		●●●			
	Insectivorous bats		●			
PRIMATES	Colobinae	XX		●		
	Other primates		●●			
EDENTATES	Anteaters		●	● ●●	●	
	Sloths	XX		●	●	
	Amadillos		● ●	●	●●●	
PANGOLINS	Pangolin		●			
LAGOMORPHS	Rabbit	X	●			
RODENTS	Sciuromorphs		●	● ●●		
	Myomorphs	XX			● ●●	
	Hystricomorphs	XX		●●	●●●●●●●	
CETACEANS	Toothed whales		●●	●●●		
	Whalebone whales			● ●		
CARNIVORES			●●●●●●●			
SEALS			● ●			
ELEPHANTS	Indian elephant*	X	●			
HYRAXES		X	●●●			
SIRENIANS	Manatee*	X	●			
ODD-TOED UNGULATES	Horses,	X		●	●●	
	Tapir, Rhinoceros, Zebra	X	●●●			
EVEN-TOED UNGULATES	Pigs	X		●	●●	
	Hippopotamus	XX			●	
	Camels	XX		●	●●●●	
	Deer	XXX			●●●●●●●●	●●●●
	Giraffids	XXX			●	●●
	Bovids	XXX			[illegible]	[illegible]

Figure 1. Pancreatic ribonuclease content in mammalian taxa. (xxx) Ruminants; (xx) ruminant-like digestion; (x) (mainly) herbivorous. (*) Pancreas badly autolysed. [Figure from Beintema *et al.* (1973), with some new data added.]

2. Trees

Amino acid sequences of pancreatic ribonucleases from 35 species are known (Table I). Evolutionary trees of these sequences were constructed by the maximum parsimony procedure (Fitch, 1977). The amino acid sequence of bovine seminal ribonuclease was found to be homologous to those of the pancreas, and therefore was also included in this investigation. These studies were done in Madison with Dr. W. M. Fitch, with frequent telephone contacts with Dr. M. Goodman in Detroit, who did the same studies independently. At both institutes equal numbers of nucleotide substitutions were found at all codon positions for trees with identical topologies. Several trees were investigated more thoroughly.

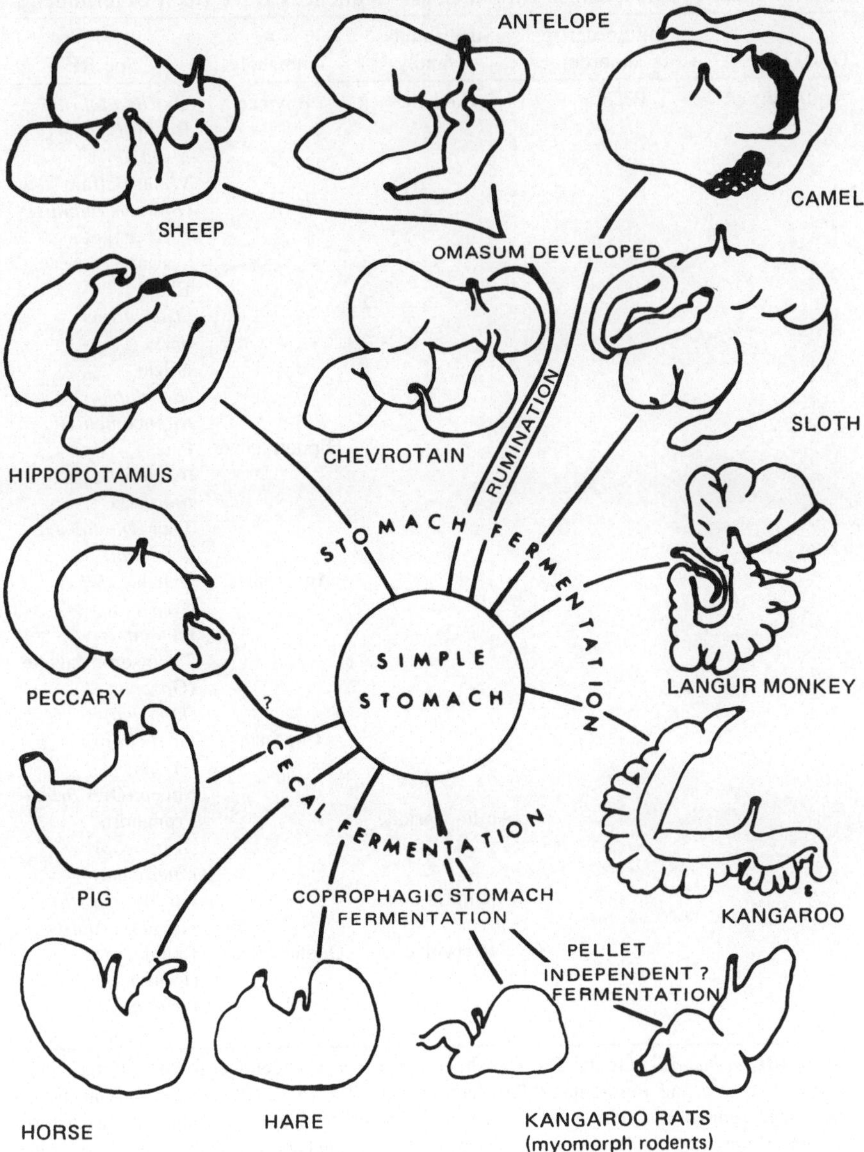

Figure 2. The radiation of stomach forms from the simple stomach. [Figure from Moir (1968).]

Table I
Mammalian Species Whose Ribonuclease Sequences Have Been Determined[a]

Order	Infraorder or suborder	Superfamily or family	Subfamily	Species
Artiodactyla	Pecora	Bovidae	Bovinae	Ox *(Bos taurus)*
				Bison *(Bison bison)*
				Water buffalo *(Bubalus bubalis)*: river type, swamp type
				Eland *(Taurotragus oryx)*
				Nilgai *(Boselaphus tragocamelus)*
			Hippotraginae	Gnu *(Connochaetus taurinus)*
				Topi *(Damaliscus korrigum)*
			Antilopinae	Impala *(Aepyceros melampus)*
				Thomson's gazelle *(Gazella thomsoni)*
			Caprinae	Goat *(Capra hircus)*
				Sheep *(Ovis aries)*
		Antilocapridae		Pronghorn *(Antilocapra americana)*
		Giraffidae		Giraffe *(Giraffa camelopardalis)*
		Cervidae	Odoicoileinae	Reindeer *(Rangifer tarandus)*

[a]Bison *(Bison bison)* is identical to ox; sheep *(Ovis aries)* is identical to goat. References: mouse (Lenstra and Beintema, 1979); hamster (Jekel *et al.*, 1979); kangaroo (Gaastra *et al.*, 1978); pronghorn (Beintema *et al.*, 1979); nilgai, water buffalo, impala, and Thomson's gazelle (Beintema, 1980); hippopotamus and sloth (Havinga and Beintema, 1980); porcupine and casiragua (Beintema *et al.*, 1982); other references can be found in Beintema *et al.* (1977). If heterogeneities were reported, only one component has been used in the determination of the most parsimonious and the biologic tree. A listing of heterogeneities can be found in Beintema *et al.* (1977). Classification and nomenclature of mammalian species is given according to Morris (1965).

Table I (*continued*)

Order	Infraorder or suborder	Superfamily or family	Subfamily	Species
				Roe deer (*Capreolus capreolus*)
				Moose (*Alces alces*)
			Cervinae	Red deer (*Cervus elaphus*)
				Fallow deer (*Dama dama*)
	Tylopoda	Camelidae		Dromedary (*Camelus dromedarius*)
				Bactrian camel (*Camelus bactrianus*)
	Ancodonta			Hippopotamus (*Hippopotamus amphibius*)[b]
	Suina			Pig (*Sus scrofa*)
Cetacea				Lesser rorqual (*Balaenoptera acutorostrata*)
Perissodactyla				Horse (*Equus caballus*)
Rodentia	Myomorpha	Muridae		Rat (*Rattus norvegicus*)
				Mouse (*Mus musculus*)
		Cricetidae		Hamster (*Mesocricetus auratus*)
				Muskrat (*Ondatra zibethica*)
	Hystricomorpha	Hystricoidea		Porcupine (*Hystrix cristata*)[c]
		Cavoidea		Guinea pig (*Cavia porcellus*)
		Chinchilloidea		Chinchilla (*Chinchilla brevicaudata*)

[b]Other component has lysine instead of glutamine at position 37 (Havinga and Beintema, 1980).

[c]Minor component; major component differs in having glycine instead of arginine at position 98 (Beintema *et al.*, 1982).

Table I (*continued*)

Order	Infraorder or suborder	Superfamily or family	Subfamily	Species
		Octodontoidea		Casiragua (*Proechimys guairae*)
				Coypu (*Myocastor coypus*)
Edentata				Two-toed sloth (*Choloepus hoffmanni*)
Marsupialia				Red Kangaroo (*Macropus rufus*)
Ribonuclease from seminal plasma		Ox (*Bos taurus*)		

Note added in proof: Three ribonuclease sequences have been determined recently: man (*Homo sapiens*), cuis (*Galea musteloides*), and capybara (*Hydrochoerus hydrochaeris*). The positions of these sequences in the trees as presented in Figs. 3 and 4 have not yet been determined. Cuis and capybara are close relatives of the guinea pig. The guinea pig pancreas contains two ribonucleases, A and B, as a result of a recent gene duplication. Cuis and capybara both have only one single pancreatic ribonuclease. However, the cuis sequence is more similar to the guinea pig B sequence and the capybara sequence is more similar to the guinea pig A sequence. This would indicate that the gene duplication has occurred in an ancestor of the three species, with both genes expressed in guinea pig, but only one of each expressed in the other two species.

Here we present the results of two of them, a most parsimonious tree requiring 451 nucleotide substitutions and a tree which requires 460 substitutions but more closely approximates current biologic opinion.

2.1. Most Parsimonious Tree

Our previous most parsimonious tree (Beintema *et al.*, 1977) was the starting point of this investigation. Sequences determined since then were added at branching positions which agree most closely with the biologic classification of the species involved. Because of its incompleteness, the ribonuclease sequence of turtle (Beintema *et al.*, 1977) was omitted from these studies. We tried to find a most parsimonious tree by interchanging neighboring branches (branch swapping) and repeating this procedure several times. However, our search did not find a tree, meanwhile found by Dr. M. Goodman in Detroit, which proved to require fewer nucleotide substitutions than all others. This tree is presented in Fig. 3.

In our previous tree (Beintema *et al.*, 1977) the main deviations from

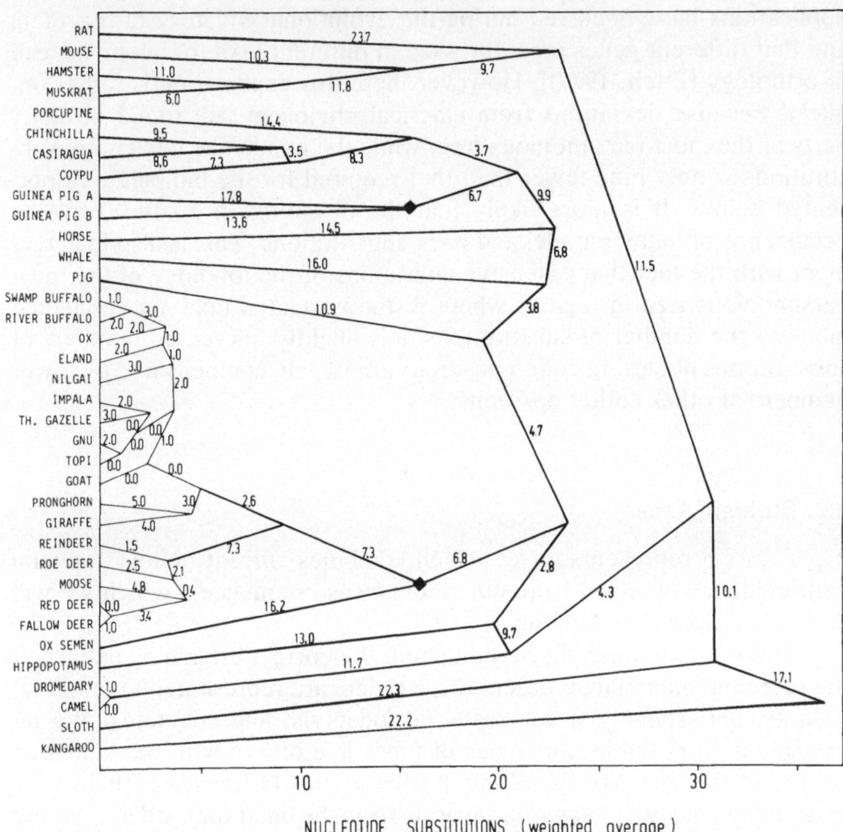

Figure 3. Most parsimonious tree of ribonucleases. The number on each leg is the minimum number of nucleotide substitutions required to account for the descent from the ancestor to its immediate descendant in the tree. Fractions result from averaging over more than one most parsimonious solution. The nodes are placed at a height equal to the weighted average number of nucleotide replacements between the node and its descendant sequences (Fitch, 1977). The species whose ribonucleases were used are given in Table I. (♦) Gene duplication resulting in two paralogous gene products.

biologic opinion were the separation of the myomorph rodents from the hystricomorph rodents and the grouping together of horse with the hystricomorph rodents. The present most parsimonious tree (Fig. 3) shows even more deviations: now not only horse, but also whale and most of the artiodactyls are connected to the hystricomorph rodents, while this entire group is connected to hippopotamus and the camels. In theory, a possible explanation for deviations from biologic opinion is that gene duplications have occurred during the evolutionary history of a protein and that different genes are expressed in different taxa [paralogy instead of orthology (Fitch, 1977)]. However, here this explanation is rather unlikely, because deviations from classical phylogeny are found in many parts of the most parsimonious tree, while the number of nucleotide substitutions is only nine fewer than that required for the biologic tree presented below. It is more likely that the deviations are caused by the occurrence of many parallel and back substitutions. This is also in agreement with the fact that extensive alterations of the topology of the most parsimonious tree in regions where distantly related taxa are connected increase the number of substitutions only slightly: increased numbers of substitutions at certain codon positions are largely compensated by lower numbers at other codon positions.

2.2. Biologic Tree

Figure 4 represents a tree which combines current biologic opinion with evidence obtained from our ribonuclease sequences, which we will discuss in more detail below.

There is still much discussion about the correct branching pattern of the placental mammalian orders. Five orders are represented in our data: rodents, perissodactyls, cetaceans, artiodactyls, and edentates. We investigated all possible topologies of these five orders with the kangaroo as root of the tree. We found that a subgroup of 11 trees, which all have sloth connected with kangaroo separate from the other four orders, require fewer substitutions than the other trees (Fig. 5). This indicates that the divergence of the edentates preceded that of the other four orders. Differences in the branching pattern of the latter have little influence on the total number of substitutions, so our data cannot be used to derive their evolutionary relationships.

Our ribonuclease data give some indications about likely positions of taxa about which there are either no opinions or conflicting biologic opinions. These are (a) the positioning of the giraffe with the pronghorn, (b) the positioning of the hippopotamus with the ruminants (camels and Pecora) and not with the pig, and (c) the positioning of the African porcupine with the South American caviomorphs in the hystricomorph suborder of the rodents (Fig. 6). These conclusions are not only supported by the parsimony criterion, but also by the fact that the shared substitutions in each of these three groupings are observed not at all or only rarely in other taxa. Our hypothesis about the grouping together of giraffe

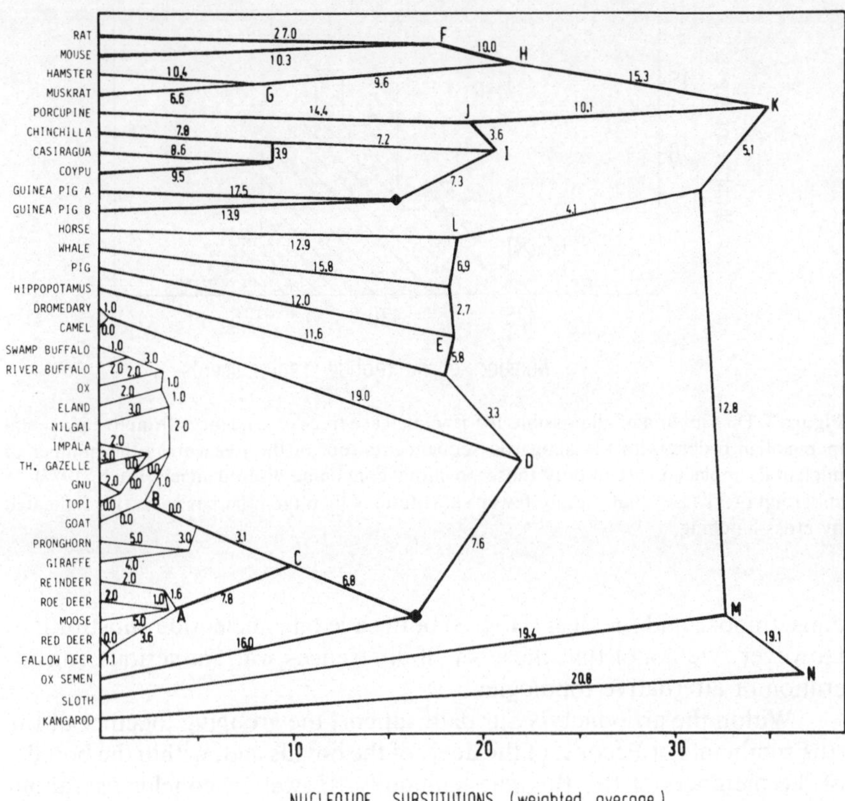

Figure 4. Ribonuclease tree which combines current biologic opinion with evidence from amino acid sequence studies of this protein (see text). For other details see legend to Fig. 3.

with pronghorn is supported by results obtained with the fibrinopeptides of these species (Goodman *et al.*, this volume, Chapter 4). In a phylogenetic study of mammalian α-crystallins, De Jong (this volume, Chapter 3) has also found that hippopotamus joins the camels and Pecora and not pig.

There are other parsimonious groupings for the ribonuclease sequences, which were not used in the biologic tree (Fig. 4) because they were not in agreement with current opinion and barely saved any substitutions. For instance, joining casiragua with chinchilla (Fig. 3) instead of with coypu (Fig. 4), or separation of reindeer from the other Odocoil-

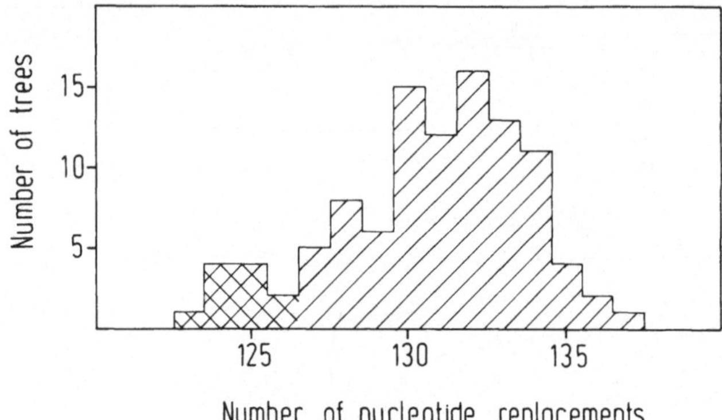

Figure 5. Distribution of all possible 105 ribonuclease trees of sequences from five placental mammalian orders with the kangaroo sequence as root of the tree versus the number of nucleotide replacements in only those positions containing discordancies (Fitch, 1977). A subgroup of 11 trees that require fewer substitutions than the others (see text) is indicated by cross-hatching.

einae (moose and roe deer) (Fig. 3) both save one nucleotide substitution. However, we doubt that these small differences warrant serious consideration of alternative topologies.

Within the artiodactyls our data support the grouping together of the true ruminants or Pecora, of the deer, of the bovids and, within the bovids, of the members of the Bovinae subfamily. However, conclusions about other relationships, between the other investigated members of the bovids, for instance, are not possible, because there are few shared substitutions in their ribonucleases.

3. Evolutionary Rates

The biologic tree was used to calculate rates of ribonuclease evolution (Table II). We did not apply corrections for leg lengths (to correct for unnoticed substitutions). Average rates were calculated for each taxon separately and these average rates were used for the calculation of the average rate in the next higher taxon. Paleontologic ages of ancestors were the same estimates as used in our previous study (Beintema *et al.*, 1977) with some new data communicated by Drs. J. J. Jaeger (Paris) and J. L. Hartenberger (Montpellier). Since there is much uncertainty in the biologic time values, we used approximate values which are multiples of

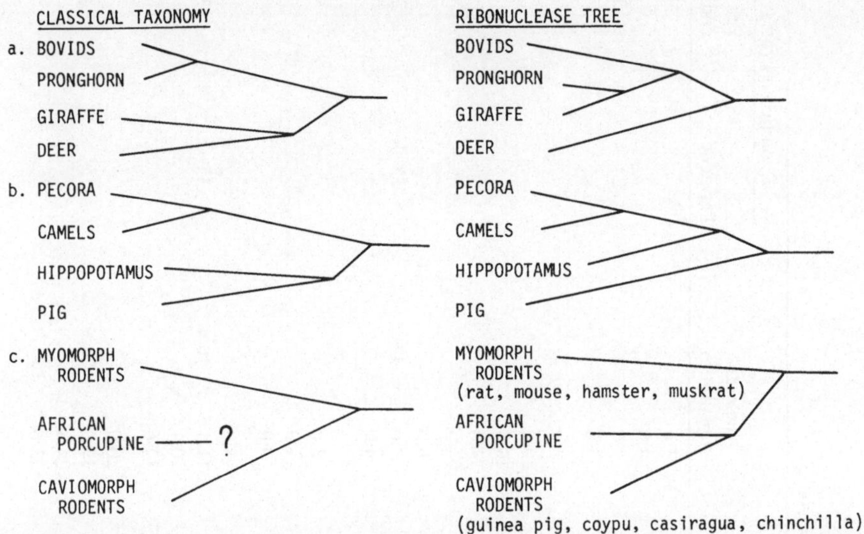

Figure 6. Differences between classical taxonomy and the ribonuclease tree as presented in Fig. 4. References to classical taxonomy: (a) position of giraffe, Simpson (1945, p. 268), Viret (1961), Romer (1966); (b) position of hippopotamus, Simpson (1945, p. 261), Morris (1965); however, see also Romer (1968); (c) position of African porcupine, Simpson (1945, p. 210), Romer (1966).

10×10^6 years. The average overall rate calculated in this way is 61 nucleotide substitutions per 10^8 years per ribonuclease gene (about 125 codons), or about 50 per 100 codons. However, there is much variation in evolutionary rate in different taxa. Values ranging from about 20 (in the Bovidea) to 150 (in the Muridea) were found. While ribonuclease on the average evolved about two times faster than hemoglobin, the ratio of the evolutionary rates in the Bovidae is the reverse, with the hemoglobin chains evolving faster than ribonuclease (Table III). The slow rate of ribonuclease evolution in the Pecora or true ruminants may have been caused by the presence of a stable symbiotic system in the rumen of ruminants, which caused a slowing down of the evolutionary rate of this enzyme. On the contrary, the high rates observed in the myomorph rodents (Table II) may have something to do with the variety in digestive systems in this taxon (see Section 8 on glycosylation and the function of carbohydrate).

In the maximum parsimony procedure of Fitch (1977) used in this study, substitutions are distributed over many branches of the tree by

Table II
Evolutionary Rates of Pancreatic Ribonuclease

Node (Fig. 4)	Ancestor: Taxon or taxa	Ancestor: Age, 10^6 years	Species (sequences): taxon or taxa	Nucleotide substitutions		Evolutionary rate[a]: Per gene	Evolutionary rate[a]: Per 100 codons
			Reindeer	3.6			
			Roe deer	4.6			
A	Cervidae	20	Moose	7.6	4.8	24	20
			Red deer	3.6			
			Fallow deer	4.7			
			Swamp buffalo	9.0			
			River buffalo	10.0			
			Ox	7.0			
			Eland	6.0			
			Nilgai	6.0			
B	Bovidae	20	Impala	3.0	4.9	25	20
			Thomson's gazelle	4.0			
			Gnu	3.0			
			Topi	1.0			
			Goat	0.0			
			Bovidae	8.0			
C	Pecora	30	Pronghorn	11.1	10.4	35	30
			Giraffe	10.1			
			Cervidae	12.6			
			Dromedary	20.0			
D	Ruminantia	40	Camel	19.0	21.9	55	40
			Pecora	24.8			
			Ox semen	23.6			

E	Artiodactyla	50	Pig Hippopotamus Ruminantia	12.0 17.4 31.0	20.1	40	30
F	Muridae	10	Rat Mouse	27.0 10.3	18.8	190	150
G	Cricetidae	10	Hamster Muskrat	10.4 6.6	8.5	85	70
H	Myomorpha	20	Muridae Cricetidae	28.8 18.1	23.5	117	90
I	Caviomorpha	30	Chinchilla Casiragua Coypu Guinea pig A Guinea pig B	15.0 19.7 20.6 24.8 21.2	20.3	68	50
J	Hystricomorpha	40	Porcupine Caviomorpha	14.4 23.9	19.2	48	40
K	Rodentia	50	Myomorpha Hystricomorpha	38.8 29.3	34.1	68	50
L	Hoofed animals + Cetacea	60	Horse Whale Artiodactyla	12.9 22.7 29.7	21.8	36	30
M	Eutheria	90	Rodentia Hoofed animals + Cetacea Sloth	52.0 38.7 19.4	36.7	41	30
N	Theria	120	Eutheria Kangaroo	55.8 20.8	38.3	32	30
			Average			61	50

[a]Nucleotide substitutions per 10^8 years.

Table III
Comparison of Evolutionary Rates (Unaugmented) Observed in Hemoglobins and Pancreatic Ribonucleases[a]

	Hemoglobin	Ribonuclease
Mammals	24	50
Bovidae	39	20
Muridae	76	150

[a]Rates are expressed as nucleotide replacements per 10^8 years per 100 codons. Hemoglobin data are from Goodman (1981, personal communication).

calculating an "average" solution, while in the procedure of Goodman *et al.* (this volume, Chapter 4) substitutions are preferentially located on the branches leading to the extant sequences (Goodman's A solution). In Table IV both methods are compared, both before and after correction for unnoticed substitutions. Dayhoff's amino acid PAMs (Fig. 9-9 of Dayhoff *et al.*, 1972) correction was applied to the data obtained with the method of Fitch. It is evident from the data presented in Table IV that both methods yield similar results.

4. Three-Dimensional Structure of Ribonuclease

The three-dimensional structure of bovine pancreatic ribonuclease is known and many functional properties of the enzyme have been studied (Richards and Wyckoff, 1971). Other pancreatic ribonucleases probably have a main-chain conformation quite similar to that of bovine ribonuclease (Lenstra *et al.*, 1977). Consequently, a comparison of the structure and the properties of the other pancreatic ribonucleases with those of the bovine enzyme may yield information about the functional roles of residues which have—or have not—remained constant during the evolution of the enzyme. Several of the relevant properties are listed in Table V and will be discussed in more detail below.

Figure 7 shows the main-chain conformation of bovine pancreatic ribonuclease. The varied and unvaried residues are indicated in this figure with filled and open circles, respectively, while positions with only two different residues are indicated with partially filled circles. Internal parts of the molecule and the active-site cleft (on the left side of the molecule) contain many unvaried residues, while the other surface parts are more varied.

Table IV
Average Number of Nucleotide Substitutions in Pancreatic Ribonuclease as Calculated with Different Methods

		Uncorrected		Corrected for unnoticed substitutions	
Node (Fig. 4)	Taxon or taxa	Data from Table II	Goodman's unaugmented values	Data from Table II[a]	Goodman's augmented values
A	Cervidae	4.8	4.7	4.8	4.9
B	Bovidae	4.9	4.9	4.9	4.9
C	Pecora	10.4	10.1	10.5	10.7
D	Ruminantia	21.9	21.8	23.4	25.6
E	Artiodactyla	20.1	19.4	21.1	22.3
F	Muridae	18.8	21.0	21.4	27.0
G	Cricetidae	8.5	8.5	8.9	9.5
H	Myomorpha	23.5	25.3	25.5	30.1
I	Caviomorpha	20.3	19.6	21.2	21.8
J	Hystricomorpha	19.2	19.0	20.1	21.4
K	Rodentia	34.1	31.7	36.5	36.3
L	Hoofed animals· + Cetacea	21.8	22.1	22.9	25.8
M	Eutheria	36.7	32.6	39.2	38.5
N	Theria	38.3	—	41.9	—

[a]After applying the correction for amino acid PAMs according to Dayhoff (Fig. 9-9 of Dayhoff *et al.*, 1972).

Lenstra *et al.* (1977) compared different methods to predict secondary structures of proteins with the pancreatic ribonucleases whose primary structures were known at that time (Fig. 8). They concluded that a number of residues are constant or are replaced conservatively in ribonucleases to conserve the hydrophobic contacts necessary for the formation of a stable secondary and tertiary structure (Lim, 1974).

Table V
Characteristics of Pancreatic Ribonucleases That May Have Evolutionary Significance

Formation of a stable secondary and tertiary structure
Enzymic activity on low-molecular-weight nucleotide substrates and on RNA
Enzymic activity on double-stranded RNA
Interaction with inhibitors
Glycosylation

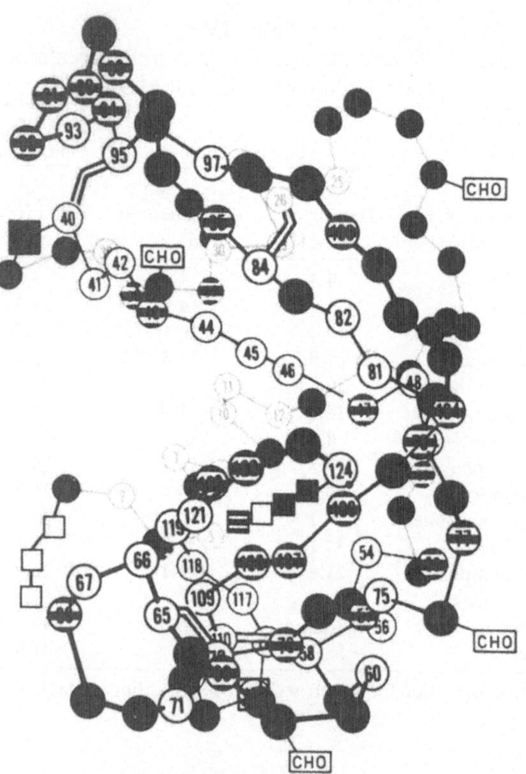

Figure 7. Three-dimensional structure of bovine pancreatic ribonuclease. Amino acid residues at unvaried positions are indicated with open circles, varied positions with closed circles, and positions with only two different residues with partially filled circles. (□) Positions involving additions or deletions. (CHO) Positions where covalently linked carbohydrate is found in one or several ribonucleases. [Figure derived from Dickerson and Geis (1969).]

5. Enzymic Activity on Low-Molecular-Weight Nucleotide Substrates and on RNA

Pancreatic ribonucleases differ little in specific activity on low-molecular-weight nucleotide substrates and on RNA. An indication that differences in amino acid sequence between pancreatic ribonucleases generally do not greatly affect the kinetic properties of these enzymes is shown by our studies on hybrid molecules obtained by mixing S-peptides and S-proteins from different ribonucleases (Welling *et al.*, 1976). For instance, the enzymic parameters and the properties of the active-site

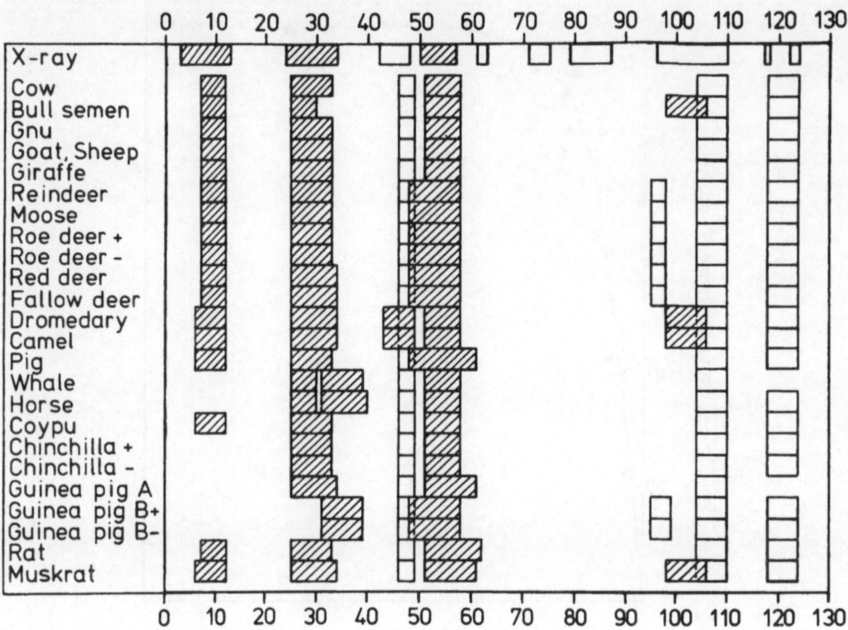

Figure 8. Prediction of ribonuclease amino acids involved in α-helices (shaded boxes) and β-sheets (open boxes) according to Lim (1974).

histidines as revealed by NMR spectroscopy were identical in bovine ribonuclease and in a hybrid formed from bovine S-protein (residues 21–124) and a synthetic rat S-peptide (residues 3–20) (Fig. 9). It should be noted that the rat peptide has only nine residues in common with the bovine S-peptide, one of which is an active-site histidine. Thus, a change of 11 residues in the first 20 positions of the molecule has no influence on either the affinity of this part to the core of the molecule (Welling *et al.*, 1976) or the properties of the resulting complex (Beintema and Lenstra, 1980).

The properties of the active-site histidines as revealed by NMR spectroscopy were found to be different between intact rat and bovine ribonuclease, but became much more similar in the presence of substrate analogs (Migchelsen and Beintema, 1973).

There are indications that amino acid replacements near the C terminus of the molecule have somewhat more influence, apparently affecting the relative preferences for low-molecular-weight nucleotide substrates

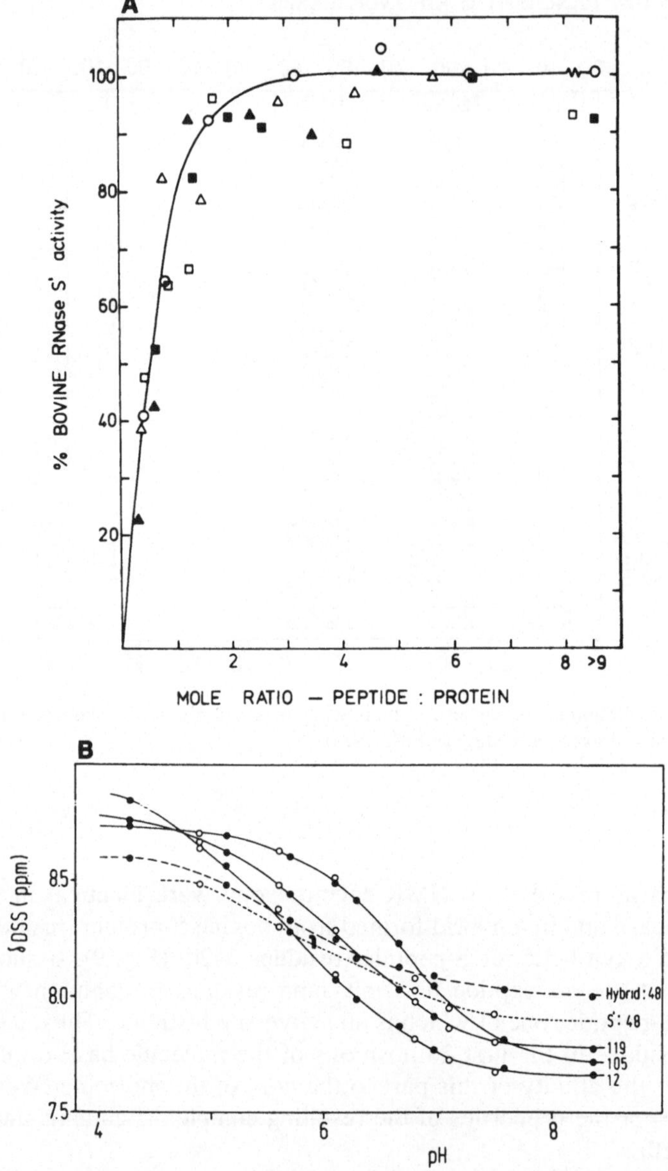

Figure 9. (A) Enzymic activity of different S-peptide–bovine S-protein complexes using cytidine 2′, 3′-cyclic phosphate as substrate. (○) Ox S-peptide; (△) kangaroo S-peptide; (▲) dromedary S-peptide; (□) rat S-peptide (residues 3–14); (■) rat S-peptide (residues 2–14). [Figure from Welling *et al.* (1976).] (B) NMR titration curves of histidines in bovine ribonuclease S′ (○) and a rat-bovine hybrid ribonuclease S′ (●).(—) Titration curves of histidines-12, -105, and -119 in both derivatives; (- - -) titration curve of histidine-48 in bovine ribonuclease S′, (– –) titration curve of histidine-48 in hybrid ribonuclease S′. [Figure from Beintema and Lenstra (1980).]

containing cytidine and uridine (Gaastra *et al.*, 1974; S. Stob, J. A. Lenstra, and J. J. Beintema, unpublished observations).

There are only minor differences between ribonucleases from different species in specific activity on RNA. Generally, yeast RNA has been used as substrate for these studies. It may be interesting to know if there are structural differences between RNAs from the microflora of different ruminants and species with ruminant-like digestion, and if there has been any coadaptation of the enzymic properties of pancreatic ribonucleases from these species.

6. Enzymic Activity on Double-Stranded RNA

Pancreatic ribonucleases differ much in net charge. There is a net change of +8 in going from moose to whale ribonuclease as the extremes (Table VI). Differences in charge have little influence on most kinetic properties of the enzyme (Ronda *et al.*, 1976). However, charge does influence the ability of ribonuclease to cleave double-stranded RNA, for the strongly positively charged enzymes from whale pancreas and bovine semen are considerably more active on double-stranded RNA than is bovine pancreatic ribonuclease, while reindeer ribonuclease, which has a lower net charge than bovine ribonuclease, also has a lower activity on this substrate (Libonati *et al.*, 1976). A second characteristic which influences the activity on double-stranded RNA is the presence of carbohydrate chains on the ribonuclease molecule. Several glycosylated ribonucleases show a higher activity on double-stranded RNA than do carbohydrate-free ribonucleases with similar charge properties of the protein part of the molecule: pig and horse ribonuclease are about ten times more active than bovine pancreatic ribonuclease, and giraffe ribonuclease shows a much higher activity on this substrate than the reindeer enzyme (Carsana *et al.*, 1981).

We do not know if the activity of a pancreatic ribonuclease on double-stranded RNA has any functional significance. Below we shall discuss the possibility that the gene for pancreatic ribonuclease is also used by other cells in the body, in which case the enzyme would not have a role in extracellular digestion exclusively. It is striking that the two enzymes with the highest activity on double-stranded RNA are those for which the biologic role is most obscure: bovine seminal ribonuclease and whale pancreatic ribonuclease. The whalebone whales are the only nonherbivores with elevated levels of ribonuclease in their pancreas (Fig. 1). Whalebone whales eat small crustaceans and possess a compartmentalized stomach, but little is known about their digestive physiology.

Table VI
Number of Charged Residues per Molecule

Species[a]	Lysine and arginine (K + R)	Aspartic acid and glutamic acid (D + E)	Difference (K + R)—(D + E)	Histidine	Sialic acid in main component[b]
Seminal ribonuclease (ox)[c]	18	10	+8	4	–
Pancreatic ribonucleases					
Whale	17	10	+7	5	–
Pronghorn, impala	14	9	+5	4	–
Mouse	15	11	+4	6	–
Hamster	14	10	+4	6	+
Rat	17	13	+4	5	–
Horse, pig	13	9	+4	5	+
Guinea pig A	15	11	+4	4	–
Ox, gnu, topi	14	10	+4	4	–
Swamp buffalo, nilgai, chinchilla	13	9	+4	4	–
Guinea pig B	14	10	+4	3	–
Kangaroo	14	11	+3	5	–

Muskrat	14	11	+3	4	−
Thomson's gazelle	13	10	+3	4	−
Hippopotamus basic component	15	13	+2	5	−
Red deer, fallow deer	13	11	+2	5	−
Porcupine, casiragua	14	12	+2	4	−
Goat	13	11	+2	4	−
River buffalo	12	10	+2	4	−
Dromedary	13	12	+1	6	−
Hippopotamus acid component	14	13	+1	5	−
Roe deer	12	11	+1	5	−
Coypu	13	12	+1	4	−
Camel, sloth	12	12	0	6	−
Reindeer	12	12	0	5	−
Giraffe	11	11	0	4	+
Moose	11	12	−1	5	+

[a]Main ribonuclease component.

[b]Present (+) or absent (−). For a review of carbohydrate in pancreatic ribonucleases see Beintema *et al.* (1976).

[c]Per monomer.

7. Interaction with Inhibitor Proteins

Although high levels of pancreatic-type ribonucleases are only found in the pancreas of a limited number of mammalian taxa and in bovine seminal plasma, very low levels of ribonucleases with similar properties are found in other organs and fluids of a wide variety of mammals. However, the literature about these is rather confusing. In this context it may be significant that proteins that effectively inhibit the enzymic activity of bovine pancreatic ribonuclease have been isolated from several sources, e.g., from *human* placenta (Blackburn and Jailkhani, 1979). A number of other pancreatic ribonucleases have been tested for inhibitor binding, and similar affinities were found (Blackburn and Gavilanes, 1980) (Table VII). Blackburn and Jailkhani (1979) have presented evidence that the interaction with the inhibitor does not affect the histidines in the active site, but that probably the surface region with the active-site lysine-41 and with tyrosine-92 is the interaction site with the inhibitor (Blackburn and Gavilanes, 1980). This region of the surface is relatively well conserved (Fig. 7) and this conservation may have something to do with the inhibitor-binding properties of the molecule.

8. Glycosylation and the Function of Carbohydrate

The most striking differences between pancreatic ribonucleases concern the presence or absence of covalently attached carbohydrate. These differences may be related to the digestive system and the diet of the

Table VII
Relative Affinity of Ribonucleases from Several Mammalian Species for the Ribonuclease Inhibitor from Human Placenta[a]

Pancreatic ribonucleases	
Ox	1.00
Dromedary	0.87
Pig	1.33
Rat	0.96
Mouse	1.37
Hamster	0.76
Seminal ribonucleases	
Ox	
Dimer	0.19
Monomer (carboxyamidomethylated)	1.83

[a]Data from Blackburn and Gavilanes (1980).

species involved. A number of pancreatic ribonucleases have carbohydrate attached to asparagine residues at positions 21, 34, 62, and 76 (Fig. 7). These positions are part of highly variant sequences at the surface of the molecule, far from the active site. Only asparagine residues in Asn-X-Ser/Thr sequences have been found to act as attachment sites, where X may be any residue. Positions 76–78 with the sequence Asn-Thr-Ser with attached carbohydrate in the enzymes from pig and whale (but not in that from hippopotamus) can be taken as an example (Fig. 10). Other species have no carbohydrate at this position because either the asparagine at position 76 or the serine at position 78 has been substituted by another residue. However, not all Asn-X-Ser/Thr sequences possess carbohydrate chains. Other features of the ribonuclease molecule and/or the effectiveness of the carbohydrate-attaching system apparently influence the glycosylation process. Some examples of how changes of the residue occupying position X or the residue directly following the Asn-X-Ser/Thr sequence influence glycosylation, together with indications of differences in the capability of the glycosylating system between related species, will

	76 77 78
rat	-Ser-Ser-Thr-
mouse	-Ser-Ser-Ala-
hamster	-His-Ser-Ala-
muskrat	-Arg-Ser-Ala-
porcupine	-Asn-Ser-Leu-
chinchilla	-Asn-Ser-Asn-
casiragua	-Thr-Ser-Asn-
coypu	-Asn-Ser-Asn-
guinea pig A	-Tyr-Ser-Ser-
guinea pig B	-Tyr-Ser-Arg-
horse	-Ser-Ser-Ser-
whale	CHO* on -Asn-Ser-Thr-
pig	CHO on -Asn-Ser-Thr-
hippopotamus	-Asn-Ser-Thr-
camels	-Ser-Thr-Ser-
bovids	-Tyr-Ser-Thr-
pronghorn	-Tyr-Ser-Thr-
giraffe	-Tyr-Ser-Ala-
deer	-Asn-Ser-Ala-
ox semen	-Lys-Ser-Thr-
sloth	-Arg-Ser-Asn-
kangaroo	-Asn-Ser-Arg-

Figure 10. Amino acid sequence of the carbohydrate attachment site 76–78 in pancreatic ribonucleases. (CHO) Carbohydrate; (*) partially glycosylated.

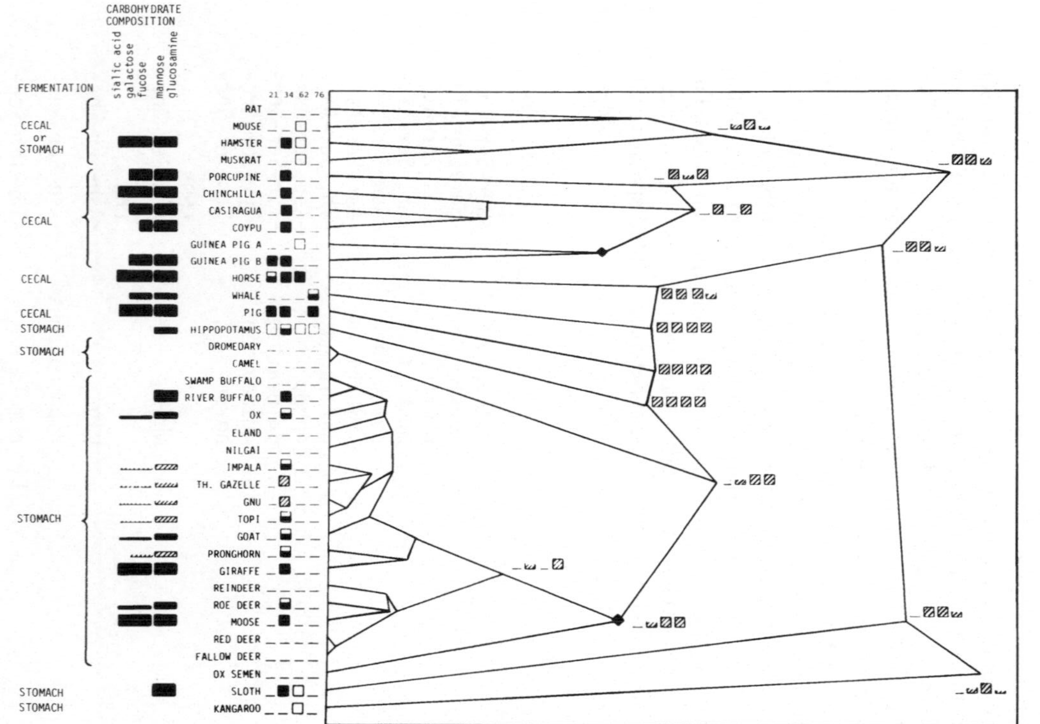

Figure 11. Occurrence of Asn-X-Ser/Thr sequences at the four carbohydrate sites (residues 21–23, 34–36, 62–64, and 76–78) in ribonucleases and the hypothetical ancestors from the biologic tree (Fig. 4) and compositions of the carbohydrate moieties. Existing sequences: ■, completely glycosylated Asn-X-Ser/Thr sequence. ⬓, part of the molecules are glycosylated. □, Asn-X-Ser/Thr sequence without carbohydrate attached. ▨, presence of carbohydrate unknown. –, no Asn-X-Ser/Thr sequence present. Ancestral sequences: ▨, Asn-X-Ser/Thr sequence present in the ancestral sequence. ▨, Asn-X-Ser/Thr sequence present in at least one of the most parsimonious solutions of the ancestral sequence. Size of the rectangle denotes the fraction of the number of solutions that code for an Asn-X-Ser/Thr sequence. –, no Asn-X-Ser/Thr sequence present. Carbohydrate compositions: ■, present in all molecules. ▬, present in all glycosylated molecules (only part of the molecules are glycosylated). ▁, only present in a minor glycosylated component; not present in major glycosylated component. ▭, carbohydrate present, but composition not determined. ﹏, presence of carbohydrate unknown.

be given below as part of a more extensive discussion of a possible function of carbohydrate in pancreatic ribonucleases.

Figure 11 summarizes, among other things, the occurrence of Asn-X-Ser/Thr sequences in the investigated ribonucleases and the glycosylation state of these sequences. It should be noticed that only horse ribonuclease has carbohydrate attached to asparagine-62, while a carbohydrate-free Asn-Val-Ser/Thr sequence at this site has been found in several other species. The sugar composition of the carbohydrate chains is also indicated schematically in Figure 11. Both simple-type chains containing only glucosamine and mannose, and complex-type chains also containing fucose, galactose, and sialic acid, occur.

We have compared the glycosylation states of the ribonucleases we investigated with the type of fermentation (stomach or cecal) used by the species involved (Fig. 11). We found that species with cecal digestion, such as pig, horse, and the hystricomorph rodents, produce ribonucleases with large carbohydrate moieties attached to several positions at the surface of the molecule (20–30% increases in molecular weight have been found). Therefore, we have suggested that the presence of carbohydrate protects ribonuclease from absorption in the gut, enabling it to be transported to the large intestine where it should hydrolyze the ribonucleic acid from the cecal microflora (Beintema *et al.*, 1976). This would be analogous to the function for ribonucleases postulated by Barnard (1969) for ruminants. (It should be noted in Fig. 1 that there are also herbivores with cecal digestion with extremely low ribonuclease contents in the pancreas, such as the lagomorphs, hyraxes, manatee, and elephant.)

Ruminants and species with ruminant-like digestion, such as the true ruminants (or Pecora), camels, hippopotamus, sloth, and kangaroo, have less carbohydrate attached to their ribonucleases. Therefore, it seems that glycosylation of pancreatic ribonuclease may not be advantageous for ruminants. This can be illustrated by substitutions occurring at positions 34 and 35 in the Pecora (Beintema, 1980). The most common site for carbohydrate attachment in pancreatic ribonucleases is the Asn-Met-Thr or Asn-Leu-Thr sequence at positions 34–36. Generally, the Asn-Met-Thr sequences are completely glycosylated (except in hippopotamus; see below), while only a minor fraction of the Asn-Leu-Thr sequences (except in moose and giraffe) are glycosylated (Fig. 12). Thus it is probable that the Asn-Met-Thr sequence is more easily glycosylated than the Asn-Leu-Thr sequence. In the ancestor of the Pecora methionine-35 has been replaced by leucine, resulting in a general decrease of glycosylation of the ribonucleases of its descendants. In two ruminant subfamilies, the Bovinae and the Cervinae, back replacements of leucine-35 by methionine have occurred. However, in these cases several types of replacements of

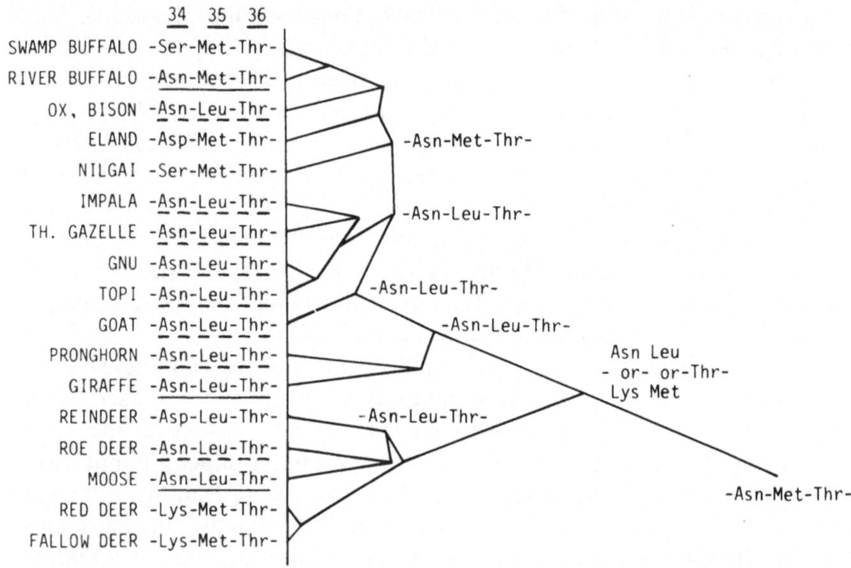

Figure 12. Amino acid sequences of the carbohydrate attachment site 34–36 in pancreatic ribonucleases from *Pecora* and in several pecoran ancestors. -Asn-Leu-Thr- or -Asn-Met-Thr-, completely glycosylated sequences; -Asn-Leu-Thr-, only part of the molecules are glycosylated.

asparagine-34 have also occurred, resulting in a complete loss of carbohydrate-accepting capability. The latter replacements may have been caused by the advantage of completely preventing glycosylation, which is suggested by the presence of the completely glycosylated ribonuclease of river-type water buffalo with the sequence Asn-Met-Thr and the reoccurrence by back replacement of the partially glycosylated Asn-Leu-Thr sequences in bovine and bison ribonuclease (Fig. 12).

Only moose and giraffe ribonuclease do possess completely glycosylated Asn-Leu-Thr sequences. This may indicate that these two species have a more effective glycosylating system in their pancreas than the other Pecora, resulting in ribonucleases with similar properties as found in species with cecal digestion. In this connection it may be significant that in a classification of Pecora based on feeding ecology and stomach structure (Hofmann, 1973, 1976; Hofmann *et al.*, 1976), moose and giraffe are grouped together as large "concentrate selectors" with a diet containing relatively little cellulose fiber and with a relatively simple (primitive) stomach (Fig. 13). A smaller "concentrate selector" is the roe deer (Fig. 13). In this species about half of the ribonuclease molecules are glycosylated at the Asn-Leu-Thr (34–36) sequence (Zwiers *et al.*,

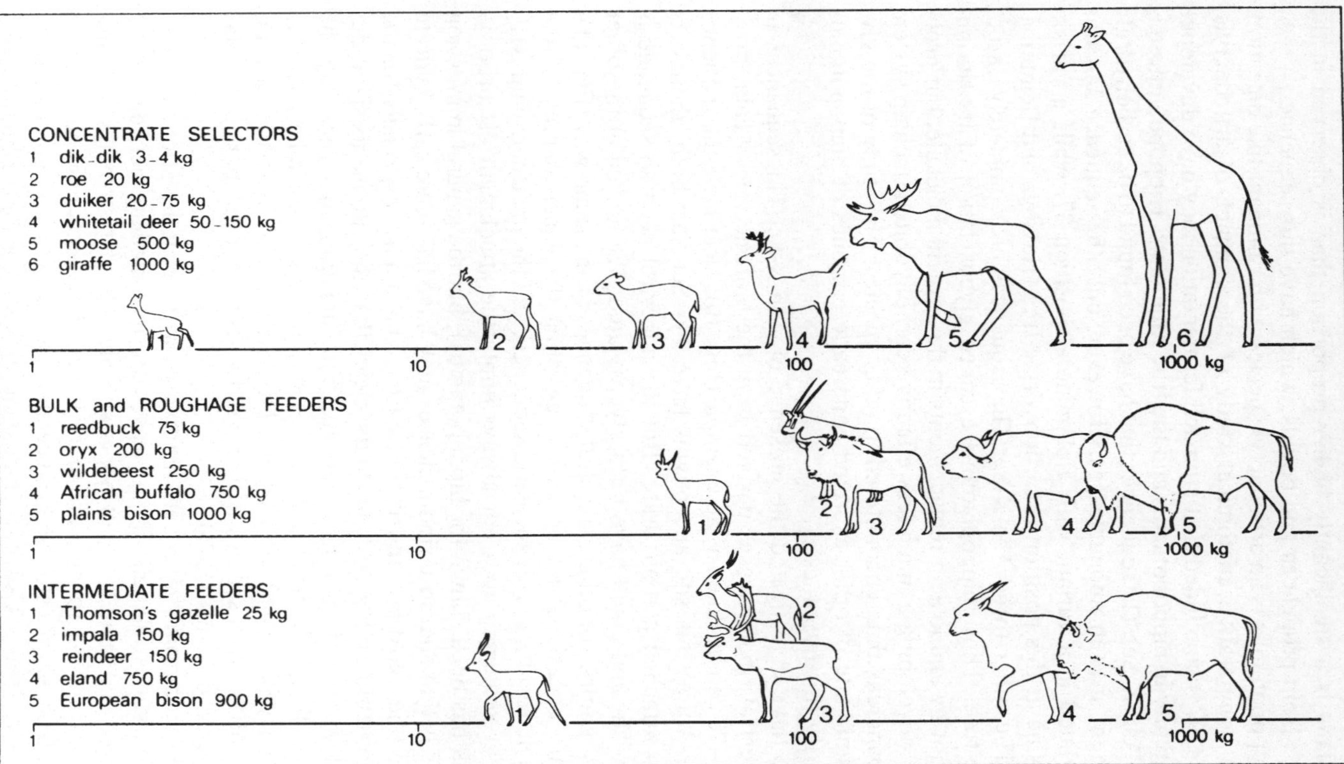

Figure 13. Classification of Pecora on basis of feeding ecology and stomach structure (Hofmann, 1973, 1976; Hofmann *et al.*, 1976). [Figure from Van de Veen (1979).]

1973). This is a significantly higher proportion than is observed in the enzymes from pronghorn and bovids, which have this sequence.

Hippopotamus is an artiodactyl species with ruminant-like digestion. From a comparative anatomical study of the stomach of artiodactyls, Langer (1974) concluded that stomach fermentation may have developed convergently in hippopotamus and the true ruminants. Hippopotamus has all four Asn-X-Ser/Thr sequences that have been found to be carbohydrate attachment sites in ribonucleases. However, only the sequence 34–36 is glycosylated in a variant with glutamine at position 37, while a variant with lysine at this position is carbohydrate-free (Havinga and Beintema, 1980). The other three Asn-X-Ser/Thr sequences are completely carbohydrate-free. Thus, hippopotamus escaped glycosylation of its ribonuclease not by amino acid replacements in the protein molecule, but probably by developing a much less effective carbohydrate-attaching system in the pancreas than encountered in other species. The study of the glycosylation state of other excreted pancreatic enzymes of hippopotamus may verify this hypothesis.

We have investigated the presence of Asn-X-Ser/Thr sequences in the ancestral ribonucleases derived from the biologic tree. Again, positions 76–78 can be taken as an example (Fig. 10). Deer have the sequence Asn-Ser-Ala at this site and bovids have Tyr-Ser-Thr; both, therefore, have no attached carbohydrate. But the ancestral pecoran sequence is Asn-Ser-Thr and could have been glycosylated. We looked at the other sites in a number of other ancestral sequences in the same way (Fig. 11). There are occasional uncertainties regarding the exact identity of the amino acid residues in an ancestral sequence, and the presence of an Asn-X-Ser/Thr sequence does not always imply the attachment of carbohydrate (as has been shown for hippopotamus), but the general impression obtained is that ancestral ribonucleases may have had more carbohydrate attached than most present ribonucleases. This applies especially for the true ruminants or Pecora and could mean that the ruminants have descended from species with cecal digestion, implying that generally the presence of carbohydrate in ruminant ribonucleases is a relic from the evolutionary history of the enzyme. Moir (1968) also suggests that ruminants evolved from species with cecal digestion, but this hypothesis has been refuted by Janis (1976).

Hystricomorph rodents are herbivores with cecal digestion. The glycosylation states of their ribonucleases are intermediate between those of the ruminants and those of horse and pig. The digestive systems and diets of myomorph rodents are more varied. Gradients exist in separate families, from simple stomach and cecum structures to very complex ones (Vorontsov, 1960). Also, the glycosylation states of their ribonucleases show variation (Fig. 11), but the data obtained are not yet sufficient to allow much more to be said. However, the high rates of ribonuclease evolution in the myomorph rodents (Table II) may be correlated with the variation shown by the digestive systems in this taxon. This contrasts with the low rates found in the Pecora as a consequence of the development of a stable symbiotic system in the rumen of this taxon.

ACKNOWLEDGMENTS. Many thanks are due to Dr. W. M. Fitch and Damon A. Smith (Department of Physiological Chemistry, University of Wisconsin, Madison, Wisconsin) for their kind hospitality and much advice and help during the determination of the ribonuclease trees and to Dr. R. N. Campagne for carefully reading the manuscript.

We gratefully acknowledge the permission obtained from Dr. R. J. Moir (University of Western Australia) to use Fig. 1, from Dr. R. E. Dickerson (California Institute of Technology, Pasadena, California), to use Fig. 7 and from Dr. H. E. van de Veen (Vrije Universiteit, Amsterdam) to use Fig. 13.

References

Barnard, E. A., 1969, Biological function of pancreatic ribonuclease, *Nature* **221**:340–344.

Beintema, J. J., 1980, Primary structures of pancreatic ribonucleases from Bovidae: Impala, Thomson's gazelle, nilgai and water buffalo, *Biochim. Biophys. Acta* **621**:89–103.

Beintema, J. J., and Lenstra, J. A., 1980, Nuclear magnetic resonance study of a hybrid of bovine and rat ribonuclease, *Int. J. Pept. Protein Res.* **15**:455–458.

Beintema, J. J., Scheffer, A. J., van Dijk, H., Welling, G. W., and Zwiers, H., 1973, Pancreatic ribonuclease: Distribution and comparisons in mammals, *Nature New Biol.* **241**:76–78.

Beintema, J. J., Gaastra, W., Scheffer, A. J., and Welling, G. W., 1976, Carbohydrate in pancreatic ribonucleases, *Eur. J. Biochem.* **63**:441–448.

Beintema, J. J., Gaastra, W., Lenstra, J. A., Welling, G. W., and Fitch, W. M., 1977, The molecular evolution of pancreatic ribonuclease, *J. Mol. Evol.* **10**:49–71.

Beintema, J. J., Gaastra, W., and Munniksma, J., 1979, Primary structure of pronghorn pancreatic ribonuclease: close relationship between giraffe and pronghorn, *J. Mol. Evol.* **13**:305–316.

Beintema, J. J., Knol, G., and Martena, B., 1982, The primary structures of pancreatic ribonucleases from African porcupine and casiragua, two hystricomorph rodent species, *Biochim. Biophys. Acta*, in press.

Blackburn, P., and Gavilanes, J. G., 1980, The role of lysine-41 of ribonuclease A in the interaction with RNase inhibitor from human placenta, *J. Biol. Chem.* **255**:10959–10965.

Blackburn, P., and Jailkhani, B. L., 1979, Ribonuclease inhibitor from human placenta: Interaction with derivatives of ribonuclease A, *J. Biol. Chem.* **254**:12488–12493.

Carsana, A., Furia, A., Gallo, A., Beintema, J. J., and Libonati, M., 1981, Degradation of double-stranded RNA by glycosylated ribonucleases, *Biochim. Biophys. Acta* **654**:77–85.

Dayhoff, M. O., Eck, R. V., and Park, C. M., 1972, A model of evolutionary change in proteins, in: *Atlas of Protein Sequence and Structure*, Volume V (M. O. Dayhoff, ed.), National Biomedical Research Foundation, Washington, D.C., pp. 89–99.

Dickerson, R. E., and Geis, I., 1969, *The Structure and Action of Proteins*, Benjamin, Menlo Park, California.

Dobson, D. E., and Wilson, A. C., 1980, Lysozyme regulation and the origin of the ruminant lifestyle, in: *Abstracts 2nd International Congress of Systematic and Evolutionary Biology*, Vancouver, Canada, p. 183.

Fitch, W. M., 1976, The molecular evolution of cytochrome c in eukaryotes, *J. Mol. Evol.* **8:**13–40.

Fitch, W. M., 1977, The phyletic interpretation of macromolecular sequence information: Simple methods, in: *Major Patterns in Vertebrate Evolution* (M. K. Hecht, P. C. Goody, and B. M. Hecht, eds.), Plenum Press, New York, pp. 169–204.

Gaastra, W., Groen, G., Welling, G. W., and Beintema, J. J., 1974, The primary structure of giraffe pancreatic ribonuclease, *FEBS Lett.* **41:**227–232.

Gaastra, W., Welling, G. W., and Beintema, J. J., 1978, The amino acid sequence of kangaroo pancreatic ribonuclease, *Eur. J. Biochem.* **86:**209–217.

Goodman, M., 1981, Decoding the pattern of protein evolution, *Prog. Biophys. Mol. Biol.* **37:**105–164.

Havinga, J., and Beintema, J. J., 1980, Pancreatic ribonucleases of mammals with ruminant-like digestion: Amino-acid sequences of hippopotamus and sloth ribonucleases, *Eur. J. Biochem.* **110:**131–142.

Hofmann, R. R., 1973, *The Ruminant Stomach—Stomach Structure and Feeding Habits of East African Game Ruminants* (T. R. Odhiambo, ed.), East African Literature Bureau, Nairobi.

Hofmann, R. R., 1976, Zur adaptiven Differenzierung der Wiederkäuer; Untersuchungsergebnisse auf der Basis der vergleichenden funktionellen Anatomie des Verdauungstrakts, *Prakt. Tierartz* **6:**351–358.

Hofmann, R. R., Geiger, G., and König, R., 1976, Vergleichend-anatomische Untersuchungen an der Vormagen-schleimhaut von Rehwild und Rotwild, *Z. Saeugetierk.* **41:**167–193.

Janis, C., 1976, The evolutionary strategy of the Equidae and the origins of rumen and cecal digestion, *Evolution* **30:**757–774.

Jekel, P. A., Sips, H. J., Lenstra, J. A., and Beintema, J. J., 1979, The amino acid sequence of hamster pancreatic ribonuclease, *Biochimie* **61:**827–839.

Langer, P., 1974, Stomach evolution in the Artiodactyla, *Mammalia* **38:**295–314.

Langley, C. H., and Fitch, W. M., 1974, An examination of the constancy of the rate of molecular evolution, *J. Mol. Evol.* **3:**161–177.

Lenstra, J. A., and Beintema, J. J., 1979, The amino acid sequence of mouse pancreatic ribonuclease; Extremely rapid evolutionary rates of the myomorph rodent ribonucleases, *Eur. J. Biochem.* **98:**399–408.

Lenstra, J. A., Hofsteenge, J., and Beintema, J. J., 1977, Invariant features of the structure of pancreatic ribonuclease: A test of different predictive methods, *J. Mol. Biol.* **109:**185–193.

Libonati, M., Furia, A., and Beintema, J. J., 1976, Basic charges on mammalian ribonuclease molecules and the ability to attack double-stranded RNA, *Eur. J. Biochem.* **69:**445–451.

Lim, V. I., 1974, Structural principles of the globular organization of protein chains. A stereochemical theory of globular protein secondary structure, *J. Mol. Biol.* **88:**857–872; Algorithms for prediction of α-helical and β-structural regions in globular proteins, *J. Mol. Biol.* **88:**873–894.

Migchelsen, C., and Beintema, J. J., 1973, Protein nuclear magnetic resonance studies of histidine residues in rat and other rodent pancreatic ribonucleases. Effects of pH and inhibitors, *J. Mol. Biol.* **79**:25–38.

Moir, R. J., 1968, Ruminant digestion and evolution, in: *Handbook of Physiology—Section 6: Alimentary Canal*, Volume V (C. F. Code, ed.), American Physiological Society, Washington, D.C., pp. 2673–2694.

Morris, D., 1965, *The Mammals*, Hodder and Stoughton, London.

Richards, F. M., and Wyckoff, H. W., 1971, Bovine pancreatic ribonuclease, in: *The Enzymes*, 3rd ed., Volume 4 (P. D. Boyer, ed.), Academic Press, New York and London, pp. 647–807.

Romer, A. S., 1966, *Vertebrate Paleontology*, 3rd ed., The University of Chicago Press, Chicago, pp. 285, 309.

Romer, A. S., 1968, *Notes and Comments on Vertebrate Paleontology*, The University of Chicago Press, Chicago.

Ronda, G. J., Gaastra, W., and Beintema, J. J., 1976, Steady-state enzyme kinetics of the pancreatic ribonucleases from five mammalian species, *Biochim. Biophys. Acta* **429**:853–859.

Simpson, G. G., 1945, *Bull. Am. Mus. Nat. Hist.* **85**:1–350.

Van de Veen, H. E., 1979, Food selection and habitat use in the red deer (*Cervus Elaphus L.*), Thesis, Groningen.

Viret, J., 1961, *Traité de Paléontologie*, Part VI, Volume I, p. 1002.

Vorontsov, N. N., 1960, The ways of food specialization and evolution of the alimentary system in Muzoidea, presented at the Symposium Theriologicum in Brünn, in: *Proceedings of the International Symposium on Methods in Mammalian Investigations*, Prague, pp. 360–377.

Welling, G. W., Lenstra, J. A., and Beintema, J. J., 1976, Activity and antigenicity of ribonuclease hybrids, *FEBS Lett.* **63**:89–94.

Zwiers, H., Scheffer, A. J., and Beintema, J. J., 1973, Amino-acid sequences of red-deer and roe-deer pancreatic ribonucleases, *Eur. J. Biochem.* **36**:569–574.

CHAPTER 3

Eye Lens Proteins and Vertebrate Phylogeny

WILFRIED W. DE JONG

1. Introduction

The vertebrate eye lens has many unique properties which make it an attractive object for a variety of molecular biologic studies (Bloemendal, 1981). In fact the eye lens was one of the earliest targets of comparative immunologic and electrophoretic studies, and a wealth of data has been collected (Clayton, 1974; De Jong, 1981). By studying and comparing the lenses of present-day vertebrates we can hope to obtain information about the evolutionary changes that have taken place in structure and composition of the lens. This will extend our knowledge of protein evolutionary processes, and at the same time provides data that can be used to infer phylogenetic relationships between the compared species.

2. Comparative Anatomy and Evolution of the Lens

Despite the similar function and gross appearance of vertebrate lenses, considerable differences in structure exist (Walls, 1963). A general scheme of the anatomy of the vertebrate lens is shown in Fig. 1. The most significant addition to this basic structure is the presence in birds and most reptiles of a ring-shaped thickening of epithelial cells around the equator, called the annular pad or *ringwulst*.

WILFRIED W. DE JONG • Department of Biochemistry, University of Nijmegen, Geert Grooteplein N21, 6525 EZ Nijmegen, The Netherlands.

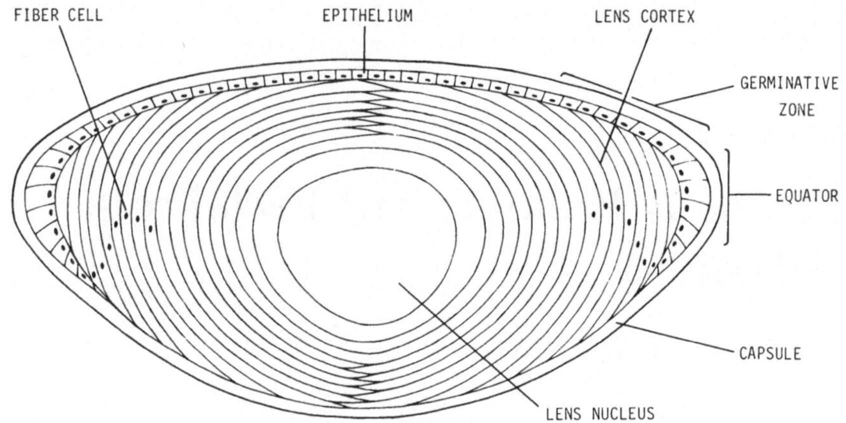

Figure 1. Schematic cross section through the vertebrate eye lens.

The anterior surface of the lens is covered with a single layer of epithelial cells. Cell division takes place mainly near the equator. At the equator the dividing cells elongate to form lens fiber cells, which cover the earlier formed fiber cells. The greatest protein synthetic activity is found in the early stages of the process of elongation. At the end of the process of elongation the cells lose their nuclei.

Since no fiber cells are shed or broken down during life, one still can find the remnants of the embryonic lens in the nucleus of the adult lens. This makes the lens a useful tool in the study of aging of cellular structures and proteins (Hoenders and Bloemendal, 1981). Lens proteins are subject to deamidation and fragmentation, and aspartyl residues are prone to racemization.

Because the lens is an avascular organ, and since the immediately surrounding tissues are devoid of blood vessels, all nutrients and metabolites have to be transported over long distances. Consequently the metabolic activity of the lens is relatively low, especially in the nucleus.

The shape of the lens in different species varies from perfectly spherical to ovoid or flattened forms. The anterior surface of the lens is usually less strongly curved than is the posterior. Spherical lenses are found in aquatic and nocturnal animals, which, for different reasons, require a high refractive power. Humans and the other diurnal primates have, together with the echidna, or spiny anteater, the flattest of all lenses. Flat lenses, providing less refraction of incident light than do rounded ones, are small relative to the size of the eyeball.

The lens is surrounded by a lens capsule of which the thickness and elasticity vary, depending on the level of the faculty to accommodate by changing the shape of the lens. The consistency of the lens ranges from very solid, as in the lamprey, to gelatinous in many birds and reptiles. The consistency usually increases from the periphery to the nucleus, being extremely pronounced in fish lenses, where a hard nucleus is often surrounded by an almost liquid cortex.

The consistency of the lens largely corresponds with its protein content. A higher protein concentration increases the refractive index, but reduces the deformability of the lens that is required for accommodation in many species. Therefore the lowest protein concentrations (±20%) are found in bird lenses, which have the highest accommodative capacities, whereas up to 50% protein can be found in fish lenses, which must be as refractile as possible to compensate for the lack of refraction at the cornea–water interface. Intermediate protein concentrations (30–40%) are found in mammalian lenses. In fact the eye lens is one of the most protein-rich organs of the body, which is one reason for its attractiveness for protein structural studies.

Although most lenses are colorless, a yellow pigment is present in the lenses of several diurnal species from widely separated taxa. Yellow lenses are found in humans and several other primates, tupaia, diurnal squirrels, snakes, gekkoes, and lampreys (Walls, 1963).

Shape and consistency of the lens are related to the accommodative properties in different species. Accommodation by displacement or deformation of the lens is not observed in all species; it is lacking in primitive fishes, in the reptile *Sphenodon*, and, most notably, in almost all mammals.

Accommodation by moving the lens as a whole is found in lampreys, fishes, amphibians, and snakes, each group having developed its own solution to reach this goal. Birds and reptiles (except snakes) do accommodate by direct muscular pressure of the ciliary body on the lens, resulting in a bulging forward of the anterior part of the lens through the iris.

A much less effective way of accommodation by deformation of the lens is found in the relatively few mammals that do accommodate actively. The curvature of the lens is changed by action of the ciliary muscle, which controls the pull of the zonule on the elastic lens capsule. At rest the lens is flattened, and by contraction of the ciliary muscle the tension of the zonule is released, allowing the plastic lens to resume a more spherical shape. This system can bring about only a very limited deformation, especially in larger lenses. Moreover, since mammalian lenses lose their plasticity rapidly with age, this accommodative capacity decreases likewise, as many middle-aged people can testify.

The evolutionary development of the vertebrate eye is shrouded in mystery. Fossil remains of this soft organ are lacking, and primitive or intermediate forms of the eye are absent among present-day vertebrates. Also, no structural or developmental homology with any type of invertebrate eye exists. Nevertheless, various hypotheses about the origin of the vertebrate eye have been put forward (Walls, 1963; Polyak, 1957; Duke-Elder, 1958). These hypotheses have not succeeded, however, in explaining the origin of the vertebrate lens in terms of gradualist Darwinian evolution. In fact, as Walls (1963) stated, it appears that "perhaps forever, the evolutionary origin of the vertebrate lens must remain a tantalizing mystery."

3. Comparative Studies of Lens Proteins

3.1. The Crystallins

The characteristic proteins of the vertebrate lens are the crystallins (Bloemendal, 1981; Harding and Dilley, 1976). They make up virtually all of the water-soluble lens protein, and are also found to a considerable degree in the water-insoluble fractions. No proteins related to the vertebrate crystallins are found in invertebrate lenses (Brahma, 1978). The crystallins are not absolutely lens-specific; small amounts have been reported to occur in chick iris, cornea, vitreous, aqueous, and retina (Bours, 1976). Interestingly, the presence of crystallins has been demonstrated by immunofluorescence in the median or parapineal "eye" of the lizard *Anolis carolinensis* (McDevitt, 1972).

The crystallins can be defined as structural proteins that serve to provide the highly concentrated and well-ordered protein contents of the lens fiber cells, required for proper refraction, transparency, and plasticity of the lens. The crystallins as a group are not set apart from the other proteins by specific common features; in fact, they vary widely in isoelectric point, aggregate size, subunit molecular weight, amino acid composition, sulfhydryl contents, and secondary structure. Table I summarizes these properties for the four main classes of crystallins that can be distinguished. The distinction between β- and γ-crystallins is not always clear; in fact, monomeric β-crystallins do occur which are not too different in amino acid composition from the γ-crystallins.

α-Crystallin, the largest and most acidic of the lens proteins, is composed of two types of polypeptide chains, αA and αB, which are both about 175 residues long and which show some 60% sequence homology

Table I
Properties of Vertebrate Lens Crystallins[a]

	α	β	γ	δ
Molecular weight	$>4 \times 10^5$	$(4–20) \times 10^4$	2×10^4	$(15–20) \times 10^4$
Quaternary structure	Aggregates	Dimers, oligomers	Monomers	Tetramers
Monomer molecular weight	20,000	21–37,000	20,000	45–50,000
N terminus	Acetylated	Acetylated	Free	Unknown
Range of native isoelectric points	pH 4.5–5.0	pH 5.7–7.0	pH 7.1–8.1	pH 5.2–5.4
Distribution	All vertebrates	All vertebrates	Not in sauropsidans?	Only in sauropsidans

[a]Data from Clayton (1974) and De Jong (1981).

(Fig. 2). The ratio of αA to αB varies from 9:1 in kangaroo to 1:4 in dogfish.

β-Crystallins occur as oligomers containing from two to six subunits; they also occur as monomers. Some 5–20 different, but related, subunits can be distinguished, depending on the species. These subunits are in part different gene products, but can also originate from post-translational modifications. The sequence of the major β-crystallin chain of the calf, βBp, is known. It shows a complete internal duplication, and reveals, moreover, a definite sequence homology with the calf γ_{II}-crystallin chain (Driessen *et al.*, 1980) (Fig. 3).

The monomeric γ-crystallins are quite heterogeneous, especially in charge, and it is uncertain to what extent the monomeric crystallins observed in all vertebrate lenses are really homologous to the mammalian γ-crystallins. Some authors therefore prefer to bring such monomeric proteins together under the general name of "low-molecular-weight lens proteins." The only γ-crystallin of which the sequence is known is the calf γ_{II} fraction. This is also the only crystallin of which the tertiary structure has been established. The molecule is an ellipsoid consisting of two globular, identical domains (Blundell *et al.*, 1981). This nicely fits with the rather weak sequence homology which, as in βBp, can be seen between the N- and C-terminal halves of the chain.

δ-Crystallin is phylogenetically interesting because it only occurs in reptiles and birds. This restricted occurrence cannot simply be correlated with the unique sauropsidan type of accommodation, because δ-crystallin occurs in snakes, which accommodate by lens displacement. δ-Crystallin is unique among the crystallins in having a high α-helical content (Kuck *et al.*, 1976).

Although the primary structure of a δ-crystallin chain has not yet

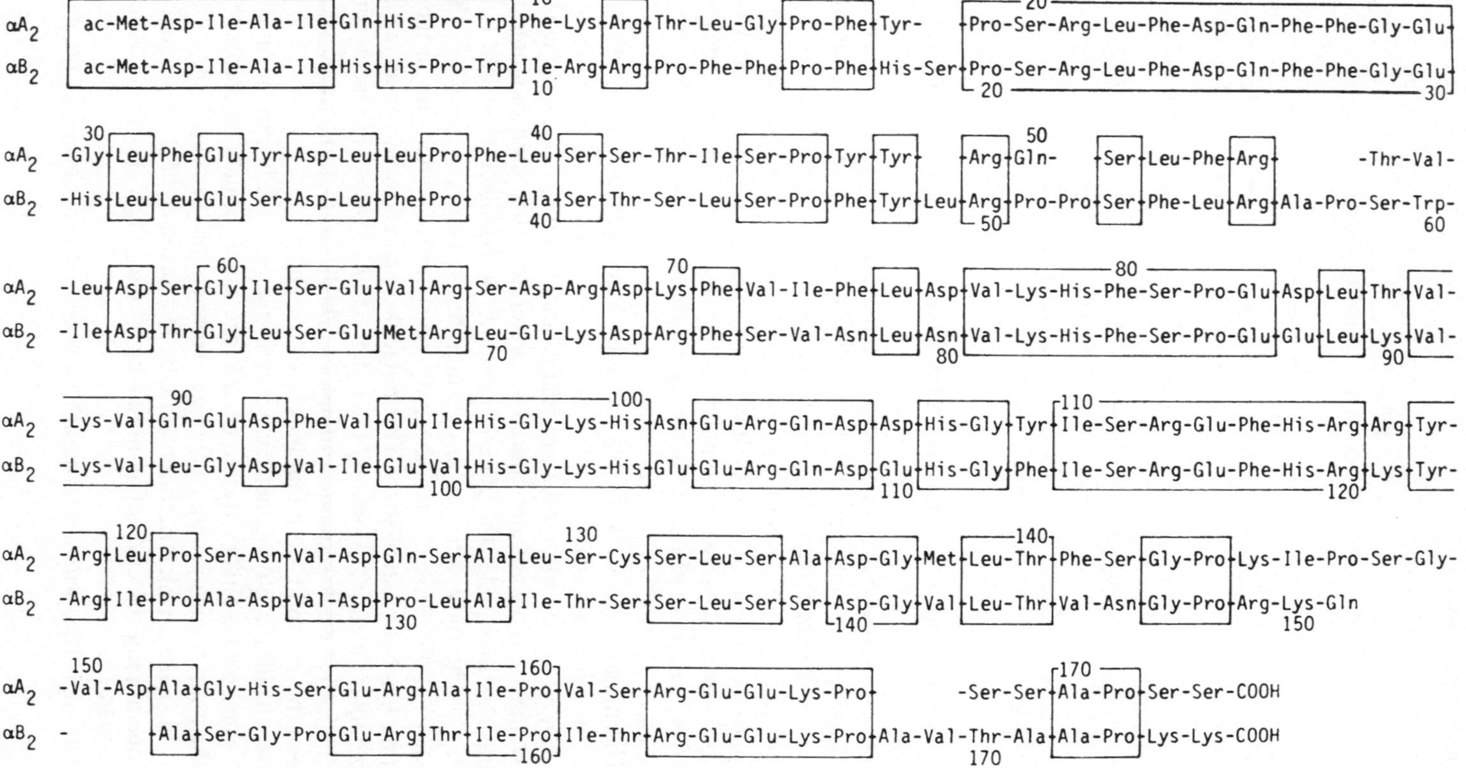

Figure 2. Sequence homology of the bovine α-crystallin A and B chains. [Van der Ouderaa *et al.* (1974).]

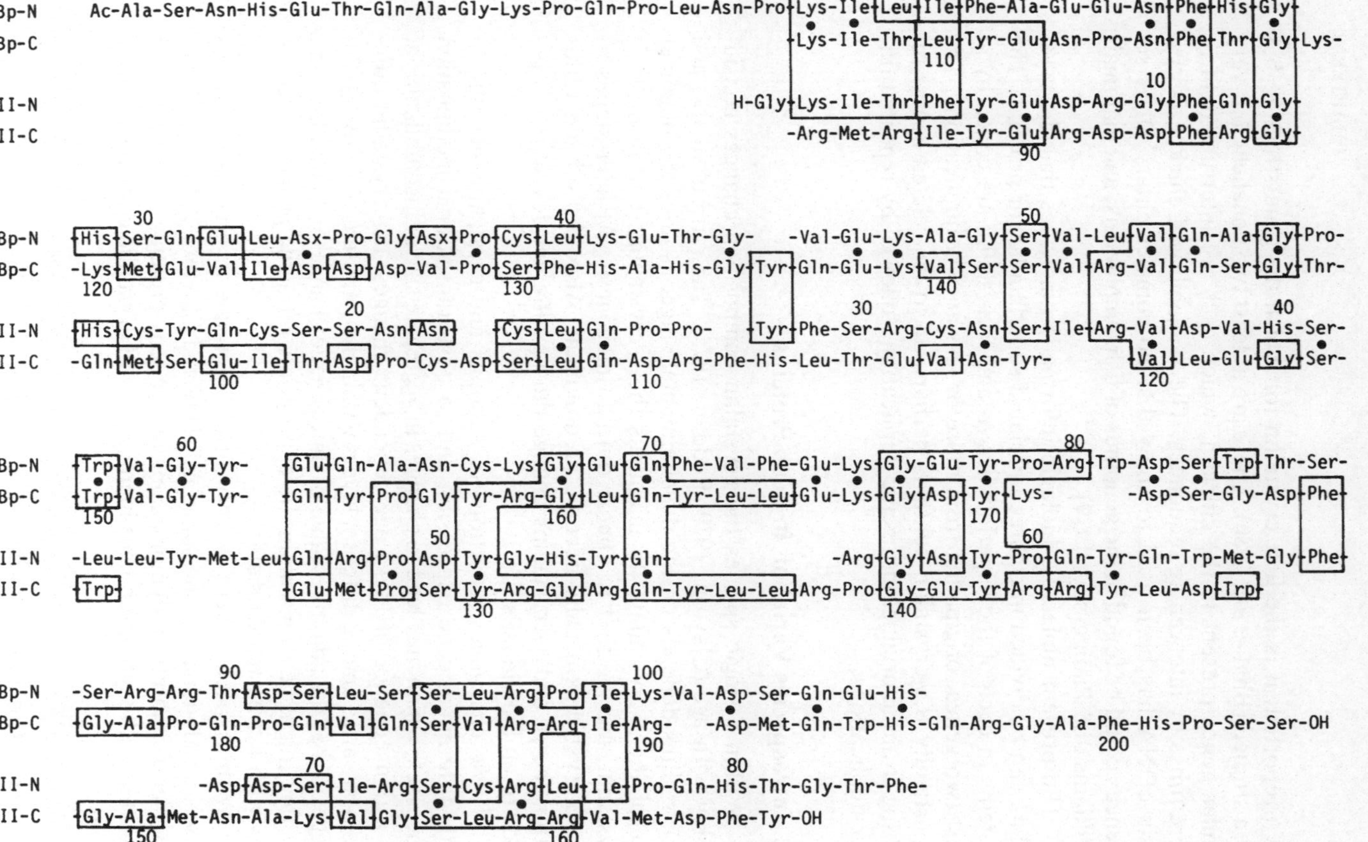

Figure 3. Sequence homology of the N- and C-terminal halves of bovine β-crystallin Bp and γ-crystallin fraction II. Intrachain sequence homology between the N- and C-terminal parts of βBp or γII is emphasized by dots. [Driessen *et al.* (1980).]

been reported, it is the first crystallin for which the gene structure has been studied (Bhat *et al.*, 1980). There are at least two nonallelic but very similar genes present in the chicken, which each are interrupted 14 or more times by intervening sequences. The availability of chicken δ-crystallin cDNA has been used to show that no complementary DNA sequences could be found in the genomes of fish, amphibians, and mammals (Williams and Piatigorsky, 1979).

The present molecular biologic information indicates that the crystallins can be divided into three unrelated superfamilies of proteins. The proteins in each of the superfamilies α- and δ-crystallin are coded by at least two genes, originating from more or less ancient duplications. The superfamily β,γ-crystallin contains perhaps up to ten related genes, resulting from an original intragenic duplication, followed by repeated further duplications.

3.2. Intraspecies Variation in Lens Proteins

Within a single species the crystallin composition changes considerably with age (Clayton, 1970; Reddan, 1975). The ratio of α-, β-, and γ-crystallins depends on the developmental stage, as does the oligomer size of the β-crystallins. In mammals the γ-crystallins, and in birds δ-crystallin, are the first to appear during ontogenesis. At later stages α- and β-crystallins usually become predominant. Also, the subunit ratios within the β- and γ-crystallins change during different development periods of an individual's life.

These changing proportions of subunits and crystallins are due to both differential gene expression and to post-translational modification processes, which may lead to changes in aggregate size or insolubilization. As a consequence of these age-dependent changes, the crystallin composition varies among different layers within the same lens. Clayton (1970) suggests that the adaptive value of ontogenetic changes in crystallin composition might be to maintain the appropriate refractile properties of the lens with increasing diameter.

Within species there is very little difference in crystallin patterns between individuals of the same age (Day and Clayton, 1973). Only very few electrophoretic polymorphisms of eye lens proteins have ever been reported, although differences in intensities of crystallin components may occur.

3.3. Crystallin Variation between Species

The variation in crystallin composition between species is very great [for reviews see Clayton (1974) and De Jong (1981)]. This variation is present at all levels of protein analysis. The proportions of monomeric, low- and high-molecular-weight aggregates are different from species to species, as is easily revealed by gel filtration chromatography of the water-soluble lens proteins (Fig. 4). Analysis of total native lens proteins from different species on the basis of their charge, by isoelectric focusing, emphasizes the variability at this level (Fig. 5). Under denaturing conditions a similar variability can be seen in the proportions between differently charged crystallin subunits (Fig. 6A), while the differences in subunit molecular weights, as demonstrated by SDS-gel electrophoresis, are equally clear (Fig. 6B).

It is the task of present and future lens research to find out which differences at the level of gene organization and regulation, and which differences in primary structure and physicochemical properties of the crystallins, are responsible for the observed variation in the lens proteins. We are still far from correlating this variation with the structural and functional characteristics of lenses in different species.

In the meantime the comparative study of vertebrate lens proteins has provided a considerable amount of evolutionary and phylogenetic information. Lens proteins have been used extensively for immunologic studies. Most impressive are the results of Manski and Halbert (1965), and Manski and Malinowski (1978). They studied the order of appearance of lens antigens in the different vertebrate classes, and could relate this to the sequence of origin of these taxa. It was found, for instance, that lenses of the shark have antigens in common with all higher vertebrates, which are lacking in lampreys, indicating that such antigens must have originated after the divergence of the ancestral gnathostomes and cyclostomes. It could also be estimated that approximately 42% of all antigenic determinants of calf α-crystallin are derived from primitive cyclostomes, 22% from subsequent aquatic vertebrates, 21% from primitive terrestrial vertebrates, and only 15% evolved with the mammals.

Electrophoretic comparison of lens proteins using different techniques yields complex patterns, which nevertheless reveal group-specific characteristics. Attempts have been made to use such electrophoretic patterns as taxonomic markers, especially in fish and birds [for review see Wright (1974)]. Above the species level the taxonomic information so obtained generally appears to be unconvincing.

More promising is comparative amino acid sequence analysis of homologous crystallins from different vertebrates. As documented in this volume, interspecies sequence comparisons of proteins such as hemoglobin, myoglobin, cytochrome *c*, and pancreatic ribonuclease provide a wealth of detailed information about protein evolutionary processes and are of considerable phylogenetic interest. Simpson (1975) indeed states about comparative protein sequence analysis that "the body of data al-

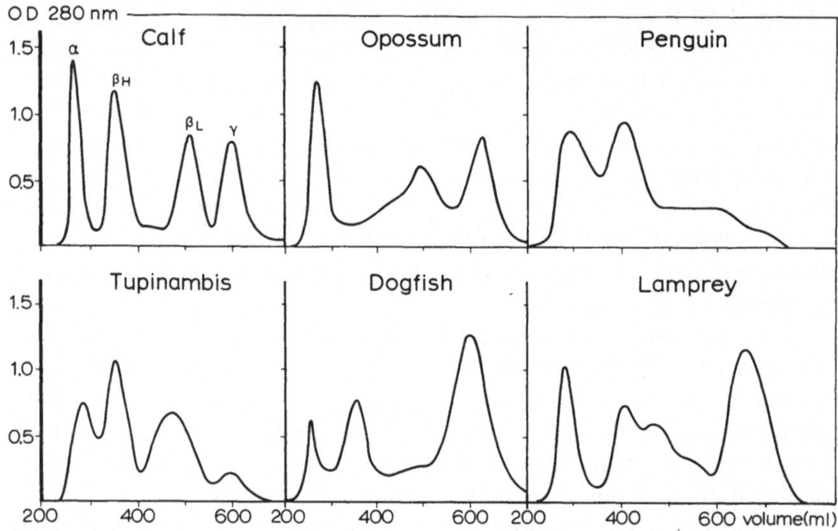

Figure 4. Variation in molecular size proportions among vertebrate water-soluble lens proteins revealed by gel filtration chromatography (Sephadex G-200). [De Jong *et al.*(1976).]

ready achieved is impressive and highly significant for taxonomy and phylogeny."

It is relevant to add the crystallins to the number of proteins already studied from an evolutionary point of view, especially since they form a unique category of proteins with properties not found in others. Most work has concentrated on the α-crystallin A chain, the results of which will be discussed in the next sections.

4. Structural Analysis of α-Crystallin

4.1. Isolation of α-Crystallin

A detailed description of the biochemical procedures involved in the sequence determination of α-crystallin chains can be found in De Jong and Terwindt (1976). We will describe here the successive steps in more general terms.

The pooled lenses from a species are homogenized in two volumes of buffer, and cell membranes and other insoluble components are removed by centrifugation. The supernatant, containing the water-soluble

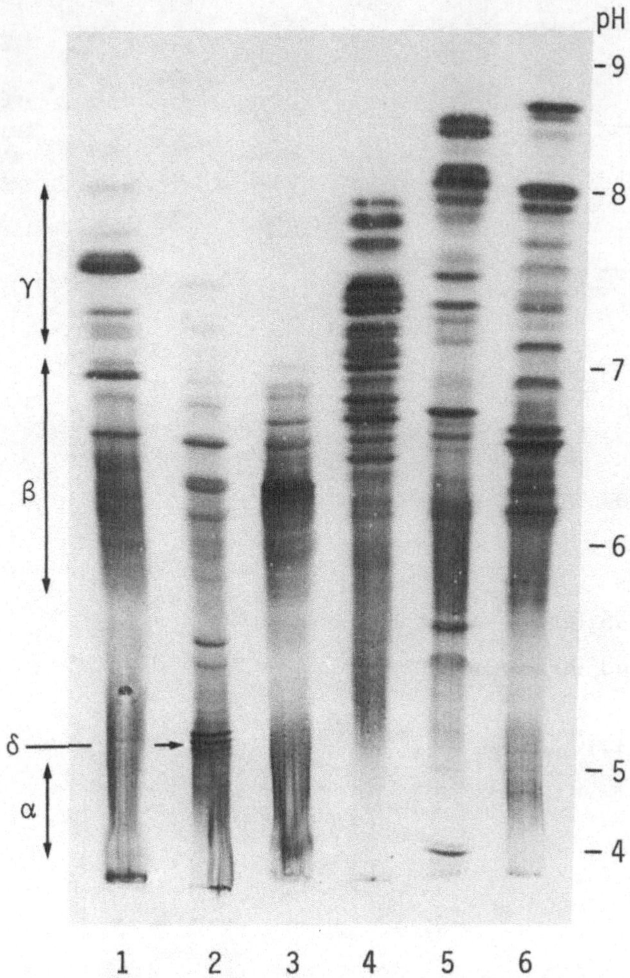

Figure 5. Variation in isoelectric points among native vertebrate lens proteins. Isoelectric focusing in polyacrylamide slabgel [according to Bours (1977)] of total lens extracts from: (1) calf; (2) chicken; (3) monitor (*Varanus exanthematicus*); (4) frog (*Rana temporaria*); (5) cod; (6) spiny dogfish. The approximate positions of bovine α-, β-, and γ-crystallins are indicated. Chicken δ-crystallin can clearly by distinguished. Note the characteristic lack of basic proteins in the sauropsidan lenses, and the abundance of very basic proteins in fish.

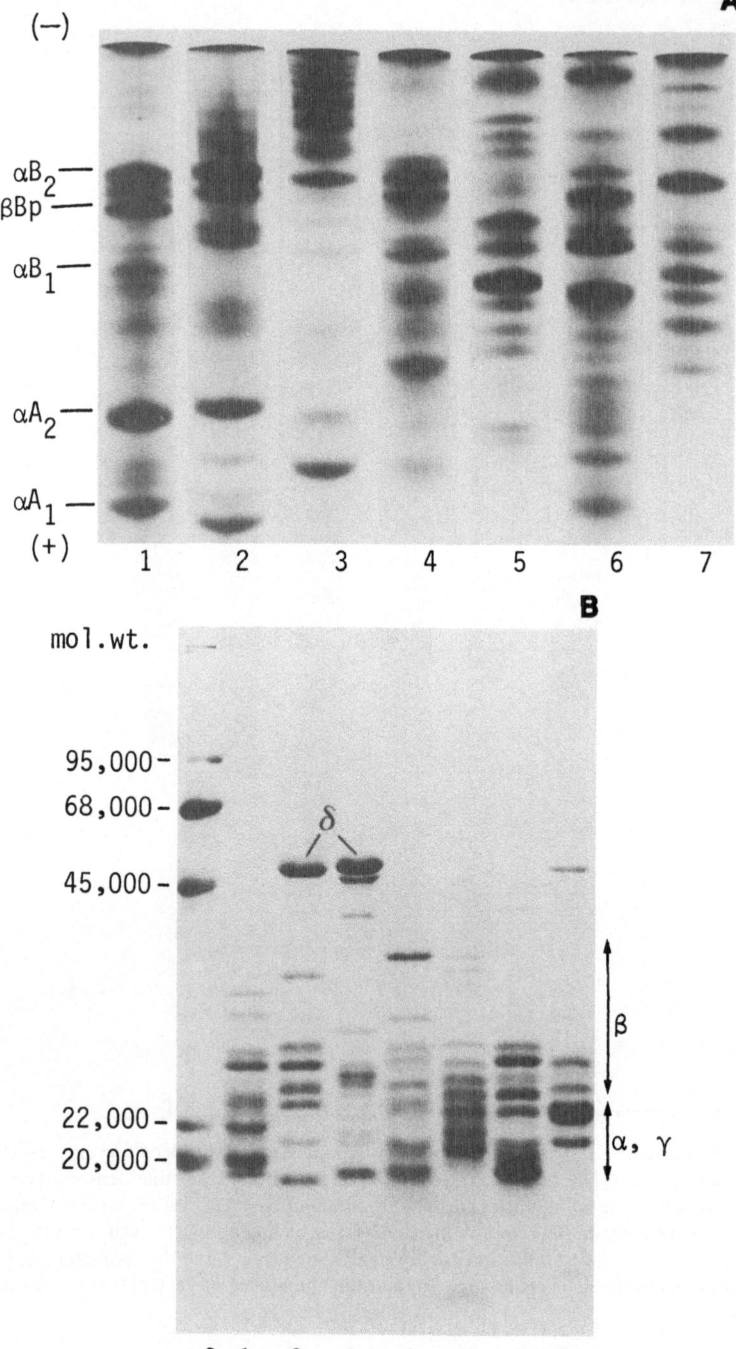
A
(–)
αB2
βBp
αB1
αA2
αA1
(+)
1
2
3
4
5
6
7
B
mol.wt.
95,000-
68,000-
45,000-
22,000-
20,000-
δ
β
α, γ
ref
1
2
3
4
5
6
7

proteins, is passed through a gel filtration column which separates the proteins on the basis of molecular weight. α-Crystallin, being the largest protein aggregate (molecular weight about 800,000) in the lens cells, elutes as the first peak from such a column, usually well separated from the lower molecular-weight β- and γ-crystallins. As can be seen in Fig. 4, the relative amount of α-crystallin varies greatly in different species. In most mammals it is about one-third of all water-soluble lens protein, which means that from one calf lens, weighing about 2.5 g, some 250 mg of α-crystallin can be obtained in one simple preparative step!

4.2. Electrophoresis of the α-Crystallin Chains

The two types of polypeptide chains, αA and αB, which compose the α-crystallin aggregate both have molecular weights of about 20,000. They can be separated from each other on the basis of their different charge, the αA chain being more acidic than the αB chain. The subunit composition of α-crystallin can thus be made visible by electrophoresis at alkaline pH, in the presence of urea in order to break hydrogen bonds which otherwise would keep the αA and αB chains together.

The patterns obtained are usually complicated by the presence of additional bands which represent post-translational modification products of the two primary gene products αA and αB. These post-translational modifications involve deamidations of asparagine and glutamine residues and breaks of specific peptide bonds, especially near the C-terminal ends of the chains (Hoenders and Bloemendal, 1981), which result in changes in net charge. In some species these modification processes are much more pronounced than in others. In particular, the human α-crystallin electrophoretic patterns already become quite complicated at a young age, whereas, for instance, the African elephant retains a simple and clear α-crystallin composition throughout life.

At the level of electrophoretic mobility only limited variability between the αA and αB chains of different species can be seen (De Jong *et al.*, 1976). Among the 41 eutherian mammalian species in which α-crystallin has been studied, only the aardvark has an αB chain which migrates

Figure 6. Variation in subunit charge and molecular weight among vertebrate lens proteins. Acrylamide gel electrophoresis in Tris-glycine buffer, pH 8.6, containing 6 M urea (panel A), and in 0.1% sodiumdodecylsulfate (panel B) of total lens extracts from: (1) calf; (2) chicken; (3) monitor; (4) frog; (5) cod; (6) spiny dogfish; (7) lamprey (*Petromyzon marinus*). The positions of the major bovine crystallin subunits are indicated in panel A. In panel B many crystallin subunits can be identified by their characteristic apparent molecular weights.

differently from all other αB chains (Fig. 7). The αA chain is slightly more variable in mobility. The mobility of the αA chain from 34 species is identical to that of the calf; in three species (mink, three-toed sloth, tree pangolin) the αA chain migrates more slowly, due to one negative charge less; in the elephant and hyrax the αA chain migrates faster, due to an additional negative charge; and the aardvark αA chain has two additional negative charges as compared to most other mammalian αA chains (Fig. 7).

It is obvious that electrophoretic analysis of α-crystallin is of no use as a taxonomic character; in fact it would lead to misleading suggestions, because the identical mobilities of mink, 3-toed sloth, and tree pangolin αA chains are the result of different amino acid substitutions.

4.3. Sequence Analysis of αA Chains

Preparative separation of αA and αB chains is performed on the basis of their charge difference, by ion exchange column chromatography, in the presence of urea (Fig. 8). The isolated chains are fragmented by

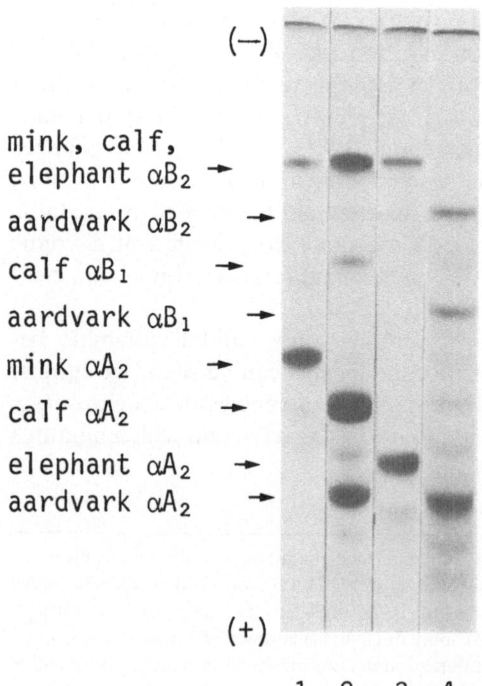

Figure 7. Variation in charge among mammalian α-crystallin chains. Alkaline urea gel electrophoresis (pH 8.6) of α-crystallin from: (1) mink; (2) calf; (3) African elephant; (4) aardvark.

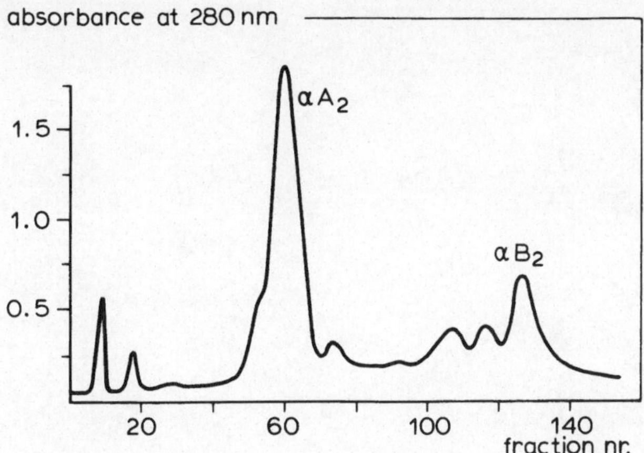

Figure 8. Separation of the α-crystallin A and B chains of the two-toed sloth by ion exchange (CM-52) column chromatography.

digestion with the proteolytic enzyme trypsin, which normally cleaves the chain at each arginine and lysine residue, and also at cysteine residues when these have chemically been converted into aminoethylcysteine. The resulting mixture of tryptic peptides can be resolved by peptide mapping, where peptides are separated by high-voltage paper electrophoresis, followed by paper chromatography in the second dimension.

Even at the level of tryptic peptide maps the differences among the α-crystallin A chains of the vertebrate species are very limited, as summarized in Fig. 9. This is understandable, since only substitutions leading to changes in charge or to great differences in hydrophobicity, or involving the presence of the proline C terminal of lysine or arginine residues (which blocks tryptic cleavage at such bonds), will give detectable changes in peptide position, and such substitutions are relatively scarce because protein evolution tends to favor conservative substitutions, which hardly change the chemical properties of amino acid side chains.

Peptides can be eluted from the peptide maps and used for amino acid analysis and sequence determination by the so-called Edman degradation. Longer tryptic peptides can be further digested with other proteolytic enzymes, such as thermolysin, chymotrypsin, and proteinase from *Staphylococcus aureus*. The same enzymes can be used to find overlapping peptides to establish the order of the tryptic peptides in the αA chain. The proposed sequences of most of the mammalian α-crystallin A chains are largely based on amino acid compositions of peptides. When amino acid compositions are found to be identical to those of corresponding

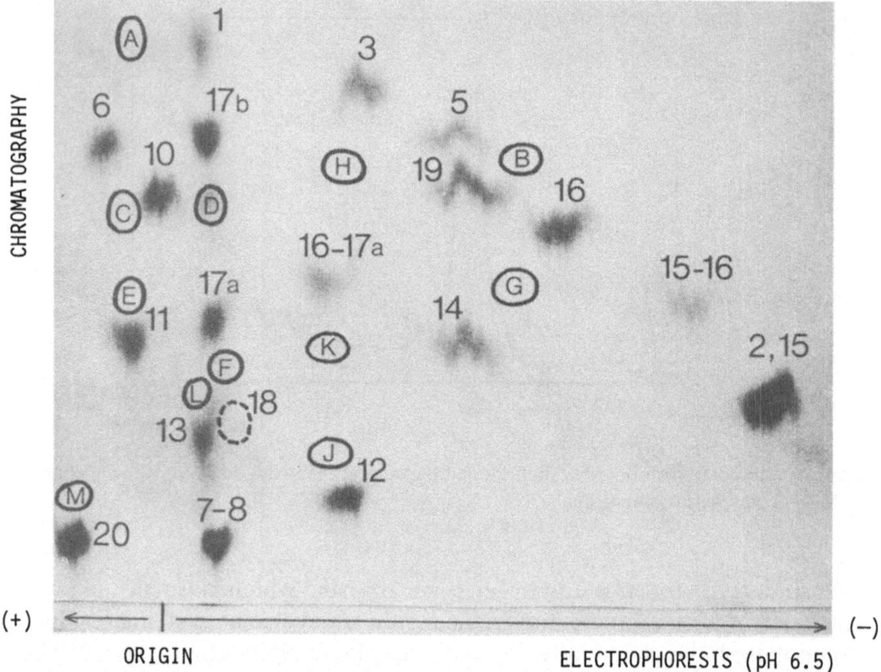

Figure 9. Fingerprint of tryptic peptides from mammalian α-crystallin A chains. The ninhydrin-stained pattern is from bovine αA chain, where peptides are numbered in the order in which they occur in the sequence from N to C terminus (cf. Fig. 10). Tryptic peptide T18 does not stain with ninhydrin, and peptides T4 and T9 are poorly soluble and not present on the peptide map. The drawn spots indicate the different positions of tryptic peptides in other mammalian αA chains due to the following substitutions (cf. Table II): (A) mink and three-toed sloth T1 (7-His→Gln); (B) prosimians T2-3 (13-Ala→Pro) and manatee T3 (19-Pro→His); (C) many species T11 (90-Gln→Leu); (D) mink and three-toed sloth T11 (90-Gln→Leu and 91-Glu→Gln or Asp→Gly, respectively) and three pangolin T11 (90-Gln→Leu and 92-Asp→Gly); (E) Malayan pangolin T11 (90-Gln→Leu and 93-Phe→Ser); (F, G) manatee T17a separated in two peptides by 126-Gln→Lys; (H) paenungulates and man T17b (142-Ser→Cys); (J) opossum T18 (147-Gln→His) and aardvark T18 (152-Ala→Asp); (K) three-toed sloth T18-19 (158-Thr→Pro); (L) manatee T20 (167-Pro→Ala); (M) marsupials T20 (169-Ser→Leu).

peptides from the previously completely determined sequence of bovine αA (Van der Ouderaa *et al.*, 1973), it is assumed that the sequence of these peptides also will be the same. When differences in composition are found with corresponding bovine peptides, an effort is made to localize the underlying amino acid substitutions by Edman degradation or further enzymatic digestions. The attempt to infer the sequence from amino acid composition bears the risk of overlooking reciprocal amino acid substitutions in peptides with the same composition. It has been demonstrated that the chance for such reciprocal substitutions in small peptides is acceptably small (Van Druten *et al.*, 1978). As an example, the sequence inference of the aardvark αA chain is shown in Fig. 10.

In addition to the bovine αA chain, the sequences of the αA chains from red kangaroo, chicken, frog (*Rana esculenta*), tegu (*Tupinambis teguixin*), and spiny dogfish (*Squalus acanthias*) have completely or almost completely been determined by Edman degradation.

Altogether the complete sequences of 41 mammalian species have now been studied, and partial sequences of two other species add additional information. The submammalian αA sequences established so far are from chicken, tegu, frog, and dogfish. In addition, the sequences of bovine and human αB chains are known, and because of their 58% sequence homology with the bovine and human αA chains, further extend the possibility of reconstructing the evolutionary history of the α-crystallin chains. All this sequence information, with appropriate species names and references, has been compiled in Table II.

4.4. Phylogenetic Reconstructions

Table II reveals, for many positions in the αA chain, which residue is most probably the ancestral, primitive one, and which residues are derived, at least if one accepts some *a priori* assumptions about the phylogenetic relationships among the included species. If all mammals and other vertebrates have serine in position 51 of the αA chain, but proline is present at this position in the seal, then it is obvious that 51-Pro is a derived character. Similarly, some shared derived (synapomorphous) amino acid substitutions in certain species can easily be recognized, such as, for instance, 13-Pro and 61-Val in the prosimians lemur, potto, and galago. The table also reveals, however, that certain substitutions, like 55-Ser, of necessity must have occurred independently in entirely unrelated taxa, and that back substitutions, for instance 101-Asn→Ser, may complicate the interpretation. Nevertheless, some relevant and important phylogenetic information can be extracted from this table, as will be shown in the next section. When the whole data set has to be brought together in one cladogram, it is advantageous to use appropriate computer algorithms to evaluate the numerous possible branching patterns.

The approach followed in constructing cladograms on the basis of amino acid sequence data is to find the most parsimonious trees—those requiring the fewest mutational events to account for the descent of the sequences. The parsimony principle is the only objective criterion that can be applied, but since protein evolution apparently does not always

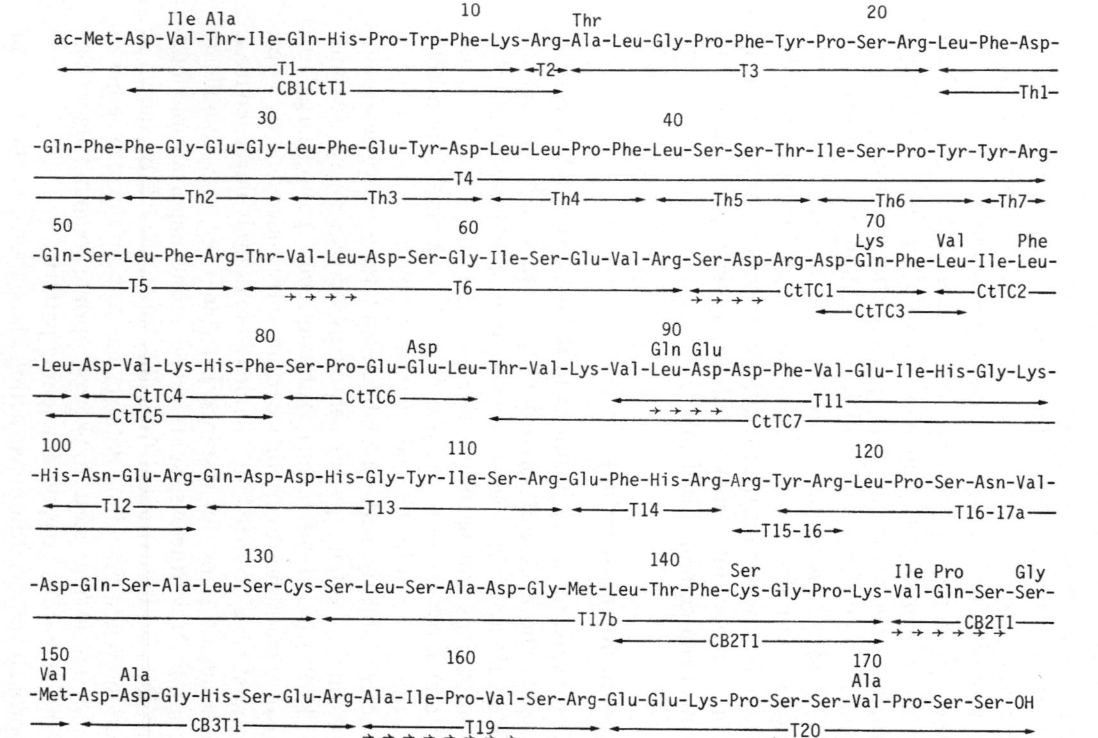

Figure 10. Sequence reconstruction of the aardvark α-crystallin A chain. Where residues are inferred to be different from the bovine sequence, the bovine residues are given above the aardvark sequence. Different types of peptides were isolated: T, tryptic; Th, thermolytic; CB CtT, tryptic peptides obtained after citraconylation of cyanogen bromide fragments; CtTC, chymotryptic peptides derived from a tryptic peptide of a citraconylated αA chain; CB T, tryptic peptides obtained from cyanogen bromide fragments.

work along most parsimonious lines (at least as far as amino acid substitutions are concerned), it follows that the most parsimonious cladograms are not necessarily the biologically most realistic ones. This problem, and how to extract in the face of it valuable phylogenetic information from sequence data, has been discussed extensively (Peacock and Boulter, 1975; Goodman *et al.*, 1979; Fitch, 1979; Foulds *et al.*, 1979).

Different algorithms have been used to construct cladograms on the basis of the αA sequences. In the earlier stages of our work, involving up to 21 mammalian species, trees were constructed by Drs. J. T. Gleaves and D. Boulter according to methods described in De Jong *et al.* (1977). More recently the extended data set of 45 αA sequences has been analyzed by a maximum parsimony approach which takes into account not only the minimum number of nucleotide replacements in a putative tree, but also the genetic cost assumed in the required number of gene duplication and gene expression events. The rationale and computer algorithms for such an approach are given in Goodman *et al.* (1979).

5. Phylogenetic Inferences from α-Crystallin A Sequences

5.1. Mammalian Phylogeny

The first phylogenetic trees of αA sequences, including 17 mammalian species, chicken, and frog, indicated that the degree of sequence difference between the chains might be especially suitable for the study of interorder and higher taxonomic relationships (De Jong *et al.*, 1977). It allowed for some relevant phylogenetic proposals. First, the obtained branching patterns of the mammalian groups nowhere violated firmly established opinions about phylogenetic relationships. The marsupials nicely diverged as the oldest offshoot from the other mammals; the ungulates grouped together, including the whale, in agreement with recent generally accepted ideas. The hyrax and elephant shared three synapomorphous substitutions, confirming their sometimes questioned common ancestry, but turned out not to show any affinity to the ungulates.

Because many uncertainties exist about the phylogenetic relationships among the mammalian orders, it was decided to extend our study of α-crystallin to phylogenetic problem cases like the orders Edentata, Pholidota, Tubulidentata, Sirenia, and Chiroptera, and to increase the overall density of the tree by including more species for which the phylogenetic affinities are not disputed. It has been shown that the denseness of a phylogenetic network bears importantly on the accuracy of evolutionary reconstructions from protein sequence data (Holmquist, 1978).

Table II
Variable Positions in the α-Crystallin A

Position	3	4	7	11	13	16	17	18	20	33	37	39	40	51	52	55	56	58	61	70	72	74	84	86	89	90	91	93	101	122	123	124
Minke whale	I	A	H	K	A	P	F	Y	S	E	L	F	L	S	L	T	V	D	I	K	V	F	D	T	V	Q	E	F	N	S	N	V
Porpoise																S																
Horse																																
Tapir					T																											
Rhinoceros					T											S			V													
Pig																			V													
Giraffe, hippopotamus																																
Ox					T																											
Camel																										L						
Dog, cat																										L						
Bear																						Y				L						
Mink			Q								M								V							L	Q					
Seal, sea lion														P	V	S										L						
Pangolin																						Y				L		S	S			
Bat		T																														
Hedgehog	V	T																								L			S	P	S	
Tupaia	V	T																								L						
Rat, hamster, gerbil	V	T																								L						
Guinea pig, springhaas	V	T																														
Pika	V	T																											S			
Rabbit	V	T			T																											
Lemur	V	T			P														V													
Galago, potto	V	T			P														V													
Rhesus monkey	V	T			T																						D					
Human	V	T			T																						D					
Elephant	V	T																		Q		L					D					
Hyrax[c]	V	T																		Q	L	L				L	D					
Manatee	V	T							N											Q	L	L				L	D					
Aardvark	V	T																		Q	L	L	E			L	D					
Two-toed sloth	V	T		R																						L	D			T	A	
Three-toed sloth	V	T	Q														A									L	G			T	A	
Tamandua	V	T															A			R		L				L	D			T	A	
Kangaroo		T				S	L											E								L	D		S			
Opossum		T				S	L											E		R		Y				L	D	Y	S			
Chicken		T					L	I				L	F			S		E			·	·		S	I	I	D		S	A		
Tegu		T					L	I				L	F					E			T			S		I						
Frog		T							N	D	F				F	G	F			R	·	·			I	L	D		S			L
Dogfish	L		Y	R	S	S				D	F		F		V	N	F			R	M	N	E	S	I	V	D	Y	A			L

[a]Positions are shown at which substitutions have been found in at least two species. The one-letter notation for amino acids has been used; (—) deletion, (·) not determined. Vertical lines are drawn where residues are identical to the topmost sequence. The αA chains are from: minke whale *(Balaenoptera acutorostrata);* common porpoise *(Phocaena phocaena);* horse *(Equus caballus);* Malayan tapir *(Tapirus indicus);* white rhinoceros *(Ceratotherium simum);* pig *(Sus scrofa);* giraffe *(Giraffa camelopardalis);* hippopotamus *(Hippopotamus amphibius);* ox *(Bos taurus);* dromedary *(Camelus dromedarius);* dog *(Canis familiaris);* cat *(Felis catus);* sloth bear *(Melursus ursinus);* American mink *(Mustela vison);* gray seal

Chains of 45 Vertebrate Species[a,b]

125	126	127	129	130	133	135	138	142	143	146	147	148	149	150	152	153	154	155	156	158	167	168	169	170	172	
D	Q	S	L	S	L	A	M	S	G	V	P	S	G	M	A	G	H	S	E	A	P	S	S	A	S	
											T															
		T			V					I												G				
		T												L												
		T																								
														V											T	
										I				V												
										I				V												
										I				V												
														V												
														V												
														V												
														V	T	S										
										L	Q			V	G											
		A									A			L			P									1[b]
											Q			L												
											Q			L												
											Q			L												
											Q			L												
											Q			L												
											Q			L												
					V						Q			L												
										I	Q	T		L		—		T								1[b]
								C		I	Q	T		L		—		T				T				
								C		I	Q					S										
								C			Q					S										
	K							C			Q					S					A				N	1[b]
								C			Q		S		D									V		
										I	—	—	—	V	P	S				T						
										I	—	—	—	V	P	S				P						1[b]
					.	.	.	.		L	—	—	—	V	P	S		G		T						
			I							I	H		D			S			D	S		T	L			
			I							I	H	T	N		S	S			D	S		T	L			1[b]
			I	T		S							D		P	S				P		T				1[b]
		A	I	T				A	A		Q		D	T	P	S		N		P		T			P	
N	E		I				I			L	M		S	L	S	S		G		P		T				3[b]
N	E		I	A		N	L	C	C	T	R	P	G	D	S	N	W	Q	D	P	Q	G	T	Q	E	12[b]

(Halichoerus grypus); California sea lion *(Zalophus californianus);* Malayan pangolin *(Manis javanica);* Jamaican fruit-eating bat *(Artibeus jamaicensis);* European hedgehog *(Erinaceus europaeus);* tree shrew *(Tupaia belangeri);* rat *(Rattus norvegicus);* golden hamster *(Mesocricetus auratus);* Mongolian gerbil *(Meriones unguiculatus);* guinea pig *(Cavia porcellus);* springhaas *(Pedetes cafer);* American pika *(Ochotona princeps);* rabbit *(Oryctolagus cuniculus);* brown lemur *(Lemur fulvus);* galago *(Galago crassicaudatus);* potto *(Perodicticus potto);* rhesus monkey *(Macaca mulatta);* human *(Homo sapiens);* African elephant (Loxodonta africana); Cape hyrax *(Procavia capensis);* Brazilian manatee *(Trichechus in-*

Table II *(continued)*

unguis); aardvark *(Orycteropus afer)*; two-toed sloth *(Choloepus hoffmanni)*; three-toed sloth *(Bradypus infuscatus)*; ant bear *(Tamandua mexicana)*; red kangaroo *(Macropus rufus)*; North American opossum *(Didelphis marsupialis)*; chicken *(Gallus domesticus)*; tegu *(Tupinambis teguixin)*; frog *(Rana esculenta)*; spiny dogfish *(Squalus acanthias)*. [References in De Jong *et al.* (1977, 1982) and unpublished data from W. W. De Jong, A. Zweers, P. J. M. Leenen, and H. G. van der Avoort.] The partially determined sequence of the giant anteater αA chain *(Myrmecophaga tridactyla)* is identical to the tamandua sequence, and the αA chain of tree pangolin *(Phataginus tricuspis)* differs in four positions from that of the Malayan pangolin (92-G; 93-F; 101-N; 152-A) (De Jong *et al.*, 1982).

[b]Positions with substitutions in a single species only: tree pangolin 92-D→G; hedgehog 23-F→L, 154-H→P; rhesus monkey 162-S→A; manatee 19-P→H; three-toed sloth 34-Y→S; opossum 151-D→E; chicken 32-F→L; frog 26-F→V, 31-L→M, 128-A→S; dogfish 67-D→E, 68-R→K, 76-D→N, 105-D→E, 107-H→Q, 109-Y→R, 110-I→V, 117-R→T, 119-R→H, 136-D→E, 141-F→L, 173-S→I, and a C-terminal extension of the chain with four residues: Arg-Ala-Asp-Pro.

[c]Hyrax position 55 showed microheterogeneity, Ala and Thr being present in equal amounts.

The addition of three edentate and one pholidote αA sequence to the earlier 17 mammalian sequences showed clearly the monophyletic origin of sloths and anteaters (De Jong *et al.*, 1982). The most parsimonious solutions placed the edentate sequences as one of the oldest offshoots of the eutherian stem, either as a separate branch or on a common base with Hyracoidea and Proboscidea. The pangolin αA sequence showed no shared derived substitutions with those of the edentates, and could most economically be connected near a common carnivore–ungulate stem.

After sequence analysis of an additional 20 mammalian species, the total data set of 41 mammalian species, plus chicken and frog, has been used to construct cladograms according to the algorithms of Goodman *et al.* (1979). The results, which will be presented and discussed in detail elsewhere (De Jong and Goodman, 1982), allow a number of interesting conclusions.

Because the principal aim of our work is to establish the most probable pattern of relationship among the mammalian orders and some disputed intraorder relationships, we limited the search for most parsimonious phylogenetic trees by including some *a priori* assumptions about cladistic relationships.

These assumptions, based on strong classical zoologic evidence, were:

1. Chicken and frog are out-groups in relation to all mammals.
2. The grouping of the investigated species into their respective orders is accepted, with tree shrew in its own higher taxonomic group [order Scandentia (Butler, 1972)] rather than assigned to either Primates or Insectivora.

3. Certain classical intraorder relationships which seem firmly established were accepted (thus, within Edentata the Bradypodidae and Myrmecophagidae are considered as sister groups).

The main freedom then allowed in the search for most parsimonious trees was the reshuffling of the mammalian orders in relation to each other as well as appreciable reshuffling of species within their orders. Two major trees were found at a lowest score of 157 NRs (nucleotide replacements). This is 5 NRs more than the most parsimonious trees that could be found when no *a priori* cladistic assumptions were introduced. However, because these lowest NR trees violate several of the well-established phylogenetic relationships, they require several hypothetical gene duplications and gene expression events, which, in terms of total genetic events, make them less parsimonious than the biologically acceptable 157 NR trees.

One of these major 157 NR trees is shown in detail in Fig. 11. The other major tree is depicted more schematically in Fig. 12. The major differences with the branching pattern in Fig. 11 are: the divergence of the edentates and paenungulates as a trichotomy with the other eutherians (which actually represents two successive but unresolvable dichotomies); the separation of the lagomorphs and primates from the insectivores and rodents (bringing them closer to the carnivore–ungulate part of the tree); the presence of pangolin and carnivores as sister groups; the unresolved trichotomy of artiodactyls, perissodactyls plus cetaceans, and carnivores plus pholidotes.

These considerable differences between equally parsimonious phylogenetic trees show that some of the depicted branching patterns are only weakly supported by the αA sequence results. Nevertheless, a number of conclusions about phylogenetic relationships seem warranted.

5.1.1. Metatherian–Eutherian Divergence

Our earlier observation that the marsupial–placental dichotomy represents the first branching event in the mammalian line (De Jong *et al.*, 1977) is confirmed by this extended data set. The marsupials and placentals clearly are sister groups, each characterized by a considerable number of autapomorphic substitutions.

5.1.2. Aardvark–Paenungulate Relationship

In agreement with prevailing opinions about mammalian phylogeny (Romer, 1966; Thenius, 1969; Simpson, 1945), the manatee, hyrax, and

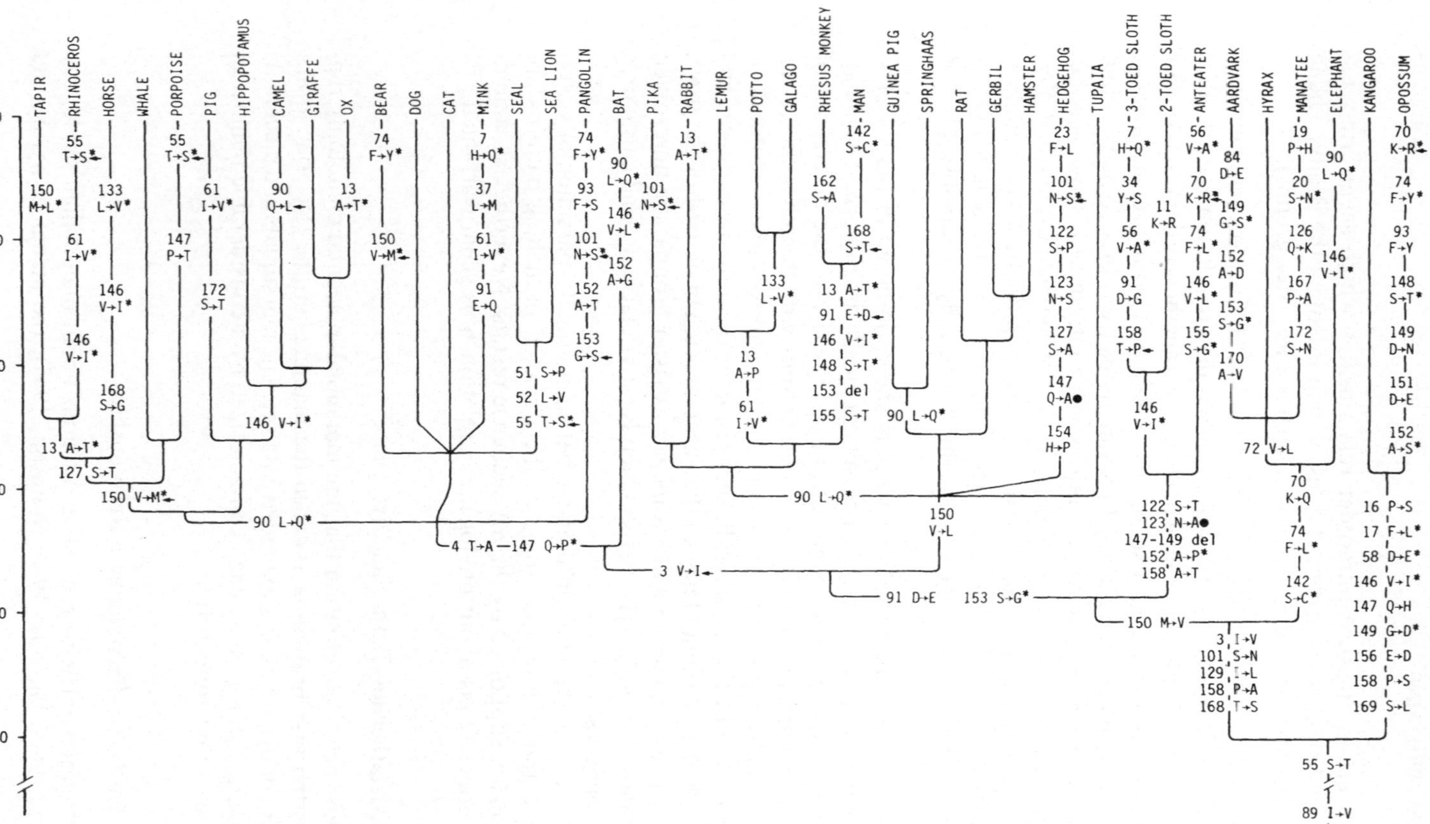

Figure 11. One of the biologically acceptable most parsimonious trees of 41 mammalian α-crystallin A chains. Chicken and frog αA chains were used as outgroups. Parallel (*) and back (←) substitutions are indicated. Two substitutions require two base replacements (●). Trees were constructed by Dr. Morris Goodman.

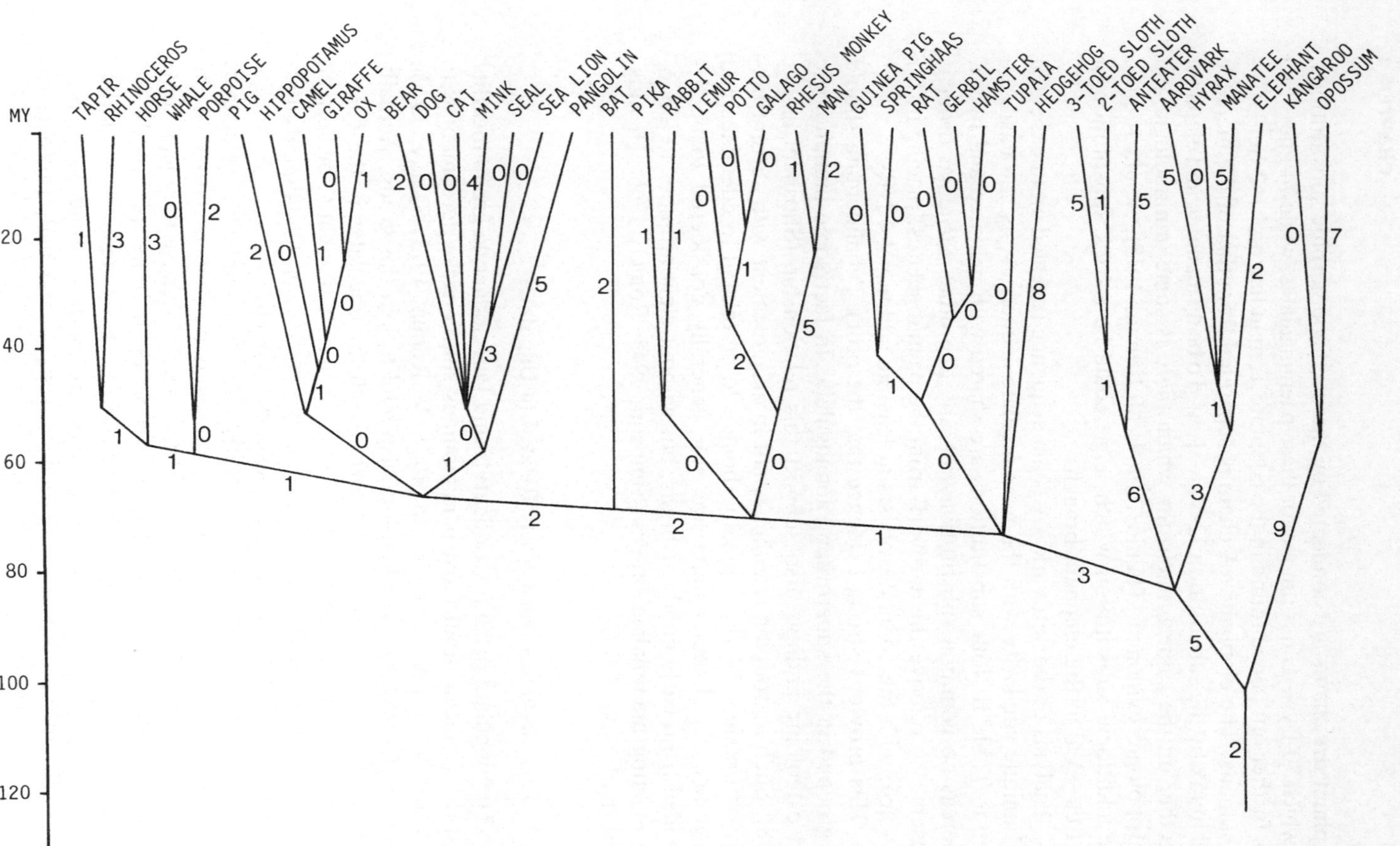

Figure 12. Alternative biologically acceptable most parsimonious tree of 41 mammalian αA chains. The numbers of nucleotide replacements are indicated for each branch. Zero branches are based on biologic evidence.

elephant αA chains are connected by three synapomorphic substitutions, of which 70-Lys→Gln is unique for these paenungulates, while 74-Phe→Leu has further only been found in the edentate Tamandua, and 142-Ser→Cys in humans. The elephant αA chain is separated from that of the manatee and hyrax by the substitution 72-Val→Leu. To bring manatee and elephant together on the same branch, for which there is some immunologic and morphologic evidence (Thenius, 1969; McKenna, 1975), would require one additional substitution which, e.g., could be a back substitution of 72-Leu→Val in the elephant branch.

Analysis of the aardvark αA-chain sequence revealed an unexpected and unique similarity with the paenungulate sequences, as can easily be seen in Table II: four substitutions are shared with manatee and hyrax, whereas no common substitution with any one other particular order is present. To remove the aardvark from the paenungulates, placing it, for instance, with the edentates, costs at least an additional 4 NRs.

This provides strong evidence that the order Tubulidentata should be grouped in the superorder Paenungulata. In fact some immunologic and osteologic findings also support this relationship (Shoshani *et al.*, 1978). Such a grouping together is not in disagreement with current taxonomic opinion. While originally thought to be related to edentates and pangolins, the Tubulidentata are now generally considered to be descended from early condylarthrans, but without clear affinities to any of the existing mammalian orders (Simpson, 1945; Romer, 1966; Thenius, 1969; Patterson, 1978).

5.1.3. Paenungulates Not Closely Related to Ungulates

The ungulate orders Artiodactyla and Perissodactyla group together with the Cetacea in all most parsimonious solutions, in agreement with most recent authors (Van Valen, 1971; McKenna, 1975; Szalay, 1977), while the Carnivora and Pholidota also tend to join this group. Actually Simpson (1945) included the Carnivora with the ungulates in the cohort Ferungulata. The paenungulate orders are generally thought to be derived from a primitive ungulate stock, although convincing synapomorphous characters with the ungulates are never mentioned.

The α-crystallin A sequences clearly place the paenungulates far apart from the ungulates. Their most parsimonious position is as one of the most ancient branches of the eutherians, mainly because they lack some of the derived substitutions shared by most of the other placentals: 91-Glu, 150-Leu or -Val, 153-Gly. To place the paenungulates (including aardvark) in the ungulate–cetacean region of the tree would cost 5 NRs more. The only other known sequences from paenungulate proteins are that of elephant myoglobin, hemoglobin β chain, and fibrinopeptides A and B. A combined parsimony analysis of these sequences and the α-crystallin A sequences for nine mammalian species also denies any special relation of the elephant to the ungulates, suggesting an early split between condylarthrans and proboscideans (Dene *et al.*, 1980a).

5.1.4. Edentates as Early Eutherian Offshoots

The three edentate species included in Figs. 11 and 12 show the most convincing synapomorphies observed among any investigated eutherian αA sequences. In the lowest NR tree the three-toed sloth should actually be placed with the tamandua, because of their common substitution 56-Val→Ala. Since the classical evidence for the division between Bradypodidae and Myrmecophagidae seems strong, one has to assume a back substitution 56-Ala→Val in the two-toed or two parallel substitutions 56-Val→Ala in three-toed sloth and tamandua.

Both in the present and in a previous analysis of αA sequences (De Jong *et al.*, 1982) the order Edentata diverges as the first or second offshoot from the main placental stem. Because equally parsimonious alternatives (Fig. 12) let edentates and paenungulates branch off as an unresolved trichotomy, it is even possible that these two taxa are sister groups. This is actually what is suggested by the results from the armadillo Hb β chain (De Jong *et al.*, 1981). The only other sequenced edentate proteins are sloth pancreatic ribonucleases, which also show them to have an anciently separated position from other mammalian orders (Beintema and Lenstra, this volume, Chapter 2).

It is obvious that the edentates have always been considered as a peculiar and in many respects primitive group among the eutherians (Montgomery, 1982). This is most explicitly expressed by McKenna (1975), who subdivided the eutherian mammals into two cohorts, one including only the edentates, and the other all remaining placentals. Our analysis of the edentate αA chains is clearly in agreement with this division.

5.1.5. The Pinnipeds Are Monophyletic

The analyzed carnivore αA sequences do not show a single synapomorphous substitution, nor do they provide much information about intraordinal relationships, apart from the fact that seal and sea lion αA sequences are identical. Both sequences are characterized by two unique substitutions, 51-Ser→Pro and 52-Leu→Val, and a third one, 55-Thr→Ser, which among mammals is also found in porpoise and rhinoceros. This clearly sheds light on the long-standing problem of the possible biphyletic origin of the Pinnipeda (Thenius, 1969). Although mostly regarded as a monophyletic group, derived from early Ursidae, it has repeatedly been argued that only the Otariidae (sea lions) might be related to the Ursidae, the Phocidae (seals) having originated independently from early Mustelidae (Tedford, 1976).

The αA sequences do not tell us anything about a possible pinniped–bear relationship, but they clearly show that the seal is more closely related to the sea lion than to the Mustelidae, represented by the mink. The only other protein sequenced both in seal and sea lion, myoglobin, provided marginal evidence in favor of a monophyletic origin (Romero-Herrera *et al.*, 1978). Immunologic evidence, using carnivore albumin, also indicated pinniped monophyly (Prager and Wilson, 1978).

5.1.6. Pholidote Relationships

Both in Figs. 11 and 12 and in our earlier analyses the Malayan pangolin αA sequence is joined to or near the carnivore branch. The partially analyzed tree pangolin αA chain differs at least at four positions from the Malayan αA sequence (Table II). After the divergence of the ancestors of the two pangolin species a unique substitution 92-Asp→Gly must have occurred in the tree pangolin line, and the Malayan pangolin line underwent the unique substitutions 93-Phe→Ser and 152-Ala→Thr and the back substitution 101-Asn→Ser.

The connection of the pholidotes to the carnivore–ungulate part of the tree is not strong. It is mainly based on the presence of the shared residues 4-Ala and 147-Pro. On the other hand, 3-Ile, 101-Ser (which, however, is Asn in the tree pangolin), and 153-Ser could be considered as primitive characters, placing the pholidotes at the base of the eutherian tree. Still, the depicted position of the pholidotes is 2 NRs more economical (or 3 NRs in the case of the tree pangolin) than having it separate as the earliest eutherian offshoot.

Some authors consider the pholidotes as the closest relatives of the edentates (Van Valen, 1971; Szalay, 1977), or at least do not exclude this possibility (Simpson, 1945), but others have entirely abandoned this idea (Romer, 1966; Thenius, 1969; McKenna, 1975). The phylogenetic relationship of the pangolins thus remains a completely open question.

The pangolin αA sequences clearly show no synapomorphies with the edentates, and it actually takes 3 NRs to force the pangolins to group with the edentates rather than with the carnivores. Lacking any other known pangolin protein sequences, and in view of the general uncertainty about the position of these animals, the information from the αA sequence might deserve serious consideration.

5.1.7. Phylogenetic Position of Chiroptera

The investigated bat αA sequence is from *Artibeus jamaicensis*, belonging to the suborder Microchiroptera. This sequence is rather weakly connected to the branch leading to carnivores, pholidotes, and ungulates, mainly based on the presence of 3-Ile and 150-Val. Such a connection is at variance with the often assumed origin of the bats from arboreal insectivores (Romer, 1966). A monophyletic bat–hedgehog branch would increase the tree length by 2 NRs only. The myoglobin of a fruit-bat (Megachiroptera) seems to support the relationship with the insectivores (Goodman *et al.*, 1979), but cytochrome *c* places it closest to the carnivores (Moore *et al.*, 1976).

5.1.8. Primate Phylogeny

The investigated αA sequences give only limited information about the relation of this order among the others. The primates tend to group with the lagomorphs, a relation for which there is no paleontologic or morphologic support, but which, strangely enough, is found also in the combined myoglobin, β-hemoglobin, fibrinopeptide, and α-crystallin A analysis (Dene *et al.*, 1980a). Within the Primates the suborders Anthropoidea and Prosimii are each characterized by specific changes, the Lorisidae further being joined by the substitution 133-Leu→Val.

5.1.9. General Remarks about Mammalian αA Sequences

Many of the questions about mammalian phylogeny remain unanswered after the analysis of 43 αA sequences, and cannot be answered by extending this number. Where no substitutions have occurred for long periods of time, as in the αA chains of tree shrew, myomorph rodents, dog, cat, and minke whale, the additional analyses of related sequences are unlikely to provide any insight into their phylogenetic relations.

Some branching patterns are marginally supported by a single, often a redundant, substitution. Sometimes this may seem to confirm an established relationship, like the substitution 13-Ala→Thr joining the Tapiridae and Rhinocerotidae. It is, however, equally parsimonious to bring the rhinoceros together with the horse on the basis of the common substitution 146-Val→Ile.

The most parsimonious branching pattern within the Artiodactyla groups the hippopotamus with the Ruminantia, again on the basis of the substitution 146-Val→Ile, whereas pig and hippopotamus both belong to the suborder Suina. This may be similar to the lungfish problem (Gardiner,

1979): while the lungfish is thought to be cladistically more closely related to the cow than to the salmon, it is still included in the class Pisces. Thus, the suborder Suina may turn out to be a grade, not a clade. Interestingly, artiodactyl ribonuclease sequences also support a cladistically closer relation between hippopotamus and ruminants (this volume, Chapter 2).

The rodent α-crystallin A chains examined so far are remarkably poor in substitutions, and actually the single substitution 90-Leu→Gln observed in guinea pig and springhaas joins these two rodents to the lagomorph–primate branch in the lowest NR trees. Just because this substitution 90-Leu→Gln has occurred several times independently in different mammalian groups, the suggested connection of the springhaas family Pedetidae, which is generally considered a phylogenetic problem case, to the Caviomorpha is very weak indeed. A further study of rodent αA sequences, especially from the families Sciuridae, Castoridae, and Hystricidae, might nevertheless prove informative, from both phylogenetic and protein evolutionary points of view.

A similar lack of substitutions prevents any conclusion about the phylogenetic affinity of the tree shrew. According to the most parsimonious solutions, it could be a sister group of both the insectivores and the rodents, but by the addition of only one NR the tree shrew can join a common stem with primates or lagomorphs. Investigations of tree shrew myoglobin and hemoglobin have been equally inconclusive (Goodman *et al.*, 1979).

Nonetheless, the considerable number of substitutions in the single investigated insectivore αA chain, that of the hedgehog, makes it promising to study αA sequences from further taxa that are thought to be derived from the primitive insectivore stock, like solenodon, tenrec, elephant shrew, flying lemurs, and Megachiroptera.

5.2. Relationships among Vertebrate Classes and Subclasses

The inclusion in our data of the αA sequences of the chicken, a lizard, a frog, and a shark enables us to make some remarks about higher order phylogenetic groupings. Figure 13 depicts the most parsimonious branching pattern, in terms of NRs, of the vertebrate classes and subclasses. In Fig. 14 a biologically more acceptable tree is shown, joining the eutherian and metatherian mammals at the cost of two additional NRs.

In both the lowest NR tree and in the biologically more acceptable tree the birds and reptiles (represented by chicken and tegu) are strongly joined by a considerable number (8 and 11, respectively) of synapomorphous substitutions. The amniotes (mammals, birds, and reptiles) are also clearly separated as a monophyletic group from the anamnia (represented by frog and dogfish).

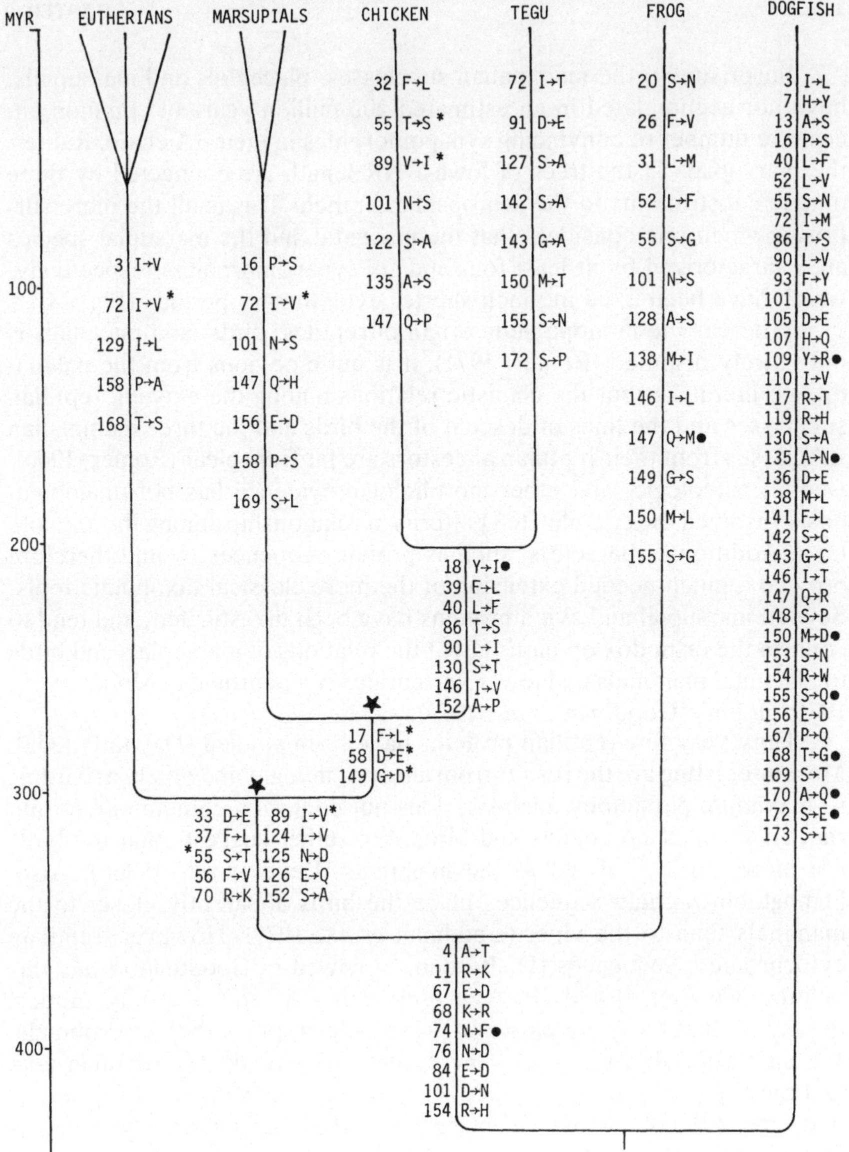

Figure 13. Most parsimonious tree for the α-crystallin A sequences from five vertebrate classes, including two mammalian subclasses. The branching points indicated by (★) are not compatible with current opinions about amniote phylogeny, and therefore do not reflect times of divergence. The sequence of the paralogous bovine α-crystallin B chain has been used to infer the distribution of substitutions over the branches from the root of the tree to dogfish and to the most recent common tetrapod ancestor. This tree requires 109 nucleotide replacements. Substitutions requiring two base replacements are indicated (●). Substitutions involved in the change of branching pattern between this figure and Fig. 14 are indicated by (*).

Surprisingly, the mammalian subclasses, placentals and marsupials, have not accumulated in an estimated 200 million years of common ancestry a number of convincing synapomorphies in their αA chain. Rather, the marsupials in the trees of lowest NR length are connected by three unique substitutions to the sauropsidan branch. This is all the more surprising when one considers that the placental and the marsupial species are characterized by at least four and six synapomorphies, respectively, which have been fixed in much shorter evolutionary periods (Fig. 14).

Whereas the monophyletic origin of reptiles, birds, and mammals is only rarely disputed (Romer, 1971), it is quite obvious from the paleontologic literature that the cladistic relations among the existing reptilian subclasses and the lines of descent of the birds and the three mammalian subclasses from their reptilian ancestors are far from clear (Romer, 1966). Dental, osteologic, and other morphologic evidence has not unambiguously resolved the complicated patterns of relationship among the amniote taxa. Additional characters, such as protein sequences, would therefore provide a much-needed extension of the more classical taxonomic tools. Several marsupial and avian proteins have been investigated, and tend to confirm the orthodox opinions about the relations of marsupials and birds to placental mammals and lower vertebrates (cytochrome *c*: Moore *et al.*, 1976; globins: Goodman *et al.*, 1979).

Only very few reptilian proteins have been studied (Dayhoff, 1978). Most interesting are the results from alligator myoglobin, which, according to maximum parsimony analysis, does not confirm a common archosaurian ancestor of crocodiles and birds, but rather suggests that the birds are more closely related to the mammals (Dene *et al.*, 1980b). Also, hemoglobin α-chain sequences place the birds apparently closer to the mammals than to the viper (Goodman *et al.*, 1979). However, reptilian cytochrome *c* sequences [D. Borden, reviewed in Goodman *et al.*, this volume, Chapter 4] and our tegu α-crystallin A sequence fully support the notion that birds are closer to present-day reptiles than to mammals. We may hope that increasing the number of investigated reptilian taxa and the types of proteins studied will provide sufficient resolving power to decide with confidence which are the most probable branching patterns in amniote evolution.

6. Molecular Aspects of α-Crystallin Evolution

It appears that the α-crystallin A and B chains, which show approximately 60% sequence homology, have different rates of evolutionary change. Whereas the bovine and human αA chains differ in ten out of 173

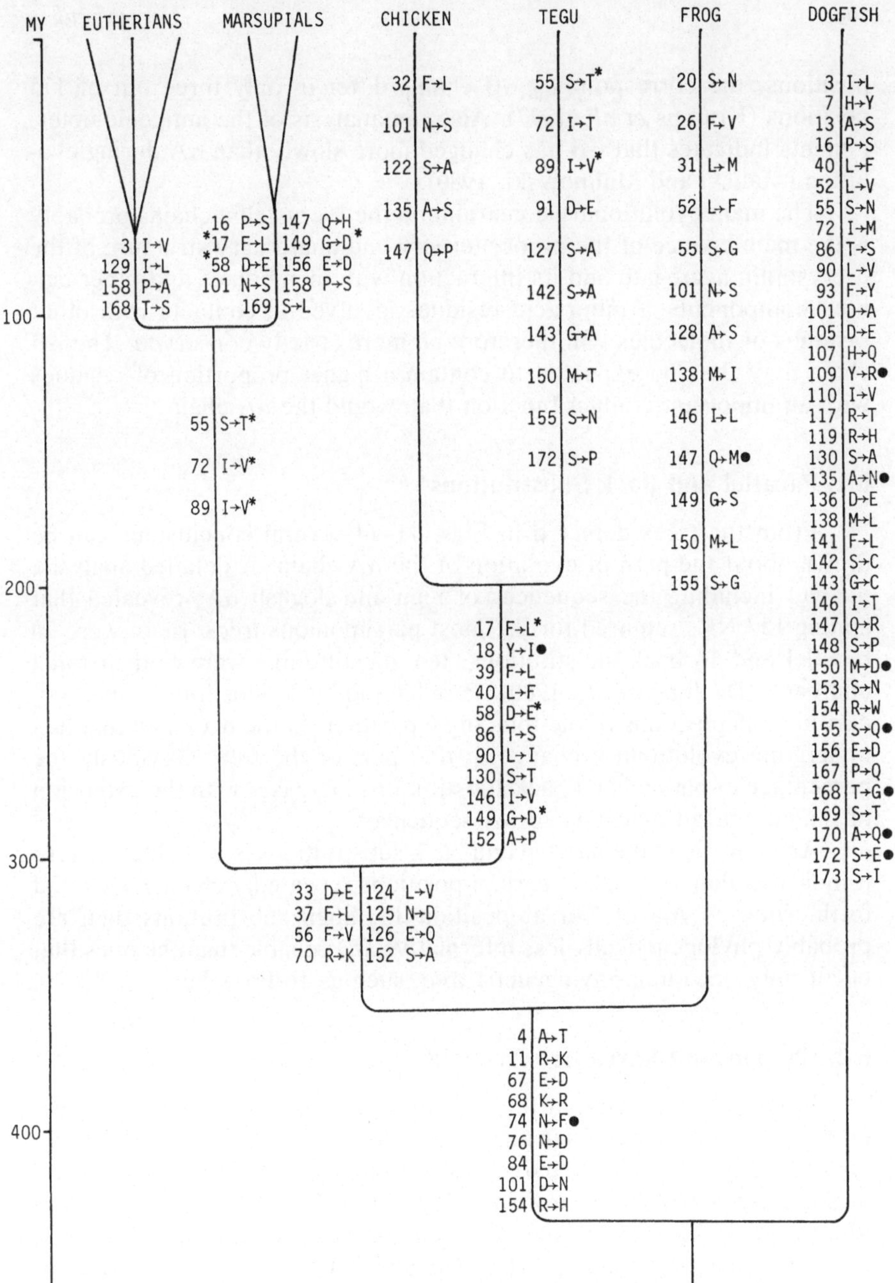

Figure 14. Biologically more acceptable tree for the vertebrate α-crystallin A sequences, requiring 111 NRs. Substitutions involved in the change of branching pattern between this figure and Fig. 13 are indicated by (*).

positions, the corresponding αB chains differ in only three out of 175 positions (Kramps *et al.*, 1977). Also, an analysis of the antigenic determinants indicates that αB has changed more slowly than αA during evolution (Manski and Malinowski, 1980).

The main evolutionary constraint on the α-crystallin chains probably is the maintenance of the proper tertiary and quaternary structure of the α-crystallin aggregate and its interaction with membranes and other cellular components. Amino acid residues involved in contacts with other residues or molecules will therefore be more strictly conserved. The αB chain may thus be expected to contain a higher proportion of residues with an important contact function than would the αA chain.

6.1. Parallel and Back Substitutions

From the trees depicted in Figs. 11–14 several conclusions can be drawn about the path of evolution of the αA chain. A detailed analysis, not yet including the sequences of tegu and dogfish αA, revealed that among 157 NRs required for the most parsimonious trees, there were 70 parallel and 16 back substitutions; ten substitutions were both parallel and back (De Jong *et al.*, 1980). Parallel and back substitutions had occurred at 26 positions of the total of 66 positions in the αA chain that had undergone evolutionary change in that part of the tree. Obviously the percentage of parallel and back substitutions increases with the extension of the number of investigated αA sequences.

An analysis of the parallel and back substitutions is valuable because it indicates that residues at certain positions repeatedly change back and forth, such as Ala or Thr at position 13. Such substitutions then are probably phylogenetically less informative and reliable than the ones that occur only once in a phylogenetic tree, such as 16-Pro→Ser.

6.2. Changes in Charge Are Avoided

In the αA tree, without the dogfish sequence, only 15 substitutions, or 9%, give rise to changes in amino acid side charge. This is considerably lower than the 32–34% of charged substitutions inferred to have occurred in the evolution of other proteins (Dayhoff, 1978, p. 346; Goodman and Moore, 1977). Moreover, nine substitutions occur in branches where they neutralize each other: mink, 7-His→Gln and 91-Glu→Gln; three-toed sloth, 7-His→Gln and 91-Asp→Gly; manatee, 70-Lys→Gln, 126-Gln→Lys, and 19-Pro→His; marsupials, 147-Gln→His and 149-Gly→Asp. It seems that selective constraints in the evolution of the αA chain have minimized changes in net charge. This is in agreement with the observation that the electrophoretic mobilities of αA chains from different vertebrate classes are remarkably similar (De Jong *et al.*, 1976).

6.3. "Covarions"

The evolution of the αA chain also provides several examples of the occurrence of concomitantly variable codons, or "covarions" (Fitch, 1971). The substitution 7-His→Gln, which has occurred independently in mink and three-toed sloth, occurs concomitantly with a substitution at position 91 of an acidic to a neutral residue (91-Glu→Gln in mink, 91-Asp→Gly in sloth). Also, substitutions at position 70 are accompanied in three different branches by changes at position 74: tamandua, 70-Lys→Arg, 74-Phe→Leu; opossum, 70-Lys→Arg, 74-Phe→Tyr; paenungulates, 70-Lys→Gln, 74-Phe→Leu. Other examples of substitutions that occur in clusters, as if local changes in the sequence of the αA chain by one substitution are compensated by the acceptance of others, can be found in seal and sea lion (51-Ser→Pro, 52-Leu→Val, 55-Thr→Ser), in higher primates (three substitutions and a deletion between positions 146 and 155), and in the branches to opossum and aardvark.

6.4. Variable Rates of Change and a Directional Trend in Substitutions

It is obvious that considerable variations in the rate of substitutions in the αA chain do occur. No substitutions have been accepted in the αA chains of dog, cat, whale, muroid rodents, tupaia, and kangaroo in the last 50–60 million years. Considerable numbers of substitutions have occurred in higher primates, hedgehog, edentates, and opossum. The rate of change was also very slow in early mammalian evolution, preceded by a much faster rate in early amniote evolution (Fig. 14).

It is clear that the "molecular clock" has run very irregularly during the evolution of the αA chain, as it has done in most other proteins. Nevertheless, an average rate of substitutions in the αA chain of 3.0% sequence change in 100 million years can be deduced from the data provided in Figs. 11–14. This places the αA chain among the moderately slowly evolving proteins (Dayhoff, 1978, p. 3). Unfortunately, the reasons for this generally slow rate of change, as well as for the conspicuous variation in different branches, can as yet only be guessed at.

Another interesting finding in the evolution of αA is a directional trend to replace residues with more generalized chemical properties by residues that have more specialized conformational and chemical properties (De Jong *et al.*, 1980) (Table III).

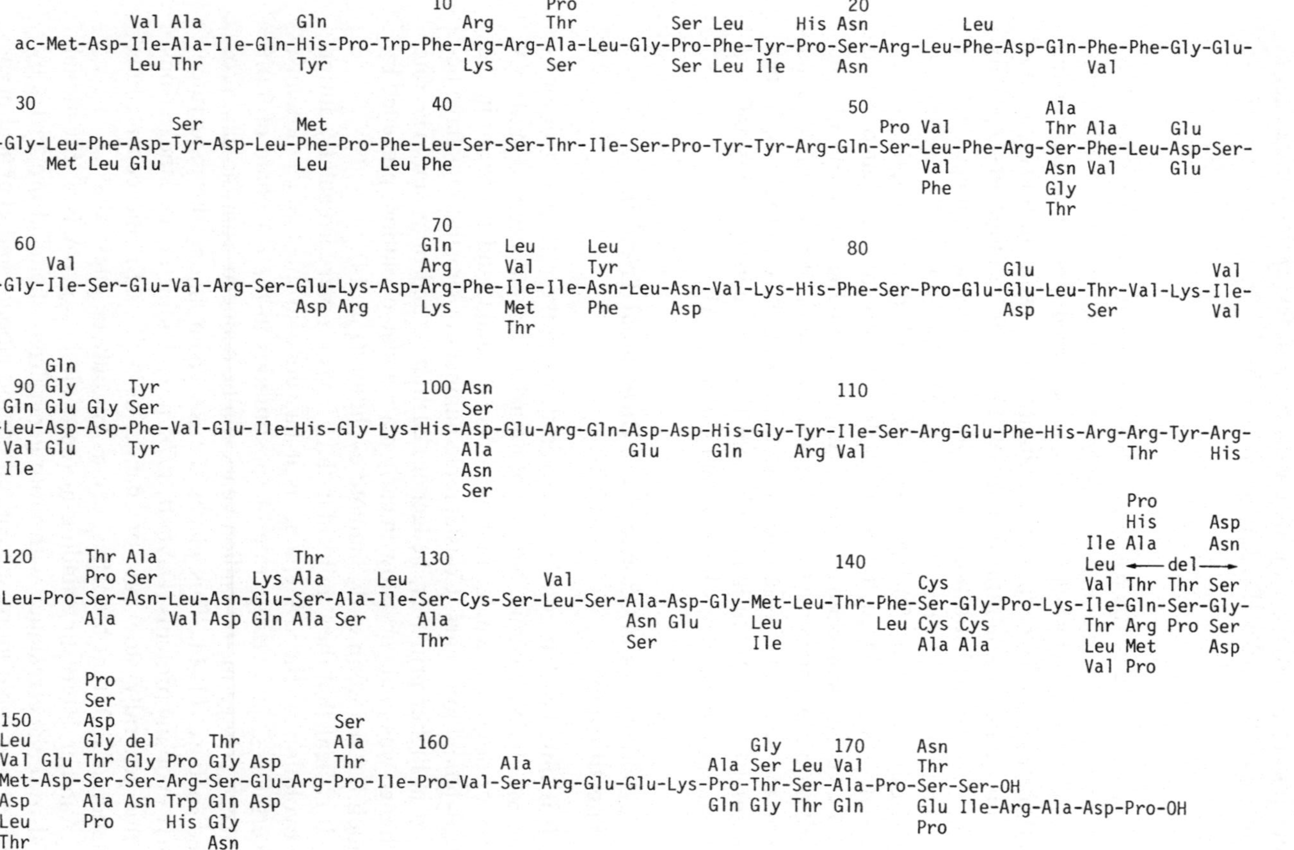

Figure 15. Reconstructed amino acid sequence of the ancestral α-crystallin A chain at the time of divergence of the ancestors of dogfish and tetrapods (the root of Fig. 13), and substitutions which have occurred in the lines to present-day mammals (Fig. 11) (above

Table III
Directional Trend among Substitutions in the Evolution of the α-Crystallin A Chain[a]

6 Val ↔ Ile 8	5 Ser ↔ Thr 9	2 Asp ↔ Glu 6	2 Ala ↔ Val 1
4 Val ↔ Leu 5	0 Ser ↔ Cys 2	1 Ala ↔ Thr 6	0 Phe ↔ Tyr 4
1 Val ↔ Met 2	6 Gly ↔ Ser 3	3 Ala ↔ Ser 4	1 Lys ↔ Arg 3
2 Val ↔ Phe 0	1 Gly ↔ Asp 2	2 Ala ↔ Pro 3	36 ← Total → 58

[a]All types of substitutions that lead to changes in stereochemical complexity and that occur more than once in the α-crystallin A tree (without tegu and dogfish) are tabulated. This shows that substitutions toward more complex residues (the right-hand ones) have occurred 58 times, those in the opposite direction 36 times. [From De Jong *et al.* (1980).]

6.5. Unequal Distribution of Substitutions over the αA Chain

All amino acid substitutions observed among vertebrate αA chains are summarized in the composite sequence in Fig. 15, where the continuous, central sequence represents the reconstructed ancestral αA sequence at the time of divergence of the lines leading to dogfish and to the tetrapods.

Above the continuous line are placed the substitutions which have occurred in the mammalian branches, and below are the substitutions in the lines to chicken, tegu, frog, and dogfish. This demonstrates that residues are mostly replaced by chemically similar ones, and that regions where substitutions are permitted are largely similar in mammalian and submammalian evolution. The unequal distribution of substitutions over the chain is clearly revealed. The highest mutability is observed between positions 146 and 158, which corresponds with the fact that this part of the chain is close to the surface of the α-crystallin aggregate (Siezen and Hoenders, 1979).

ACKNOWLEDGMENTS. Large parts of the phylogenetic and molecular evolutionary inferences are based on unpublished maximum parsimony analyses of αA sequences kindly made available by Dr. Morris Goodman, Detroit. The electrophoretic experiments shown in Figs. 6 and 7 were expertly carried out by A. Zweers. We thank E. Tolhuizen (Department of Ophthalmology, University of Nijmegen) for making the electropherogram shown in Fig. 5. The many persons who have made available lens material are acknowledged in our previous publications. This work was supported in part by the Netherlands Foundation for Chemical Research (SON) and by financial aid from the Netherlands Organization for Pure Research (ZWO).

References

Bhat, S. P., Jones, R. E., Sullivan, M. A., and Piatigorsky, J., 1980, Chicken lens crystallin DNA sequences show at least two δ-crystallin genes, *Nature* **284**:234–238.

Bloemendal, H. (ed.), 1981, *Molecular and Cellular Biology of the Eye Lens*, Wiley—Interscience, New York.

Blundell, T., Lindley, P., Miller, L., Moss, D., Slingsby, C., Tickle, I., Turnell, B., and Wistow, G., 1981, The molecular structure and stability of the eye lens: X-ray analysis of γ-crystallin II, *Nature* **289**:771–777.

Bours, J., 1976, Isoelectric focusing in free solution, in: *Isoelectric Focusing* (N. Catsimpoolas, ed.), Academic Press, New York, pp. 209–228.

Bours, J., 1977, The crystallins of the aging lens from five species studied by various methods of thin-layer isoelectric focusing, in: *Electrofocusing and Isotachophoresis* (B. J. Radola, and D. Graesslin, eds.), De Gruyter, Berlin, pp. 303–312.

Brahma, S. K., 1978, Ontogeny of lens crystallins in marine cephalopods, *J. Embryol. Exp. Morphol.* **46**:111–118.

Butler, P. M., 1972, The problem of insectivore classification, in: *Studies in Vertebrate Evolution* (K. A. Joysey and T. R. Kemps, eds.), Oliver and Boyd, Edinburgh, pp. 253–265.

Clayton, R. M., 1970, Problems of differentiation in the vertebrate lens, in: *Current Topics of Developmental Biology*, Volume 5 (A. A. Moscona and A. Monroy, eds.), Academic Press, New York, pp. 115–180.

Clayton, R. M., 1974, Comparative aspects of lens proteins, in: *The Eye*, Volume 5 (H. Davson and L. T. Graham, eds.), Academic Press, New York, pp. 399–494.

Day, T. H., and Clayton, R. M., 1973, Intraspecific variation in lens proteins, *Biochem. Genet.* **8**:187–203.

Dayhoff, M. O. (ed.), 1978, *Atlas of Protein Sequence and Structure*, Volume 5, Supplement 3, National Biomedical Research Foundation, Washington, D.C.

De Jong, W. W., 1981, Evolution of lens and crystallins, in: *Molecular and Cellular Biology of the Eye Lens* (H. Bloemendal, ed.), Wiley—Interscience, New York, pp. 221–278.

De Jong, W. W., and Goodman, M., 1982, Mammalian phylogeny studied by sequence analysis of eye lens protein α-crystallin, *Z. Säugetierk.*, in press.

De Jong, W. W., and Terwindt, E. C., 1976, The amino acid sequences of the α-crystallin A chains of red kangaroo and Virginia opossum, *Eur. J. Biochem.* **67**:503–510.

De Jong, W. W., Terwindt, E. C., and Groenewoud, G., 1976, Subunit compositions of vertebrate α-crystallins, *Comp. Biochem. Physiol.* **55B**:49–56.

De Jong, W. W., Gleaves, J. T., and Boulter, D., 1977, Evolutionary changes of α-crystallin and the phylogeny of mammalian orders, *J. Mol. Evol.* **10**:123–135.

De Jong, W. W., Zweers, A., and Goodman, M., 1980, Trends in the molecular evolution of α-crystallin, in: *Protides of the Biological Fluids*, Volume 28 (H. Peeters, ed.), Pergamon Press, Oxford, pp. 161–164.

De Jong, W. W., Leunissen, J. A. M., and Cuijpers, H. T., 1981, Primary structure of the major β-chain of armadillo (*Dasypus novemcinctus*) haemoglobin, *Biochim. Biophys. Acta* **668**:57–62.

De Jong, W. W., Zweers, A., Joysey, K. A., Gleaves, J. T., and Boulter, D., 1982, Protein sequence analysis applied to xenarthran and pholidote phylogeny, in: *The Evolution and Ecology of Sloths, Anteaters, and Armadillos* (G. G. Montgomery, ed.), Smithsonian Institution Press, Washington, D.C.

Dene, H., Goodman, M., and Romero-Herrera, A. E., 1980a, The amino acid sequence of elephant (*Elephas maximus*) myoglobin and the phylogeny of Proboscidea, *Proc. R. Soc. Lond. B* **207**:111–127.

Dene, H., Sazy, J., Goodman, M., and Romero-Herrera, A. E., 1980b, The amino acid sequence of alligator (*Alligator mississippiensis*) myoglobin, *Biochim. Biophys. Acta* **624**:397–408.

Driessen, H. P. C., Herbrink, P., Bloemendal, H., and De Jong, W. W., 1980, The β-crystallin Bp chain is internally duplicated and homologous with γ-crystallin, *Exp. Eye Res.* **31**:243–246.

Duke-Elder, S., 1958, *System of Ophthalmology*, Volume 1, *The Eye in Evolution*, H. Krimpton, London.

Fitch, W. M., 1971, Rate of change of concomitantly variable codons, *J. Mol. Evol.* **1**:84–96.

Fitch, W. M., 1979, Cautionary remarks on using gene expression events in parsimony procedures, *Syst. Zool.* **28**:375–379.

Foulds, L. R., Penny, D., and Hendy, M. D., 1979, A general approach to proving the minimality of phylogenetic trees illustrated by an example with a set of 23 vertebrates, *J. Mol. Evol.* **13**:151–166.

Gardiner, B. G., Janvier, P., Patterson, C., Forey, P. L., Greenwood, P. H., Miles, R. S., and Jefferies, R. P. S., 1979, The salmon, the lungfish and the cow: A reply, *Nature* **277**:175–176.

Goodman, M., and Moore, G. W., 1977, Use of Chou–Fasman amino acid conformational parameters to analyze the organization of the genetic code, *J. Mol. Evol.* **10**:7–47.

Goodman, M., Czelusniak, J., Moore, G. W., Romero-Herrera, A. E., and Matsuda, G., 1979, Fitting the gene lineage into its species lineage, A parsimony strategy illustrated by cladograms constructed from globin sequences, *Syst. Zool.* **28**:132–163.

Harding, J. J., and Dilley, K. J., 1976, Structural proteins of the mammalian lens: A review, *Exp. Eye Res.* **22**:1–73.

Hoenders, H. J., and Bloemendal, H., 1981, Ageing of lens proteins, in: *Molecular and Cellular Biology of the Eye Lens* (H. Bloemendal, ed.), Wiley—Interscience, New York, pp. 279–326.

Holmquist, R., 1978, A measure of the denseness of a phylogenetic network, *J. Mol. Evol.* **11**:225–231.

Kramps, J. A., De Man, B. M., and De Jong, W. W., 1977, The primary structure of the B_2 chain of human α-crystallin, *FEBS Lett.* **74**:82–84.

Kuck, J. F. R., East, E. J., and Yu, N. T., 1976, Prevalence of α-helical form in avian lens proteins, *Exp. Eye Res.* **23**:9–14.

Manski, W., and Halbert, S. P., 1965, Immunochemistry of the lens with special reference to phylogeny, *Invest. Ophthalmol.* **4**:539–559.

Manski, W., and Malinowski, K., 1978, The evolutionary sequence and quantities of different antigenic determinants of calf lens α-crystallin, *Immunochemistry* **15**:781–786.

Manski, W., and Malinowski, K., 1980, Molecular evolution and subunit structure of cattle lens α-crystallin, *J. Mol. Evol.* **15**:219–230.

McDevitt, D. S., 1972, Presence of lateral eye lens crystallins in the median eye of the American chameleon, *Science* **175**:763–764.

McKenna, M. C., 1975, Toward a phylogenetic classification of the Mammalia, in: *Phylogeny of the Primates* (W. P. Luckett and F. S. Szalay, eds.), Plenum Press, New York, pp. 21–46.

Montgomery, G. G. (ed.), 1982, *The Evolution and Ecology of Sloths, Anteaters, and Armadillos*, Smithsonian Institution Press, Washington, D.C.

Moore, G. W., Goodman, M., Callahan, C., Holmquist, R., and Moise, H., 1976, Stochastic versus augmented maximum parsimony method for estimating superimposed mutations, *J. Mol. Biol.* **105**:15–38.

Patterson, B., 1978, Pholidota and Tubulidentata, in: *Evolution of African Mammals* (V. J. Maglio and H. B. S. Cooke, eds.), Harvard University Press, Cambridge, Massachusetts, pp. 268–278.

Peacock, D., and Boulter, D., 1975, Use of amino acid sequence data in phylogeny and evaluation of methods using computer simulation, *J. Mol. Biol.* **95:**513–527.

Polyak, S., 1957, *The Vertebrate Visual System*, University of Chicago Press, Chicago.

Prager, E. M., and Wilson, A. C., 1978, Construction of phylogenetic trees for proteins and nucleic acids: Empirical evaluation of alternative matrix methods, *J. Mol. Evol.* **11:**129–142.

Reddan, J. R., 1975, Molecular embryology of the lens, in: *Cataract and Abnormalities of the Lens* (J. G. Bellows, ed.), Grune & Stratton, New York, pp. 29–42.

Romer, A. S., 1966, *Vertebrate Paleontology*, 3rd ed, University of Chicago Press, Chicago.

Romer, A. S., 1971, Unorthodoxies in reptilian phylogeny, *Evolution* **25:**103–112.

Romero-Herrera, A. E., Lehmann, H., Joysey, K. A., and Friday, A. E., 1978, On the evolution of myoglobin, *Phil. Trans. R. Soc. B* **283:**61–163.

Shoshani, J., Goodman, M., and Prychodko, W., 1978, Cladistic analysis of Paenungulata by computer (Abstract), *Am. Zool.* **18:**601.

Siezen, R. J., and Hoenders, H. J., 1979, The quaternary structure of bovine α-crystallin, *Eur. J. Biochem.* **96:**431–440.

Simpson, G. G., 1945, Principles of classification and a classification of mammals, *Bull. Am. Mus. Nat. Hist.* **85:**1–350.

Simpson, G. G., 1975, Recent advances in methods of phylogenetic inference, in: *Phylogeny of the Primates* (W. P. Luckett and F. S. Szalay, eds.), Plenum Press, New York, pp. 3–20.

Szalay, F. S., 1977, Phylogenetic relationships and a classification of eutherian mammals, in: *Major Patterns in Vertebrate Evolution* (M. K. Hecht, P. C. Goody, and B. Hecht, eds.), Plenum Press, New York, pp. 315–374.

Tedford, R. H., 1976, Relationship of pinnipeds to other carnivores (Mammalia), *Syst. Zool.* **25:**363–374.

Thenius, E., 1969, *Phylogenie der Mammalia*, De Gruyter, Berlin.

Van der Ouderaa, F. J., De Jong, W. W., and Bloemendal, H., 1973, The amino acid sequence of the αA_2 chain of bovine α-crystallin, *Eur. J. Biochem.* **39:**207–222.

Van der Ouderaa, F. J., De Jong, W. W., Hilderink, A., and Bloemendal, H., 1974, The amino acid sequence of the αB_2 chain of bovine α-crystallin, *Eur. J. Biochem.* **49:**157–168.

Van Druten, H. A. M., Peer, N. G. M., Bos, F. A. B. H., and De Jong, W. W., 1978, Reciprocal amino acid substitutions in the evolution of homologous peptides, *J. Theor. Biol.* **73:**549–561.

Van Valen, L., 1971, Adaptive zones and the orders of mammals, *Evolution* **25:**420–428.

Walls, G. L., 1963, *The Vertebrate Eye and its Adaptive Radiation*, Hafner, New York.

Williams, L. A., and Piatigorsky, J., 1979, Comparative and evolutionary aspects of δ-crystallin in the vertebrate lens, *Eur. J. Biochem.* **100:**349–357.

Wright, C. A. (ed.), 1974, *Biochemical and Immunological Taxonomy of Animals*, Academic Press, New York.

CHAPTER 4

Amino Acid Sequence Evidence on the Phylogeny of Primates and Other Eutherians

MORRIS GOODMAN, ALEJO E. ROMERO-HERRERA, HOWARD DENE, JOHN CZELUSNIAK, and RICHARD E. TASHIAN

1. Introduction

The biomolecular approach to systematic and evolutionary biology is in a state of transition. Laboratories that had been determining the amino acid sequences of proteins are now caught up by the excitement of the new recombinant DNA gene cloning and sequencing technology. The possibilities for advancing knowledge in systematic and evolutionary biology by application of this new technology seem almost boundless. It is obvious that knowing the actual nucleotide sequences of genes, rather than having to infer them from the amino acid sequences of encoded proteins, allows more accurate data to be used in figuring out the genealogic relationships of organisms (see Hewett-Emmett *et al.*, this volume, Chapter 9; also Scott and Smith, this volume, Chapter 8). During the transition, while laboratories engaged in studying molecular evolution are retooling in order to engage in nucleotide sequencing, it is worth preparing for the impending flood of these gene sequence data by taking

MORRIS GOODMAN, ALEJO E. ROMERO-HERRERA, and HOWARD DENE • Department of Anatomy, Wayne State University School of Medicine, Detroit, Michigan 48201. *JOHN CZELUSNIAK* • Departments of Biology and Anatomy, Wayne State University, Detroit, Michigan 48202. *RICHARD E. TASHIAN* • Department of Human Genetics, University of Michigan Medical School, Ann Arbor, Michigan 48104.

stock of what has already been learned about phylogeny from the substantial body of amino acid sequence data. With that objective in mind, this chapter focuses attention on the phylogeny of the order Primates, both on the subbranching within the order and on the genealogic position of Primates within the subclass Eutheria as well as on the broader pattern of vertebrate branching. We will concentrate on these groups because more species are represented in them by amino acid sequence data than in any other eukaryotic branch.

We also focus primarily on results obtained from genealogic trees constructed by the maximum parsimony method (Moore, 1976; Moore *et al.*, 1973; Goodman *et al.*, 1974, 1979a), since this method, in our experience, produces more accurate and informative phylogenetic reconstructions than those obtained by other tree-building algorithms. On the basis of the parsimonious reconstructions, we question the general validity of the molecular evolutionary clock hypothesis (Zuckerkandl and Pauling, 1962; Wilson *et al.*, 1977), in which protein sequences are supposed to evolve at approximately constant rates. From evidence for nonconstant rates of amino acid substitutions we support the Darwinian thesis that natural selection is the principal force behind the evolution of molecules and organisms. A recent review (Goodman, 1981a) of this topic has amassed many of the facts and concepts which are explored further in the present study.

2. Genealogic Evidence from Amino Acid Sequences

2.1. Tree Construction Strategy

2.1.1. Rationale

The aim of the maximum parsimony method applied to amino acid sequences is to account for the evolutionary descent of these sequences by the fewest possible genic changes. To show such minimum evolution, a genealogic arrangement must be found which maximizes the genetic likenesses associated with common ancestry while minimizing the incidence of parallel and back mutations. This can be done by utilizing the genetic code to represent the amino acid sequences as messenger RNA sequences and by then seeking a tree with the minimum number of nucleotide replacements, i.e., lowest NR length. Since common ancestry rather than convergent evolution is the most probable explanation for any extensive matching of nucleotide sequences between species, tree reconstruction algorithms based on maximum parsimony offer a way of using Occam's razor to find the preferred genealogic hypothesis.

2.1.2. Search for Lowest NR Length Trees

The strategy developed in previous studies (Goodman, 1976; Goodman *et al.*, 1979a) for finding the most parsimonious genealogic tree for a collection of related sequences was followed in the present study. In this strategy, as the first step, a matrix of minimum mutation distances is calculated for the amino acid sequences by the method of Fitch and Margoliash (1967). Then an unweighted pair group tree is constructed from this matrix by the clustering algorithm of Sokal and Michener (1958). Another initial dendrogram, the distance Wagner tree, is constructed from the same matrix by the algorithm of Farris (1972). The unweighted pair group and distance Wagner trees provide a means to start the search for the parsimonious genealogy without being biased by preconceived ideas on the phylogeny of the species from which the sequence data come. This search is conducted with a branch-swapping maximum parsimony algorithm (Goodman *et al.*, 1979a). Input data are the starting tree and the original file of amino acid sequences. The algorithm determines the NR length of the starting tree and of each alternative tree produced by the branch exchanges. The alternative having the lowest NR count is the start of the next round of exchanges. The rounds of exchanges continue until no new trees of lower NR count are found. Since the NR count reached may represent only a local minimum, trees which differ extensively from those previously examined, and which test a wide range of phylogenetic possibilities, can now be submitted to the branch-swapping algorithm. This continuation of the heuristic search sometimes reaches a deeper valley approaching the true minimum for that set of sequences in that further searching fails to reach a lower NR score.

2.1.3. Fitting the Gene Phylogeny into the Species Phylogeny

An important finding is that the lowest NR length trees contain fewer branching errors than the unweighted pair group and distance Wagner trees when judged in terms of accumulated knowledge on the phylogeny of the species represented by the sequences (Goodman *et al.*, 1979a). Nevertheless, despite this improvement in phylogenetic accuracy, the lowest NR length trees constructed for different proteins often yield nonconcordant branching arrangements on the same animal species and usually violate in each tree some of the features of the species phylogeny strongly supported by prior evidence. Such violations are indicative of incorrect groupings of sequences that happen to have excesses of convergent residues. Alternatively, the violations could be due to real differences between the gene phylogeny and the species phylogeny. Indeed, this later possibility opens the way to the construction of more accurate genealogic trees by an extension of the parsimony criterion. Not only are base replacements counted, but so are the additional genetic events needed to fit the putative gene phylogeny into well-established features of the species phylogeny.

While the sequence file for a gene phylogeny can have *orthologous* sequences aligned against *paralogous* sequences, the sequence file for a species phylogeny should only have orthologous sequences aligned against orthologous sequences. *Orthologous* sequences represent gene lineages that split from common ancestors simultaneously with separation of the species lineages, whereas *paralogous* sequences represent gene lineages that arose from gene duplication prior to species separation (Fitch, 1970). Inasmuch as the ancestral splitting of gene lineages can never postdate the ancestral splitting of their corresponding species lineages, it is always possible to fit the gene lineages within their species lineages. This is done by counting the events of gene duplication (GD) and gene expression (GE), either gene loss or inactivating and activating regulator mutations, needed for fitting one tree topology into another. The most parsimonious tree then becomes the one with the lowest NR + GD + GE score—in other words, the tree with the least number of genic changes for the group of sequences. The number of GDs that have to be postulated is equivalent to the number of bifurcations of gene lineages that must precede in time the bifurcations of the species lineages in order that the gene phylogeny can fit within the species phylogeny (Goodman *et al.*, 1979a). The number of GEs that then have to be hypothesized is invariably several times larger than the number of GDs and can be precisely determined by procedures described in Goodman *et al.* (1979a).

In pursuing the search for the parsimonious genealogy by this approach, it is necessary to single out features of the species phylogeny which are strongly indicated by prior evidence. Each such *a priori* feature can be represented in a dendrogram as a monophyletic portion of the total tree. In turn, the interior nodes which serve in the dendrogram as ancestors for these monophyletic subtrees can be so designated. The search for the tree that minimizes the summed NR + GD + GE score can then be aided by a modification of the maximum parsimony branch-swapping algorithm [Appendix I-E in Goodman *et al.* (1979a)]. The modification prevents branch swaps across any interior node designated as the ancestor of a monophyletic subtree, but allows branch-swapping for all undesignated regions within either the subtrees or the total tree. This helps find the lowest NR score for trees that do not require hypothetical GDs and

GEs. This score can then be compared to that of the trees that have the minimum NR count but in having it require hypothetical GDs and GEs.

The results obtained on proteins such as α- and β-hemoglobins and myoglobin (Goodman *et al.*, 1979a), cytochrome *c* (Baba *et al.*, 1981), and lens α-crystallin A (De Jong and Goodman, 1981) suggest that the gene lineages for each kind of protein deviate relatively little from established species relationships, i.e., most of the lineages among different species may be regarded as orthologous rather than paralogous. The real paralogous gene lineages are best identified by genetic evidence of duplicated loci, now bolstered by recombinant DNA and nucleotide gene sequencing data (Hewett-Emmett *et al.*, this volume, Chapter 9; also Scott and Smith, this volume, Chapter 8), although occasionally, as illustrated later, presumptive evidence for paralogous lineages may be obtained by the parsimony strategy recapitulated here.

It seems, therefore, that most differences between a putative gene phylogeny based on the sole criterion of minimum NR length and the well-established features of the species phylogeny result from mistakes made in the parsimony reconstruction when the statistical sample of aligned sequence positions is too small to override an erratic distribution of parallel and back mutations. However, when the number of sequence positions in the alignment is greatly expanded by combining different protein chains in tandem, the trees of lowest NR length found by the branch-swapping maximum parsimony algorithm should violate very few well-established species relationships.

Even if the traditional, nonmolecular evidence on species relationships were ignored, the molecular evidence itself would require that hypothesized events of gene duplication and expression be accounted for as well as base substitutions when constructing gene phylogenies from sequence data by the parsimony method. The problem is comparable to that encountered when seeking an alignment that must use gaps and insertions to maximize homology among evolutionarily related sequences. The genetic cost of any postulated intragenic deletions and insertions must be counted in the parsimony reconstruction and weighed against the savings in base replacements in order to decide the positioning and number of the extra genic events. Another comparable problem stems from the domain structure of proteins and the finding that different functional regions in a protein appear to be coded for by corresponding exons within the gene (Gilbert, 1978; Blake, 1979; Hewett-Emmett *et al.*, this volume, Chapter 9, Section 3.3; Scott and Smith, this volume, Chapter 8, Section 3.4). Exons, or coding portions of the gene, which get expressed, are separated from one another within the gene by introns, or nonexpressed intervening sequences. This finding raises the possibility that the genealogic histories of the different domains of a protein would not invariably coincide with one another, i.e., all fit the same way into the species phylogeny. Thus, if separate parsimonious trees are constructed for these different domains the cost in extra duplications and deletions or crossover events has to be added to the NR cost in order to evaluate where the trees should coincide and where they should vary. This in fact is the problem faced when dealing with different proteins, each of which is represented by sequences from the same group of species. The true gene phylogenies represented by these sequences will either coincide (all sequences orthologous) or will vary in branching topology (some sequences paralogous). Thus, when the evidence on phylogeny comes from more than one protein, the parsimony reconstructions must logically take into account possible events of gene duplication and expression as well as base substitutions in reconciling the gene phylogenies to one another.

2.1.4. Levels of Phylogenetic Inference

It should be pointed out that the necessity of taking account of possible events of gene duplication and gene expression in the search for most parsimonious trees limits the scope for phylogenetic inference when such trees are found. Nevertheless, the inferences that may be drawn, although limited, are likely to be more accurate provided judicious use has been made of prior evidence on phylogeny. For example, there is strong prior evidence that the order Primates is monophyletic and that the lemurs of Madagascar, including the sportive lemur (*Lepilemur mustelinus*), evolved within the Primates. Yet, the trees of lowest NR length found for myoglobin sequences (see Section 2.3) group *Lepilemur* with elephants rather than with other primates. However, when constrained to stay within a monophyletic order Primates, *Lepilemur* yields the least expensive NR arrangement by grouping with the branch containing species of the infraorder Lorisiformes. Furthermore, the unconstrained arrangement (*Lepilemur* joins elephants) is more expensive in genetic events because the number of NRs saved is less than the added cost of GDs and GEs needed to reconcile such a putative gene phylogeny with the primate status of *Lepilemur* in the species phylogeny. Limiting the scope for phylogenetic inference enhanced the chances of constructing accurate gene and species phylogenies. The fact that fewer total genetic events are needed for the *Lepilemur* sequence to be orthologously related to other primate sequences agrees with the expectation that most corresponding genes of different species (e.g., myoglobin genes) are likely to be orthologous genes rather than paralogous ones. Thus, a parsimony reconstruction of the genealogy of myoglobin sequences that minimizes the total NR + GD + GE count is probably more accurate than the reconstruction that minimizes only the NR count. If so, such a lowest NR + GD + GE length tree is the more reliable reconstruction for analyzing evolutionary trends in the descent of myoglobin sequences. Moreover, while no evidence is provided that *Lepilemur* is a primate (this has been assumed *a priori*), the lowest NR + GD + GE length tree of myoglobin sequences does provide evidence that in the species phylogeny the infraorder Lemuriformes, which contains the genus *Lepilemur*, is genealogically closest to Lorisiformes. This phylogenetic inference is confirmed in fact by the aggregate body of amino acid sequence data, in particular, by the lowest NR length trees found (see Section 2.5) for a combined tandem alignment spanning seven different polypeptide chains, including the four represented in lemurs (α- and β-hemoglobin, myoglobin, and lens α-crystallin A). Lemuriform and lorisiform branches join together first and then group with other Primates.

Some further comments on the strategy of constructing species phylogenies from sequence data may be useful before turning to a fuller review of our maximum parsimony results. In theory, if the statistical sample of aligned orthologous sequence positions representing the different species being compared is sufficiently large, the parsimony tree that minimizes only the NR count should yield the correct genealogic arrangement for these species. In practice, the available sequence data constitute a rather meager statistical sample. Very few species can be compared by several or more orthologous sequences. Nor can we always be sure that we employed only orthologous sequences in these comparisons. Thus, with the present aggregate body of sequence data, minimizing NR length alone does not guarantee reconstruction of the correct species phylogeny. Even when the sequences of different kinds of proteins are combined in tandem as though they were products of a single giant gene, it is still helpful at some stage in the search for the correct tree to specify the genealogic relationships among the species lineages for which there is strong *a priori* evidence. For this purpose, we treat the extended sequence alignment as if it were indeed encoded by a single gene, i.e., we pursue the search as though we were constructing a gene phylogeny. Then each violation of a well-established species relationship in a tree of lowest NR length becomes evident as a gene duplication, a bifurcation in the gene phylogeny preceding the splitting of the species containing these expressed genes. In other words each violation of a prespecified species relationship can be treated as costing on the average 1 GD and about 3 GEs, a small price considering the large number of sequence positions at which NRs can accumulate in the extended alignment. Thus, if a violation of a prespecified species relationship in the tree minimizing NR length alone does not save more than about 4 NRs, the prespecified relationship, i.e., the orthologous arrangement, is considered to be correct, although of course we have not provided independent evidence for it. Clearly, we might leave too little room for phylogenetic inference if the prior identification of monophyletic subtrees is excessive. This is not a problem, however, in the present study, since most questions of mammalian phylogeny are still wide open (Novacek, this volume, Chapter 1). Furthermore, even prespecified species relationships can be challenged if trees that violate them save a large number of NRs. Then, if the species at issue were represented in the extended alignment by more than several types of sequences, we could choose to conclude for these species that the prior evidence on phylogeny was mistaken rather than that we were dealing with paralogous gene lineages from duplicated loci. Obviously any new genealogic arrangement suggested by the maximum parsimony reconstruction would be subject to further testing once additional proteins or genes were sequenced in these species.

2.2. Sequences Analyzed

The genealogic reconstructions were carried out on amino acid sequence data from 553 polypeptide chains of 244 species. The polypeptide chains consist of 94 cytochromes *c*, 46 lens α-crystallins, 47 combined fibrinopeptide A and B sequences, 18 carbonic anhydrases, 29 members of the protein family containing calmodulin, and 268 globins, of which those from the jawed vertebrates subdivide into myoglobin, β-hemoglobin, and α-hemoglobin. The vast majority of the sequences (488 out of 553) and a majority of the species (181 out of 244) are from vertebrates. The vertebrate class best represented by the sequences is the Mammalia (138 species), in particular its subclass Eutheria (130 species), within which the order Primates is especially well sampled (37 species). A catalog of almost all these sequences and the species from which they come can be found in Goodman (1981a). The maximum parsimony approach was used to construct gene phylogenies for the separate collections of sequences (each collection consisting only of evolutionarily related sequences) and species phylogenies from sequences of separate collections combined in an extended tandem alignment. In this connection many of these sequences come from the same mammalian taxa, as evident from the information compiled in Table I.

Table I
Species List of Amino Acid Sequences Analyzed by the Maximum Parsimony Method[a]

Common name	Species	αHb	βHb	Mb	Cry	Fib	Cyt	Car	Cal
Primates									
Human	*Homo sapiens*	1	5	1	1	2	1	3	1
Chimpanzee	*Pan troglodytes*	2	4	1		2	1	1	
Gorilla	*Gorilla gorilla*	2	2	1		2		1	
Orangutan	*Pongo pygmaeus*	2	1	1		2		1	
Gibbon	*Hylobates lar*		3						
Gibbon	*Hylobates agilis*		1	1		2			
Siamang gibbon	*Symphalangus syndactylus*			1		2			
Rhesus	*Macaca mulatta*	1	2		1	2	1	2	
Crab-eating macaque	*Macaca fascicularis*	2		1					
Pit tail macaque	*Macaca nemestrina*	2	1						
Japanese macaque	*Macaca fuscata*	1	1						
Baboon	*Papio cynocephalus*	1	2						
Baboon	*Papio anubis*			1					
Drill	*Papio leucophalus*					2			
Gelada baboon	*Theropithecus gelada*	1	1						
Mangabey	*Cercocebus atys*	3	1						
Vervet	*Cercopithecus aethiops*	1	1			2			
Patas	*Erythrocebus patas*	1	1	1					
Colobus	*Colobus badius*	1	1						
Langur	*Presbytis entellus*	1	1	1					
Spider monkey	*Ateles geoffroyi*	1	2			2	1		
Woolly monkey	*Lagothrix lagothricha*			1					
Capuchin monkey	*Cebus apella*	1	1	1		2	1	1	
Squirrel monkey	*Saimiri sciureus*		2	1					
Night monkey	*Aotes trivirgatus*		2						
Marmoset	*Saguinus nigricollis*		2						
Marmoset	*Saguinus mystax*		2						
Marmoset	*Saguinus fusciollis*	1	1						
Marmoset	*Callithrix jacchus*			1					
Tarsier	*Tarsius bancanus*	1	1						
Slow loris	*Nycticebus coucang*	1	1	1		2	1		
Slender loris	*Loris tardigradus*	1	1						
Potto	*Perodicticus potto*			1	1				
Bush baby	*Galago crassicaudatus*	1		1	1				
Lemur	*Lemur fulvus*	1	1		1				
Sportive lemur	*Lepilemur mustelinus*			1					

[a]αHb, α-hemoglobin; βHb, β-hemoglobin, including ε-, γ-, and δ-hemoglobins; Mb, myoglobin; Cry, α-crystallin; Cyt, cytochrome *c*; Car, carbonic anhydrases I, II, or III; Cal, proteins of the calmodulin family. References to compilations of these sequences up through 1979 and to the individual sequences obtained since 1979 are given in Goodman (1981a). In addition, goose-beaked whale myoglobin (Lehman et al., 1980) and Hubb's-beaked whale myoglobin (Dwulet *et al.*, 1980) are included in the present study.

Table I *(continued)*

Common name	Species	αHb	βHb	Mb	Cry	Fib	Cyt	Car	Cal
Sifaka	*Propithecus verreauxi*	1	1						
Lagomorpha									
Rabbit	*Oryctolagus cuniculus*	1	1	1	1	2	1	2	5
Pika	*Ochotona princeps*			1	1				
Scandentia									
Tree shrew	*Tupaia glis*	1	1	1					
Tree shrew	*Tupaia belangeri*				1				
Rodentia									
Mouse	*Mus musculus*	2	3				1		
Mouse	*Mus cervicolor*		2						
Rat	*Rattus norvegicus*	1	1		1	2	1		
Hamster	*Mesocricetus auratus*				1				
Gerbil	*Meriones unguiculatus*				1				
Lemming	*Lemmus sibiricus*	1	1						
Muskrat	*Ondatra zibethicus*	1	1						
Red-backed mouse	*Clethrionomys rutilus*	1	1						
Vole	*Microtus xanthognathus*	1	1						
Guinea pig	*Cavia porcellus*	1	1		1	1	1		
Springhaas	*Pedetes cafer*				1				
Carnivora									
Dog	*Canis familiaris*	2	1	1	1	2	1		
Coyote	*Canis latrans*	1	1						
Fox	*Urocyon cineroargenteas*	1	1						
Bat-eared fox	*Otocyon megalotis*			1					
Fox	*Vulpes vulpes*					2			
Hunting dog	*Lycaon pictus*			1					
Bear	*Thalarctos maritimus*	1	1						
Sloth bear	*Melursus ursinus*				1				
Mink	*Mustela vison*				1				
Badger	*Meles meles*	1	1	1		1			
Racoon	*Procyon lotor*	1	1						
Coatimundi	*Nasua narica*	1	2						
Coatimundi	*Nasua nasua*		1						
Sea lion	*Zalophus californianus*			1	1				
Seal	*Mirounga leonina*						1		
Gray seal	*Halichoerus grypus*				1	1			
Harbor seal	*Phoca vitulina*			1					
Cat	*Felis felis*	1	2		1	2			
Lion	*Panthera leo*	1	1			2			
Pholidota									
Pangolin	*Manis javanica*				1				
Artiodactla									
Sheep	*Ovies aries*	1	4	1		2	1	1	
Barbary sheep			1						

(continued)

Table I *(continued)*

Common name	Species	αHb	βHb	Mb	Cry	Fib	Cyt	Car	Cal
Goat	*Capra sp.*	2	2			2			
Persian gazelle						2			
Ox	*Bos taurus*	1	3	1	2	2	1	3	5
Bison	*Bison bonasus*					2			
Water buffalo	*Bubalus bubalus*					2			
Cape buffalo	*Syncerus caffer*					2			
Red deer	*Cervus elephas*			1		2			
Elk	*Cervus canadensis*					2			
Sika deer	*Cervus nippon*					2			
Muntjak	*Muntiacus muntjak*					2			
Black-tailed deer	*Odocoileus hemionus*					2			
Reindeer	*Rangifer tarandus*					2			
Giraffe	*Giraffa camelopardalis*				1	2			
Pronghorn	*Antilocapra americana*					2			
Llama	*Lama peruana*	1	1			2			
Guanaco	*Lama guanicoe*						1		
Vicuna	*Vicugna vicugna*					2			
Camel	*Camel dromedarius*	1	1		1	2	1		
Pig	*Sus scrofa*	1	1	1	1	2	1		1
Hippopotamus	*Hippopotamus amphibius*								
Cetacea									
Killer whale	*Orcinus orca*			1					
Black sea dolphin	*Delphinus delphis*			1					
Common dolphin	*Delphinus delphis*			1					
Pilot whale	*Globicephala melaena*			1					
Bottlenosed dolphin	*Tursiops truncatus*			1					
Pacific spotted dolphin	*Stenella attenuata*			1					
Common porpoise	*Phocaena phocaena*			1	1				
Doll porpoise	*Phocoenoides dalli*			1					
River dolphin	*Inea geoffrensis*			1					
Sperm whale	*Physter catadon*			1					
Dwarf sperm whale	*Kogia simus*			1					
Goose-beaked whale	*Ziphius cavirostris*			1					
Hubb's-beaked whale	*Mesoplodon carlhubbsi*			1					
Minke whale	*Balaenoptera acutorostrata*			1	1				
Finback whale	*Balaenoptera physalus*			1					
Sei whale	*Balaenoptera borealis*			1					
Humpback whale	*Megoptera novaeangliae*			1					
Gray whale	*Eschrichtius gibbosus*			1					
Whale	*Eschrichtius glaucus*						1		
Perissodactyla									
Horse	*Equus caballus*	2	1	1	1	2	1	1	
Zebra	*Equus burchelli*			1					

(continued)

Table I *(continued)*

Common name	Species	αHb	βHb	Mb	Cry	Fib	Cyt	Car	Cal
Zebra	*Equus quagga*						1		
Zebra	*Equus sp.*					4			
Donkey	*Equus asinus*	1				2	1		
Mule	*E. caballus x asinus*					4			
Tapir	*Tapirus indicus*				1				
Tapir	*Tapirus terrestris*					2			
Rhinoceros						2			
White rhinoceros	*Ceratotherium simon*				1				
Insectivora									
Hedgehog	*Erinaceus europaeus*	1	1	1	1				
Chiroptera									
Fruit bat	*Rousettus aegyptiacus*			1					
Bat	*Artibeus jamaicensis*				1				
Bat	*Miniopteris schreibersi*				1				
Edentata									
Armadillo	*Dasypus novemcinctus*		1						
Two-toed sloth	*Choloepus hoffmanni*				1				
Three-toed sloth	*Bradypus infuscatus*				1				
Anteater	*Tamandera mexicana*				1				
Probosidea									
African elephant	*Loxodonta africana*			1	1				
Asian elephant	*Elephas maximus*		1	1		2			
Sirenia									
Manatee	*Trichechus inungius*				1				
Hyracoidea									
Cape hyrax	*Procavia capensis*				1				
Tubulidentata									
Aardvark	*Orycteropus afer*				1				
Marsupialia									
Kangaroo	*Macropus giganteus*	1	1						
Kangaroo	*Macropus rufus*				1				
Kangaroo	*Macropus cangura*						1		
Kangaroo						2			
Kangaroo	*Megaleia rufa*			1					
Potorous	*Potorous tridactylus*		1						
Opossum	*Didelphis marsupialis*	1	1	1	1				
Monotremata									
Platypus	*Ornithorhynchus anatinus*	1	1	1					
Echidna	*Tachyglossus aculeatus*	2	1	1					
Aves									
Chicken	*Gallus gallus*	4	1	1	1	1	1	1	2
Turkey	*Meleagris gallopano*						1		

(continued)

Table I *(continued)*

Common name	Species	αHb	βHb	Mb	Cry	Fib	Cyt	Car	Cal
Penguin	*Aptenodytes patagonica*						1		
Penguin	*Aptenodytes forsteri*			1					
Emu	*Dromaius novae-hollandiae*						1		
Ostrich	*Struthio camelus*	1	1				1		
Pigeon	*Columba livia*						1		
Pekin duck	*Anas platyrhynchos*						1		
Goose	*Anser anser*	2	1						
Barheaded goose	*Anser indicus*	1	1						
Reptilia									
Alligator	*Alligator mississippiensis*			1					
Viper	*Viper aspis*	1							
Rattlesnake	*Crotalus adamanteus*						1		
Lizard	*Varanus varanus*			1			1		
Lizard	*Tupinombis teguixin*				1				
Turtle	*Graptemys geographica*			1					
Turtle	*Chelydra serpentina*						1		
Turtle	*Malaclemys terrapin*							1	
Amphibia									
Bullfrog	*Rana catesbeiana*		2				1		
Frog	*Rana esculentia*		1		1				2
Newt	*Taricha granulosa*	1							
Oxolote	*Ambystoma mexicanum*	1							
Crossopterygii									
Coelacanth	*Latimeria cholumnae*								2
Teleostei									
Carp	*Cyprinus carpio*	1	2	1			1		2
Sucker	*Catostomus clarkii*	1							
Chub									1
Goldfish	*Carassius auratus*	1	1						
Whiting	*Cadus merlangus*								1
Cod	*Cadus callarias*								1
Hake	*Merlucius merlucius*								1
Pike	*Esox lucius*								2
Trout	*Salmo irideus*								1
Tuna	*Thunnus thynnus*						1		
Tuna	*Thunnus albacares*			1					
Bonito	*Katsuwonus vagrans*						1		
Elasmobranchii									
Shark	*Heterodontus portusjacksoni*	1	1	1					
Shark	*Mustelus antarcticus*			1					
Dogfish	*Squalis sucklii*						1		
Dogfish	*Squalis acanthias*				1				
Thornback ray	*Raja clavata*								1

(continued)

Table I *(continued)*

Common names	Species	Globin	Cyt	Cal
Agnatha				
Lamprey	*Lampetra fluviatilis*	1		
Lamprey	*Petromyzon marinus*	1		
Lamprey	*Entosphenus tridentatus*		1	
Hagfish	*Myxine glutinosa*	1		
Echinodermata				
Starfish	*Asterias ruberis*		1	
Mollusca				
Gastropod	*Aplysia limacina*	1		
Gastropod	*Busycon canaliculatum*	1		
Scallop				1
Snail	*Helix aspersa*		1	
Arthropoda				
Prawn	*Macrobrachium malcomsonii*		1	
Chironomus	*Chironomus thummi*	4		
Fruit fly	*Drosophila melanogaster*		1	
Blowfly	*Lucilia cuprina*		1	
Screw-worm	*Haemotobia irritans*		1	
Ceratitis	*Ceratitis capitata*		1	
Silkworm moth	*Samia synthia*		1	
Hornworm moth	*Manduca septa*		1	
Locust	*Schistocerca gregaria*		1	
Annelida				
Bloodworm	*Glycera dibranchiata*	1		
Brandling worm	*Eisenia foetida*		1	
Coelenterata				
Renilla	*Renilla reniformes*			1
Plantae				
Enteromorpha	*Enteromorpha intestinalis*		1	
Ginkgo	*Ginkgo biloba*		1	
Soybean	*Glycine max*	2		
Kidney bean	*Phoseolus vulgaris*	1		
Mung bean	*Phoseolus aureus*		1	
Broad bean	*Vicia faba L.*	1		
Lupin	*Lupinus luteus*	1		
Parsnip	*Pastinaca olerocea*		1	
Niger	*Guizotia abyssinica*		1	
Sunflower	*Helianthus annuus*		1	
Elder	*Sambucus nigra*		1	
Pumpkin	*Cucurbita maxima*		1	
Arum	*Arum maculatum*		1	
Leek	*Allium porrum*		1	
Maize	*Zea mays*		1	

(continued)

Table I *(continued)*

Common names	Species	Globin	Cyt	Cal
Wheat	*Triticum aestivum*		1	
Cauliflower	*Brassica oleracea*		1	
Rape	*Brassica napus*		1	
Sesame	*Sesamum indicum*		1	
Castor	*Ricinus communis*		1	
Abutilon	*Abutilon theophrasti*		1	
Cotton	*Gossypium barbadense*		1	
Box elder	*Acer negundo*		1	
Hemp	*Cannabis sativa*		1	
Love-in-the-mist	*Nigella damascena*		1	
Nasturtium	*Tropaeolum majus*		1	
Spinach	*Spinacea oleracea*		1	
Buckwheat	*Fagopyrum esculentum*		1	
Potato	*Solanum tuberosum*		1	
Tomato	*Lycopersicum esculentum*		1	
Fungi				
Ustilago	*Ustilago sphaerogena*		1	
Humicola	*Humicola languginosa*		1	
Neurospora	*Neurospora crassa*		1	
Bakers Yeast	*Saccharomyces oviformes*		1	
Bakers Yeast	*Saccharomyces cerevisiae*		1	
Candida	*Candida krusei*		1	
Debaryomyces	*Debaryomyces kloeckeri*		1	
Physarum	*Physarum polycephalum*		1	
Euglena	*Euglena gracilis*		1	
Crithidia	*Crithidia oncopelti*		1	
Tetrahymena	*Tetrahymena pyriformis*		1	
Prokaryota				
	Rhodomicrobium vannielii		1	
	Rhodopseudomonas palustris		1	
	Rhodospirillum rubrum		1	
	Rhodopseudomonas spheroides		1	
	Rhodopseudomonas capsulata		1	
	Paracoccus dentrificans		1	

2.3. Gnathostome Myoglobin Genealogy

2.3.1. Inferences from Lowest NR Length Trees

The construction of gene phylogenies and the kinds of phylogenetic inferences on organisms that they provide may be illustrated by results obtained on 68 gnathostome myoglobin sequences. After extensive searching, in which thousands of alternative trees were examined by the branch-swapping algorithm, the lowest NR score reached for these myoglobin sequences was 758. A typical tree at this lowest NR length is shown in Fig. 1. This tree, in common with the others at the lowest NR length, captures a number of well-established phylogenetic relationships. The two sharks group together, as do the two teleosts. All the amniotes (reptiles, birds, and mammals) cluster as a monophyletic assemblage, within which the class Aves and the class Mammalia are monophyletic groups. Within Mammalia, the Monotremata, Anthropoidea, Pinnipedia, Lagomorpha, Lorisoidea, Cetacea, Elephantinae, and Pecora (deer and bovids) are each monophyletic. Similarly, within Cetacea the families Rhachianectidae and Balaenopteridae of the suborder Mysticeti, the families Ziphiidae and Physeteridae of the superfamily Physeteroidea, and the superfamilies Platanistoidea and Delphinoidea, the latter containing Delphinidae and Phocaenidae, all appear as monophyletic branches. Furthermore, if we view the cetacean suborder Odontoceti (toothed whales) as a grade rather than clade, the grouping in the lowest NR length tree of Physeteroidea with Mysticeti (the baleen whales) rather than with other Odontoceti, the two grouped superfamilies Platanistoidea and Delphinoidea, is not with any certainty in disagreement with the correct species phylogeny. Anthropoidea also branches as expected, first into Platyrrhini and Catarrhini, then Catarrhini into Cercopithecoidea and Hominoidea, and within Hominoidea—in agreement with data from other proteins, from DNA, and from chromosomes—*Gorilla* and *Pan* share a closer kinship with *Homo* than with *Pongo* and Hylobatinae (*Hylobates* and *Symphalangus*). Finally, certain of the lowest NR length trees (e.g., the tree in Fig. 1) have a monophyletic Eutheria and a monophyletic Carnivora.

These lowest NR length trees, however, show some major differences between the branching patterns of gene lineages and species lineages with regard to the well-established features of the species phylogeny. For the mammalian region of the tree shown in Fig. 1 it takes 29 further genic events (7 GDs + 22 GEs) to reconcile the differences. Of course these genic events are hypothetical; nevertheless, they are required to fit the putative gene phylogeny into the assumed features of the species phylogeny. Starting in the early mammalian region of this putative gene phylogeny and going toward the present, the first duplication explains why kangaroo does not group with opossum, the other marsupial. On the assumption that the therian mammals (marsupials and eutherians) are monophyletic, the second duplication is needed because opossum comes out closer genealogically to the monotreme branch than to the eutherian branch. The third duplication accounts for the separate descent of the anthropoid lineage from a lineage which groups other primates with non-primates. Similarly, the fourth duplication accounts for the wide separation of pig from its order Artiodactyla. The fifth duplication explains why the lorisoid branch (one of the two remaining primate branches) is no closer to sportive lemur than to *Equus*, elephants, pecorans, and cetaceans. The sixth duplication results from the assumption that the ungulate orders Perissodactyla and Artiodactyla are closer genealogically to one another than either is to Primates; this duplication allows the equine branch to be closer to sportive lemur than to other ungulates. Finally, the seventh duplication explains why sheep (a bovid) groups first with deer rather than with its closer species relative, ox (the other bovid). All these assumed features of the species phylogeny (which the putative gene phylogeny of Fig. 1 takes account of) are supported by traditional phylogenetic evidence, and almost all are also supported by the analysis of the combined body of amino acid sequence data described in Section 2.5.

Another feature of the myoglobin lowest NR length trees is a closer grouping of Aves to Mammalia than to Crocodilia and other reptiles. This disagrees with the traditional view that among extant vertebrates birds are closest in kinship to crocodilians and also closer to snakes and lizards than to mammals. Moreover, 8–12 NRs are added to the cost of the reconstruction if the birds are first joined either to alligator alone or to a branch containing alligator and other reptiles. Thus, even if the traditional view is correct, the most parsimonious myoglobin gene phylogeny must still have a lineage to reptile sequences separate from a lineage to bird and mammal sequences, because even after counting the extra events of gene duplication and expression it is cheaper to have the reptile myoglobin sequences paralogously related to the bird and mammal myoglobin sequences. In the case of α-hemoglobin sequences, in which Reptilia is represented by the snake *Viper aspis*, it is also cheaper to have this reptile lineage separate from a lineage to birds and mamals. Reptiles are further represented in the total body of amino acid sequence data by cytochromes *c* of a turtle, a snake, and a lizard and by α-crystallin A of a lizard. In contrast to the myoglobin and α-hemoglobin results, birds group with reptiles rather than with mammals in the maximum parsimony trees constructed, respectively, for cytochrome *c* sequences and α-crystallin sequences. The fact that the cytochrome *c* and α-crystallin results

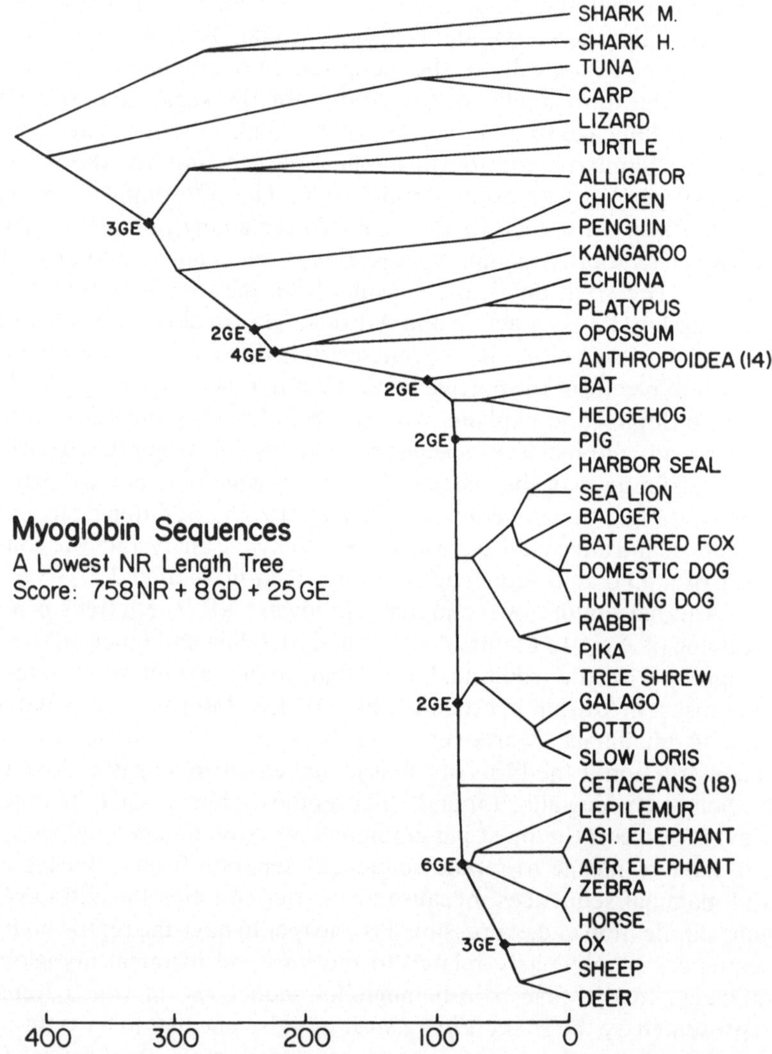

Figure 1. A lowest NR length putative gene phylogeny found for 68 myoglobin sequences. The subbranchings within Anthropoidea and within Cetacea are shown above. (♦) Hypothetical gene duplication. These hypothetical GDs and their accompanying GEs (gene expression events) are needed to reconcile this putative gene phylogeny to well-established features of the species phylogeny. Each such GD must always precede in time the ancestral splitting

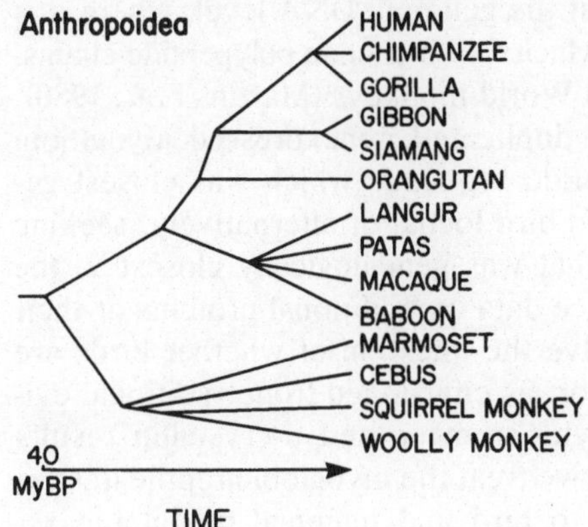

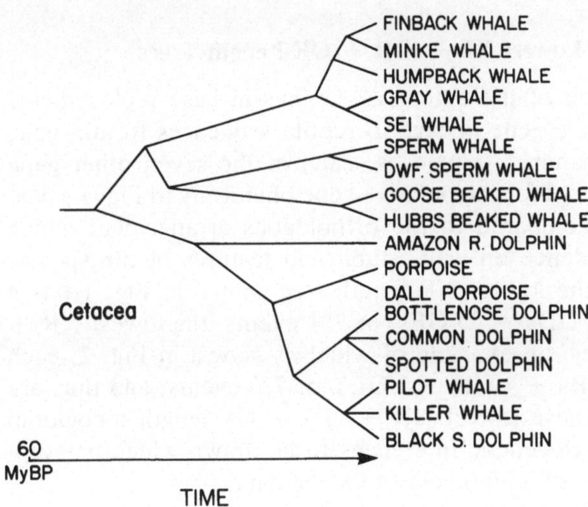

of the species exhibiting the hypothesized paralogous sequences. The time scale for the branch points in the tree is based on paleontologic views concerning the ancestral separations of the species represented by the sequences. Thus the branching times of orthologous sequences coincide with the corresponding branching times in the species phylogeny.

support the traditional view strengthens the hypothesis that the reptile myoglobins are paralogously related to bird and mammal myoglobins.* This hypothesis is open to testing at the genomic DNA level, where it is possible to detect duplicated genes which fail to encode polypeptide chains, e.g., the δ-hemoglobin genes of Old World monkeys (Martin *et al.*, 1980). The test would involve seeking a duplicated nonexpressed myoglobin locus in reptiles that had a nucleotide sequence which was closest genealogically to that of the expressed bird locus, or alternatively, seeking a silent myoglobin locus in birds that was genealogically closest to the expressed locus in reptiles. Sequence data on additional proteins or their encoding genes can also help resolve the question of whether birds are closest genealogically to mammals or, as envisioned from traditional evidence, to reptiles. Because the cytochrome *c* and α-crystallin results agree with the traditional evidence, we treat the myoglobin reptile lineage in Fig. 1 as paralogously related to bird and mammal myoglobin sequences.

2.3.2. Inferences from Lowest NR + GD + GE Length Trees

With the exception of the extensive savings in base replacements obtained by separating a gene lineage to reptile sequences from a gene lineage to bird and mammal sequences, each of the seven other gene duplications hypothesized for the putative gene phylogeny in Fig. 1 saved only 1 or 2 NRs as compared to the orthologous arrangement which conforms to prior evidence on well-established features of the species phylogeny. Whereas the lowest NR length tree shown in Fig. 1 has a score of 758 NRs + 8 GDs + 25 GEs, or 791 events, the lowest NR + GD + GE length trees, an example of which is shown in Fig. 2, each have a score of 771 NRs + 1 GD + 3 GEs, or 775 events, and thus are more parsimonious. These lowest NR + GD + GE length myoglobin trees allow several phylogenetic inferences to be drawn which are supported by the full body of amino acid sequence data.

Within the order Primates, Lemuriformes and Lorisiformes are not simply similar due to their primitive features as reflected in primate grade classifications (Simpson, 1945; Le Gros Clark, 1959) which include tree

*The α- and β-hemoglobin sequences of Nile crocodile, caiman, and alligator have recently been published (Leclercq *et al.*, 1981). Thus Crocodilia is now represented in our newest maximum parsimony trees constructed for α- and β-hemoglobin sequences (M. L. Weiss and M. Goodman, unpublished data): a caiman-alligator branch is joined by Nile crocodile and then the Crocodilia branch attaches to the stem of the Aves branch. This is consistent with the genealogic grouping of Aves and Crocodilia in a monophyletic superclass Archosauria (an arrangement proposed by cladists on the basis of morphologic characters).

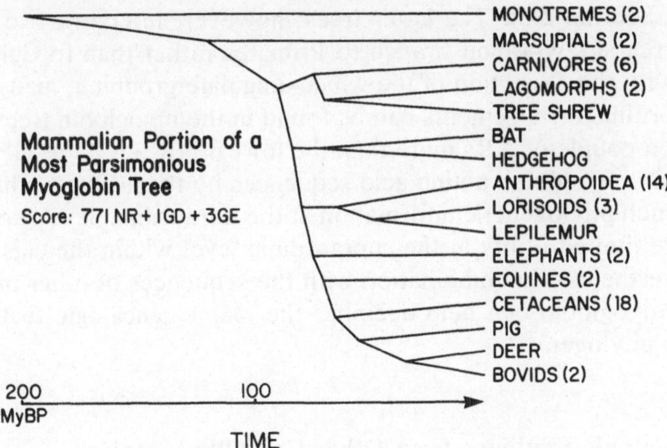

Figure 2. A most parsimonious (lowest NR +GD +GE length) tree found for the 68 myoglobin sequences. Only those arrangements of branches that differ from the tree in Fig. 1 are shown.

shrews in Lemuriformes and place Lemuriformes, Lorisiformes, and Tarsiformes in the suborder Prosimii. Rather, Lemuriformes (minus tree shrews) and Lorisiformes can be grouped together in a genealogic branch which is distinct from Tarsiformes and Anthropoidea, a branch called Strepsirhini in the classification of Hill (1953). The grouping of *Lepilemur* and the lorisoid branch was found in the computer search and was not dictated by any prior assumption about relationships within Primates.

Although each eutherian order represented by myoglobin sequences was designated as a monophyletic subtree during the computer search for the trees of lowest NR + GD + GE length, the only interordinal constraint on the grouping of these orders with one another was that they constitute a monophyletic Eutheria. Under these conditions the close grouping of Perissodactyla, Artiodactyla, and Cetacea in the trees of lowest NR + GD + GE length provides evidence for the monophyletic origin of these three extant orders, in agreement with newer paleontologic views of Van Valen (1966), McKenna (1969), Romero-Herrera *et al.* (1978) and Novacek (this volume, Chapter 1), but not the older views of Simpson (1945) and Romer (1966). Maximum parsimony trees constructed for cytochrome *c* and α-crystallin also group whales with ungulates.

Another possibly significant phylogenetic relationship depicted by the tree in Fig. 2 is the grouping of Scandentia (tree shrews) with Lagomorpha (pika and rabbit). The grouping of these two orders is also found in the maximum parsimony trees constructed from combined sequence data (see Section 2.5). The latter trees, however, in contrast to Fig. 2, join the tree shrew–rabbit branch to Primates rather than to Carnivora. Indeed, with the exception of the whale–ungulate grouping, many different interordinal arrangements can be found in the myoglobin trees which cost only a couple of NRs more than the lowest NR + GD + GE length trees. Thus, myoglobin amino acid sequences by themselves, while containing much phylogenetic information at the infraordinal level, appear to have quite limited utility at the supraordinal level within the class Eutheria. Nevertheless, in combination with the sequences of other proteins, those of myoglobin can help decipher the major genealogic features of eutherian phylogeny.

2.4. Genealogic Evidence from Other Gene Phylogenies

2.4.1. Interclass Relationships

The lowest NR length tree results for β-hemoglobin support the myoglobin results in grouping all Mammalia examined into a monophyletic branch and all Aves examined into another monophyletic branch. Other single protein results have been obtained on α-hemoglobin α-crystallin A, and cytochrome *c*. Each fails to depict the Mammalia as monophyletic; instead, each yields a different pattern of polyphyletic origin of mammalian lineages. Nevertheless, the lowest NR length tree results from combined sequence data encompassing these various proteins (Section 2.5) does depict both a monophyletic Mammalia and a monophyletic Aves.

As already mentioned, the gene phylogenies for α-crystallin and cytochrome *c*, in contrast to the myoglobin results, group Aves first with reptilian lineages rather than with Mammalia. In the case of α-crystallin A there are eight synapomorphic amino acid substitutions which group the bird (chicken) and reptile (lizard) sequences (De Jong, this volume, Chapter 3). In the case of cytochrome *c*, the lowest NR length trees group the Aves–lizard branch with turtle; rattlesnake, however, emerges as a much more ancient branch in the Metazoa. Nevertheless, evidence was obtained during the search for the lowest NR + GD + GE length tree of cytochrome *c* sequences that snakes belong with lizards in the reptile order Squamata and that Aves is genealogically closer to Squamata than to Mammalia. In conducting the search for this most parsimonious tree (the vertebrate region of which is shown in Fig. 3), the major vertebrate branches were treated as orthologously related, with the exception of the rattlesnake branch. Its origin was examined at many different nonmammalian positions on the assumption that it might be paralogous to other

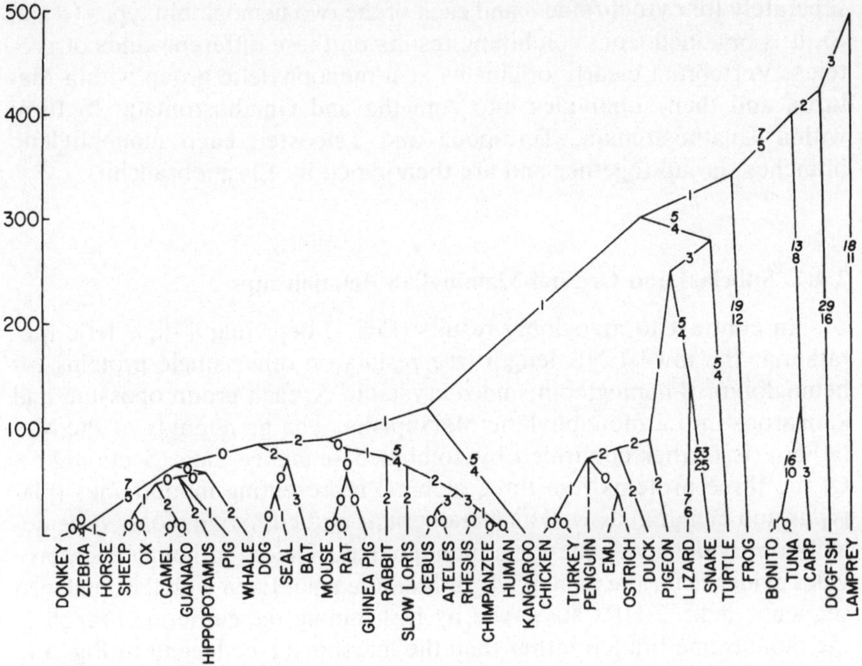

Figure 3. The vertebrate region of a most parsimonious tree found for 94 cytochrome *c* sequences. The complete tree requires 1319 NRs and no hypothetical GDs and GEs. Unaugmented and augmented (italicized numbers) NR values for the codon positions shared between ancestral and descendant sequences are shown on the links. The ordinate scale, in millions of years, is based on paleontologic views concerning the ancestral separations of the organisms represented by the sequences.

amniote branches. Under these conditions, the tree costing the fewest NRs (Fig. 3) joins rattlesnake first to lizard, next joins Squamata to Aves, and then groups Chelonia (turtle) with the Aves–Squamata branch.

Just as the lowest NR length trees for myoglobin sequences group the amniote (mammal, bird, reptile) branches closer to one another than to nonamniotes (in this case teleosts and sharks), the lowest NR length trees for α-crystallin sequences also group the amniote branches closer to one another than to nonamniotes (in this case frog and dogfish). Moreover, the latter trees depict a monophyletic origin of the tetrapods, frog and amniote A-chain sequences being closer to one another than to dogfish. While such good phylogenetic discrimination for lower vertebrate branches is not obtained from the lowest NR length trees constructed separately for cytochrome c and each of the two hemoglobin types (α and β), it is obtained from combining results on these different kinds of proteins. Vertebrata clearly originates as a monophyletic group within Metazoa and then subdivides into Agnatha and Gnathostomata. In turn, within Gnathostomata, Tetrapoda and Teleostei, each monophyletic branches, group together and are then joined by Elasmobranchii.

2.4.2. Subclass and Ordinal Mammalian Relationships

In contrast to myoglobin results (Fig. 1) depicting a diphyletic Metatheria, the lowest NR length tree results on other single proteins (α-hemoglobin, β-hemoglobin, and α-crystallin A) each group opossum and kangaroos into a monophyletic Marsupialia. The monophyly of the Metatheria is further confirmed by combined sequence data (Section 2.5). Of the three proteins (the three globins) representing monotremes (platypus and echidna) as well as metatherians and eutherians, only α-hemoglobin provides evidence for a monophyletic Theria, although each provides evidence for the monophyly of Monotremata. Even with the combined sequence data, 2 NRs are saved by first joining the eutherian branch to the monotreme branch rather than the marsupial (see legend to Fig. 5 in Section 2.5). However, this savings is not considered large enough to challenge the traditional evidence for a monophyletic origin of therian (metatherian and eutherian) mammals.

Turning to relationships within Eutheria, sequence evidence is beginning to emerge for several large cladistic groupings. One consists of Simpson's (1945) extant Paenungulata or the orders Proboscidea (elephants), Sirenia (manatee), and Hyracoidea (hyrax) enlarged to include Tubulidentata (aardvark), with this huge branch possibly being joined by Edentata. A second consists of Perissodactyla, Artiodactyla, and Cetacea. A third consists of Lagomorpha (rabbit and pika) and Scandentia (tree shrews) joined by Primates. The positioning of these three large branches with respect to one another and other eutherian orders such as Rodentia, Insectivora (hedgehog), Chiroptera (bats), Carnivora, and Philodota (pangolins) is not yet clearly indicated. Carnivora, however, from the α-crystallin results is closest to Philodota.

The amino acid sequence evidence for a monophyletic origin of an enlarged Paenungulata (the three traditional extant orders plus aardvark) comes from α-crystallin A and is reviewed by De Jong (this volume, Chapter 3) as well as elsewhere (De Jong and Goodman, 1981). Among these four orders only Proboscidea is represented by amino acid sequence data on other proteins.* Thus it has not yet been possible to either validate or invalidate the α-crystallin grouping of the four extant orders into Paenungulata by maximum parsimony analyses of additional proteins. However, dendrograms constructed from immunologic data (Shoshani *et al.*, 1981) support the grouping together of Proboscidea, Sirenia, Hyracoidea, and Tubulidentata.

In the α-crystallin A maximum parsimony results (De Jong, this volume, Chapter 3; De Jong and Goodman, 1981), Edentata (represented by anteater and two- and three-toed sloths) and the enlarged Paenungulata both descend from the earliest Eutheria either as sister groups or separately from the stem to the rest of the Eutheria. The most parsimonious arrangements for β-hemoglobin sequences (lowest NR length and lowest NR + GD + GE length trees) always group Edentata (armadillo) and Proboscidea (Asian elephant) together. However, the positioning of this paenungulate–edentate branch is not the same as in the α-crystallin tree. The lowest NR + GD + GE length β-hemoglobin tree joins the elephant–armadillo branch to the rodent branch and has the following branches separate in succession from the eutherian stem to the Primates: hedgehog, ungulate, tree shrew, rodent–armadillo–elephant, carnivore, and rabbit. While Edentata is not yet represented by any other proteins, Proboscidea is represented by myoglobin and fibrinopeptides A and B in addition to α-crystallin A and β-hemoglobin. The myoglobin genealogy joins Proboscidea to the ungulate–cetacean branch (Fig. 2), whereas fibrinopeptides A and B join Proboscidea to Artiodactyla. In the combined sequence analysis, the Paenungulata branch (Proboscidea) descends from the earliest Eutheria, as it does in the α-crystallin tree (De Jong, this volume, Chapter 3). Obviously, sequence data on more proteins or genes are needed to determine if this is the correct position for the Paenungulata.

The myoglobin results (Fig. 2) provide evidence, as already discussed, for placing Cetacea with ungulates. Further evidence for this second major cladistic branch of eutherians comes from the maximum parsimony analysis of α-crystallin A sequences (De Jong, this volume, Chapter 3; De Jong and Goodman, 1981) and cytochrome *c* sequences

*The amino acid sequence of aardvark myoglobin has recently been completed (Dene *et al.*, 1982). We find in the lowest NR length trees constructed for myoglobin sequences that aardvark groups with a bat-hedgehog branch. However, when dendrograms are employed which follow branching arrangements obtained with seven polypeptide chains (see Fig. 5) the lowest NR length trees place aardvark either with the elephant branch or with the bat-hedgehog branch. Moreover the lowest NR length trees for combined myoglobin and α-crystallin A sequence data on 27 taxa group aardvark with elephant as the most anciently separated branch of Eutheria.

(Baba *et al.*, 1981). Although whales and ungulates are represented in the sequence data only by these three proteins, the two orders of ungulates—Artiodactyla and Perissodactyla—are much more extensively represented. Artiodactyl and perissodactyl branches occur in the genealogic trees constructed for myoglobin, α hemoglobin, β hemoglobin, α-crystallin, cytochrome *c*, fibrinopeptides A and B, and carbonic anhydrases. It is apparent in the combined protein analysis (Section 2.5) that both Artiodactyla and Perissodactyla along with Cetacea constitute a monophyletic branch. However, certain of the single protein analyses do not provide their own evidence for this phylogenetic conclusion. For example, in the lowest NR length trees for fibrinopeptides A and B, Perissodactyla does not appear as a sister group of Artiodactyla; it originates at a more ancient position in the Eutheria, which costs 1 NR less than when it is near Artiodactyla in the lowest NR + GD + GE length trees. Similarly, to have the lowest NR score for carbonic anhydrase I sequences, ox and horse cannot be joined together. In this case, however, it proves more parsimonious not to group these two sequences, since the added cost in GDs and GEs is more than balanced by the lowered cost in NRs.

As shown in Fig. 4, in the most parsimonious genealogic tree of carbonic anhydrase sequences, horse carbonic anhydrase I (CA I) joins a primate–rabbit branch, then ox CA I joins this cluster. To first group together the horse and ox sequences costs an additional 5 NRs. Thus, even if we accept from the overall evidence on mammalian phylogeny that Perissodactyla is closer genealogically to Artiodactyla than to Primates and Lagomorpha, it is cheaper to have the lower NR length arrangement of the gene phylogeny in Fig. 4 with 1 GD plus 3 GEs that this necessitates than to treat the horse and ox sequences as orthologously related. The fact that ox CA I is not found in red blood cells but comes from rumen epithelium, whereas the horse, rabbit, and primate CA I sequences all come from red blood cells, lends credence to the possibility that the ox CA I sequence is paralogously related to the CA I of other species. Certainly this hypothesis is worth proposing, because it is subject to testing at the DNA level by searching in the horse and ox genomes for duplicated but unexpressed CA I genes. If such genes exist, one could then find out if a horse silent gene is more closely related to the expressed ox gene, or conversely if an ox silent gene is more closely related to the expressed horse gene.

The third major cladistic grouping within Eutheria is partially depicted in the most parsimonious myoglobin tree (Fig. 2) by the sister group relationship between Scandentia (*Tupaia*) and Lagomorpha. The myoglobin results (Fig. 2), however, place Primates closer to ungulates than to Lagomorpha or *Tupaia*, whereas in the most parsimonious trees for other proteins Lagomorpha and *Tupaia* are consistently closer to Primates than are ungulates. Lagomorpha and Scandentia are both represented in the trees for myoglobin, α-hemoglobin, β-hemoglobin, and α-crystallin. In addition, the rabbit is represented in the fibrinopeptides A and B, cytochrome *c*, and carbonic anhydrase trees.

The β-hemoglobin results, as already indicated, place rabbit, carnivores, rodents (and elephants and armadillo), tree shrew, ungulates, and hedgehog at increasing genealogic distances from the primate lineage. The lowest NR + GD + GE length tree results for α-hemoglobin join a rabbit–tree shrew branch to a rodent–carnivore branch and then group these lineages with the primate branch. The ungulate branch is at the next further distance and the hedgehog at the farthest distance. In the α-crystallin results (De Jong, this volume, Chapter 3; De Jong and Goodman, 1981), one of two sets of parsimony solutions has Lagomorpha closest to Primates; next closest are *Tupaia*, Rodentia, and hedgehog; then further away is the large branch to ungulates and cetaceans, which also contains carnivores, pangolins, and Chiroptera; farthest away are edentates and paenungulates. The other set of parsimony solutions has the Lagomorpha just as far away from Primates as the large branch containing ungulates, cetaceans, carnivores, pangolins, and Chiroptera; even further away are *Tupaia*, Rodentia, and hedgehog; and farthest away, as before, are Edentata and Paenungulata. Both the lowest NR length and lowest NR + GD + GE length trees for fibrinopeptides A and B join together Rodentia and Lagomorpha and then group these two orders with Primates. Next, carnivores (felids in the lowest NR length trees and felids plus canids in the lowest NR + GD + GE length trees) join the branch of rabbit, rat, and primates. In the most parsimonious cytochrome *c* results, Primates is closest to rabbit, whereas Carnivora is closest first to Chiroptera and then the branch containing ungulates and Cetacea. That Primates is closer to rabbit than to ungulates is further demonstrated in the most parsimonious gene phylogeny of carbonic anhydrases (Fig. 4).

2.4.3. Subordinal Eutherian Relationships

The single-protein genealogies provide valuable evidence on relationships among major branches within such eutherian orders as Lagomorpha, Primates, Rodentia, Artiodactyla, Perissodactyla, and Cetacea. The myoglobin evidence on relationships within Cetacea has already been reviewed (Section 2.3). The only additional piece of evidence comes from α-crystallin; it reveals in the trees of lowest NR length several synapomorphic amino acid substitutions grouping representatives of the two major extant divisions of the order, a porpoise (Odontoceti) and the minke whale (Mysticeti). The myoglobin evidence in which several synapomorphic amino acid substitutions grouped the two major extant branches of Lagomorpha represented by pika (Ochotonidae) and rabbit (Leporidae) may also be referred to again because it overrides the α-crystallin grouping of rabbit closer to Primates than to pika, which saved only 1 NR.

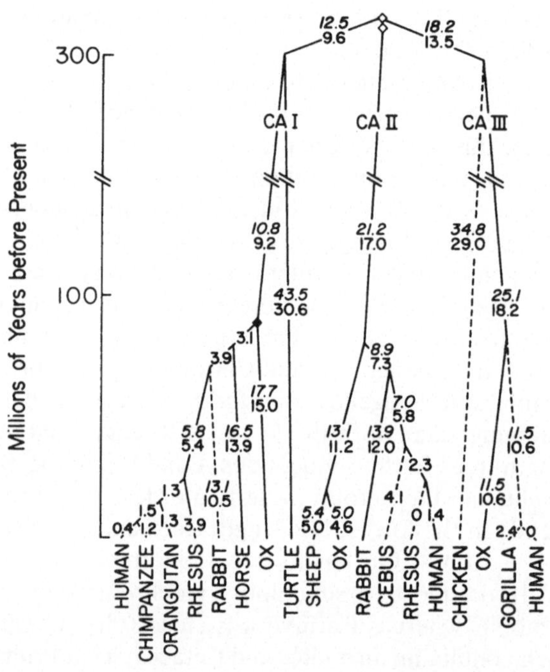

Figure 4. Most parsimonious genealogic tree of the carbonic anhydrase isozymes CA I, CA II, and CA III constructed from the available sequence data (Tashian *et al.*, 1980a,b; Goodman 1981a). Unaugmented and augmented (italicized numbers) NR values (normalized to numbers of base changes per 100 codons) are shown on the links. The ordinate scale, in millions of years, is based on paleontologic views concerning the ancestral separations of the organisms represented by the sequences. The order of occurrence of the two events of gene duplication, each represented by the symbol ◇, cannot yet be determined. Thus, the rooting of the tree is at present ambigious; sequence information from nonamniote vertebrates and nonvertebrates should establish the order of these duplications. Broken lines indicate CA sequences less than 85% complete. As can be seen, the gene duplications that produced the three isozymes probably occurred prior to the divergence of the mammalian and avian–reptilian lineages. Critical to this conclusion is the fact that two distinct isozymes of carbonic anhydrase are present in the red cells of the turtle (Hall and Schraer, 1979). About 88% of the amino acid sequence of one of these isozymes has now been completed (Tashian *et al.*, 1981), and a comparison of this sequence with those of mammalian CA I, CA II, and CA III isozymes shows that the turtle sequence is more homologous to the

The monophyletic origin of Anthropoidea (platyrrhines and catarrhines) within Primates is demonstrated not only by myoglobin sequences, but also by α- and β-hemoglobins, fibrinopeptides A and B, cytochrome *c*, and carbonic anhydrase II. In addition, each of the two types of hemoglobin sequences group Anthropoidea and tarsier together into a monophyletic branch which can be designated as the taxon Haplorhini using the classification of Hill (1955). Hill's other major taxon of Primates, the Strepsirhini for lemurs and lorises, also appears to be monophyletic. The evidence for this comes from myoglobin, α-crystallin, and combined sequence data (Section 2.5).

Evidence that a monophyletic Rodentia embracing among its major branches Caviomorpha (guinea pig) and Myomorpha (murids and microtines) comes from α- and β-hemoglobin, cytochrome *c*, and the combined sequence data. Evidence that Caviomorpha and Histricomorpha share a common ancestor after their divergence from Myomorpha is provided by ribonuclease (Beintema and Lenstra, this volume, Chapter 2) and by α-crystallin A (De Jong, this volume, Chapter 3; De Jong and Goodman, 1981).

Evidence for a monophyletic Pecora (Cervoidea and Bovoidea) within Artiodactyla comes from the proteins representing the two pecoran branches, namely myoglobin, ribonuclease (Beintema and Lenstra, this volume, Chapter 2), and fibrinopeptides A and B. Within Pecora the lowest NR length trees of both ribonuclease and fibrinopeptides A and B group Giraffidae (giraffe) and Antilocapridae (pronghorn) into a monophyletic branch, although ribonuclease joins this branch to the Bovoidae, whereas fibrinopeptides A and B first join Bovoidea and Cervoidea before adding the giraffe–pronghorn branch. The grouping of Tylopoda (llama and camel) with Pecora into a monophyletic Ruminantia is found in the lowest NR length trees constructed for fibrinopeptides A and B and α-crystallin, but

---

mammalian CA I isozymes. In particular, this is supported by comparing the turtle residues with those residues that are invariant and unique to each of the three mammalian isozymes. When this is done, ten of the turtle residues are identical to the unique CA I residues of mammals, four are identical to the unique CA II residues, and none are identical to the unique CA III residues. A cleavage fragment of the other turtle isozyme has also been sequenced (Hewett-Emmett, unpublished data); and although the data are not conclusive, it appears that this isozyme is homologous to the CA II isozymes of mammals. This evidence of both a CA I-like and CA II-like isozyme in a reptile suggests that the gene duplication that produced these two isozymes occurred over 300 million years ago. Examination of the values in Table VI (derived from this figure) reveals that the CA III genes have been evolving in mammals at an average rate of 15 NR% compared to the rates of 24 and 18 NR% for the CA II and CA I isozymes, respectively. This somewhat slower rate suggests that selection has been acting to conserve an important role for the CA III isozymes. Also, the apparent slowdown in the rates of nucleotide substitutions of the CA I and CA II isozymes in the higher primates previously reported (Tashian *et al.*, 1980a) still appears to be true.

not in those constructed for cytochrome *c*, ribonuclease (Beintema and Lenstra, this volume, Chapter 2), and α- and β-hemoglobins. Since these latter four proteins each vary in where they place the Tylopoda, we tend to give more credence to the fibrinopeptides A and B and α-crystallin results, which support the traditional taxonomic grouping of Tylopoda and Pecora. Suiformes (pig) poses a problem similar to that encountered with Tylopoda. There is no consistent trend from the lowest NR length tree results for single proteins as to where Suiformes should be positioned. The best sequence evidence for grouping it with other artiodactyls comes from fibrinopeptides A and B and cytochrome *c*. However, in the results on proteins such as myoglobin and ribonuclease a few NRs are saved by removing Suiformes from a monophyletic Artiodactyla. Nevertheless, these savings are far too small in our opinion to challenge the traditional evidence.

Extant Perissodactyla is usually subdivided into two suborders, Ceratomorpha for tapir and rhinoceros, and Hippomorpha for horses and their close relatives. Fibrinopeptide A and B and α-crystallin provide the only sequence data on these two suborders. They depict a monophyletic Perissodactyla. In the case of α-crystallin A, it is as parsimonious for either tapir or rhinoceros to group first with horse as to group first with each other. However, in the case of fibrinopeptides A and B, it proves more parsimonious to first group together tapir and rhinoceros before joining them to the equine branch. Thus these latter sequence data support the division of Perissodactyla into Ceratomorpha and Hippomorpha.

2.4.4. Relationships within Hominoidea

The lowest NR length tree results for β-hemoglobin and for fibrinopeptides A and B, like myoglobin results, place chimpanzee and gorilla closer to man than to either orangutan or gibbons. These proteins yield seven synapomorphic amino acid substitutions on the stem to *Homo*, *Pan*, and *Gorilla*, but fail to distinguish if two of these three genera share a more recent common ancestor, i.e., they depict a trichotomy. However, in the case of α-hemoglobin, one synapomorphic substitution groups *Pan* and *Homo*, separating these two genera from *Gorilla* and *Pongo*. Further evidence for the closer grouping of *Pan* to *Homo* than to *Pongo* comes from carbonic anhydrase I. This protein yields four synapomorphic amino acid substitutions on the stem of the *Pan–Homo* branch.

Although with myoglobin sequences it saves one NR to place the Hylobatinae branch (gibbon and siamang gibbon) rather than *Pongo* next to the gorilla–human–chimpanzee branch, with fibrinopeptides A and B sequences just the reverse is found. On the basis of evidence from nuclear DNA (Benveniste and Todaro, 1976) as well as from karyotypes (Dutrillaux, 1975) the closer grouping between the *Homo–Pan–Gorilla* branch and *Pongo* would seem to be correct. This can be represented in a genealogic classification by subdividing the extant Hominoidea into two families, Hylobatidae and Hominidae, and by then dividing the Hominidae into Ponginae for *Pongo* and Homininae for *Pan*, *Homo*, and *Gorilla*. It is worth noting that high resolution analysis of chromosomal fine genetic organization has the lineage to *Gorilla* separate from the common Homininae ancestor before the final divergence of *Pan* from *Homo* (Yunis and Prakash, 1982).

2.5. Species Phylogeny from Combined Sequence Data

Figure 5 depicts a most parsimonious branching arrangement found for 49 vertebrate taxa, using an alignment for seven polypeptide chains combined in tandem. Each taxon was represented in the extended tandem alignment by at least two polypeptide chains and usually by three or four, but because of gaps in the available sequence data only nine species could be represented by all seven chains. Of the 49 taxa, 20 are from the order Primates and another 20 OTUs (operational taxonomic units) are from ten other eutherian orders. The eutherian branching arrangement agrees generally with the trend of results from the single-protein genealogies. The Primates are joined first by Lagomorpha (rabbit) and Tupaioidea (tree shrew), followed by Rodentia and Carnivora. This cluster then joins a monophyletic branch leading to Perissodactyla (horse), Cetacea (whales), and Artiodactyla (bovids, camelids, pig). The three remaining eutherian orders represented in the data, Insectivora (hedgehog), Chiroptera (bat), and Proboscidea (elephant), emerge out of the earliest eutherian stem after its separation from marsupials and monotremes. Within the order Primates, Strepsirhini (lorises and lemurs) separates from Haplorhini (tarsier and Anthropoidea). Within Anthropoidea, Platyrrhini (marmosets and New World monkeys) separates from Catarrhini. The latter divides into Cercopithecoidea (Old World monkeys) and Hominoidea, and within Hominoidea the two most closely related genera are *Homo* and *Pan*. This human–chimpanzee branch is joined by gorilla to form the genealogic subfamily Homininae (Goodman and Moore, 1971), which in turn groups with Ponginae (*Pongo* or orangutan), the two extant branches of the genealogic family Hominidae, whose sister group is Hylobatidae (represented here by *Hylobates* or gibbon).

The close grouping of Primates to Lagomorpha is supported, as already reviewed, by the individual most parsimonious trees constructed

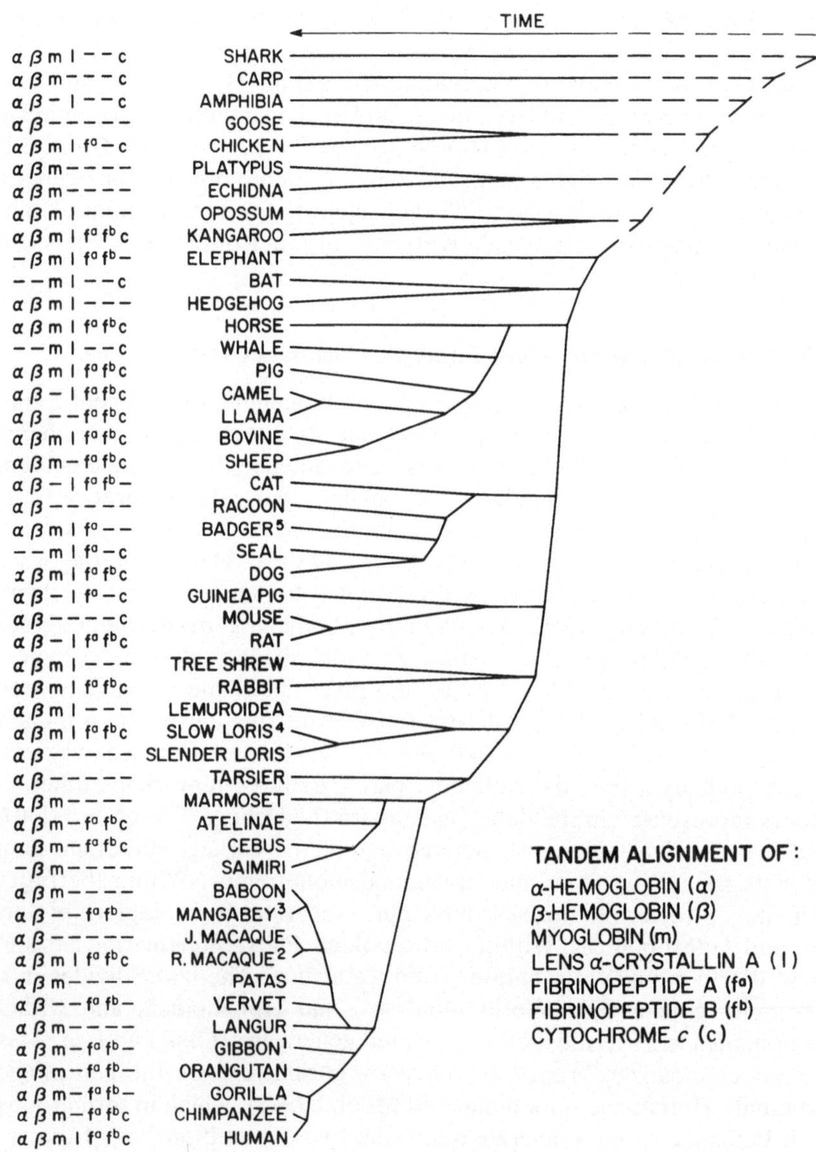
TIME
α β m l – – c SHARK
α β m – – – c CARP
α β – l – – c AMPHIBIA
α β – – – – – GOOSE
α β m l f^a – c CHICKEN
α β m – – – – PLATYPUS
α β m – – – – ECHIDNA
α β m l – – – OPOSSUM
α β m l f^a f^b c KANGAROO
– β m l f^a f^b – ELEPHANT
– – m l – – c BAT
α β m l – – – HEDGEHOG
α β m l f^a f^b c HORSE
– – m l – – c WHALE
α β m l f^a f^b c PIG
α β – l f^a f^b c CAMEL
α β – – f^a f^b c LLAMA
α β m l f^a f^b c BOVINE
α β m – f^a f^b c SHEEP
α β – l f^a f^b – CAT
α β – – – – – RACOON
α β m l f^a – – BADGER[5]
– – m l f^a – c SEAL
α β m l f^a f^b c DOG
α β – l f^a – c GUINEA PIG
α β – – – – c MOUSE
α β – l f^a f^b c RAT
α β m l – – – TREE SHREW
α β m l f^a f^b c RABBIT
α β m l – – – LEMUROIDEA
α β m l f^a f^b c SLOW LORIS[4]
α β – – – – – SLENDER LORIS
α β – – – – – TARSIER
α β m – – – – MARMOSET
α β m – f^a f^b c ATELINAE
α β m – f^a f^b c CEBUS
– β m – – – – SAIMIRI
α β m – – – – BABOON
α β – – f^a f^b – MANGABEY[3]
α β – – – – – J. MACAQUE
α β m l f^a f^b c R. MACAQUE[2]
α β m – – – – PATAS
α β – – f^a f^b – VERVET
α β m – – – – LANGUR
α β m – f^a f^b – GIBBON[1]
α β m – f^a f^b – ORANGUTAN
α β m – f^a f^b – GORILLA
α β m – f^a f^b c CHIMPANZEE
α β m l f^a f^b c HUMAN
TANDEM ALIGNMENT OF:
α-HEMOGLOBIN (α)
β-HEMOGLOBIN (β)
MYOGLOBIN (m)
LENS α-CRYSTALLIN A (l)
FIBRINOPEPTIDE A (f^a)
FIBRINOPEPTIDE B (f^b)
CYTOCHROME c (c)

for six of the seven polypeptide chains and by the genealogic reconstruction for carbonic anhydrases (Fig. 4), in which it proved much cheaper for rabbit and primates to join together first and then group with ungulates. Similarly, in the maximum parsimony tree constructed for 29 nucleotide sequences of α- and β-hemoglobin genes from humans, rabbits, mice, goats, and chickens, human and rabbit branches among orthologous sequences always join each other first (see Hewett-Emmett *et al.*, this volume, Chapter 9). The close primate–lagomorph relationship suggested by this sequence evidence is neither contradicted nor supported by any overriding paleontologic views on eutherian phylogeny (Simpson, 1945; Romer, 1966; McKenna, 1969; Novacek, this volume, Chapter 1). In fact, the degrees of genealogic relationship among most eutherian orders do not appear to be clearly delineated by the fossil record, perhaps because these orders seem to have diverged in a bush-like radiation from their common ancestor over a relatively short period of evolutionary time. More recent interpretations of the fossil record, as already noted in Section 2.3.2, suggest that Cetacea belongs with Artiodactyla and Perissodactyla in a monophyletic branch of Eutheria (Van Valen, 1966; McKenna, 1969; Romero-Herrera *et al.*, 1978; Novacek, this volume, Chapter 1). This view is supported by the sequence evidence from the species phylogeny (Fig. 5) as well as from the individual trees for those proteins (cytochrome *c*, myoglobin, and α-crystallin A) upon which sequence data on Cetacea exist.

Figure 5. A parsimonious genealogic tree for 49 vertebrate taxa using a tandem alighment of up to seven polypeptide chains. This tree requires 2425 NRs. The lowest NR length tree found costs 2415 NRs and can be reached from the tree shown by (a) switching the monotreme and marsupial branches and (b) joining the whale to the Camelidae branch and the horse to the bovid branch. The overall evidence on mammalian phylogeny argues strongly for the arrangement shown in the figure: Marsupialia, rather than Monotremata, closer to Eutheria, and similarly Camelidae closer to Bovidae than to Cetacea, with horse not as close as pig to Camelidae and Bovidae. Thus, even if the tandem alignment were treated as representing a single genetically encoded polypeptide chain rather than seven encoded chains, a minimum of 1 GD plus 3 GEs would be required for the 2 NRs saved by change (a) in branching arrangement to reach the lowest NR length tree, and similarly 2 GDs plus 7 GEs would be required for the 8 NRs saved by change (b). Consequently the tree shown requires fewer genetic events than the lowest NR length tree. Notes: (1) The only portion of the gibbon α-hemoglobin chain (the first N-terminal 31 positions) which has been sequenced (Boyer *et al.*, 1972) is included as part of the sequence data for this OTU. (2) The myoglobin sequence for this OTU comes from the crab-eating macaque; all other sequences come from the rhesus monkey. (3) Fibrinopeptide A and B sequences for this OTU come from the drill; all other sequences are from the mangabey. (4) The α-crystallin A sequence for this OTU comes from galago; all other sequences are from slow loris. (5) The α-crystallin A sequence for this OTU comes from mink; all other sequences are from badger.

The evidence from proteins for the ordinal positions of Carnivora, Insectivora, Chiroptera, and Proboscidea depicted in Fig. 5 is very weak. It may be recalled (Section 2.4) that α- and β-hemoglobins place carnivores closer to primates than to ungulates, whereas cytochrome *c* and α-crystallin A do just the opposite. Moreover, with the extended alignment of the seven polypeptide chains, equally good solutions to the one shown in Fig. 5 join the bat–hedgehog branch to the Carnivora and then unite this cluster either with the whale–ungulate branch or with the primate–rabbit–tree shrew–rodent branch. In the case of Proboscidea, we have seen that similar discrepancies exist. The ancient divergence of Proboscidea from the other eutherian orders in the species tree shown in Fig. 5 is found in the maximum parsimony tree for α-crystallin A sequences, but is not found in the individual maximum parsimony trees for the sequence data on β-hemoglobin chains, myoglobins, and fibrinopeptides A and B. Each of these latter trees has a different position for Proboscidea. Indeed, in a species tree constructed from a tandem alignment including just the three types of jawed vertebrate globins (myoglobin and α- and β-hemoglobin) the elephant branch is closest to the primate branch, whereas in the tree constructed for fibrinopeptides A and B, elephant is closest to Artiodactyla. Moreover, in an earlier maximum parsimony species tree constructed for 40 vertebrate taxa utilizing the extended alignment of the seven polypeptide chains, elephant grouped with the whale–ungulate branch. It is apparent that more extensive sequence data on a range of proteins will be needed to determine the preferred branching arrangement for the orders within Eutheria. Nevertheless, in spite of the uncertainty as to the correct branching arrangement, the data reviewed in the next two sections show that a meaningful analysis of the tempo and mode of molecular evolution in eutherian phylogeny can still be carried out.

3. Utilizing the Clock Model of Protein Evolution

3.1. Rationale and Procedure

The molecular clock hypothesis (Zuckerkandl and Pauling, 1962; Wilson *et al.*, 1977) can be used to estimate rates of molecular evolution for different proteins. This hypothesis assumes that each type of protein evolves at its own characteristic rate and that such a rate is relatively constant throughout the different periods of descent and on comparing one lineage to another in the genealogic tree for that protein. Fossil evidence is drawn upon to date, in millions of years before the present (my bp), a strategically placed ancestral node in the genealogic reconstruction, strategic in the sense of setting the "clock" for the range of dates to be calculated from it. Next, this node's assigned time span to the present (its my bp) is equated to its number of nucleotide replacements to the present (its NR span) averaged over the lineages descending from it. Then each remaining node in the tree is dated by extrapolation from the ratio of each such node's NR span over the *a priori* dated node's NR span. Using these molecular clock dates for the ancestral nodes, rates of molecular evolution can be designated in units of NR% (number of NRs per 100 codons per 10^8 years).

For the molecular clock dates to have validity for systematists and evolutionary biologists, they should agree with the dates deduced from the best evidence of the fossil and geologic records on the nodes where such evidence can apply, i.e., where the splitting times of species lineages represented by the sequence data are considered to be well established by paleontologic evidence. The better the agreement, the more confidence one can have in the accuracy of the clock dates in those regions of the tree where fossil evidence on splitting times is grossly lacking. Another consideration in judging the validity of dates calculated by the molecular clock procedure is that they be consistent with the branching structure of the rooted genealogic tree. To be consistent, each offspring node should not have a more ancient clock date than its nearest ancestor node. An offspring must either have a more recent date than its nearest ancestor or roughly the same date if both are part of a multichotomous splitting, where several or more lineages arise at about the same evolutionary time.

Clock dates passing the test of internal consistency could still be grossly in error in some regions of the tree. For example, if certain lineages show a marked deceleration in rates of molecular evolution, the branch times within these lineages would have clock dates much closer to the present than the branch times actually were. Yet the clock dates would be perfectly consistent with the branching structure of the rooted tree, in that each offspring in the region of the tree where evolutionary rates slowed down would have a more recent clock date than its nearest ancestor. Moreover, if a lineage experienced in its ancestry earlier periods of accelerated evolution followed by later periods of decelerated evolution, its overall amount of change could be roughly the same as that of a lineage that always evolved at an average rate of change. Thus, just because comparable numbers of mutations accumulate in different lineages in a genealogic tree does not prove, although Wilson *et al.* (1977) imply that it does, the existence of a reliable molecular clock. The dates for branch times calculated by the molecular clock procedure could still be grossly in error in some of these lineages. Indeed, this appears to be so among higher primate lineages as well as some other lineages when molecular clock dates are evaluated in terms of the paleontologic evidence on mammalian phylogeny. The data we consider demonstrate that certain of these clock dates are pushed so close to the present as to not be acceptable.

The data used to evaluate the molecular clock hypothesis are from Goodman (1981a). They involve the genealogic trees constructed by the parsimony method for 518 sequences of the several types of globins, α-crystallin, fibrinopeptides A and B, cytochrome c, and the several types of carbonic anhydrases [see Tables 1–5 in Goodman (1981a)]. With each

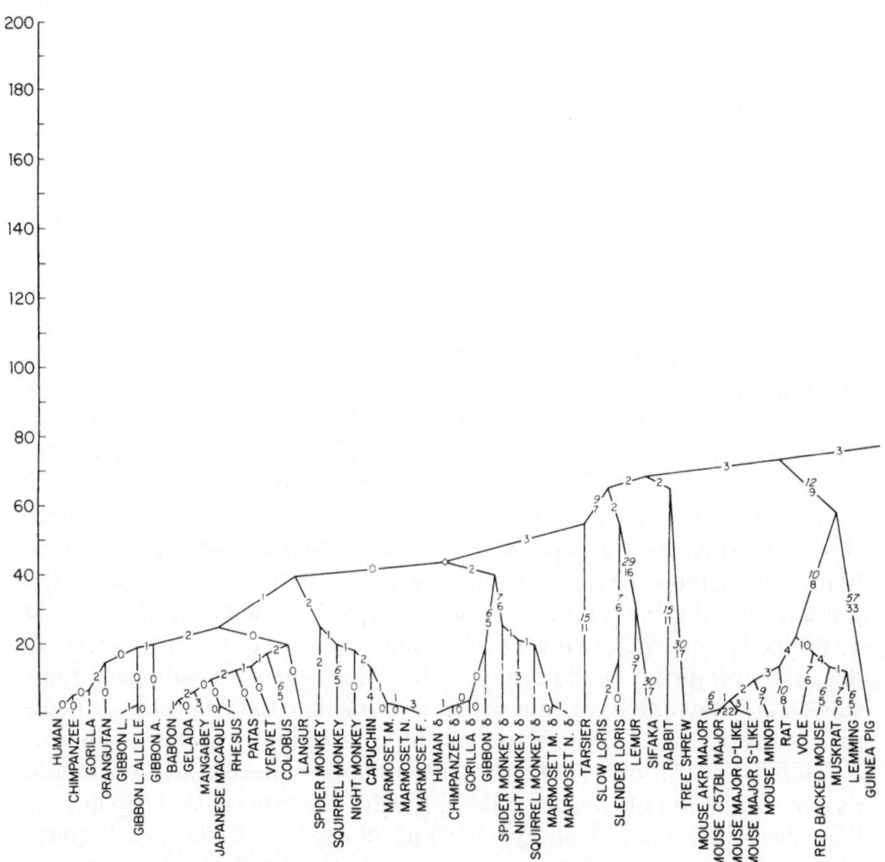

Figure 6. Mammalian portion of the genealogic tree for 103 β-type hemoglobin sequences. This tree costing 1173 NRs follows the species phylogeny that was found for the tandem alignment of seven polypeptide chains (Fig. 5) and that was used for the calculations of Tables II–IV and VI. Unaugmented and augmented (italicized numbers) NR values for the codon positions shared between ancestral and descendant sequences are shown on the links. The ordinate scale is in millions of years based on paleontologic views concerning the ancestral separations of the organisms represented by the sequences. (◇) An obvious gene

of the trees constructed to follow the branching arrangement of the species phylogeny depicted in Fig. 5 (e.g., the β-hemoglobin tree shown in Fig. 6) dates of branching were calculated by the clock procedure for 26 ancestral nodes within this species phylogeny. From over 200 interior nodes depicting the splitting of different species lineages, these 26 were chosen because they were the most extensively represented by sequence data. Each chosen node was represented by an average of six polypeptide chains and by never less than four chains (or five if fibrinopeptides A and B were

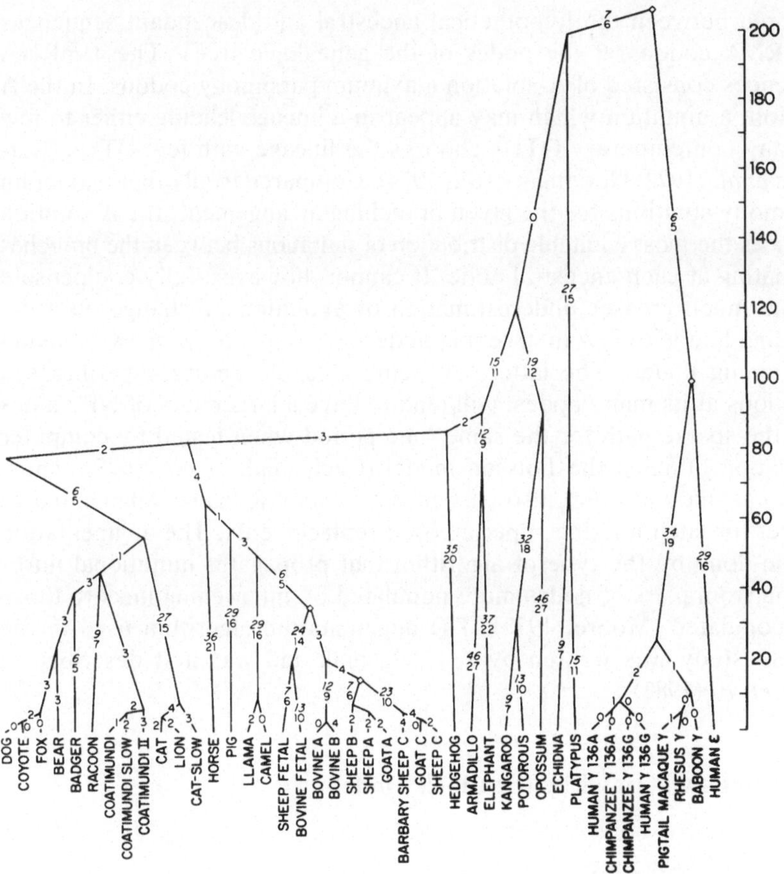

duplication. (The cheapest β-hemoglobin tree found requiring no hypothetical GDs and GEs cost 1167 NRs. It joins the elephant–armadillo branch to the rodent branch and has the following succession of branches departing from the stem descending from the eutherian to primate ancestor: hedgehog, ungulates, tree shrew, rodents–armadillo–elephant, carnivores, and rabbit. This 1167-NR tree and a range of other phylogenetically plausible trees show the same trends in evolutionary rates as depicted in Table XI for the given 1173-NR tree.)

in the group of sequences representing the node). Of the various lineages depicted in the species phylogeny, those of eutherian mammals provided not only the densest body of sequence data (23 nodes within Eutheria) but also the best fossil record. Thus, the results obtained on eutherian lineages were especially relevant for judging both how well clock dates pass the test of internal consistency and how well they agree with dates based on paleontologic evidence.

The clock calculations were carried out with NR spans obtained from using the numbers of base replacements (unaugmented NR values) which occurred between the hypothetical ancestral and descendant sequences of mRNA codons at the nodes of the genealogic trees. These mRNA sequences consisted of A-solution maximum parsimony codons. In the A solution, a mutation which may appear in a lineage leading either to few or many contemporary OTUs chooses the lineage with few OTUs (Barnabas *et al.*, 1972; Goodman *et al.*, 1974). Compared to all other maximum parsimony solutions for the given branching arrangement, the A solution produces the most equitable distribution of mutations between the branches originating at each ancestral node. It cannot, however, fully compensate for the much grosser underestimation of evolutionary change on a descending line with few intervening nodes as compared to one with many intervening nodes. The latter, by being able to capture superimposed mutations at its many nodes, will tend to have a larger sum of NR values than the sparse path for the same time period when tested by computer simulation utilizing the Poisson model (Czelusniak *et al.*, 1978). Thus, clock calculations were also carried out using NR values augmented to correct for such missing superimposed replacements. The augmentation can be done by the type of algorithm that propagates mutational information from pairs of nodes more populated by intervening links to those less populated (Moore, 1977). The augmentation algorithm used in the present study was written by J. Czelusniak and was first described in Baba *et al.* (1981).

3.2. Results Obtained by the Clock Model

3.2.1. The Clock Dates

Tables II and III list for each protein the clock dates of the 26 ancestral nodes most extensively represented by sequence data in the species phylogeny. Unaugmented NR values were employed for calculating the clock dates in Table II and augmented NR values in Table III. An averaged clock date per node was also obtained by equally weighting the results

Table II

Clock Dates of 26 Ancestral Nodes Calculated from Unaugmented Nucleotide Replacement Values

Ancestral node	βHb, my bp	Mb, my bp	αCry, my bp	Fib A&B, my bp	αHb, my bp	Cyt *c*, my bp	CA I, my bp	CA II, my bp	CA III, my bp	Averaged clock date, my bp
x. Aves–Mammalia	160.3	176.7	148.8		150.7	150.8	143.9			155.2
y. Theria	119.6	105.4	123.0		104.7	85.2				107.6
z. Marsupalia	72.9	39.1	37.2		82.9					58.0
A. Eutheria	90	90	90	90						90
B. Insectivora–Primates	84.5	81.4	55.5		81.5	72.9				75.2
C. Ungulata–Primates	76.7	74.6	54.4	86.4	73.0	73.0	73.0	73.0	73.0	73.0
D. Ungulata	69.9	62.4	28.7	68.7	49.4	38.5				52.9
E. Artiodactyla	68.1	58.7	17.0	73.1	48.6	21.3				47.8
F. Tylopoda–Pecora	62.0		7.1	60.6	46.5	18.9				39.0
G. Bovidae	41.0	17.4		30.4	22.5	0		20.9		22.0
H. Carnivora–Primates	70.3	63.2	50.3	86.3	64.9	95.9				71.8
I. Carnivora	32.6		19.1	47.8	45.6					36.3
J. Mustelidae–Canidae	18.0	42.4	21.3		25.2					26.7
K. Rodentia–Primates	67.9		31.9	65.4	63.1	87.6				63.2
L. Rodentia	66.8		5.3		63.2	0				33.8
M. Lagomorpha–Primates	53.0	62.1	39.4	55.5	58.1	105.1	63.7	42.8		60.0
N. Lagomorpha–Tupaia	38.3	39.1	14.1		52.3					36.0
O. Primates	48.0	65.4	47.8	45.1	46.2	63.9				52.7
P. Strepsirhini	47.8	50.0	5.3		32.6					33.9
Q. Anthropoidea	17.7	30.1		19.4	29.0	48.3		14.7		26.5
R. Cebus–Atelinae	14.8	10.1		7.3	6.8	0				7.8
S. Catarrhini	13.8	14.0	15.9	11.8	26.2	9.5	18.8	3.1		14.1
T. *Macaca–Papio*	4.9	2.2		12.9	19.6					9.9
U. Hominidae	4.8	8.7		7.3	3.5		10.9			7.0
V. Homininae	0.9	2.9		0	1.9				7.5	2.6
W. *Homo–Pan*	0	2.2		0	0	0	4.2			1.1

Table III

Clock Dates of 26 Ancestral Nodes Calculated from Augmented Nucleotide Replacement Values

Ancestral node	βHb, my bp	Mb, my bp	αCry, my bp	Fib A&B, my bp	αHb, my bp	Cyt *c*, my bp	CA I, my bp	CA II, my bp	CA III, my bp	Averaged clock date, my bp
x. Aves–Mammalia	177.2	193.2	149.8		156.4	166.5	152.9			166.0
y. Theria	124.1	104.4	124.1		105.8	84.6				108.6
z. Marsupalia	100.3	46.9	49.9		110.5					76.9
A. Eutheria	90	90	90	90						90
B. Insectivora–Primates	84.4	81.0	53.5		80.8	73.3				74.6
C. Ungulata–Primates	77.3	75.0	51.5	86.4	72.6	72.6	72.6	72.6	72.6	72.6
D. Ungulata	76.4	61.3	25.6	68.3	51.1	40.0				53.8
E. Artiodactyla	73.7	71.2	14.8	73.3	50.4	18.3				50.3
F. Tylopoda–Pecora	67.1		6.2	60.0	49.4	16.3				39.8
G. Bovidae	40.7	17.8		29.8	23.3	0		19.4		21.8
H. Carnivora–Primates	70.5	66.3	47.9	87.6	64.5	89.8				71.1
I. Carnivora	35.1		20.2	50.7	46.9					38.2
J. Mustelidae–Canidae	16.0	56.6	23.2		25.9					30.4
K. Rodentia–Primates	68.7		31.0	65.5	62.4	85.2				62.6
L. Rodentia	71.1		4.6		71.5	0				36.8
M. Lagomorpha–Primates	53.4	63.0	38.5	55.7	55.9	102.3	62.7	41.7		59.2
N. Lagomorpha–Tupaia	50.9	38.9	12.3		63.6					41.4
O. Primates	48.5	66.6	49.0	45.4	43.5	59.6				52.1
P. Strepsirhini	65.0	52.5	4.6		33.3					38.9
Q. Anthropoidea	15.3	29.0		18.2	26.3	47.8		12.6		24.9
R. Cebus–Atelinae	12.9	9.2		6.6	5.9	0				6.9
S. Catarrhini	12.0	12.8	13.8	10.8	24.3	8.2	17.2	2.6		12.7
T. *Macaca–Papio*	4.3	2.0		11.8	19.0					9.3
U. Hominidae	4.1	7.9		6.6	2.9		10.0			6.3
V. Homininae	0.8	2.7		0	1.6				6.9	2.4
W. *Homo–Pan*	0	2.0		0	0	0	3.9			1.0

on each protein. These averaged dates are shown in the last columns of Tables II and III. Another way to calculate clock dates was by combining the results on the ten polypeptide chains as if the mutational replacements over these different chains had accumulated at the residue positions of a single extended alignment. These latter clock dates, from both unaugmented and augmented NR spans, are recorded in Table IV. The hypothesized protein clock provided by this aggregate body of amino acid sequence data was set by assigning a date of 90 my bp for the earliest ancestral node of the extant Eutheria (node A in Tables II–IV). The resulting time-dependent tree using the dates (Table IV) from augmented NR spans is depicted in Fig. 7.

Individual proteins prove to be rather poor clocks. They occasionally provide dates which are more ancient for offspring nodes than for ancestor nodes, e.g., nodes H, K, and M compared to C and B in the cytochrome *c* genealogy. Moreover, different proteins can yield markedly different clock dates for the same ancestral node. For example, dates for the rodent ancestor (node L) are very recent (in the range of 0–5 my bp) when calculated from cytochrome *c* and α-crystallin sequences and are ancient (in the range of 63–71 my bp) when calculated from α- and β-hemoglobin sequences. Similarly, cytochrome *c* and α-crystallin sequences, in contrast to globins and fibrinopeptides A and B, yield far too recent dates for the Artiodactyla and Tylopoda–Pecora ancestors (nodes E and F).

When the clock dates for the 26 ancestral nodes are compared to the possible dates for these nodes suggested by the fossil record the individual proteins again are found to be poor time-keepers. However, when results from the different proteins are combined (Table IV), the hypothesized protein clock does not perform too badly, in accord with conclusions previously reached by Fitch and Langley (1976). The clock dates for branch times pass the test of internal consistency (see Fig. 7) and for a majority of nodes are within or near the range of dates suggested from paleontologic evidence. Nevertheless, several major disagreements with this evidence consistently present themselves. Regardless of whether the clock dates are derived from individual proteins or combined results or whether derived from unaugmented or augmented NR values, the branch times depicted within Anthropoidea are too close to the present. A split within Anthropoidea between Catarrhini and Platyrrhini is already evident in Oligocene strata in the range of 30 million years ago (Szalay and Delson, 1979; Ciochon and Chiarelli, 1981). Yet the clock date for this split (node Q) is in the range of 20–25 my bp. Similarly, by 19–20 million years ago the two major catarrhine branches, Hominoidea and Cercopithecoidea, can both be distinguished in the fossil record (Simons, 1976; Szalay and Delson, 1979); yet this catarrhine node (S) has a clock date of about 13

Table IV
Comparison between Clock Dates and Paleontologic Dates on Calculating the Clock Date for Each of the 26 Ancestral Nodes by Equally Weighting the Nucleotide Replacements over All Sequences Represented at That Node

	Clock date from		
Ancestral node	Unaugmented NR span, my bp	Augmented NR span, my bp	Paleontologic date,[a] my bp
x. Aves–Mammalia	161.9	173.8	290–320
y. Theria	115.3	117.3	80–120
z. Marsupalia	68.5	92.2	65–100
A. Eutheria	90	90	70–90
B. Insectivora–Primates	83.4	83.2	65–85
C. Ungulata–Primates	76.4	76.4	65–85
D. Ungulata	59.5	61.4	55–65
E. Artiodactyla	57.2	61.0	50–60
F. Tylopoda–Pecora	50.0	52.7	45–55
G. Bovidae	25.7	25.5	15–20
H. Carnivora–Primates	71.5	71.5	65–85
I. Carnivora	39.5	41.7	45–60
J. Mustelidae–Canidae	27.4	29.8	40–50
K. Rodentia–Primates	64.8	65.0	65–85
L. Rodentia	56.3	62.1	40–60
M. Lagomorpha–Primates	57.7	57.0	65–85
N. Lagomorpha–Tupaia	42.9	51.7	65–85
O. Primates	51.5	51.0	50–65
P. Strepsirhini	41.2	48.4	20–50
Q. Anthropoidea	22.5	20.5	40–55
R. Cebus–Atelinae	9.4	8.3	30–40
S. Catarrhini	14.7	13.4	25–40
T. *Macaca–Papio*	10.2	9.7	4–10
U. Hominidae	7.2	6.3	15–20
V. Homininae	2.1	1.8	5–15
W. *Homo–Pan*	1.5	1.3	5–15

[a]The ranges of paleontologic dates used for these 26 nodes are derived from Løvtrup (1977), Romer (1966), Romero-Herrera *et al.*, (1978), Perutz *et al.*, (1981), Novacek (this volume, Chapter 1), Simons (1976), Ciochon and Chiarelli (1981), Szalay and Delson (1979), Walker (1976), Pilbeam (1979), Johanson and White (1979), McHenry and Corruccini (1980), and Lovejoy (1981).

my bp. Furthermore, although the fossil evidence on Miocene hominoids and on *Australopithecus* strongly indicates that the ancestral divergence between *Homo* and *Pan* preceded 5 my bp (Walker, 1976; Pilbeam, 1979; Johanson and White, 1979; McHenry and Corruccini, 1980; Lovejoy, 1981), the protein clock places this divergence point (node W) at only about 1.3 my bp. Since this much too recent separation of *Pan* and *Homo* is shown as clearly by clock dates based on unaugmented NR spans as those based on augmented NR spans, it cannot be attributed to the augmentation algorithm overestimating superimposed mutations. Instead, it is indicative of inaccurate time-keeping by the amino acid-changing base substitutions in protein evolution.

Clock dates within Carnivora, such as the date of 30 my bp for the last common Mustelidae–Canidae ancestor (node J), also appear to be contradicted by fossil evidence favoring a more ancient split (Romer, 1966;

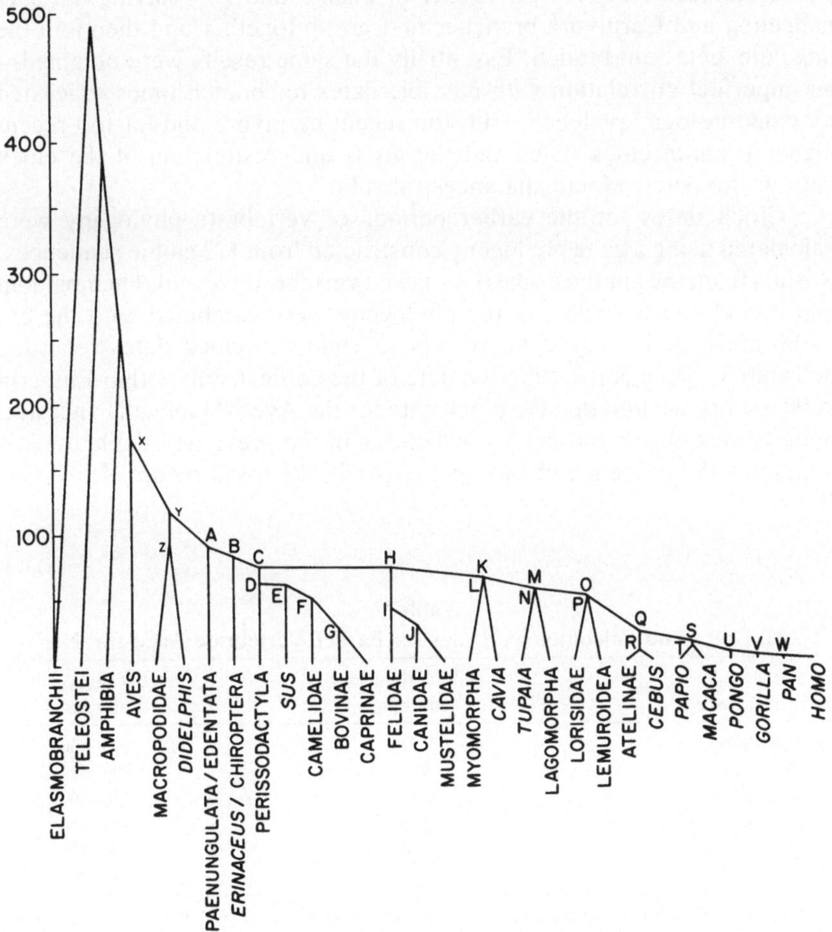

Figure 7. A tree consisting of 30 contemporary taxa and their 29 ancestral nodes with the ordinate representing the molecular clock times shown in Table IV for 26 of these ancestral nodes and in Table V for the remaining three nodes (for the tetrapod, teleost–tetrapod, and gnathostome ancestral nodes).

Romero-Herrera *et al.*, 1978; Novacek, personal communication). Outside Eutheria, but within Amniota, there is a serious disagreement between clock- and fossil-derived dates for branching times. The clock date of 174 my bp found for the last common Aves–Mammalia ancestor (node x) appears to be too recent by over 100 million years, since paleontologists place this split in the early reptiles close to 300 my bp (Romer, 1966; Løvtrup, 1977; Romero-Herrera *et al.*, 1978).

Further extensive clock calculations were carried out on genealogic reconstructions in which the species phylogeny followed by the ten protein chains differed from the phylogeny in Figs. 5 and 7 by having the bat–hedgehog and Carnivora branches first group together and then join the ungulate–cetacean branch. Essentially the same results were obtained—an imperfect correlation with possible dates for branch times suggested by paleontologic evidence, with too recent carnivore and far too recent higher primate clock dates and the gross underestimation of the clock date of the Aves–Mammalia ancestral split.

Clock dates for the earlier periods of vertebrate phylogeny were calculated using a gene phylogeny constructed from 172 globin sequences. Results from the gnathostome (i.e., jawed vertebrate) α- and β-hemoglobin and myoglobin branches in the phylogeny were combined with the cytochrome *c* and α-crystallin results to yield the clock dates recorded in Table V. On again setting the date of the earliest split within Eutheria at 90 my bp, we find that the clock date for the Aves–Mammalia ancestral node comes out to 180 my bp, which, as in the previous calculations, is long after the evidence of this split exists in the fossil record. However,

Table V
Clock Dates and Paleontologic Dates for Earlier Vertebrate Ancestral Nodes

Ancestral node[a]	Clock date,[b] my bp	Paleontologic date,[c] my bp
Vertebrate (B)	790	440–500
Gnathostome	485	395–425
Teleost–tetrapod	387	370–400
Amphibian–amniote	260	340–350
Bird–mammal (C)	180	290–320
Eutherian (D)	90	70–90

[a]B, C, and D (the letters in parentheses) refer to nodes so labeled in Fig. 8.
[b]These clock dates were calculated from augmented NR spans, using the α-crystallin tree, cytochrome *c* tree, and the vertebrate myoglobin and α- and β-hemoglobin regions of the broad globin genealogic tree.
[c]Paleontologic views such as those of Romer (1966) and Løvtrup (1977) were used in assigning this range of approximate dates for earlier vertebrates.

the clock date of 387 my bp for the teleost–tetrapod ancestor agrees with paleontologic evidence. On the other hand, the clock date of 790 my bp of the last common vertebrate ancestor is far too ancient.

Natural selection could be the cause of the disagreements between clock and fossil dates. Alternating between a transforming force and a stabilizing force, it could have caused speedups and slowdowns in rates of protein evolution during the course of vertebrate phylogeny. If so, a less erratic clock should be obtained from gene nucleotide sequence data. In particular, the silent base replacements, those from one synonymous codon to another, by tending to evade selection, could well accumulate at a more uniform rate than the amino acid-changing nucleotide replacements. Perhaps the extremely recent clock dates for the rodent ancestor (node L) from cytochrome *c* and α-crystallin results would become ancient dates, comparable to those obtained from hemoglobin results, once the genes for cytochrome *c* and α-crystallin A are sequenced in a range of rodents and other mammals. Moreover, it should be possible to tell from such gene sequence data if silent substitutions accumulate as a function of the number of cell divisions in the germ line of a lineage or as the ticking of a clock measuring elapsed sidereal time. The conclusion of Wilson and his colleagues (Wilson *et al.*, 1977) that the accumulation of amino acid substitutions bears no relationship to generation time, i.e., does not go faster in the lineages with shorter generation times, could be true in those situations where stabilizing selection preserves the structures of proteins in their minute detail. Indeed, the genetic load entailed by strong selection against mutant proteins could readily be tolerated in mice and other short-generation rodents which replenish and expand their population numbers by rapid rates of reproduction. By the same reasoning, the finding (Wilson *et al.*, 1977) of lack of correlation between generation time and rate of evolution of such a conservative protein as cytochrome *c*, which yields erroneous clock dates for many taxa, provides no evidence for a sidereal-time-dependent clock.

3.2.2. Even Clock Dates Yield Nonuniform Rates

The idea underlying the clock model that each type of protein evolves at a relatively constant rate in the different lines of descent is not well supported by the results of this model. For example, the rates calculated for the conservatively evolving proteins cytochrome *c* and α-crystallin A show that these two proteins evolved at extremely decelerated rates in their descent from the clock-dated rodent ancestor at 62 my bp (Table IV) to the present. As compared to average eutherian rates of 6.9 and 6.3 NR%, the rodent rates are 0 and 0.5 NR%, respectively, for cytochrome *c* and α-crystallin. Conversely, both α- and β-hemoglobin chains evolved at highly accelerated rates in the caviomorph lineage to guinea pig. As compared to average eutherian rates of 28.9 and 30.1 NR%, the caviomorph rates are 61.2 and 63.0 NR%, respectively, for α- and β-hemoglobin chains.

The α-hemoglobin sequences of the Old World monkey tribe Papionini provide an especially interesting example of nonuniformity of rates and of accelerated evolution. Within this tribe the baboon sequences evolved at an extremely elevated rate. From the clock-dated papionin ancestor at 9.7 my bp to the present, the baboon rate is 138.2 NR%, almost five times faster than the average eutherian rate and 16 times faster than the rhesus macaque rate. The overall rate from the papionin ancestor to the present for the 11 α sequences representing this tribe is twice as fast as the average eutherian rate. Hewett-Emmett *et al.* (1976), who first observed this accelerated α-hemoglobin evolution within the Papionini, suggested that it was caused by natural selection. They discovered a nonrandom pattern of amino acid substitutions. Instead of 36–38% of the substitutions being electrophoretically charge-altering, as would be expected from the genetic code if the substitutions occurred randomly, 100% of the substitutions were charge-altering. The substitutions in the papionin α chains leading to new positive charges were exactly balanced by substitutions leading to new negative changes. At least three new intrachain salt bridges could have resulted from these substitutions (Hewett-Emmett *et al.*, 1976) which would have affected the functional behavior of baboon hemoglobins. This possibility suggests that positive natural selection caused the accelerated rate and nonrandom pattern of substitutions.

Aside from the nonuniform rates of evolution of single proteins, there is at least one example of decelerated rates encompassing a range of proteins. The slow rates are found within Homininae in the lineage to present-day *Homo* in spite of the very abbreviated time allowed by the molecular clock dates for this lineage to evolve (1.3 and 1.8 my bp, respectively, for the *Homo–Pan* and Homininae ancestral nodes). For the seven polypeptide chains represented by sequence data in both *Homo* and *Pan* (β- and α-hemoglobin, myoglobin, fibrinopeptides A and B, cytochrome *c*, and carbonic anhydrase I) the evolutionary rate in descent from the *Homo–Pan* ancestor to man is 9.2 NR% compared to an average eutherian rate of 23.5 NR%. Similarly, for the six polypeptide chains represented by sequence data in both *Homo* and *Gorilla*, the evolutionary rate in descent from the hominine ancestor to man is 9.7 NR% compared to an average eutherian rate of 27.2 NR%. Thus, within Homininae the rate of molecular evolution in the lineage to *Homo* is two and one-half to three times slower than the average eutherian rate as well as slower than *Pan* and *Gorilla* rates. In fact there was probably a much more profound deceleration of evolution of these proteins in the origin of man, because, as noted earlier, the fossil evidence on *Australopithecus*, the parent of *Homo*, strongly indicates that the lineages to *Homo* and *Pan* must have diverged from a common ancestor by at least 5 my bp. Not only did protein evolution decelerate but apparently chromosomal evolution too; Yunis and Prakash (1982) found that from their reconstructed common Homininae ancestral karyotype far fewer changes in the lineage to *Homo* than in either the lineage to *Gorilla* or to *Pan*.

4. The Tempo and Mode of Protein Evolution

4.1. Pattern of Rate Variations

We can obtain a sharper picture of variations in the rate of molecular evolution by not biasing the results with the initial assumption that molecular evolution must proceed in a clocklike manner. To escape this assumption, we can use paleontologic evidence on the branch times of the species lineages with well-documented fossil histories to assign dates to a number of ancestral nodes—not just one—in the genealogic tree. The rate estimations are then confined to the lines of descent from the ancestral nodes with assigned dates for the splitting times of these lineages. Other nodes, such as those depicting gene duplications that gave rise to paralogous gene lineages, can be dated if they fall between nodes with assigned dates by extrapolation from the NR values of these particular internodal links.

We realize, of course, that gaps in the fossil records and mistakes in classifying the fossils into clades can lead to erroneous dates being assigned to some of the nodes in the genealogic tree. Similarly, mistakes in the branching arrangement of the tree constructed from amino acid sequence data as well as mistakes in identifying which sequences are orthologously related and which are paralogously related can compound the errors in calculating rates. These pitfalls, however, can largely be avoided by judicious use of the fossil evidence and by the care taken in searching for the correct branching arrangement of the geneaologic tree.

When we follow this heuristic approach, which does not try to fit, *a priori*, the data to any particular model, we find a pattern of accelerations and decelerations in rates of molecular evolution. Highly accelerated rates are found in the genealogic trees between the nodes for the vertebrate and bird–mammal ancestors and also between the nodes for the Eutheria and Anthropoidea ancestors, whereas sharply decelerated rates are found between the bird–mammal and eutherian ancestors and again between the Anthropoidea ancestor and present-day *Homo*.

This acceleration–deceleration pattern is illustrated in Figs. 8–10.

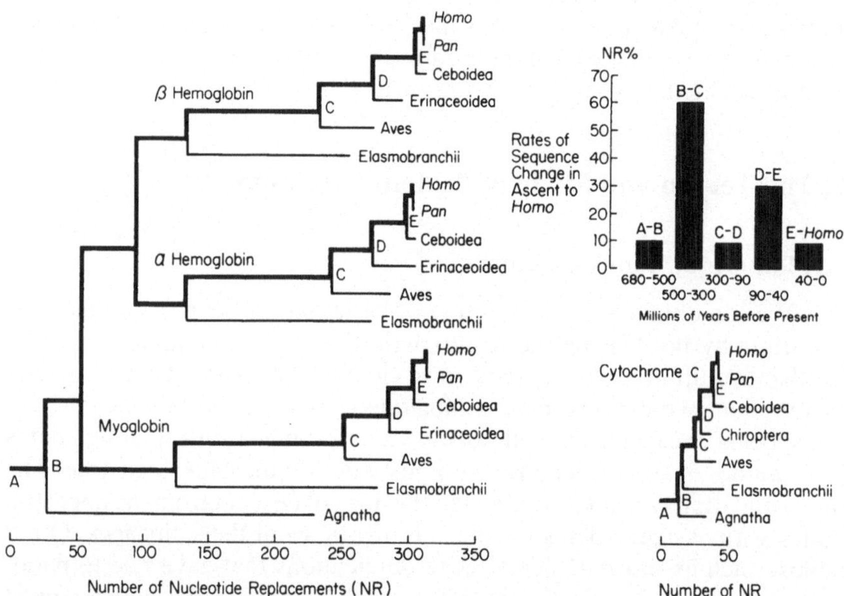

Figure 8. Acceleration–deceleration pattern in rates of sequence evolution. The portions of the globin and cytochrome *c* genealogic trees shown depict the phyletic line to *Homo* from the early vertebrates to the present. The length of each branch equals the average number of NRs, corrected by the augmentation algorithm for superimposed replacements, calculated to have occurred along that line of evolution, with the main branches shown being those that circumscribe the accelerated evolution depicted in the bar graph between the vertebrate and bird–mammal ancestor and again between the Eutheria and Anthropoidea ancestor. In finding the lowest NR +GD +GE length tree for the full set of 159 globin sequences, the species phylogeny that had yielded the lowest score for a tandem alignment of gnathostome myoglobin and α- and B-hemoglobin sequence was followed for the orthologous sequences in each of the three gnathostome regions of the globin gene phylogeny. Separate parsimonious trees were constructed for the three sets of exons, yielding a summed length of 3745 NRs (unaugmented results), and then, because almost the same branching pattern was found for these three separate trees, the NR values of the links in the three trees after correction for superimposed replacements by the augmentation algorithm were added together for the globin branch results shown in this figure. The full cytochrome *c* tree consisted of 87 sequences (seven less than in the more recent tree used for the calculations in Tables II–VI) and cost 1306 NRs (unaugmented results).

The portions of the genealogic trees charted in these figures focus on the amounts of sequence evolution on the way to present-day *Homo*. Figure 8 illustrates that accelerated rates of myoglobin, α-hemoglobin, β-hemoglobin, and cytochrome *c* evolution occurred first between about 500 and 300 my bp during the emergence of jawed vertebrates and tetrapods and then, considerably later in vertebrate phylogeny, between about 90 and 40 my bp during the emergence of primates from basal eutherians. The much faster rate of sequence evolution in the early eutherians and early primates as compared to the much slower rate from the ancestor of the Anthropoidea to humans is further illustrated by the eutherian regions of the most parsimonious species trees constructed from α- and β-hemoglobin sequences combined in an extended tandem alignment (Fig. 9) and from fibrinopeptide A and B sequences (Fig. 10).

It is apparent from the information summarized in Table VI that there was a general trend toward highly elevated rates of protein evolution between 500 and 300 my bp during the emergence of Gnathostomata and Tetrapoda. Myoglobin, α-hemoglobin, β-hemoglobin, cytochrome *c*, and α-crystallin represent this early period of major vertebrate progress, and each of these five proteins evolved 5–7 times faster than they did in the following 210 or so million years in the lineage from the bird–mammal ancestor to the basal eutherian mammals. Then, during the emergence of primates, α-crystallin A and cytochrome *c* evolved 10–20 times faster than in the previous period. Furthermore, myoglobin, α- and β-hemoglobin, and carbonic anhydrase I evolved 2–5 times faster. After this upsurge in rates, the globins, α-crystallin A, fibrinopeptides A and B, and carbonic anhydrases eventually all evolved at markedly decelerated rates in the lineage to man. The deceleration became especially rapid within Homininae, with the evolutionary rate of the sequences in the protohuman lineage being at least 9–10 times slower than the average eutherian rate. In fact, no compelling evidence rules out the paleontologic view that not only *Australopithecus* but also *Ramapithecus* is closer genealogically to *Homo* than to any other extant hominoids (Simons, 1976). If so, since some *Ramapithecus* fossils date back to 14 my bp, the protohuman rate would be at least 27–30 times slower than the average eutherian rate. Similarly, if we follow the paleontologic deductions of Simons (1976) and have the hominoid–cercopithecoid divergence date in the range of 35 my bp rather than 25 my bp and the catarrhine–platyrrhine divergence date in the range of 55 my bp rather than 40 my bp, the earlier accelerated rate of protein evolution in the primate stem to the Anthropoidea would be all the faster and the later decelerated rate from the Anthropoidea ancestor to the present would be all the slower.

It should be emphasized that these trends are shown by a range of

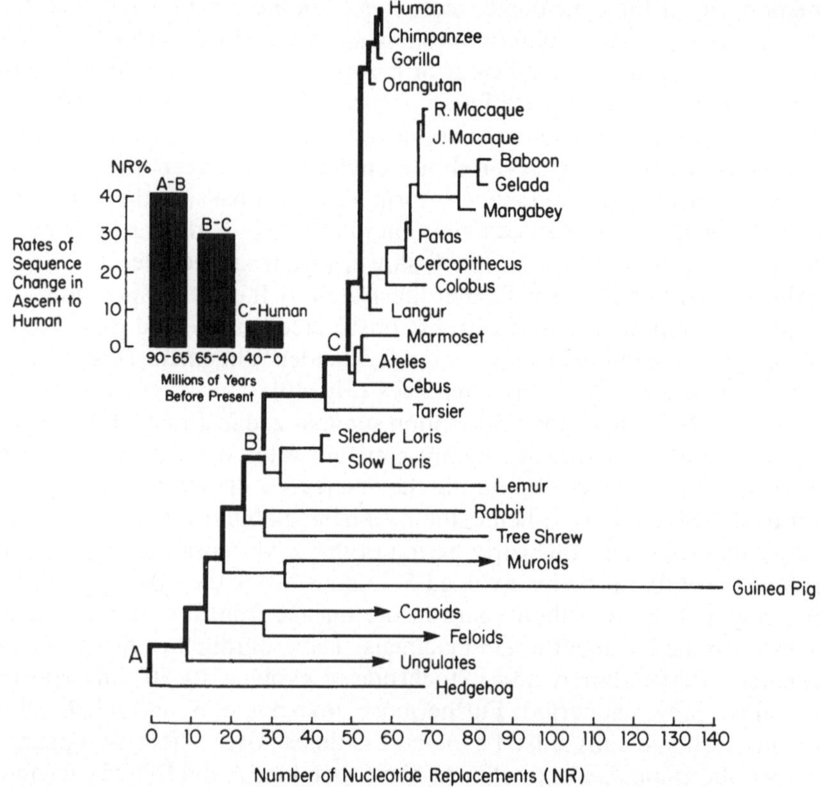

Figure 9. Portions of the eutherian region of the most parsimonious (lowest NR + GD + GE length) tree (costing 1613 NRs) constructed for 55 contemporary vertebrate taxa from orthologous sets of β- and α-hemoglobin sequences combined in a tandem alignment. The length of each branch shown equals the average number of NRs, corrected by the augmentation algorithm for superimposed replacements, calculated to have occurred along that line of evolution. The corresponding bar graph depicts the rapid rate of sequence evolution in the early Eutheria and early Primates followed by the slow rate in the Anthropoidea on the phyletic line ascending to *Homo*.

alternative genealogic reconstructions. The acceleration–deceleration pattern in rates is obtained when the genealogic trees follow the species phylogeny depicted in Fig. 5, as revealed by results listed in Table VI. This basic pattern is obtained when the genealogic trees follow the species phylogeny in which the carnivore and bat–hedgehog branches group together first and then group with the ungulate–cetacean branch, or alternatively when trees of lowest NR + GD + GE length are constructed separately for the different sets of protein sequences without necessarily following a common branching arrangement for the species. It is also obtained in the broad globin gene phylogeny (Figs. 8 and 11), where the orthologous sequences in each of the tree's three gnathostome regions follow the branching arrangement of the most parsimonious species phylogeny constructed from a tandem alignment of gnathostome myoglobin and α- and β-hemoglobin sequences. In this arrangement Proboscidea groups first with Primates, and at increasingly further distances are Carnivora, rabbit, tree shrew, Rodentia, the ungulate–cetacean branch, and the hedgehog–bat branch.

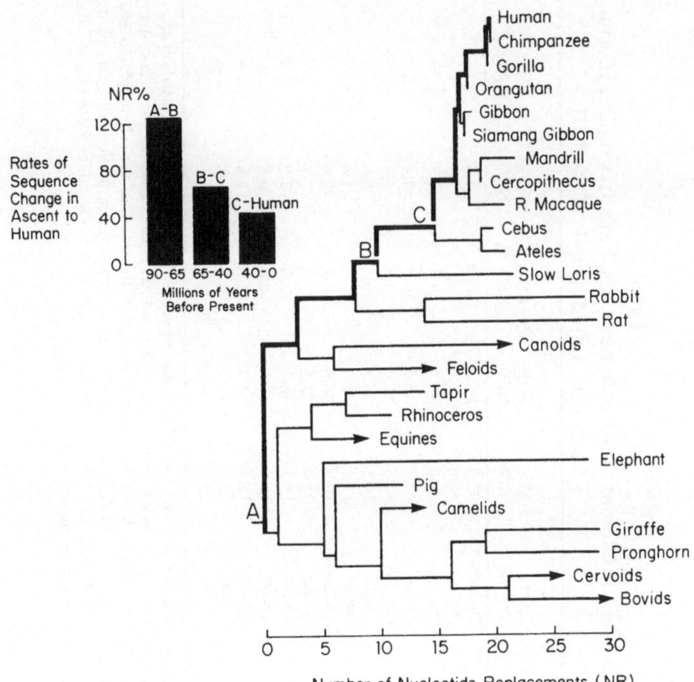

Figure 10. Portions of the eutherian region of the most parsimonious (lowest NR +GD +GE length) tree (costing 260 NRs) constructed for 47 OTUs from tandemly combined fibrinopeptide A and B sequences. The length of each branch shown equals the average number of NRs, corrected by the augmentation algorithm for superimposed replacements, calculated to have occurred along that line of evolution. The corresponding bar graph depicts the rapid rate of sequence evolution in the early Eutheria and early Primates followed by the slow rate in the Anthropoidea on the phyletic line ascending to *Homo*. The tree used for the calculations in Tables II–IV and Table VI (costing 264 NRs) follows the species genealogy depicted in Fig. 5 and shows an even more rapid rate of sequence evolution in the early Eutheria.

Table VI

Rates of Evolution for Globins, Cytochrome *c*, αA Lens Crystallin, Fibrinopeptides A and B, and Carbonic Anhydrases I, II, and III[a]

Evolutionary period	Age, my bp	Mb, NR%	αHb, NR%	βHb, NR%	Cyt *c*, NR%	αCry, NR%	Fib A&B, NR%	CA I, NR%	CA II, NR%	CA III, NR%
Vertebrate to gnathostome ancestor	500–425	87.0	107.0	113.4	4.5	—	—	—	—	—
Gnathostome ancestor[b] to present	425–0	29.0	26.8	30.5	5.3	5.6	—	—	—	—
Gnathostome to amniote ancestor	425–300	71.8	53.2	65.3	6.7	11.0	—	—	—	—
Amniote ancestor[b] to present	300–0	12.1	15.5	16.8	4.4	3.1	—	10.0	—	—
Amniote to eutherian ancestor	300–90	10.3	11.7	12.8	1.3	1.5	—	6.3	—	—
Eutherian ancestor[b] to present	90–0	16.5	28.3	30.1	6.8	6.3	75.2	17.7	24.1	15.3
Eutherian to primate ancestor	90–65	13.1	53.1	43.2	41.6	14.8	200.0	—	—	—
Eutherian to Anthropoidea ancestor	90–40	18.7	36.5	38.3	22.8	—	154.1	—	38.6	—
Primate to Anthropoidea ancestor	65–40	24.4	23.2	33.5	7.7	—	74.8	—	—	—
Primate to Catarrhini ancestor	65–25	18.5	18.1	22.6	13.2	9.6	55.1	—	—	—
Anthropoidea ancestor to humans	40–0	11.4	7.1	10.6	10.8	—	33.3	—	9.4	—
Catarrhini ancestor to humans	25–0	13.1	5.7	14.2	3.8	4.6	40.0	12.8	5.6	—
Hominine ancestor to humans	5–0	0	14.2	0	—	—	0	—	—	0
Homo–Pan ancestor to humans	5–0	0	0	0	0	—	0	7.6	—	—

[a]The myoglobin and α- and β-hemoglobin rates for the first five evolutionary periods listed were obtained from the broad globin genealogic tree. Otherwise rates were calculated from the individual genealogic trees constructed to follow for their orthologous sequences the branching arrangement of the species phylogeny depicted in Fig. 5. The eutherian ancestor was placed as listed at 90 my bp except for three of the trees; in the α-hemoglobin and cytochrome *c* tree it was placed at 85 my bp and in the carbonic anhydrase tree it was placed for CA I, CA II, and CA III at 81 my bp (my bp, millions of years before the present as estimated from the paleontologic record). NR%, nucleotide replacements per 100 codons per 100 million years (augmented NR values were employed).

[b]Average of all lineages descending from this node to the present.

Clearly such results, as exemplified by the data in Table VI, falsify Kimura's (1968, 1969, 1979) constant rate hypothesis and thereby undermine his version of the neutral theory of protein evolution in which each kind of protein is supposed to evolve at its own constant rate. The parsimony reconstructions show that Kimura's Poisson estimates of rates of evolution from pairwise comparisons of contemporary amino acid sequences are seriously in error because they miss most of the superimposed replacements which occurred during the descent of the sequences. In the case of hemoglobin evolution, for instance, Kimura's Poisson estimates yield a rate of only 10^{-9} per amino acid site per year, or ten fixed mutations per 100 amino acid residues per 10^8 years when the pairwise comparisons are between eutherian orders (Kimura, 1969, 1979). This is three times slower than the corresponding eutherian rate calculated in the parsimony reconstructions. By grossly undercounting the number of superimposed replacements, the Poisson estimates tend to obscure what is so obvious in the parsimony reconstructions, namely that each kind of protein has a history of marked nonuniformity of evolutionary rate during different periods of descent.

4.2. The Central Role of Natural Selection

4.2.1. Synchronized Evolutionary Tempos

The two periods of highly accelerated protein evolution during the past half billion years, first in the earlier vertebrates and second in the earlier eutherians, were periods of large-scale adaptive radiation. In the first period the Tetrapoda and other major branches of Gnathostomata emerged; in the second period, the Primates and other orders of eutherian mammals arose. The bursts of rapid protein evolution observed in the genealogic reconstructions correspond to bursts of rapid morphologic evolution observed in the fossil record. This synchrony in evolutionary tempos conforms to expectations of Darwinian theory. Intensified selection should occur for new organismal adaptations at molecular and morphologic levels in lineages invading new ecologic zones. After the accelerated molecular evolution in the successful lineages, decelerated evolution should ensue from stabilizing selection holding perfected adaptations constant.

Positive natural selection as the dominant force directing change in protein molecules should cause the amino acid substitutions during periods of accelerated evolution to concentrate in the residue positions that acquired new or altered functions. This hypothesis is supported by the genealogic reconstructions for globins, cytochromes *c*, and the calmodulin family. In each case, a detailed correlation has been found between the rates of evolution in different parts of the protein molecule and the change of function in these parts. If the bursts of rapid protein evolution were due simply to an accumulation of selectively neutral mutations from a relaxation of constraints as might occur in duplicated genes that were nonexpressed for a period of their history, the mutations should have been randomly distributed over the encoded protein chains. The fact that they were not so distributed in the genealogic reconstructions but instead were concentrated in functionally important residue positions (thus indicating that they were adaptively advantageous) provides evidence for positive selection. The key examples of such a pattern in protein evolution from a recent review of this topic (Goodman, 1981a) can be recapitulated here.

4.2.2. Globin Evolution

It will help in presenting the examples from the genealogic history of globin genes if the origins and relationships of the major branches of myoglobin and hemoglobin sequences are first described. The main features of the most parsimonious genealogic tree encompassing five leghemoglobins, seven invertebrate globins, three agnathan globins, and 157 gnathostome myoglobin and α- and β-hemoglobin chains are shown in Fig. 11; the alignment used in this parsimony reconstruction is given for representative sequences in Fig. 12. It may be noted (Fig. 11) that branches to globins, which functionally can be described as myoglobins, have arisen more than once. The mollusc myoglobins (those of *Aplysia* and *Buscyon*) descend from a branch in the invertebrate region of the tree, whereas the gnathostome myoglobin branch originates within the vertebrates and is therefore closer to gnathostome α- and β-hemoglobin branches than to the mollusc myoglobin branch. Moreover, the myoglobin isolated from lamprey heart muscle by Romero-Herrera *et al.* (1979) appears to be genealogically closest to lamprey hemoglobin. On the basis of a tentative tryptic peptide alignment (portions of which are shown in Fig. 13), this lamprey myoglobin shows 60% similarity (amino acid homology) with lamprey hemoglobin and 33% similarity with hagfish hemoglobin. In contrast, each of the two sequenced lamprey hemoglobins (Braunitzer and Fujuki, 1969; Li and Riggs, 1970), while showing 40–43% similarity with hagfish hemoglobin (Liljeqvist *et al.*, 1979), shows only 25–26% similarity with gnathostome myoglobins and hemoglobins. These comparisons certainly indicate that a gene duplication within the lamprey lineage after divergence of the hagfish lineage produced the gene for lamprey myoglobin, but of course, for a rigorous analysis of whether or not this is true, the actual amino acid sequence of lamprey myoglobin is needed.

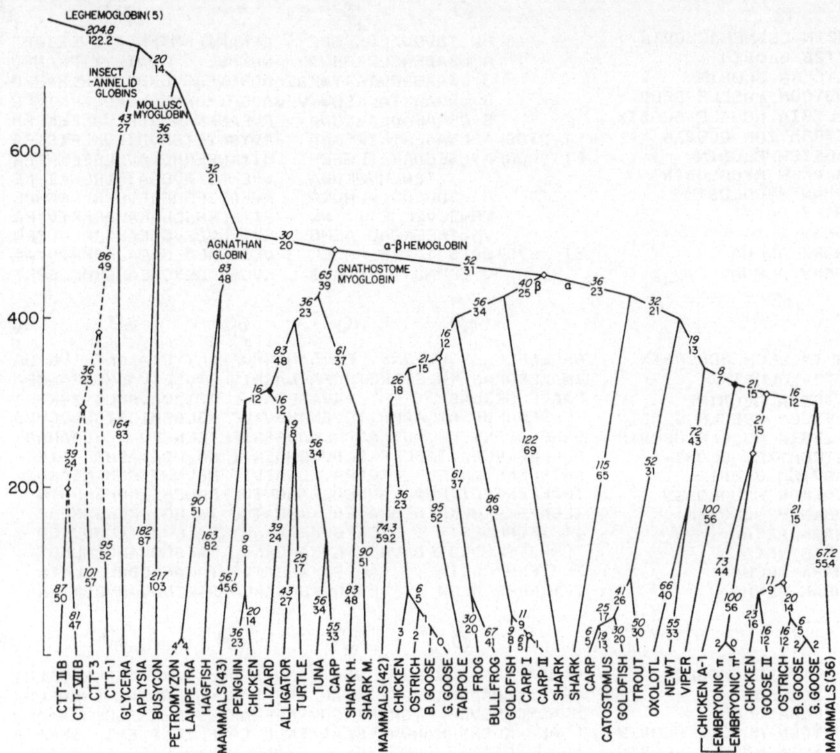

Figure 11. The parsimonious globin gene phylogeny costing 4048 NRs (unaugmented results), found for 172 globin sequences. Unaugmented and augmented (italicized numbers) NR values for the codon positions shared between ancestral and descendant sequences are shown on the links, with averaged NR values shown for the leghemoglobin branch and the three mammalian (myoglobin and α- and β-hemoglobin) branches. The ordinate scale, in millions of years, is based on paleontologic views concerning the ancestral separations of the organisms from which the globins came. In the case of the separations among CTT globins, dashed lines are used because the geologic time level of these ancestral nodes is unknown. (◇) An obvious gene duplication. (◆) Only circumstantial evidence for gene duplication.

The most striking finding which can be deduced from the most parsimonious tree of the globin sequences (Fig. 11) is that the gene duplication separating gnathostome myoglobin from gnathostome hemoglobin occurred in the early vertebrate stem to the Gnathostomata. This vertebrate origin of the gnathostome myoglobin branch, as compared to the much

```
                                10        20        30        40        50
LUPIN LEGHEMOGLOBIN             GVLTDVQVALVKSSFE   EFNANIPKNTHRFFTLVLEIAPG
CTT2B GLOBIN                    APLSADEASLVRGSWA   QVKHSE    VDILYYIFKANPD
GLYCERA GLOBIN                  G LSAAQRQVIAATWKDI AGNDNGAGVGKDCLIKHLSAHPQ
BUSYCON MUSCLE GLOBIN           G LDGAQKTALKESWKVLGADGPTMMKNGSLLFGLLFKTYPD
APLYSIA MUSCLE GLOBIN           S LSAAEADLAGKSWA   PVFANKNANGADFLVALFEKFPD
PETROMYZON GLOBIN       PIVDTGSVAPLSAAEKTKIRSAWA   PVYSNYETSGVDILVKFFTSTPA
HAGFISH GLOBIN          PITDHGQPPTLSEGDKKAIRESWP   QIYKNFEQNSLAVLLEFLKKFPK
SHARK H MYOGLOBIN                    TEWEHVNKVWA   VVEPDIPAVGLAILLRLFKEHKE
HUMAN MYOGLOBIN                 G LSDGEWQLVLNVWG   KVEADIPGHGQEVLIRLFKGHPE
SHARK BETA                      VHWSEVELHEITTTWK   SI  DKHSLGAKALARMFIVYPW
HUMAN BETA                      VHLTPEEKSAVTALWG   KV  NVDEVGGEALGRLLVVYPW
SHARK ALPHA             ST  STSTSDYSAADRAELAALSK   VLAQNAEAFGAEALARMFTVYAA
HUMAN ALPHA                     V LSPADKTNVKAAWG   KVGAHAGEYGAEALERMFLSFPT

                                60        70        80        90       100
LUPIN LEGHEMOGLOBIN     AKDLFSFLKGSSEVPQNN  PDLQAHAGKVF KLTYEAAIQLE  VN GA
CTT2B GLOBIN            IMAKFPQFAG KDLETLKGTGQFATHAGRIV GFVSEIVALMGNSANMPA
GLYCERA GLOBIN          MAAVF GFSGASD       PAVADLGAKVL AZIGVAVSHLGDZGK
BUSYCON MUSCLE GLOBIN   TKKHFKHFDD ATFAAMDTTGVGKAHGVAVF SGLGSMICSIDDDDCVBG
APLYSIA MUSCLE GLOBIN   SANFFADFKG KSVADIKASPKLRDVSSRIF TRLNEFV   NDAANAGK
PETROMYZON GLOBIN       AQEFFPKFKGLTTADELKKSADVRWHAERII NAVNDAVASMD  DTEKM
HAGFISH GLOBIN          AQDSFPKFSAKKSH  LEQDPAVKLQAEVII NAVNHTIGLMDKEAAMKK
SHARK H MYOGLOBIN       TKDLFPKFKEI PVQQLGNNEDLRKHGVTVL RALGNILKQKGKHSTN
HUMAN MYOGLOBIN         TLEKFDKFKHLKSEDEMKASEDLKKHGATVL TALGGILKKKGHHEAE
SHARK BETA              TTRYFGNLKEFTA     CSYGVKEHAKKVT GALGVAVTHLGDVKSQ
HUMAN BETA              TQRFFESFGDLSTPDAVMGNPKVKAHGKKVL GAFSDGLAHLDNLKGT
SHARK ALPHA             TKSYFKDYKDFTA     AAPSIKAHGAKVV TALAKACDHLDDLKTH
HUMAN ALPHA             TKTYFPHF DLSH     GSAQVKGHGKKVA DALTNAVAHVDDMPNA

                               110       120       130       140       150
LUPIN LEGHEMOGLOBIN     VASDA TLKSVHVSKGV  VDAH FPVVKEAILKTIKEVVGDKWSEELNT
CTT2B GLOBIN            METLIKDMAANHKARGI  PKAQ FNEFRASLVSYLQSKV  SWNDSLGA
GLYCERA GLOBIN          MVAQMKAVGVRHKGYGNKHIKGQYFEPLGASLLSAMEHRIGGKMNAAAKD
BUSYCON MUSCLE GLOBIN   L AK  KLSRNHLARGV  SAAD FKLL EAVFKZFLD EAT  QRKATD
APLYSIA MUSCLE GLOBIN   MSAMLSQFAKEHVGFGV  GSAQ FENVR SMFPGFVASVAA  PPAGAD
PETROMYZON GLOBIN       S   MKNLSGKHAKSFQ  VDPQYFKVLAAVIADTV         AAGDA
HAGFISH GLOBIN          Y   LKDLSTKHSTEFQ  VNPDMFKELSAVFVSTM          GGKA
SHARK H MYOGLOBIN           VKELADTHINKHK  IPPKNFVLITNIAVKVLTEMYPSDMTGPMQE
HUMAN MYOGLOBIN             IKPLAQSHATKHK  IPVKYLEFISECIIQVLQSKHPGDFGADAQG
SHARK BETA                  FTDLSKKHAEELH  VDVESFKLLAKCFVVELGILLKDKFAPQTQA
HUMAN BETA                  FATLSELHCDKLH  VDPENFRLLGNVLVCVLAHHFGKEFTPPVQA
SHARK ALPHA                 LHKLATFHGSELK  VDPANFQYLSYCLEVALAVHL TEFSPETHC
HUMAN ALPHA                 LSALSDLHAHKLR  VDPVNFKLLSHCLLVTLAAHLPAEFTPAVHA

                               160       170
LUPIN LEGHEMOGLOBIN     AWTIAYDELAIIIKKEMKDAA
CTT2B GLOBIN            AWTQGLDNVFNMMFSYL
GLYCERA GLOBIN          AWAAAYADISGALISGLQS
BUSYCON MUSCLE GLOBIN   AQKDADGALLTMLIKAHV
APLYSIA MUSCLE GLOBIN   AWTKLFGLIIDALKAAGK
PETROMYZON GLOBIN       GFEKLRMICIL LRSAY
HAGFISH GLOBIN          AYEKLFSIIATLLRSTYDA
SHARK H MYOGLOBIN       SFSKVFTVICSDLETLYKEANFQG
HUMAN MYOGLOBIN         AMNKALELFRKDMASNYKELGFQG
SHARK BETA              IWEKYFGVVVDAISKEYH
HUMAN BETA              AYQKVVAGVANALAHKYH
SHARK ALPHA             ALDKFLTNVCHGLSSRYR
HUMAN ALPHA             SLDKFLASVSTVLTSKYR
```

Figure 12. Globin sequences illustrating the alignment of 174 positions employed for constructing the broad globin gene phylogeny.

		•	•		•		•	•		•		•	•	•	•	•	•				•
		1	2	3	4	5	6	7	8	9	10	11	12	13	14	15	16	17	18	19	20
HAGFISH	Hb	Pro	Ile	Thr	Asp	His	Gly	Gln	Pro	Pro	Thr	Leu	Ser	Glu	Gly	Asp	Lys	Lys	Ala	Ile	Arg
P. MARINUS	Hb	Pro	Ile	Val	Asp	Thr	Gly	Ser	Val	Ala	Pro	Leu	Ser	Ala	Ala	Glu	Lys	Thr	Lys	Ile	Arg
P MARINUS	Mb	Pro	Ile	Ser	Asp	Ser	Gly	Ser	Asp	Ala	Gly	Leu	Ser	Ala	Ala	Glu	Lys	Ser	Ala	Val	Arg
											1	2	3	4	5	6	7	8	9	10	11
S. WHALE											Val	Leu	Ser	Glu	Gly	Glu	Trp	Gln	Leu	Val	Leu

		•		•			•	•		•		•	•		•		•	•	•		•
		21	22	23	24	25	26	27	28	29	30	31	32	33	34	35	36	37	38	39	40
HAGFISH	Hb	Glu	Ser	Trp	Pro	Gln	Ile	Tyr	Lys	Asn	Phe	Glu	Gln	Asn	Ser	Leu	Ala	Val	Leu	Leu	Glu
P. MARINUS	Hb	Ser	Ala	Trp	Ala	Pro	Val	Tyr	Ser	Asn Thr	Tyr	Glu	Thr	Ser	Gly	Val	Asp	Ile	Leu	Val	Lys
P. MARINUS	Mb	Ser	Asn	Trp	Gly	Lys	Val	Tyr	His	Thr	Pro	Glu	Thr	His	Gly	Ala	Asp	Ile	Leu	Ile	Lys
		12	13	14	15	16	17	18	19	20	21	22	23	24	25	26	27	28	29	30	31
S. WHALE	Mb	His	Val	Trp	Ala	Lys	Val	Glu	Ala	Asp	Val	Ala	Gly	His	Gly	Gln	Asp	Ile	Leu	Ile	Arg

					•	•		•		•	•								
		136	137	138	139	140	141	142	143	144	145	146	147	148					
HAGFISH	Hb	Ser	Ile	Ile	Ala	Thr	Leu	Leu	Arg	Ser	Thr	Tyr	Asp	Ala					
		138	139	140	141	142	143	144	145	146	147	148							
P. MARINUS	Hb	Arg	Met	Ile	Cys	Ile	Leu	Leu	Arg	Ser	Ala	Tyr							
P MARINUS	Mb	Thr	Thr	Leu	Cys	Ile	Gly	Leu	Lys	Ser	Ala	Phe							
		136	137	138	139	140	141	142	143	144	145	146	147	148	149	150	151	152	153
S. WHALE	Mb	Glu	Leu	Phe	Arg	Lys	Asp	Ile	Ala	Ala	Lys	Tyr	Lys	Glu	Leu	Gly	Tyr	Gln	Gly

HAGFISH Hb vs. P. MARINUS Hb = 40% SIMILARITY
HAGFISH Hb vs. P. MARINUS Mb = 33% SIMILARITY
P. MARINUS Hb vs. P. MARINUS Mb = 60% SIMILARITY

Figure 13. Portions of the tentative tryptic alignment of lamprey myoglobin against hagfish and lamprey hemoglobins and sperm whale myoglobin. The portions shown are those that are the most readily inferred from the preliminary evidence. Dots are placed at positions that have identical residues between lamprey myoglobin and hemoglobin.

more ancient origins favored by other investigators (Dayhoff, 1978; Kimura, 1979), is not only cheaper in base replacements but also in the number of insertions and deletions required to account for the evolution of these globin genes. A phylogeny minimizing the number of these insertions and deletions is shown in Fig. 14, and an alternative scheme which agrees with Fig. 11 in having the teleost myoglobin branch orthologously related to the tetrapod myoglobin branch is shown in Fig. 15.

The next important event in the phylogeny of vertebrate globins, after the gnathostome myoglobin–hemoglobin gene duplication, was the gene duplication that gave rise to separate α- and β-hemoglobin loci. The most parsimonious genealogic tree in base replacements (Fig. 11) and in insertions and deletions (Fig. 16) places the occurrence of this duplication in the early gnathostomes just before the divergence of Elasmobranchii from boney fish and tetrapods. Thus, the duplication occurred at a time when the vertebrates evolved into larger bodied, faster moving animals, and needed a more efficient hemoglobin for delivery of oxygen. This

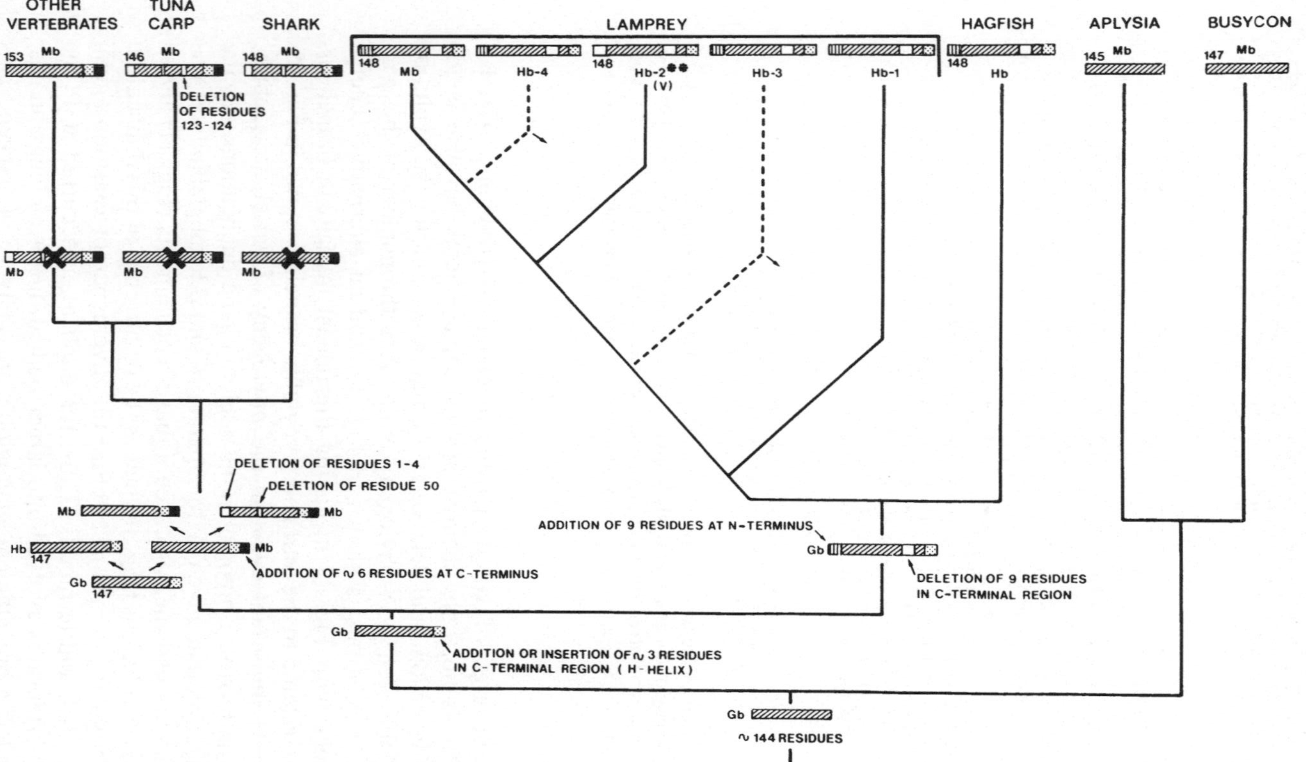

Figure 14. A scheme for insertions and deletions in the descent from the vertebrate–mollusc globin ancestor to three major branches of vertebrate myoglobins. Although this scheme costs two insertions less than the alternative scheme in Fig. 15, to do so it has one gene duplication of myoglobin in the early jawed vertebrates plus three subsequent gene losses or inactivations (as signified by the heavy ×), one each in the lines to shark, to teleosts, and to other vertebrates. In this scheme, while the myoglobins of shark and teleosts (carp and tuna) are orthologously related to one another, they are paralogously related to the myoglobins of other vertebrates.

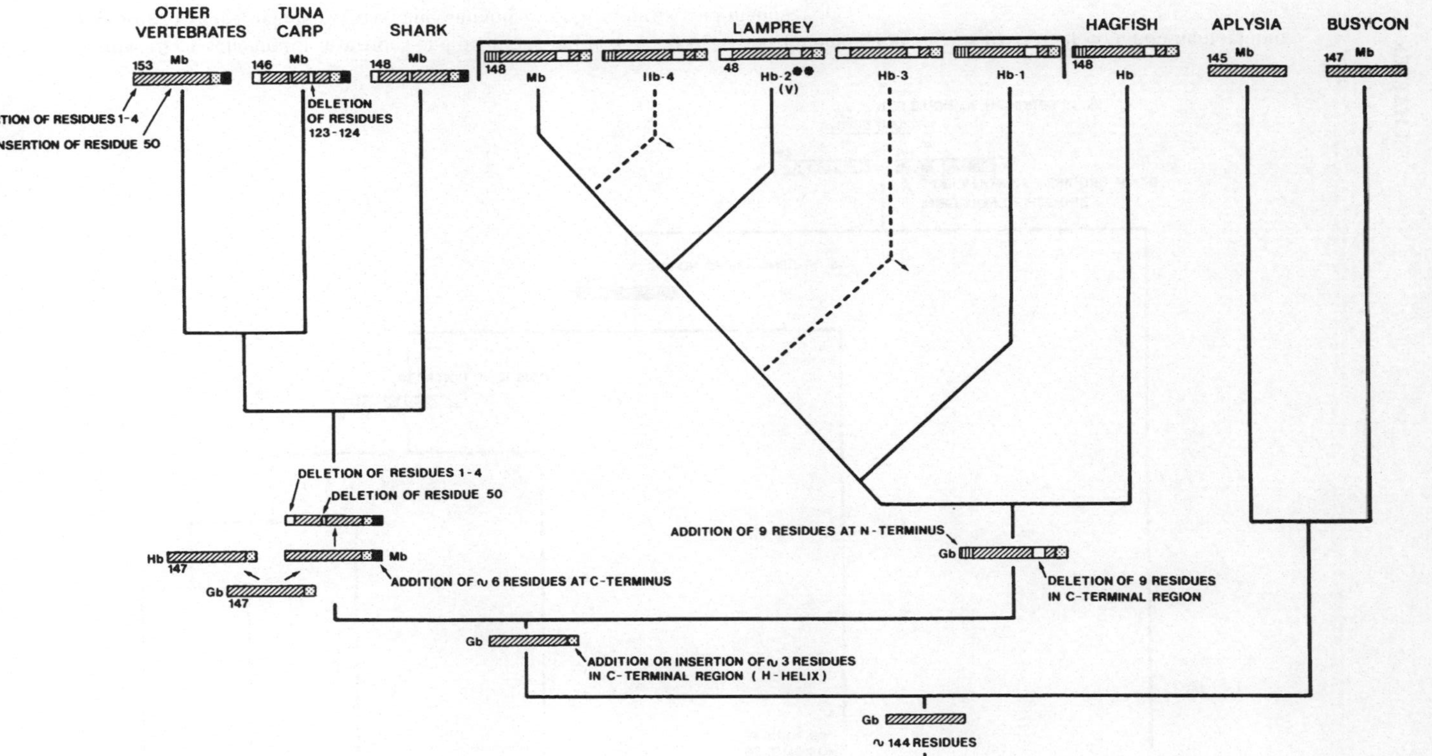

Figure 15. The alternative scheme for insertions and deletions in the descent from the vertebrate–mollusc globin ancestor to three major branches of vertebrate myoglobins. In this scheme, in contrast to that of Fig. 14, the myoglobins of shark, teleosts, and other vertebrates are all orthologously related.

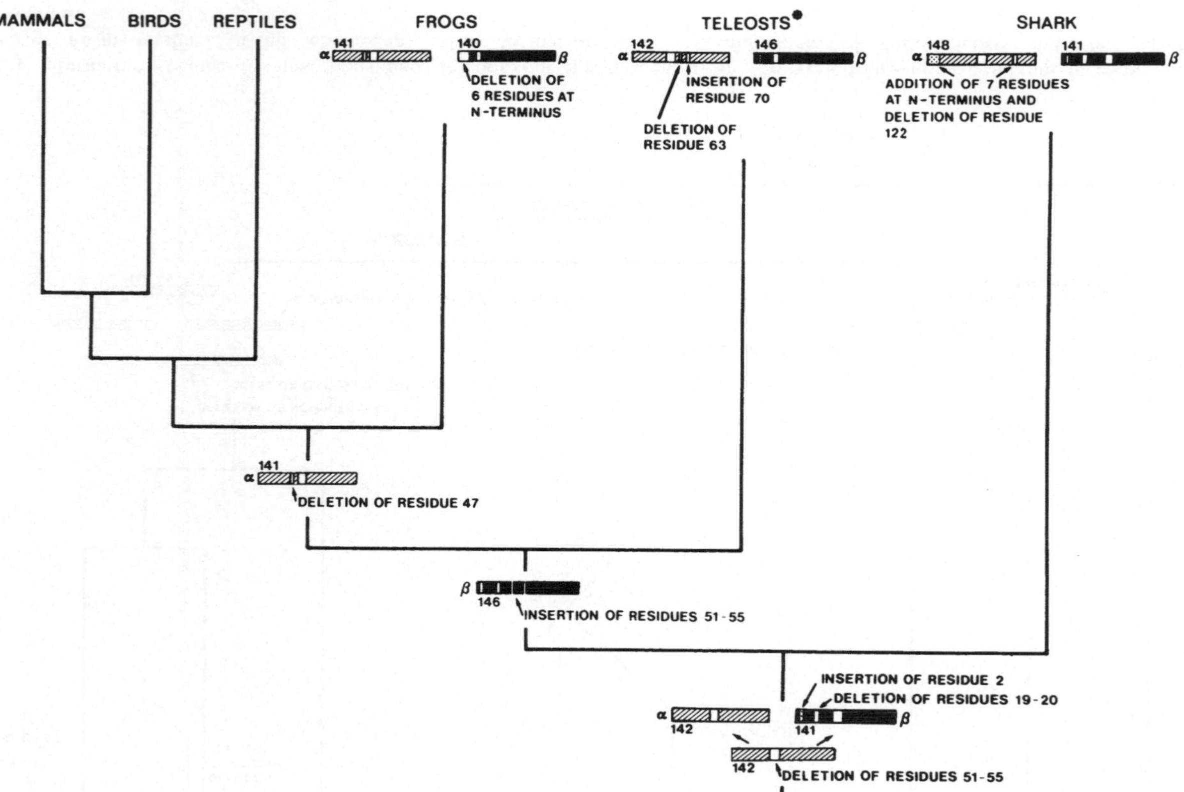

Figure 16. A scheme for insertions and deletions in the descent of α- and β-hemoglobin chains from their common hemoglobin ancestor to shark, teleosts, frogs, and amniotes (reptiles, birds, and mammals).

hypothesis is supported by the pattern of mutations between ancestral and descendant sequences in the parsimony reconstructions. During the early vertebrate period of accelerated globin evolution, when monomeric hemoglobin evolved into a tetramer with a sigmoid oxygen equilibrium curve, there was first an elevated rate of substitutions in the prospective $\alpha_1\beta_1$ contact sites. Then, following upon the α–β gene duplication, the sequence positions in the hemoglobin molecule that became responsible for efficient oxygen delivery evolved more rapidly than other positions (Table VII). These more rapidly evolving positions were the $\alpha_1\beta_2$ contact sites, the salt-bridge-forming sites associated with the Bohr effect, and the 2,3-diphosphoglycerate (DPG)-binding sites. Their rapid rate of evolution can reasonably be attributed to selection of adaptive substitutions. In contrast to their rapid evolution in the early gnathostomes, the $\alpha_1\beta_2$ contacts and Bohr-effect sites along with the heme contacts have been extremely conservative from the bird–mammal ancestor to the present. The early elevated rate at the positions acquiring cooperative functions is about four times greater than the rate at the external positions without sharply defined functions. This difference in rate is highly significant ($\chi^2 = 18.515$, $p < 0.001$, 1 d_f). Conversely, the later depressed rate at the positions with cooperative functions is about four times less than the rate at the external positions without sharply defined functions. Again this difference is highly significant ($\chi^2 = 93.009$, $p << 0.001$, 1 d_f).

It was also found during this early vertebrate period of rapid globin evolution that on the myoglobin stem to the gnathostome ancestor the positions that serve as salt bridges in the myoglobin molecule evolved at an elevated rate. To a lesser extent this was also true on the myoglobin stem for the heme contacts and for the positions corresponding to $\alpha_1\beta_2$ contacts in heterotetrameric hemoglobin. In myoglobin the residues at these latter positions, as well as those responsible for the intramolecular salt bridges, may have been selected to repel any interchain contacts among myoglobin molecules. Between the bird–mammal ancestor and present-day myoglobins the three categories of functional sites that had previously evolved in the primitive vertebrates at an elevated rate now evolved at a very slow rate (Table VIII).

If we accept the proposition that natural selection acting to conserve perfected adaptations caused the slow rate of evolution at the functionally important sites in the myoglobin molecule, it logically follows that the earlier faster rate at these functional sites was also due to natural selection. Clearly, at some time in the past positive selection must have spread the amino acid substitutions which later were conserved by stabilizing selection. The distinction, of course, between the two forms of natural selection concerns the frequency of beneficial genes. Positive selection acts on

Table VII

Nucleotide Replacement Lengths for Groups of Hemoglobin Residue Positions in Several Evolutionary Periods in Descent from the Myoglobin–Hemoglobin Ancestor to Present-Day Amniote αs and βs[a]

Type of position	Mb–Hb to α–β-Hb ancestor		α–β to α gnathostome ancestor		α amniote ancestor to αs		α–β to β gnathostome ancestor		β amniote ancestor to βs	
	N_P	NR_{av}	N_P	NR_{av}	N_P	NR/100 my	N_P	NR_{av}	N_P	Av. 100 my
Heme contacts	22	0.18	19	0.11	19	0.02	23	0.04	23	0.03
Nonheme $\alpha_1\beta_2$ contacts	10	0.20	11	0.27	11	0.02	13	0.62	13	0.06
Bohr effect	4	0.00	5	0.80	5	0.03	2	0.50	2	0.04
DPG binding	3	1.00	—	—	—	—	4	0.50	4	0.13
Non-Bohr $\alpha_1\beta_1$	16	0.38	15	0.07	15	0.20	14	0.07	15	0.19
Remaining interior	19	0.21	21	0.05	21	0.12	18	0.11	19	0.08
Remaining exterior	67	0.15	71	0.11	70	0.20	67	0.16	70	0.21
All	141	0.21	142	0.13	141	0.14	141	0.18	·146	0.14

[a] N_P, number of positions. Positions are grouped according to their functional roles in present-day mammalian hemoglobin chains as may be identified from Table 12 in Goodman (1981a), which updated the previous assignment of residue positions to functional groups (Goodman *et al.*, 1975) on the basis of further high-resolution x-ray crystallographic analyses (Fermi, 1975; Frier and Perutz, 1977; Ladner *et al.*, 1977; Dickerson and Geis, 1981). NR_{av} is number of unaugmented NRs in the functional group divided by number of positions in the group. Between Mb–Hb and α–β-Hb ancestors the group of prospective $\alpha_1\beta_1$ contacts include those residue positions common to α and β chains and also include B16 and C1, which are $\alpha_1\beta_2$ contacts in β chains but $\alpha_1\beta_1$ contacts in α.

mutations that produce beneficial genes to spread them through an evolving lineage, whereas stabilizing selection acts to conserve the beneficial genes that have long been established at high frequency throughout the lineage.

The effects of alternating positive and stabilizing selection are also illustrated by hemoglobin genes in their later periods of descent. For example, there was selection for loss of DPG-binding capacity in the gene line to higher primate γ chains after the gene duplication that separated this line from ϵ chains. Of 19 base replacements on the γ line in the genealogic reconstruction (Fig. 6) three are at DPG-binding sites, valine mutating to glycine through one base replacement at NA1 and histidine mutating to serine through two base replacements at H21. Since hemoglobin has a lowered affinity for oxygen when DPG is bound to it, the mutations that caused the loss of DPG-binding capacity in fetal hemoglobin favored the ability of this hemoglobin to compete with adult hemoglobin for oxygen. The advantage of having such a redesigned fetal hemoglobin may have contributed to the marked deceleration in γ-chain evolution that ensued after the substitutions in the DPG-binding sites had taken place. From the catarrhine ancestor to the present, amino acid-changing base replacements occurred in γ lineages at the slow rate of 5.6 NR%; moreover, from the *Homo–Pan* ancestor to the present no amino acid substitutions at all occurred.

Another example of the effects of alternating positive and stabilizing selection on hemoglobin evolution is provided by the avian β-hemoglobin lineages. This example pertains to the six residue positions in bird β chains that bind inositol pentaphosphate (IPP), the phosphate compound that functions in bird erythrocytes as DPG does in mammalian erythrocytes (Arnone and Perutz, 1974). The six IPP sites consist of the four that are also DPG sites in mammals (NA1, NA2, EF6, and H21) and two additional sites, H13 and H17. Of 23 base replacements on the line from the bird–mammal ancestor to the bird ancestor, three occurred at IPP sites; glycine at H13 mutated to arginine, asparagine at H17 mutated to histidine, and histidine at H21 mutated to arginine. Yet from the bird ancestor to the present not only did no amino acid substitutions occur at the six IPP sites, but few occurred over all 146 alignment positions, an average of just 5.75 NRs in the descending bird lineages. During this same period 22.8 NRs occurred on the average α bird lineage. It is as though the amino acid residues over much of the β chain had to be rigorously selected in order that the residues at the six IPP-binding positions have the correct geometry to execute their binding function. This would explain why natural selection, after it perfected this IPP binding function, acted strongly to prevent any further amino acid substitutions in the bird β chains.

Table VIII
Comparison of Functional Groups in Earlier and Later Stages of Vertebrate Myoglobin Evolution

Type of position	Mb–Hb to Mb gnathostome ancestor		Mb amniote ancestor to Mbs	
	N_P	NR_{av}	N_P	NR/N_P 100 My
Heme contacts	22	0.32	22	0.03
$\alpha_1\beta_2$ contacts in both α and β chains	10	0.30	10	0.03
Salt bridges	20	0.40	20	0.06
Remaining interior	20	0.20	20	0.06
Remaining exterior	73	0.25	81	0.14
All	145	0.28	153	0.09

[a]N_P, number of positions. Assignment of positions to functional groups is based on Takano (1977) and Romero-Herrera *et al.* (1978). Heme contacts are the same as for β-Hb, except F1 is considered only an interior position in Mb. The group of $\alpha_1\beta_2$ contacts in both α- and β-Hb chains is included because it is thought (Romero-Herrera *et al.*, 1973) that selection acts on these positions in myoglobin evolution to prevent them from becoming subunit contacts; moreover, five of these ten positions, C3, C6, CD2, G3, and HC3 (H24 in myoglobin), are salt bridges in myoglobin. There are 25 salt bridges in myoglobin; the 20 not in the $\alpha_1\beta_2$ contact group are A2, A14, A16, B1, B8, CD5, D2, D6, E3, E20, EF1, EF2, EF5, G6, G19, GH4, H10, H16, H18, and HC4. The remaining interior positions are the same as for β-Hb with the inclusion of F1. NR_{av} is number of unaugmented NRs in the functional group divided by number of positions in the group. The time between the vertebrate Mb–Hb gene duplication and the gnathostome Mb ancestor may be less than 75 million years if, as indicated by the maximum parsimony reconstruction (Figs. 8 and 11), this duplication occurred in the protognathostomes after separation of Agnatha. The times of the connecting links over the Mb amniote (bird–mammal) region add up to 2.59 billion years.

4.2.3. Cytochrome *c* Evolution

The maximum parsimony reconstruction of the genealogic history of cytochrome *c* provides a further indication that natural selection, alternating between its positive and stabilizing forms, caused accelerations and decelerations in rates of protein evolution. During the burst of rapid cytochrome *c* evolution which occurred between the eutherian ancestor and Anthropoidea ancestor (Table VI and Figs. 3 and 8), most of the amino acid substitutions in the early primate cytochrome c molecule were concentrated at sites concerned with the critically important function of interaction with mitochondrial cytochrome oxidase (Baba *et al.*, 1981). In all other vertebrate regions of the genealogic tree, these strong functional sites showed no amino acid substitutions. Yet in the early primate lineage they evolved six times faster than the exterior sites not implicated in strong functions, a nonrandom distribution of mutations found to be significant at the $p < 0.02$ level ($\chi^2 = 5.72$, 1 d_f). Moreover, the mutations produced a functional divergence of higher primate cytochrome *c* from other mammalian cytochromes *c*. It has been established (Margoliash, 1980) that Old World monkey and hominoid cytochromes *c* bind more tightly and incorrectly than ox cytochrome *c* to ox heart oxidase and thus inhibit electron transport actively in the ox system. Conversely, when the heart mitochondria are from a catarrhine primate, the higher primate cytochrome *c* is highly efficient in transporting electrons to the primate oxidase, whereas nonprimate cytochromes *c* are less efficient. These findings, therefore, suggest that adaptive substitutions caused the accelerated rate of cytochrome *c* evolution in the early primate lineage. This hypothesis is supported by the deceleration of cytochrome *c* evolution in the higher primates. Such decelerated evolution can be attributed to stabilizing selection preserving the perfected adaptations that had been brought about by positive selection.

Probably every amino acid sequence position of cytochrome *c* is involved to some extent in the functional behavior of the protein. Aside from the positions directly involved in particular functions, such as interactions with cytochrome reductase and cytochrome oxidase, other positions may be indirectly involved through the effects of their amino acid residues on the folding pattern and overall three-dimensional shape of the protein. Specific stereochemical interactions with other molecular surfaces might well exist, but have simply not yet been identified. Every region over the surface of cytochrome *c* would have to be exquisitely tailored for this molecule to interact only with certain specific proteins or molecular constituents in its cellular environment and not with any others normally present, i.e., there was selection to prevent physiologically harmful molecular interactions as well as selection for physiologically beneficial interactions. The fact that for the last 300 million years, from the bird–mammal ancestor to the present, cytochrome *c* evolved at a slow average rate indicates that stabilizing selection acted on almost all positions of the molecule. If true, it follows that at some earlier period in the evolution of the protein, positive selection acting within the stem lineage of later taxa must have spread the adaptive amino acid substitutions which thereafter were preserved by stabilizing selection.

4.2.4. Calmodulin Evolution

Some of the best evidence for this connection between positive and stabilizing selection shaping and preserving the fine detailed features of a protein comes from the genealogic history of calmodulin (Goodman *et al.*, 1979b; Goodman, 1980). Extensive amino acid sequence homologies demonstrate that calmodulin is a member of a family of proteins which include cardiac and skeletal muscle troponin C, alkali and regulatory light chains of myosin, parvalbumins, intestinal calcium binding protein, and the brain S-100 phenylalanine-rich acidic protein. Calmodulin and other members of this family mediate the messenger role played by calcium in the regulation of eukaryotic cellular activities such as energy-linked movement processes (Kretsinger, 1977, 1980; Klee *et al.*, 1980; Jamieson *et al.*, 1980).

The genealogic tree constructed by the maximum parsimony method for these important proteins is shown in Fig. 17. Rates of evolution in different regions of the tree are listed in Table IX. As can be noted in this table, from 700 or so million years ago (the time of the coelenterate–vertebrate ancestor) to the present, calmodulin evolved at the exceedingly slow rate of about only 0.3 NR%. If earlier calmodulins and their precursors had evolved at the same rate (i.e., just accumulated two base replacements per 100 codons every 700 million years) the date for the ancestral divergence between calmodulin and troponin C (node F in Fig. 17) would have to be placed at 15.7 billion years ago and the date for the origin of the primordial four-domain calcium-binding protein (node A in Fig. 17) at 40.5 billion years ago. These are far too early dates inasmuch as the earth itself has only existed for about 4.5 billion years. Thus, clearly, calmodulin in its earlier stages of descent evolved at a much more rapid rate than in its later stages. Positive selection can be implicated in the early rapid evolution, for it was then that adaptive substitutions must have started to shape the protein–protein interaction sites that allow the structural domains of calmodulin to recognize and interact advantageously with a large number of proteins involved in regulating a wide range of cellular processes (Kretsinger, 1980; Klee, 1980; Jamieson *et al.*, 1980). This sculpturing of calmodulin was essentially completed by the late Precambrian. From then on, stabilizing selection preserved the structure of calmodulin in its minute detailed features as well as basic design. The extreme conservatism of metazoan calmodulin is further indicated by the finding that sea anemone calmodulin differs from bovine brain calmodulin at only three amino acid residue positions (Takagi *et al.*, 1980).

Since the early rapid evolution of the gene lineage to the Precambrian ancestor of metazoan calmodulins occurred in conjunction with gene duplications that gave rise to other lineages of the calmodulin family, advocates of Kimura's neutral theory (Kimura, 1981) might claim that removal of constraints rather than the natural selection of Darwinian theory (Goodman, 1981b) caused the rapid evolution. In the neutral theory, random drift of selectively neutral mutations accounts for the vast majority

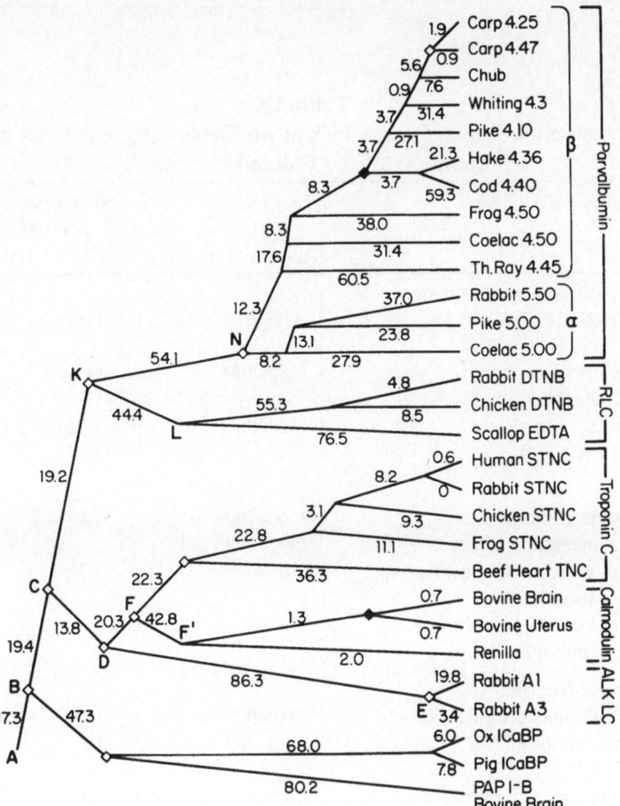

Figure 17. Gene phylogeny of calmodulin and its relatives. Numbers on the links are NRs per 100 codons. As described in Goodman *et al.* (1979b), the NR values are corrected for superimposed replacements and were obtained by putting together the results for the most parsimonious tree constructed for the separate domains as well as from any N-terminal chain positions coming before the first domain. (◆) An obvious gene duplication. (◇) Only circumstantial evidence for gene duplication. Sequence errors rather than gene duplication might account for the two amino acid difference between uterus and brain ox calmodulins.

of amino acid substitutions in proteins, while Darwinian selection of adaptive substitutions accounts for only a minute portion. This clearly, however, cannot be so for the early rapid evolution of calmodulin. Of the 148 positions of calmodulin, 141 must have been subjected to intense stabilizing selection from about at least 700 my bp to the present since they are invariant between the coelenterate and ox brain proteins. Yet the genealogic reconstruction shows that the preceding period of rapid evolution led at these 141 positions to 41 amino acid differences between the

Table IX
Rates of Nucleotide Replacements in Known Evolutionary Periods of Different Members of the Calmodulin Family

	Age, my bp	Nucleotide replacements per 100 codons per 100 my
Regulatory light chains of myosin		
Mollusc–vertebrate ancestor to scallop EDTA RLC	680–0	11.3
Mollusc–vertebrate to amniote DTNB RLC ancestor	680–300	14.6
Amniote ancestor to chicken and rabbit DTNB RLCs	300–0	2.2 (1.6–2.8)
Calmodulin		
Primordial to coelenterate–vertebrate ancestor	3500–700[a] 1500–700	4.1–14.2
Coelenterate–vertebrate ancestor to Renilla calmodulin	700–0	0.3
Coelenterate–vertebrate ancestor to bovine calmodulin	700–0	0.3
Skeletal muscle troponin C		
Tetrapod ancestor to frog STNC	340–0	3.3
Tetrapod to amniote STNC ancestor	340–300	7.8
Amniote ancestor to chicken STNC	300–0	3.1
Amniote to eutherian STNC ancestor	300–75	3.6
Eutherian to human and rabbit STNCs	75–0	0.4 (0.0–0.8)
Parvalbumins		
Gnathostome ancestor to ray β	425–0	14.2
Gnathostome to teleost–tetrapod β ancestor	425–400	103.6
Teleost–tetrapod β ancestor to teleost and frog βs	400–0	9.4 (5.8–17.8)
Teleost–tetrapod α ancestor to pike and rabbit αs	400–0	7.6 (6.0–9.3)

[a]The primordial ancestor of the calmodulin family could not be any more ancient than the origin of life on the earth and might even be as late as the origin of Eukaryota, i.e. about one and a half billion years or so ago.

troponin C–calmodulin ancestral sequence (node F) and the coelenterate–mammalian calmodulin ancestral sequence (node F″) out of the 63 base replacements detected in the reconstruction. Similarly, at these 141 positions, which became extremely conservative, 81 amino acid differences occur between the primordial four-domain calmodulin-like ancestor of the family (node A) and the coelenterate–mammalian calmodulin ancestral sequence (node F″) out of the 157 base replacements detected in the reconstruction. Thus, a substantial fraction of these base replacements (in fact a majority of those counted in the reconstruction) produced amino acid substitutions that proved to be adaptive and were selected.

4.2.5. Conservative Chromosomal Proteins

It seems likely that other conservative proteins that arose through gene duplication, such as histones H3 and H4, also evolved much more rapidly after they originated than later in descent. According to calculations of Hunt and Dayhoff (this volume, Chapter 5), histone H4 evolved at the extremely slow rate of 0.10 PAMs per 10^8 years from the plant (pea)–animal (bovine) ancestor of one billion years ago to the present. Yet while pea and bovine histone H4s differ in amino acid sequence by only 2%, their difference from *Tetrahymena* histone H4 is 22% and from histone H3 about 80% (see Tables VI and VII and Figs. 16 and 19 of Hunt and Dayhoff, this volume, Chapter 5). The earth's age of 4.5 billion years is far too young to account for the 22% and 80% differences by the constant-rate hypothesis. It is likely, therefore, that histone H4 showed much more rapid rates of evolution in the earlier period of eukaryotic phylogeny than during the phylogeny of higher eukaryotes. As can be seen from Tables VI and VII and Figs. 15 and 19 of Hunt and Dayhoff (this volume, Chapter 5), similar conclusions can also be drawn for histone H3. Thus, positive natural selection probably acted on these two proteins in their earlier period of evolution to fix a number of adaptive amino acid substitutions, whereas strong stabilizing selection has acted for the past billion years to prevent almost all further amino acid substitutions.

5. Conclusions

Analysis of a large body of amino acid sequence data by the maximum parsimony method has provided a phylogenetic framework for investigating the tempo and mode of protein evolution. Because the sequence data come from a broad range of species and almost invariably for each of the sequenced proteins represent *Homo sapiens*, the geneaologic reconstructions prove to be especially informative about trends during the history of life in the eukaryotic lineage that ultimately descended to present-day human beings. The trends observed suggest that rates of protein evolution were very rapid early in the history of life when the domains and active sites of the major classes of proteins were first evolving. Natural selection was initially a positive transforming force which acted on proteins to coordinate their interactions in ways that improved cell metabolism. As a result, complicated networks of interacting molecular surfaces with complementary specificities evolved. Then natural selection became a stabilizing force and slowed rates of protein evolution. However, natural selection could still be a transforming force and cause upsurges in rates of protein and polynucleotide evolution in those lineages at the forefront of metazoan phylogeny that were mastering new ecologic frontiers.

After the emergence of the jawed vertebrates and tetrapods and again after the emergence of the primates, as higher degrees of homeostasis evolved in the organisms of these lineages, pervasive decelerations in evolutionary rates occurred. This is what is to be expected from anagenesis, that form of progress in which the organisms that evolved had with their more highly organized internal environments greater independence from and control over perturbing forces of the external environment (Huxley, 1942; Rensch, 1959; Goodman, 1963, 1976; Zuckerkandl, 1976; Wicken, 1980). As a process of molecular evolution, anagenesis increased the density of lock-and-key interaction sites among macromolecules and therefore the number of structural constraints that cause most mutations in modern-day proteins to be detrimental. Not only did the force of stabilizing selection become stronger, but due to improvements in homeostasis, the pressure from the external environment for selection of new mutations in genes was reduced. Such a reduction in the opportunities for positive selection to transform genes may be greatest in the human species, where the development of advanced technological cultures erected a further buffer against external perturbing forces. This can help explain why the genealogic reconstructions detect remarkably little protein sequence change during the hominine radiation in the lineage descending to modern human beings.

Almost certainly, the significance of our present findings on tempo and mode in the evolution of macromolecular sequences will become better understood when the data analyzed are actual nucleotide sequences of many corresponding genes from a large number of primates and other vertebrates.

Epilogue: New Cladistic Findings on Globin Phylogeny

Rapid growth in amino acid sequence data on globin chains and in nucleotide sequence data on hemoglobin genes and pseudogenes has occurred during the last several years and continues unabated. Our most recent analysis of globin phylogeny by the maximum parsimony approach has now been carried out on sequence data encompassing *in toto* 296 globins (Goodman *et al.*, 1982; Goodman and Beeber, 1982), of which the most exciting new findings are revealed by a tree constructed for 195 of these globin sequences from 87 vertebrate and 19 nonvertebrate species. The cladistic branching pattern of α- and β-hemoglobin lineages in this tree (some main features of which are shown in Fig. 18), indicate that avian and mammalian embryonic α-like chains (π- and ζ-chains respectively) had a common monophyletic origin involving an α-gene duplication which occurred about 400 million years ago . . . not long after the gene duplication which separated α- and β-gene lineages. This lineage to embryonic α-like chains of birds and mammals illustrates Haekel's biogenetic "law" of ontogeny recapitulating phylogeny, in that the embryonic chains appear to be cladistically closer to the teleost α-lineage than to the adult amniote α-lineage. Such differentiation in the stem line to land vertebrates between hemoglobin genes expressed in embryos and those expressed in subsequent stages of ontogeny presumably allowed further homeostatic adaptations in oxygen transport mechanisms to evolve in tetrapod clades that invaded uncharted environs.

In the β-hemoglobin line of descent, a similar duplication occurred in the common stem of teleosts and tetrapods (Fig. 18) as evidenced by the divergence from this stem of the lineage to a tadpole β-hemoglobin gene. This locus for expressed fetal β-chains was lost in amniotes; however, much later in phylogeny independent duplications of the major β-locus occurred in Aves and Mammalia to produce separate loci for embryonic β-(ρ-) chains of birds and embryonic/fetal β-(ϵ-/γ-) chains of mammals. This duplication in mammals appears to have immediately preceded the prototherian-therian divergence, i.e., occurred in the range of about 180–220 million years ago. Somewhat later it was followed in the ϵ-/γ-line by a further duplication that led to differentiation between ϵ-genes expressed in embryonic life and γ-genes expressed in either fetal or embryonic life. Both gene lines are widespread in mammals. For example, the ϵ-hemoglobin chain of humans is cladistically closer to the ϵ-hemoglobin chains of goats and mice than to human δ- or β-hemoglobin chains; similarly, human γ-loci (and their counterparts in other catarrhine primates) are cladistically closer to γ-loci in rabbits and mice than to ϵ-, δ-, or β-

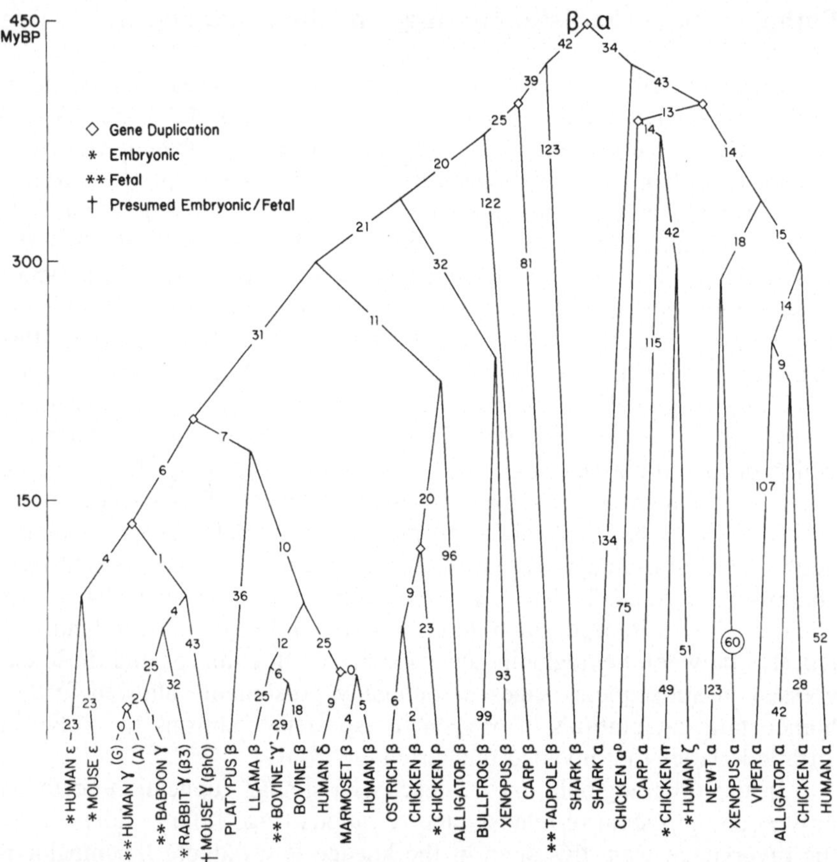

Figure 18. Representative lineages from the β- and α-hemoglobin region of the parsimony globin phylogeny found for 195 globin amino acid sequences. NR values corrected for superimposed replacements are shown on the lines of descent. The NR value on the *Xenopus* α branch is circled because this branch represents a partial sequence consisting of 114 amino acid positions as opposed to the full complement of 141 positions.

loci of humans (Czelusniak *et al.*, 1982). The original γ-locus was apparently lost in Artiodactyla; however, after separation of Camelidae from Pecora, in the line to Bovoidea, a new "γ"-locus arose from duplication of the major β-locus. That β-like hemoglobin chains specialized for fetal life in eutherian mammals, i.e., γ-chains, evolved more than once out of generalized β-hemoglobin ancestors parallels the finding previously reviewed (Section 4.2.2.)—that globin chains specialized for muscle cells evolved more than once out of generalized globin ancestors. Clearly, the force directing such evolution was positive (transformational) selection acting on mutations in duplicated genes.

ACKNOWLEDGMENTS. This work was supported by NSF grant DEB7810717 and NIH grant GM24681.

References

Arnone, A., and Perutz, M. F., 1974, Structure of inositol hexaphosphate–human deoxyhaemoglobin complex, *Nature* **249**:34–36.

Baba, M. L., Darga, L. L., Goodman, M., and Czelusniak, J., 1981, Evolution of cytochrome c investigated by the maximum parsimony method, *J. Mol. Evol.* **17**:197–213.

Barnabas, J., Goodman, M., and Moore, G. W., 1972, Descent of mammalian alpha globin chain sequences investigated by the maximum parsimony method, *J. Mol. Biol.* **69**:249–278.

Benveniste, R. E., and Todaro, G. J., 1976, Evolution of type C viral genes: Evidence for an Asian origin of man, *Nature* **261**:101–108.

Blake, C. C. F., 1979, Exons and the structure, function and evolution of haemoglobin, *Nature* **291**:616.

Boyer, S. H., Noyes, A. N., Timmons, C. F., and Young, R. A., 1972, Primate hemoglobins: Polymorphisms and evolutionary patterns, *J. Hum. Evol.* **1**:515–543.

Braunitzer, G., and Fujuki, H., 1969, Zur evolution der vertebraten die konstitution und tertiärstruktur des hämoglobins des flussneunauges, *Naturwissenschaften* **56**:322–323.

Ciochon, R. L., and Chiarelli, A. B. (eds.), 1981, *Evolutionary Biology of the New World Monkeys and Continental Drift*, Plenum Press, New York.

Czelusniak, J., Goodman, M., and Moore, G. W., 1978, On investigating the statistical properties of the populous path algorithm by computer simulation, *J. Mol. Evol.* **11**:75–85.

Czelusniak, J., Goodman, M., Hewett-Emmett, D., Weiss, M. L., Venta, P. J., and Tashian, R. E., 1982, Phylogenetic origins and adaptive evolution of avian and mammalian haemoglobin genes, *Nature*, in press.

Dayhoff, M. O., 1978, *Atlas of Protein Sequence and Structure*, Volume 5, National Biomedical Research Foundation, Washington, D.C.

De Jong, W. W., and Goodman, M., 1981, Mammalian phylogeny studied by sequence analysis of eye lens protein α-crystallin, *Z. Saugetierkd.*, in press.

Dene, H., Goodman, M., Walz, D., and Romero-Herrera, A. E., 1982, The phylogenetic position of aardvark (*Orycteropus afer*) as suggested by its myoglobin, in preparation.

Dickerson, R. E., and Geis, I., 1981, *Hemoglobin: Structure, Function, Evolution, and Pathology*, Benjamin Cummings, Menlo Park.

Dutrillaux, B., 1975, Sur lar nature el l'origine des chromosomes humains, *Monogr. Ann. Génét. Expansion Sci. Fr.* **1975**:41–71.

Dwulet, J. A., Dwulet, F. E., and Gurd, F. R. N., 1980, Complete amino acid sequence of the major component myoglobin from Hubb's beaked whale, *Mesoplodon carlhubbsi*, *Biochim. Biophys. Acta* **624**:121–129.

Farris, J. S., 1972, Estimating phylogenetic trees from distance matrices, *Am. Nat.* **106**:645–668.

Fermi, G., 1975, Three-dimensional fourier synthesis of human deoxyhaemoglobin at 2.5 Å resolution refinement of the atomic model, *J. Mol. Biol.* **97**:237–256.

Fitch, W. M., 1970, Distinguishing homologous and analogous proteins, *Syst. Zool.* **19**:99–113.

Fitch, W. M., and Langley, C. H., 1976, Evolutionary rates in proteins: Neutral mutations and the molecular clock, *Molecular Anthropology* (M. Goodman and R. E. Tashian, eds.), pp. 197–219, Plenum Press, New York.

Fitch, W. M., and Margoliash, E., 1967, The construction of phylogenetic trees—A generally applicable method utilizing estimates of the mutation distance obtained from cytochrome c sequences, *Science* **155**:279–284.

Frier, J. A., and Perutz, M. F., 1977, Structure of human foetal deoxyhaemoglobin, *J. Mol. Biol.* **112**:97–112.

Gilbert, W., 1978, Why genes in pieces?, *Nature* **271**:501.

Goodman, M., 1963, Man's place in the phylogeny of the primates as reflected in serum proteins, in: *Classification and Human Evolution* (S. L. Washburn, ed.), Aldine, Chicago, pp. 204–234.

Goodman, M., 1976, Towards a genealogical description of the Primates, in: *Molecular Anthropology* (M. Goodman and R. E. Tashian, eds.), Plenum Press, New York, pp. 321–353.

Goodman, M., 1980, Molecular evolution of the calmodulin family, in: *Calcium-Binding Proteins: Structure and Function* (F. L. Siegel, E. Carafoli, R. H. Kretsinger, D. H. MacLennan, and R. H. Wasserman, eds.), Elsevier/North-Holland, New York, pp. 347–354.

Goodman, M., 1981a, Decoding the pattern of protein evolution, *Progr. Biophys. Mol. Biol.* **37**:105–164.

Goodman, M., 1981b, Globin evolution was apparently very rapid in early vertebrates: A reasonable case against the rate-constancy hypothesis, *J. Mol. Evol.* **17**:114–120.

Goodman, M., and Beeber, J. E., 1982, Molecular evolution above the species level: Cladogenesis and anagenesis revisited. Proceedings of C.N.R.S. International Colloquium on "Les Modalites, Rythmes Et Mecanismes De L'Evolution: Gradualisme Phyletique On Equilibres Ponctues?" Dijon, France, May 9–14, 1982.

Goodman, M., and Moore, G. W., 1971, Immunodiffusion systematics of the Primates, I. The Catarrhini, *Syst. Zool.* **20**:19–62.

Goodman, M., Moore, G. W., Barnabas, J., and Matsuda, G., 1974, The phylogeny of human globin genes investigated by the maximum parsimony method, *J. Mol. Evol.* **3**:1–48.

Goodman, M., Moore, G. W., and Matsuda, G., 1975, Darwinian evolution in the genealogy of haemoglobin, *Nature* **253**:603–608.

Goodman, M., Czelusniak, J., Moore, G. W., Romero-Herrera, A. E., and Matsuda, G., 1979a, Fitting the gene lineage into its species lineage, A parsimony strategy illustrated by cladograms constructed from globin sequences, *Syst. Zool.* **28**:132–163.

Goodman, M., Pechère, J.-F., Haiech, J., and Demaille, J. G., 1979b, Evolutionary diversification of structure and function in the family of intracellular calcium-binding proteins, *J. Mol. Evol.* **13**:331–352.

Goodman, M., Weiss, M. L., and Czelusniak, J., 1982, Molecular evolution above the species level: branching pattern, rates, and mechanisms, *Syst. Zool.*, in press.

Hall, G. E., and Schraer, R., 1979, Purification and partial characterization of high and low activity carbonic anhydrase isoenzymes from *Malaclemys terrapin contrata*, *Comp. Biochem. Physiol.* **63B**:561–567.

Hewett-Emmett, D., Cook, C. N., and Barnicot, N. A., 1976, Old World monkey hemoglobins: Deciphering phylogeny from complex patterns of molecular evolution, in: *Molecular Anthropology* (M. Goodman and R. E. Tashian, eds.), pp. 257–275, Plenum Press, New York.

Hewett-Emmett, D., Czelusniak, J., Goodman, M., Venta, P. J., and Tashian, R. E., 1981, Evolution of nucleotide sequences coding for hemoglobin chains, *Fed. Proc.* **40**:1591.

Hill, W. C. O., 1953, *Primates—Comparative Anatomy and Taxonomy*, Volume I: *Strepsirhini*, University Press, Edinburgh.

Hill, W. C. O., 1955, *Primates—Comparative Anatomy and Taxonomy*, Volume II: *Haplorhini: Tarsioidea*, University Press, Edinburgh.

Huxley, J. S., 1942, *Evolution, the Modern Synthesis*, Allen and Unwin, London.

Jamieson, G. A., Hayes, A., Blum, J. J., and Vanaman, T. C., 1980, Structure and function relationships among calmodulins from divergent eukaryotic organisms, in: *Calcium-Binding Proteins: Structure and Function* (F. L. Siegel, E. Carafoli, R. H. Kretsinger, D. H. MacLennan, and R. H. Wasserman, eds.), Elsevier/North-Holland, New York, pp. 165–172.

Johanson, D. C., and White, T. D., 1979, A systematic assessment of early African hominids, *Science* **203**:321–330.

Kimura, M., 1968, Evolutionary rate at the molecular level, *Nature* **217**:624–626.

Kimura, M., 1969, The rate of molecular evolution considered from the standpoint of population genetics, *Proc. Natl. Acad. Sci. USA* **63**:1181–1188.

Kimura, M., 1979, The neutral theory of molecular evolution, *Sci. Am.* **241**(5):94–104.

Kimura, M., 1981, Was globin evolution very rapid in the early stages? A dubious case against the rate constancy hypothesis, *J. Mol. Evol.* **17**:110–113.

Klee, C. B., Crouch, T. H., and Richman, P. G., 1980, Calmodulin, *Annu. Rev. Biochem.* **49**:489–515.

Kretsinger, R. H., 1977, Evolution of the informational role of calcium in eukaryotes, in: *Calcium-Binding Proteins and Calcium Function* (R. H. Wasserman, R. A. Corradino, E. Carafoli, R. H. Kretsinger, D. H. MacLennan, and F. L. Siegel, eds.), North-Holland, New York, pp. 63–72.

Kretsinger, R. H., 1980, Structure and evolution of calcium-modulated proteins, *CRC Crit. Rev. Biochem.* **1980**:119–174.

Ladner, R. L., Heidner, E. J., and Perutz, M. F., 1977, The structure of horse methaemoglobin at 2.0 Å resolution, *J. Mol. Biol.* **114**:385–414.

Leclercq, F., Schnek, A. G., Braunitzer, G., Stangl, A., and Schrank, B., 1981, Direct reciprocal allosteric interaction of oxygen and hydrogen carbonate sequence of the haemoglobins of the caiman (*Caiman crocodylus*), the Nile crocodile (*Crocodylus niloticus*) and the Mississippi crocodile (*Alligator mississippiensis*), *Hoppe-Seyler's Z. Physiol. Chem.* **362**:1151–1158.

Le Gros Clark, W. E., 1959, *The Antecedents of Man*, Edinburgh University Press, Edinburgh.

Lehman, L. D., Jones, B. N., Dwulet, F. E., Bogardt, R. A., and Gurd, F. R. N., 1980, Complete amino acid sequence of the major component myoglobin from the goose-beaked whale, *Ziphius cavirostris*, *Biochim. Biophys. Acta* **625**:221–229.

Li, S. L., and Riggs, A., 1970, The amino acid sequence of hemoglobin V from the lamprey *Petromyzon marinus*, *J. Biol. Chem.* **245**:6149–6169.

Liljeqvist, G., Braunitzer, G., and Paléus, S., 1979, Hämoglobine, XXVII Die sequenz der monomeren hämoglobine III von *Myxine glutinosa* L: ein neurer hämkomplex: E7 glutamin, E11 isoleucin, *Hoppe-Seyler's Z. Physiol. Chem.* **360**:125–135.

Lovejoy, L. O., 1981, The origin of man, *Science* **211**:341–350.

Løvtrup, S., 1977, *The Phylogeny of Vertebrata*, Wiley, New York.

Margoliash, E., 1980, Evolutionary adaptation of mitochondrial cytochrome c to its functional milieu, in: *The Evolution of Protein Structure and Function* (D. S. Sigman and M. Brazier, eds.), Academic Press, New York, pp. 299–321.

Martin, S. L., Zimmer, E. A., Kan, Y. W., and Wilson, A. C., 1980, Silent δ-globin gene in old world monkeys, *Proc. Natl. Acad. Sci. USA* **77**:3563–3566.

McHenry, H. M., and Corruccini, R. S., 1980, Late tertiary hominoids and human origins, *Nature* **285**:397–398.

McKenna, M. C., 1969, The origin and early differentiation of therian mammals, *Ann. N.Y. Acad. Sci.* **167**(1):217–240.

Moore, G. W., 1976, Proof for the maximum parsimony ("red king") algorithm, in: *Molecular Anthropology* (M. Goodman and R. E. Tashian, eds.), Plenum Press, New York, pp. 117–137.

Moore, G. W., 1977, Proof of the populous path algorithm for missing mutations in parsimony trees, *J. Theor. Biol.* **66**:95–106.

Moore, G. W., Barnabas, J., and Goodman, M., 1973, A method for constructing maximum parsimony ancestral amino acid sequences on a given network, *J. Theor. Biol.* **38**:459–485.

Moore, G. W., Goodman, M., Callahan, C., Holmquist, R., and Moise, H., 1976, Stochastic versus augmented maximum parsimony method for estimating superimposed mutations in the divergent evolution of protein sequences. Methods tested on cytochrome c amino acid sequences, *J. Mol. Biol.* **105**:15–37.

Perutz, M. F., Bauer, C., Gros, G., Leclercq, F., Vandecasserie, C., Schnek, A. G., Braunitzer, G., Friday, A. E., and Joysey, K. A., 1981, Allosteric regulation of crocodillian haemoglobin, *Nature* **291**:682–684.

Pilbeam, D., 1979, Recent finds and interpretations of Miocene hominoids, *Annu. Rev. Anthropol.* **8**:333–352.

Rensch, B., 1959, *Evolution above the Species Level*, Columbia University Press, New York.

Romer, A. S., 1966, *Vertebrate Paleontology*, University of Chicago Press, Chicago.

Romero-Herrera, A. E., Lehmann, H., Joysey, K. A., and Friday, A. E., 1973, Molecular evolution of myoglobin and the fossil record: A phylogenetic synthesis, *Nature* **246**:389–395.

Romero-Herrera, A. E., Lehmann, H., Joysey, K. A., and Friday, A. E., 1978, On the evolution of myoglobin, *Phil. Trans. R. Soc. Lond. B* **283**:61–163.

Romero-Herrera, A. E., Lieska, N., and Nasser, S., 1979, Characterization of the myoglobins *Petromyzon marinus*, *J. Mol. Evol.* **14**:259–266.

Shoshani, J., Goodman, M., Barnhart, M., Prychodko, W., Vereshchagin, N. K., and Mikhelson, V. M., 1981, Blood cells and proteins in the Magadan mammoth calf: Immunodiffusion comparisons of *Mammuthus* to extant paenungulates and tissue ultrastructure, in: *The Magadan Mammoth* (N. K. Vereshchagin, ed.), Nauka, Leningrad.

Simons, E., 1976, The fossil record of primate phylogeny, in: *Molecular Anthropology* (M. Goodman and R. E. Tashian, eds.), Plenum Press, New York, pp. 35–62.

Simpson, G. G., 1945, The principles of classification and a classification of mammals, *Bull. Am. Mus. Nat. Hist.* **85**:1–350.

Sokal, R. R., and Michener, C. D., 1958, A statistical method for evaluating systematic relationships, *Univ. Kans. Sci. Bull.* **38**:1409–1438.

Szalay, F. S., and Delson, E., 1979, *Evolutionary History of the Primates*, Academic Press, New York.

Takagi, T., Nemoto, T., and Konishi, K., 1980, The amino acid sequence of the calmodulin obtained from sea anemone (*Metridium senile*) muscle, *Biochem. Biophys. Res. Commun.* **96**:377–381.

Takano, T., 1977, Structure of myoglobin refined at 2.0 Å resolution, *J. Mol. Biol.* **110**:537–584.

Tashian, R. E., Hewett-Emmett, D., Stroup, S. K., Goodman, M., and Yu, Y.-S. L., 1980a, Evolution of structure and function in the carbonic anhydrase isozymes of mammals, in: *Biophysics and Physiology of Carbon Dioxide* (C. Bauer, G. Gros, and H. Bartels, eds.), Springer-Verlag, Berlin, pp. 165–176.

Tashian, R. E., Hewett-Emmett, D., and Goodman, M., 1980b, Evolutionary diversity in the structure and activity of carbonic anhydrase, in: *Protides of the Biological Fluids* (H. Peeters, ed.), Volume 28, Pergamon Press, Oxford, pp. 153–156.

Tashian, R. E., Stroup, S. K., and Hall, G. E., 1981, Primary sequence of turtle low-activity red cell carbonic anhydrase: Homology with mammalian low-activity CA I isozymes, *Fed. Proc.* **40**:1677.

Van Valen, L., 1966, Deltatheridae, A new order of mammals, *Bull. Am. Mus. Nat. Hist.* **132**:1–126.

Walker, A., 1976, Splitting times among hominoids deduced from the fossil record, in: *Molecular Anthropology* (M. Goodman and R. E. Tashian, eds.), Plenum Press, New York, pp. 63–77.

Wicken, J. S., 1980, A thermodynamic theory of evolution, *J. Theor. Biol.* **87**:9–23.

Wilson, A. C., Carlson, S. S., and White, T. J., 1977, Biochemical evolution, *Annu. Rev. Biochem.* **46**:573–639.

Yunis, J. J., and Prakash, Om., 1982, The origin of man: A chromosomal pictorial legacy, *Science* **215**:1525–1530.

Zuckerkandl, E., 1976, Programs of gene action and progressive evolution, in: *Molecular Anthropology* (M. Goodman and R. E. Tashian, eds.), Plenum Press, New York, pp. 387–447.

Zuckerkandl, E., and Pauling, L., 1962, Molecular disease, evolution, and genetic heterogeneity, in: *Horizons in Biochemistry* (M. Kasha and N. Pullman, eds.), Academic Press, New York, pp. 189–225.

CHAPTER 5

Evolution of Chromosomal Proteins

LOIS T. HUNT and MARGARET O. DAYHOFF

1. Introduction

We use the term "chromosomal" proteins to include various kinds of proteins closely associated with the chromosomal DNA. In eukaryotes these would be: histones, including nuclear protein A24; sperm histones, including protamines; and nonhistone chromosomal proteins. Some of the eukaryote proteins have the most highly conserved sequences known, which should be an indication of their functional importance. In prokaryotes the type so far characterized from several species is called DNA-binding protein.

In addition, for eukaryotes, prokaryotes, and viruses there are other proteins that may interact in some way with the genome nucleic acid during replication, transcription, recombination, and repair, but that may not be as closely and as continuously associated with this nucleic acid as are the "chromosomal" proteins. Examples could include deoxyribonucleases and ribonucleases, polymerases, repressors, and helix-destabilizing proteins, and perhaps also those proteins binding to the 5′ end of viral nucleic acids. Lastly, a few other eukaryote chromatin-associated or nuclear proteins have been purified but not fully characterized by both structural and functional studies, so their relationships are not yet known. In general we will discuss only the "chromosomal" proteins listed in the first paragraph.

The structures of eukaryote chromosomes, of chromatin, of nucleosomes, and of DNA and its associated proteins are currently being studied

LOIS T. HUNT and MARGARET O. DAYHOFF • National Biomedical Research Foundation, Georgetown University Medical Center, Washington, D.C. 20007.

on several levels of organization and by an array of techniques, some very new, as many questions remain to be answered about structural–functional relationships. Among the excellent reviews that appeared recently are these on chromatin and nucleosome structure (Felsenfeld, 1978; Lilley and Pardon, 1979; Kornberg, 1977; Finch *et al.*, 1977; McGhee and Felsenfeld, 1980), nonhistone chromosomal proteins (Goodwin *et al.*, 1978), primary and secondary structure of histones (Isenberg, 1979), structure of histone variants (von Holt *et al.*, 1979), and histone evolution (Temussi, 1975; Reeck *et al.*, 1978). So rapid are advances in this field, however, that some of the information is already out of date.

We are specifically concerned with two related aspects of the protein structures: (1) the evolutionary history of proteins and species, especially distant relationships, as derived by computer analyses of the sequences and displayed as evolutionary trees; and (2) the organization of related protein sequences into families (sequences less than 50% different) and into superfamilies (sequences more than 50% different) where similarity of sequences among different families can be recognized by statistical procedures (Dayhoff, 1979). Additional similarities in secondary and tertiary structure and/or in function(s) increase the probability of relationship among proteins whose sequences are more than 50% different.

2. Nucleosome Structure

2.1. The Chromatin Fiber

The structure of eukaryote chromatin will be briefly described to provide a background for later discussion of the histones. According to proposals based on biochemical data, x-ray crystallography, light and electron microscopy, and neutron scattering, the chromatin fiber, about 10 nm in diameter, consists of a flexible chain of nucleosomes (the basic repeating unit) (Kornberg, 1977; Felsenfeld, 1978; Lilley and Pardon, 1979; McGhee and Felsenfeld, 1980). It is suggested that the 10-nm fiber may be coiled into a superhelix of approximately 30 nm diameter and having six or seven nucleosomes per turn (McGhee and Felsenfeld, 1980; Thoma *et al.*, 1979; Suau *et al.*, 1979).

According to the model proposed by Thoma *et al.* (1979), a fiber with nonameric nucleosomes can condense from an open zigzag configuration into a closed zigzag, helical configuration in which the disk-shaped nucleosomes are opposed side to side (not face to face); the number of nucleosomes per turn increases to six to eight and the pitch of the superhelix is about one nucleosome diameter (10–11 nm). Suau *et al.* (1979)

proposed a similar model, based on neutron scattering studies, in which the superhelix had an outside diameter of 34 nm, a central space of diameter 9.5 nm, a pitch of 11 nm, and seven nucleosomes, edge to edge, per turn; they suggest that the central space may contain nucleosomes, and that perhaps every seventh one in the coil has "flipped" into the center of the supercoil. Other models with somewhat different orientation of nucleosome core components and DNA have been described (Worcel and Benyajati, 1977; Trifonov, 1978).

2.2. The Nucleosome and Nucleosome Core

A nucleosome is composed of a segment of DNA about 200–250 base pairs (bp) in length, one molecule of histone H1 (usually), an octamer of histones H2A, H2B, H3, and H4 (two each), and possibly certain ones of the nonhistone chromosomal proteins (McGhee and Felsenfeld, 1980; Thoma *et al.*, 1979; Kornberg, 1977; Mardian *et al.*, 1980). The wedge-shaped nucleosome core particle (Fig. 1), about 11 nm in diameter and 5.5 nm high, contains the histone octamer and 146–166 bp of the DNA; the histone octamer forms a flattened cylinder around which is coiled 1.75–2 turns of the DNA in a flat superhelix having a pitch of about 3 nm (McGhee and Felsenfeld, 1980; Finch *et al.*, 1977; Finch and Klug, 1978; Lilley and Pardon, 1979; Thoma *et al.*, 1979). The remaining DNA (15–100 bp, depending on cell type and on species) forms the "linker" region (McGhee and Felsenfeld, 1980; Thoma *et al.*, 1979). New information as to the arrangement of their binding to each other (Klug *et al.*, 1980) and to the core DNA (Mirzabekov *et al.*, 1980) is now available. Klug *et al.* (1980) have deduced, from electron microscopy and image reconstruction

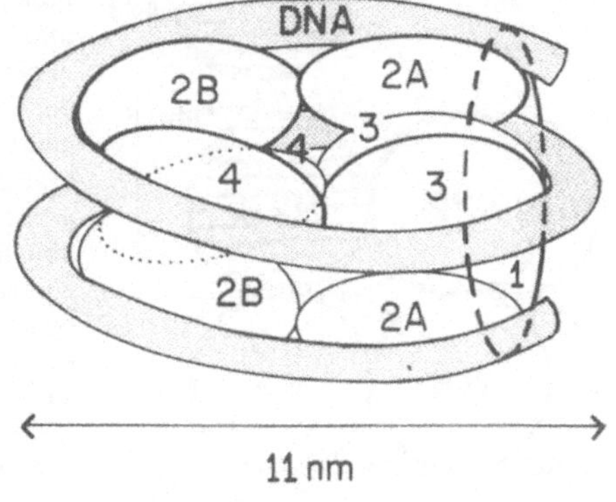

Figure 1. Diagrammatic representation of a nucleosome core particle. The relative positions of the four types of histones in the octamer and of histone H1 outside the core, where the DNA begins and ends the coil about the octamer, are indicated. The coil makes 1.75 turns, with a pitch of 3 nm.

combined with the histone–DNA cross-linking studies of Mirzabekov *et al.* (1978, 1980), a detailed structure for the histone octamer. The wedge-shaped nucleosome core particle has 166 bp of DNA wound in 1.75 turns around it. The histone octamer, also wedge-shaped, is composed of an H4–$(H3)_2$–H4 tetramer and two H2B–H2A heterodimers and has a two-fold axis of symmetry; the order of the eight histones (within and along the DNA coil) is then H2A-H2B-H4-H3-H3-H4-H2B-H2A (see Fig. 1). Figure 2 [from Goldknopf *et al.* (1977)] is a diagrammatic representation of the histone octamer, showing an arrangement of the four kinds of histones and also of nuclear protein A24, the histone H2A–ubiquitin dimer found in 5–10% of nucleosomes (Goldknopf *et al.*, 1980; Goldknopf and Busch, 1980). All four histones are primarily globular proteins, with approximately 50% α-helical secondary structure; however, the terminal portions of the chains appear to be either outside of the globular portion of the molecules or exposed at the surface (Lilley and Pardon, 1979).

Strong interchain binding occurs between the two histones H4 (residues 38–102) and H3 (residues 42–120) to form a tetramer and between histones H4 and H2B (residues 65–125) and histones H2B (residues 37–114) and H2A (residues 31–95) to form dimers, as indicated in Fig. 3 and summarized in Klug *et al.* (1980), Isenberg (1979), and McGhee and Felsenfeld (1980). Other, less strong interchain binding occurs between H3 and H2A, H2A and H4, and H3 and H2B (Isenberg, 1979). The existence *in vivo* of H3 dimers, through the formation of a disulfide bond involving Cys-110, is possible but not yet established (Garrard *et al.*, 1977; Camerini-Otero and Felsenfeld, 1977). The more basic amino-terminal region in all of the core histones is a region of secondary modifications: acetylation,

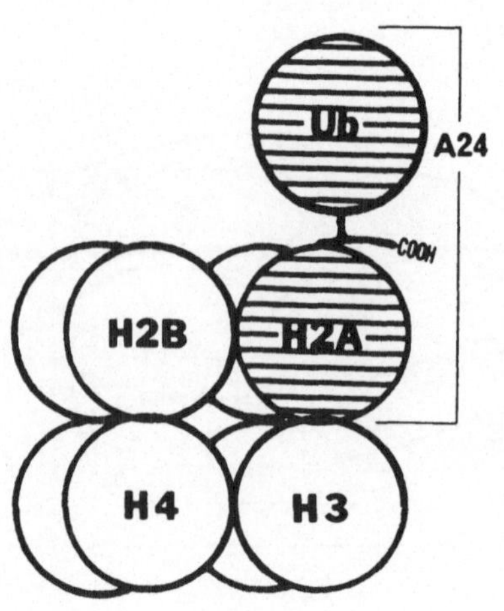

Figure 2. A possible way in which a branched protein A24 molecule (hatched spheres) could fit into an octameric nucleosome core composed of dimers of H2A, H2B, H3, and H4 interacting according to the cross-complexing pattern of D'Anna and Isenberg (1974). In this nucleosome, the histone 2A portion of protein A24 takes the place of a histone 2A molecule in the core, whereas the ubiquitin (Ub) is outside the core. The ubiquitin-containing polypeptide of protein A24 is linked by a Gly–Gly isopeptide bond to the ϵ-NH_2 group of Lys-119 of histone H2A (Goldknopf and Busch, 1977). [This figure with its legend (slightly modified) is reproduced, with permission, from Fig. 5 in Goldknopf *et al.* (1977).]

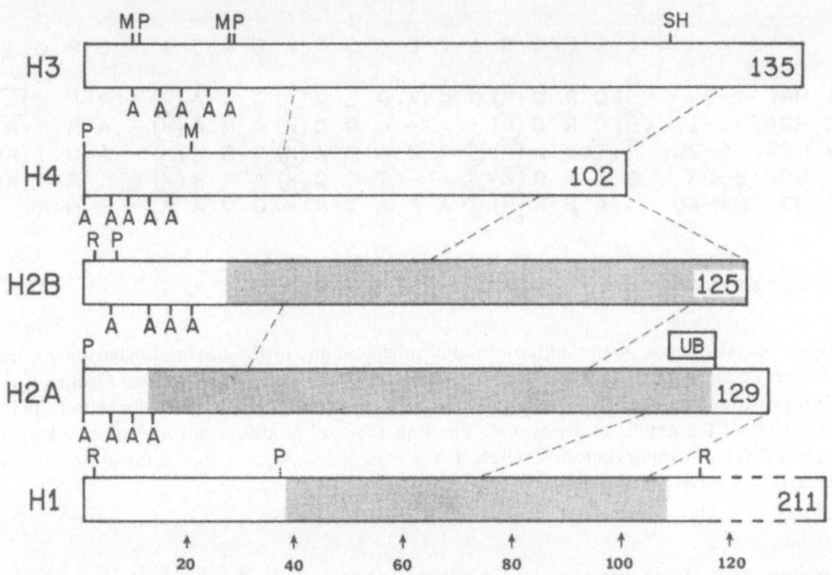

Figure 3. Comparison of features of the five histone types. Lengths of the chains and residue positions are given. Dashed lines between proteins indicate regions of interaction between histones, and shaded areas indicate conserved regions in those histones with variable regions. Chain modifications *in vivo* (according to McGhee and Felsenfeld, 1980) are identified by P (phosphorylation), A (acetylation), M (methylation), and R (ADP-ribosylation); the site of ubiquitin (UB) binding to H2A and the location of the potential interchain disulfide bond (SH) in H3 are also shown. Some sites may have more than one type of modification, depending on the species.

methylation, phosphorylation, and ADP-ribosylation (Isenberg, 1979; McGhee and Felsenfeld, 1980; Ogata *et al.*, 1980); we show one possible alignment of these modification sites in the four bovine core histones in Fig. 4. The formation of nuclear protein A24 (mentioned above) may be considered a fifth type of modification; it involves residues at the carboxyl end of each chain. Figure 5 diagrams this binding, discussed in Fig. 2, in greater detail.

Histone H1 has a middle, 72-residue, somewhat hydrophobic globular region of which the carboxyl half can bind to the carboxyl-terminal region (residues 58–129) of histone H2A to form a nonamer (see Figs. 1 and 3) (Boulikas *et al.*, 1980; Reudelhuber *et al.*, 1980; McGhee and Felsenfeld, 1980) and nonglobular hydrophilic terminal regions that are proposed to bind to DNA (Thoma *et al.*, 1979). Residues in the amino-terminal region may be acetylated, phosphorylated, and ADP-ribosylated (see Fig. 3), and

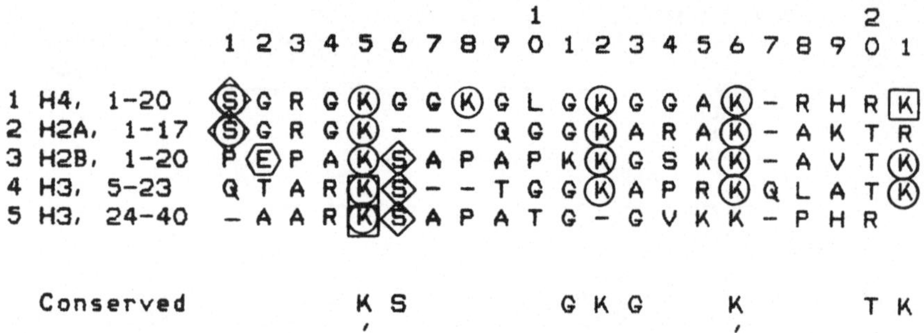

Figure 4. Alignment of the amino-terminal modifications in the bovine nucleosome core histones, according to McGhee and Felsenfeld (1980), Isenberg (1979), and Ogata *et al.* (1980). Residue numbers for each sequence are given. Conserved residues are present in at least 60% of the chains at a position. The four types of modifications are indicated by a diamond for phosphorylation, a circle for acetylation, a square for methylation, and a hexagon for ADP-ribosylation.

serines and threonines in the carboxyl-terminal half may be phosphorylated (Langan *et al.*, 1971; Hohmann *et al.*, 1976; Iwasa *et al.*, 1980; Isenberg, 1979). Histone H1 has not been considered a part of the core particle and it is not always present, although recent studies suggest it may play an integral role in the nucleosome (Thoma *et al.*, 1979; Reudelhuber *et al.*, 1980). In the model of nucleosome structure proposed by Thoma *et al.* (1979), the DNA superhelix forms two full turns (166 bp) about the octamer and the H1 binds to the DNA coil where it enters and leaves the core particle (see Fig. 1), thus stabilizing the nucleosome and allowing higher-order coiling of the chromatin.

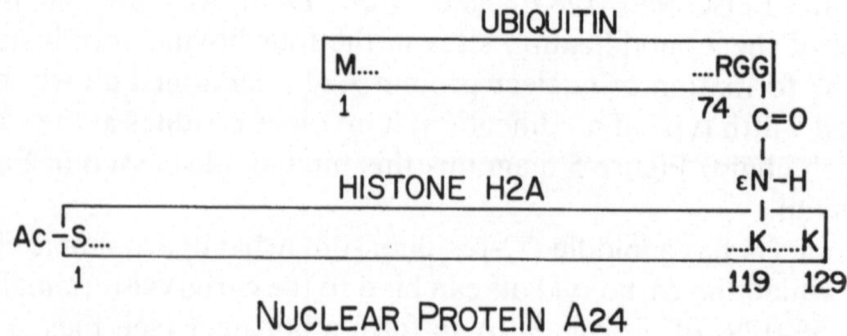

Figure 5. Diagrammatic representation of the structure of nuclear protein A24, a Y-shaped dimer of histone H2A and ubiquitin (Goldknopf and Busch, 1980). The chains are drawn proportional in length. The isopeptide link between Gly and Lys is indicated.

Nonhistone chromosomal proteins (NHCP) may also be closely associated with the nucleosome. For example, ubiquitin (part of protein A24) has also been called NHCP HMG-20 (Walker *et al.*, 1978b). More recently, Mardian *et al.* (1980) have identified two binding sites on the DNA of the nucleosome core for NHCP HMG-14 and/or HMG-17; it was already known that these two proteins were structurally and functionally related (see also Section 4.4) (Walker *et al.*, 1979c; Reeves and Candido, 1980; Gazit *et al.*, 1980) and had the same binding sites (Albright *et al.*, 1980; Weisbrod *et al.*, 1980). Furthermore, the related NHCP HMG-6 from trout testis chromatin has been found closely associated with a certain subset of transcriptionally active nucleosomes in the ratio of two molecules of HMG-6 per nucleosome core octamer (Levy-Wilson *et al.*, 1980).

3. Computer Methods

3.1. Detection of Protein Relationships

We have three programs that we use to detect and/or to quantitate relationships between and within proteins (especially distant proteins): SEARCH, ALIGN, and RELATE (Dayhoff, 1979).

We examine new sequences of unknown relationship for evolutionary similarity to other sequences in the data base with program SEARCH. We select a test piece of, for example, at least 25 residues from a region of the new sequence containing a concentration of the less mutable amino acids, or we may search all consecutive 25-residue segments. The test piece (or pieces) is compared with all 25-residue segments and the shorter end segments of each known sequence in our data base (Dayhoff, 1979). An input matrix is supplied to the program. For each segment comparison, a search score is computed by summing the pair scores for each pair of amino acids matched in the comparison. For detecting distant relationships, we use our mutation data matrix (Dayhoff, 1979). However, other matrices, such as the unitary matrix, the genetic code matrix, or a specially constructed matrix, may be used, depending on the structure or peculiarities of the sequence to be searched.

The distribution of scores from unrelated sequences is approximately normal. Typically, for a 25-residue piece, all corresponding sequences of the same family (sequences <50% different) appear above the distribution of scores of unrelated segments. About half of the more distantly related sequences in the same superfamily are also above this distribution, whereas the rest are in the upper tail. The sequences within the same family will still stand out clearly above the distribution of scores of unrelated segments. If a search produces no scores above the distribution, in all probability there are no other sequences from the same family in the data collection.

For comparing a pair of sequences that may be related (for example, a pair with a high SEARCH score for matching segments), we use program ALIGN, which calculates the best alignment between any pair of sequences (see also Needleman and Wunsch, 1970), given a matrix of amino acid pair scores and a penalty for breaking a sequence (gap) (Dayhoff, 1979). In this work we have used our mutation data matrix, usually with a bias of 6 and a gap penalty of 6. This program can also be used with nucleotide sequences. The maximum score that can be achieved by an alignment of a pair of real sequences is compared with the distribution of maximum scores for a large number (usually 100) of random permutations of the two sequences. The mean and standard deviation of this approximately normal distribution are calculated. The alignment score (AS) is the number of standard deviations by which the maximum score for the real sequences exceeds the average maximum score for the random permutations. The probability that a score as high as that from the real sequences could have been obtained in a comparison of randomized sequences can be determined from the cumulative standardized normal distribution table.

Program RELATE can be used to detect unusual similarity between a pair of sequences, especially when of unequal length, or to show the existence and identity of single or multiple repeated segments (Dayhoff, 1979); the ALIGN program is somewhat more sensitive for comparing sequences of similar length without repetitions. This computer method compares all possible segments of a given length from one sequence with all segments of the same length from the second sequence (see also Fitch, 1966). A segment score is accumulated from the pair scores of the amino acids occupying corresponding positions within the two segments. A scoring matrix is supplied as before. For detecting distant relationships, we use the mutation data matrix and segments of at least 20 residues. When looking for small, recent duplications, we use the unitary matrix and a length of five or ten residues.

3.2. Protein Alignments and Evolutionary Trees

Information derived from the above three programs can be used to construct (or to add a sequence to) alignments and evolutionary trees. Program DISP (for "display") produces alignments of two or more related sequences, matrices of differences, and a matrix of PAM (accepted point mutations/100 residues) values (Dayhoff, 1979).

The output from program ALIGN is especially useful as a guide in optimizing an alignment, particularly in selecting gap positions (part of the input data for DISP), which represent insertion/deletion events. There may have been a number of insertions and deletions during the evolution of very distantly related sequences and there are often a number of alignments that are almost equally good. From the limited sequence information available, it is often impossible to recreate all of the insertion/deletion events correctly. Sometimes there has been a repetition of several residues in one sequence that is still clearly visible and can be reflected in the alignment.

Program DISP also calculates the conserved line below the alignments. The sequences are divided into groups of closely related types, and two parameters A and B (>50) are assigned. A group may contain a single sequence. An amino acid is selected as a "group" amino acid if it appears in at least A percent of the sequences within the group. It is selected as a conserved residue if it is the group amino acid in at least B percent of the groups. An apostrophe appears below residues that are common to all sequences.

Each difference matrix, derived from an alignment, shows the number of differences (above the diagonal line) and the percent difference (below the diagonal line). For a given sequence, each alignment position can contain an amino acid, a gap represented by a dash, an undetermined amino acid (X), or nothing (that is, it is blank). The percent difference between two sequences is derived from positions where both sequences contain amino acids or where one contains an amino acid and the other contains a dash. Positions where either sequence is blank or contains an X, or where both contain dashes, are ignored. In calculating the matrices or in listing the conserved residues or alternatives beneath the alignments, a residue that may be either aspartic acid or asparagine, or glutamic acid or glutamine, is not considered different from either possible alternative.

Two main methods can be used to derive evolutionary trees from protein sequence data; one is based on a matrix of differences between aligned sequences, and the other is based on generating ancestral sequences. Our program MATTOP, a modified version of the matrix method first described by Fitch and Margoliash (1967), was used for the trees presented here; the matrix method is superior to the ancestral sequence method for distantly related sequences (Dayhoff, 1976b).

We start with the alignment PAM matrix (the matrix of the total amount of evolutionary change between each pair of sequences), which has been derived from a matrix of percent differences by correcting for inferred parallel and superimposed mutations (see Table 36 in Dayhoff, 1979). For each possible topology, our program determines a set of branch lengths to give the weighted least-squares fit of the terms of the reconstructed and original matrices. Typically the weights are inversely proportional to the size of the matrix element (approximating the variance of the number) and to a factor that partially compensates for the nonindependence of the matrix elements, Z^n, where Z is a number between 1 and 2, and n is the number of nodes on the tree between the two sequences that generated the matrix element. We generally use no more than eight sequences or (averaged) sets of sequences in a single tree run because of the impractically larger number of topologies to be compared if there are more than eight branches; we can also run a single, selected topology of many branches to get their lengths. We may also select a set of complete sequences from within an alignment, leave out partial sequences, or select a specific region (for example, a conserved region) of an alignment from which to derive a tree. MATTOP allows the user to delete sequences (by removing the matrix elements) and to lump groups of branches, to give a single branch, by averaging the matrix elements.

The best topology is usually the one with minimal total (absolute value) branch length, all (or nearly all) positive branches, and consistency with biologic data. The branch lengths are expressed in PAMs.

The trees presented here should be considered as tentative for several reasons: the sample size is generally small, with only a few sequences, from a few species, being available; some of the proteins are multigenic, making recognition of closest homologs between species difficult; the degree of conservation in some of the histones and nonhistone chromosomal proteins may be less than present estimates; for very conserved and ancient proteins, with few variant positions, mutation events can obscure the order of divergence of evolutionary lineages; some of the proteins have subregions of different degrees of conservation and with internal duplications. Furthermore, as the point of earliest time is not known for any of these trees, the trunk is always arbitrarily placed. Only the tree for histone H4 has a branch based on a partial sequence.

3.3. Protein Families and Superfamilies

The statistical information provided by the search scores (SEARCH), alignment scores (ALIGN), and segment comparison scores (RELATE) and the difference matrices produced by DISP are used to assign protein sequences into family and superfamily groups (Dayhoff, 1976a, 1979; Dayhoff *et al.*, 1975). Proteins within a family are less than 50% different and they are either homologs in various species or products of gene duplication; their similarity of function was usually recognized before the sequences were determined and they may have identical or very similar names. Proteins and protein families are combined into superfamilies if they are more than 50% different but similarity of sequences can be statistically demonstrated, that is, if the probability that two proteins (in a superfamily) could have such similar sequences by chance is less than 10^{-6}, or less than 10^{-3} if they are known to share a similar function (Dayhoff, 1979).

3.4. Mutation Acceptance Rates

In order to calculate the rate of mutation acceptance, the observed number of amino acid differences between two sequences must be corrected to estimate the actual number of mutations, some superimposed, that have occurred since the proteins diverged. This actual number, given in PAMs (accepted point mutations/100 residues) is the evolutionary distance and is derived from data describing the point mutations that have occurred (Dayhoff, 1979). A PAM matrix for each alignment is calculated by program DISP. Briefly, the evolutionary distance in PAMs between two species (or two proteins) on an evolutionary tree is divided by the total time on both branches since the suggested divergence time to obtain the mutation acceptance rate in PAMs per 100 million years (Dayhoff, 1979).

4. Chromosomal Proteins

4.1. Prokaryote DNA-Binding Proteins

At least three DNA-binding proteins have been found in *Escherichia coli*. Two of these, NS1 and NS2, are related proteins that probably exist primarily as heterotypic dimers, are closely associated with the bacterial nucleoid, and also bind specifically to the 30S ribosomal subunit (Mende *et al.*, 1978). NS1 and NS2, also called HUβ and HUα (Rouvière-Yaniv and Kjeldgaard, 1979), and probably histone-like protein II (Lathe *et al.*, 1980), are 62% identical in sequence. Proteins NS1 and NS2, as well as histone-like protein I (mol. wt. 17,000), are said to resemble histones in several physical properties (Lathe *et al.*, 1980); it has been suggested that dimers (or tetramers) of NS1 and NS2 may play a role in DNA condensation analogous to that of the H3–H4 histones (Rouvière-Yaniv and Kjeldgaard, 1979).

Other prokaryotes possess DNA-binding proteins related structurally to NS1 and NS2 (see Fig. 6). The partial sequences of two DNA-binding proteins from the blue-green alga *Synechocystis* (Aitken and Rouvière-Yaniv, 1979) are nearly 50% identical with the two *E. coli* proteins, yet the time of divergence for these prokaryotes was probably well over two billion years ago. In their highly conserved sequences and probable slow rate of evolution these prokaryote proteins also resemble the eukaryote histones.

The DNA-binding protein of the free-living mycoplasma *Thermoplasma acidophilum* is 27% identical with NS1 and NS2 (Searcy and DeLange, 1980; DeLange *et al.*, 1981; Stein and Searcy, 1978) and shares two short conserved regions of seven and 12 residues (see Fig. 6). Searcy and Stein (1980) reported that this protein condenses DNA into globular

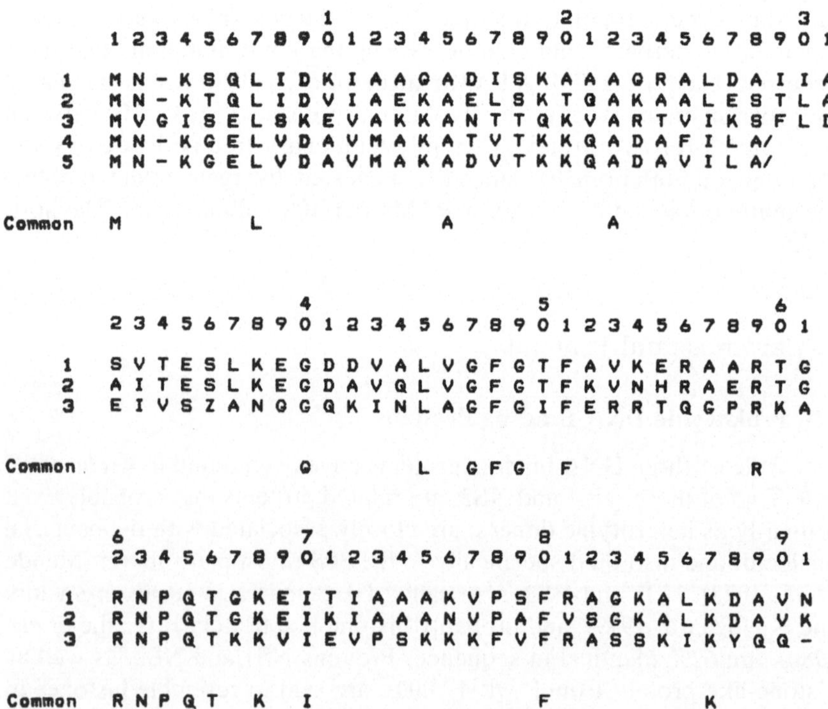

Figure 6. Alignment of prokaryote DNA-binding proteins. Sequences 1 and 2 are *Escherichia coli* NS1 and NS2 (Mende *et al.*, 1978; Rouvière-Yaniv and Kjeldgaard, 1979); sequence 3 is from *Thermoplasma acidophilum* (Searcy and DeLange, 1980; DeLange *et al.*, 1981); and sequences 4 and 5 are NS1 and NS2 from *Synechocystis* (Aitken and Rouvière-Yaniv, 1979). Residue 36 is Glu in the *Thermoplasma* sequence.

particles, each containing 40 bp of DNA wound about four of the protein molecules, similar to eukaryote nucleosome cores. Searcy and DeLange (1980) compared the amino-terminal 24 residues from *Thermoplasma* with several regions in histones H2A and H3, combined identical positions and residues related by single-base substitutions, and calculated that there were statistically significant homologies. They proposed that the *Thermoplasma* protein was intermediate between the *E. coli* proteins and histones H2A and H3; it also resembles eukaryote histones in that it inhibits RNA synthesis and its initiator Met is not formylated (Searcy and Stein, 1980). Our computer analyses did not demonstrate any similarities that were statistically significant by our criteria between eukaryote histones and any of these DNA-binding proteins. Nor were matches to the seven- and 12-residue conserved regions found in any histones. However, we did find some suggestive similarities between the *Thermoplasma* protein and histones H2A and H3; for example, the segments with residues 1–61 (*Thermoplasma* protein) and 58–118 (histone H2A) are 24% identical. On the other hand, our comparison of the *Thermoplasma* protein and NS1 with program ALIGN gives a significant AS of 9.86 SD units.

4.2. Viral Nucleic Acid-Binding Proteins

At least one animal DNA virus, simian virus 40, has a minichromosome with over 20 nucleosomes, in which the viral genome DNA is complexed with the five kinds of histones from the host cell (Griffith, 1975; Varshavsky *et al.*, 1976, 1978; Bellard *et al.*, 1976); similar chromatin structure is found for polyoma virus and others (Sergeant *et al.*, 1979; Schaffhausen and Benjamin, 1976).

In adenovirus the basic unit of viral chromatin, analogous to the nucleosome, appears to be a complex of viral DNA with two kinds of virus-coded, basic, histone-like proteins, protein VII (mol. wt. 18,000) and protein V (mol. wt. 45,000) (Corden *et al.*, 1976; Lischwe and Sung, 1977; Sung *et al.*, 1977). Corden *et al.* (1976) have calculated that there are 180 units per viral chromosome and that each complex could contain six copies of protein VII, one of protein V, and 200 bp of DNA. The synthesis of the precursor form (mol. wt. 20,000) of protein VII begins simultaneously with that of viral DNA; both the protein VII precursor and protein V appear to be acetylated during synthesis (Fedor and Daniell, 1980). A small amount of sequence data for protein VII precursor is available (Sung *et al.*, 1977; Lischwe and Sung, 1977); the amino-terminal 40 residues of protein VII somewhat resemble protamine sequences, according to results from our computer search. Adenovirus, in addition, has a DNA-binding protein of mol. wt. 72,000, coded by a gene in the E2 region of the genome, that covalently binds to the 5′ CMP of both single-stranded and double-stranded DNA; it is one of the components controlling DNA replication and is important for both initiation and elongation (Schechter *et al.*, 1980; Arens *et al.*, 1977; Horwitz, 1978; van der Vliet *et al.*, 1978). Green *et al.* (1980) found an apparent second protein covalently bound to the 5′ end of the DNA; this protein, of mol. wt. 55,000, did not correspond to any other known virus-coded protein and was possibly coded by a host cell gene. No sequence data are available for either of these proteins.

Such proteins that covalently bind to the 5′ end of the genome and are involved in nucleic acid replication have been described from other animal DNA viruses, from animal and plant RNA viruses, and from bacteriophages (Salas and Viñuela, 1980). These proteins may be in a functional category different from those complexed in nucleosomes or in nucleosome-like particles.

Another example is the small virus-coded protein VPg that covalently binds the single-stranded RNA of poliovirus through a phosphodiester link between the single tyrosine and the 5′ UMP (Kitamura *et al.*, 1980); the protein probably acts as a primer for elongation of RNA strands (Salas and Viñuela, 1980). A fragment of poliovirus RNA that contains the coding region for VPg has recently been sequenced (Kitamura *et al.*, 1980), although the exact length of the protein is not yet determined. However, it lies within the segment of the genome specifying the RNA polymerase precursor. The same RNA-binding protein has been reported from encephalomyocarditis virus (Vartapetian *et al.*, 1980). Other picornaviruses and some plant viruses possess a similar RNA-binding protein (Kitamura *et al.*, 1980).

4.3. Protamines

Protamines, or sperm histones, are very basic polypeptides associated with nuclear DNA that replace histones in maturing sperm of vertebrates. Protamine-like peptides have also been identified in some invertebrates. All have a high percentage of arginine. Coelingh and Rozijn (1975) have proposed that there are three main types: cysteine protamines (mammalian sperm histones), tyrosine protamines (galline, tuna thynnins), and arginine protamines (other fish protamines).

Black and Dixon (1967) presented a scheme for the evolution of the herring clupeines from an ancestral pentapeptide (Ala-Arg-Arg-Arg-Arg) by a series of partial gene duplications with subsequent point mutations. It was later proposed (Fitch, 1971) that clupeine Z is the result of a crossover between the genes for clupeines YI and YII. Thynnin Z2 may also be a crossover product. Recent evidence suggests that trout may have as many as six protamine genes per haploid set, including some crossover products (Sakai *et al.*, 1978).

We constructed several alignments of the 13 fish protamine sequences having different arrangements of gaps; the one that best clustered the Arg residues, shown in Fig. 7, in general had the lowest percent differences. This alignment, which best clustered the arginines and aligned the prolines, may have some experimental support. A model derived from x-ray diffraction and circular dichroism studies (using a salmine AI–yeast transfer RNA crystal complex) suggests that the conformation of a protamine may change, upon binding to double-helical nucleic acid, from a random coil to a structure composed of three to four α-helical domains alternating with flexible "joints" (see Fig. 7); such a segmented molecule could lie along the grooves of double-helical DNA and "condense" adjacent DNA molecules by hydrogen-bonding between arginines and phosphates (Warrant and Kim, 1978). According to predictions of secondary structure in the protamines (Toniolo, 1980), most of the sequences have one or two α-helices that fall within these same domains, as well as one or two β-turns correlated with phosphorylatable serines (see Fig. 7).

Figure 8 is an evolutionary tree that we constructed from this alignment, with the two possible crossover sequences and one of the sturgeon sequences omitted. We selected this tree, as it has one of the shortest topologies and also may correspond best to generally accepted evolutionary relationships for these fishes. Each species probably has at least two major protamines, but because of the small number of sequence positions at which differences occur and the types of point mutations, it is not clear how to match the two types among the species. Furthermore, the evolutionary distance between sturgeons and the other fishes is great enough (and the branch lengths so long) that the order of branching at nodes becomes uncertain when both sturgeon sequences are included. Both protein divergences, including gene duplications for three of the species, and species divergences are shown.

4.4. Nonhistone Chromosomal Proteins

Nonhistone chromosomal proteins (NHCP) are accessory proteins associated with chromatin in nuclei of animals, plants, and fungi (yeasts) (Goodwin *et al.*, 1978; Spiker *et al.*, 1978; Weber and Isenberg, 1980). The high-mobility-group (HMG) NHCP are generally characterized by low molecular weights, a high percentage of both acidic and basic residues,

```
                                    1                   2                   3
                  1 2 3 4 5 6 7 8 9 0 1 2 3 4 5 6 7 8 9 0 1 2 3 4 5 6 7 8 9 0 1 2 3 4 5 6

 1 Tuna Y2        P R R R R - - Q A S R P V R R R R R Y R R S T A A R R R R R V V R R R R
 2 Tuna Z2        P R R R R - - R S S R P V R R R R R Y R R S T A A R R R R R V V R R R R
 3 Tuna Z1        P R R R R - - R S S R P V R R R R R Y R R S T V A R R R R R V V R R R R
 4 Salmon AII    (P.R.R.R.R.R,R,S.S.S.R.P.I.R.R.R.R.- P.R.R.A.S.- - R.R.R.R.R.G.G.R.R.R.R)
 5 Trout IB       P R R R R R R S S S R P I R R R R - P R R V S - - R R R R R G G R R R R
 6 Salmon AI      P R R R R - - S S S R P V R R R R R P R - V S R - R R R R R G G R R R R
 7 Trout IA       P R R R R - - S S S R P V R R R R R P R R V S R - R R R R R G G R R R R
 8 Trout II       P R R R R - - S S S R P V R R R R - A R R V S R - R R R R R G G R R R R
 9 Herring YII    P R R R - T R R A S R P V R R R R - P R R V S - - R R R R - - A R R R R
10 Herring Z      A R R R R S R R A S R P V R R R R - P R R V S - - R R R R - - A R R R R
11 Herring YI     A R R R R S - - S S R P I R R R R - P R R R T T - R R R R - A G R R R R
12 Sturgeon B     A R R R R R - - S S R P Q R R R R R - R R H G - - R R R R - - G R R - -
13 Sturgeon A     A R R R R R H A S T K L K R R R R R - R R H Q - - K K - - - - S H K - -

   Conserved      P R R R R       S S R P V R R R R R P R R   S     R R R R R   G R R R R
                    ' ' '                   ' ' ' '     '
```

Figure 7. Alignment of fish protamines. Coiled lines indicate potential α-helical domains in one or more protamines; positions 8–11, 9–12, and 23–28 are potential locations of β-turns in some protamines. Most serines, and especially the one at position 10, can be phosphorylated. Sequences and species are: 1–3, thynnins Y2, Z2, and Z1 from *Thunnus thynnus* (Bretzel, 1972, 1973b, 1973a); 4 and 6, salmines AII and AI from *Oncorhynchus keta*, and 5, 7, and 8, iridines IB, IA, and II from *Salmo gairdneri* (Ando and Watanabe, 1969); 9–11, clupeines YII, Z, and YI from *Clupea pallasii* (Suzuki and Ando, 1972a; Iwai *et al.*, 1971; Suzuki and Ando, 1972b); 12, sturine B from *Acipenser guldenstadti* (Yulikova *et al.*, 1976); 13, stellin A from *Acipenser stellatus* (Yulikova *et al.*, 1979).

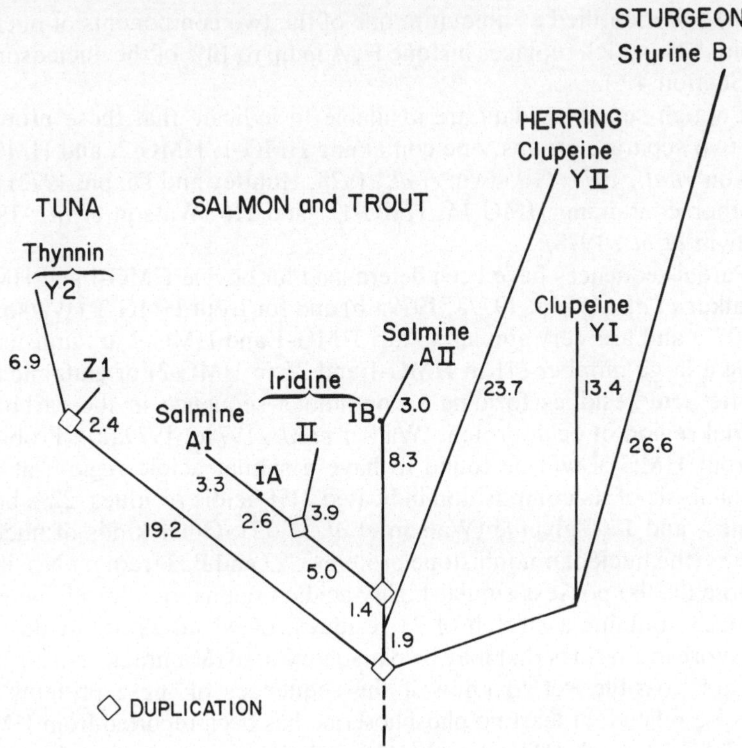

Figure 8. Evolutionary tree of fish protamines, derived from the alignment of Fig. 7, with sequences 2 (thynnin Z2), 10 (clupeine Z), and 13 (stellin A) omitted. Both protein and species divergences are shown. The trunk (dashed line) is arbitrarily positioned.

extractability from chromatin with dilute salt solutions, and solubility in 2% trichloroacetic acid (Goodwin *et al.*, 1978; Kuehl, 1979; Weber and Isenberg, 1980); however, there are exceptions to this definition among known HMG proteins (Weber and Isenberg, 1980). At least four distinct HMG proteins have been described from various tissues of several mammals, including calf (Goodwin *et al.*, 1978), pig (Walker *et al.*, 1978a), rat (Kuehl, 1979; Inoue *et al.*, 1980), rabbit (Mathew *et al.*, 1979; Goodwin *et al.*, 1977), and mouse (Saffer and Glazer, 1980), and from chicken erythrocytes (Goodwin *et al.*, 1978). These four proteins are designated HMG-1, HMG-2, HMG-14, and HMG-17 (Goodwin *et al.*, 1978). At least two NHCP, HMG-T and H6, have been reported from rainbow trout (Huntley and Dixon, 1972; Watson *et al.*, 1977, 1979; Kuehl *et al.*, 1980). A fifth protein, HMG-20, was isolated from calf thymus by Walker *et al.*

(1978b) and identified as ubiquitin, one of the two components of nuclear protein A24, which replaces histone H2A in up to 10% of the nucleosomes (see Section 4.6).

Enough sequence data are available to indicate that these proteins form two separate groups, one containing HMG-1, HMG-2, and HMG-T (Watson *et al.*, 1977; Goodwin *et al.*, 1978; Huntley and Dixon, 1972) and the other containing HMG-14, HMG-17, and H6 (Watson *et al.*, 1979; Goodwin *et al.*, 1978).

Partial sequences have been determined for bovine HMG-1 and HMG-2 (Walker *et al.*, 1976a, 1977a, 1979a,b) and for trout HMG-T (Watson *et al.*, 1977) and are very similar so far. HMG-1 and HMG-2 are unusual in having a large number (41 in HMG-1 and 35 in HMG-2) of glutamic and aspartic acid residues forming a continuous sequence in the carboxyl-terminal region of each protein (Walker *et al.*, 1978c, 1979a,b). Probably the trout HMG-T will be found to have a similar acidic region; at any rate, analysis of its composition indicated 21% acidic residues, 22% basic residues, and 17% glycine (Watson *et al.*, 1977). Other kinds of nuclear proteins (the nucleolar nonhistone proteins C23 and B23 from rat Novikoff hepatoma cells) possess similar highly acidic regions; one tryptic peptide from C23 contains a stretch of 27 residues, of which 23 are acidic and three more are serines that may be phosphorylated (Mamrack *et al.*, 1979). It is not possible yet to know if the sequences of these proteins are otherwise related; at least no phosphoserine has been reported from HMG-1 or HMG-12. Both HMG-1 and HMG-2 do, however, contain dimethylarginine (Boffa *et al.*, 1979).

The complete sequences of bovine HMG-14 (Walker *et al.*, (1979c) and HMG-17 (Walker *et al.*, 1976b, 1977b), of chicken HMG-17 (Walker and Johns, 1980; Walker *et al.*, 1980), and of rainbow trout H6 (Watson *et al.*, 1979) have been determined and are structurally related. Partial sequences of pig HMG-14 (Walker *et al.*, 1978a) and of chicken HMG-14 (Walker and Johns, 1980) are also available. All of these sequences share a highly conserved 15-residue region near the amino end that contains two serines, which may be the sites for the reported phosphorylation (Inoue *et al.*, 1980; Saffer and Glazer, 1980). In addition, the four complete sequences share another seven-residue conserved region near the carboxyl end that begins Ala-Glu-Asn-Gly; in trout H6 the Glu residue binds ADP-ribosyl (Wong *et al.*, 1977; Watson *et al.*, 1979). Although the function(s) of these proteins has not been established, they appear to be preferentially associated with transcriptionally active genes (Weisbrod *et al.*, 1980; Gazit *et al.*, 1980; Kuehl *et al.*, 1980; Levy-Wilson *et al.*, 1980) and in some way they partially inhibit endogenous histone deacetylase (Reeves and Candido, 1980); furthermore, HMG-17 and HMG-14 can bind interchangeably to the two binding sites on the nucleosome core (Mardian *et al.*, 1980; Albright *et al.*, 1980).

The evolutionary tree in Fig. 9 is derived from one possible alignment of these proteins. A number of gaps are required to align HMG-6 and HMG-14 with HMG-17; consequently, these branches on the tree are somewhat distorted. HMG-6 is distant enough to be barely within the same family as HMG-17, but HMG-14 is 55% different from bovine HMG-17 and therefore in another family. The mutation acceptance rate for HMG-17 since the mammal–bird divergence 300 million years before the present (my bp) is 0.97 PAM/100 my.

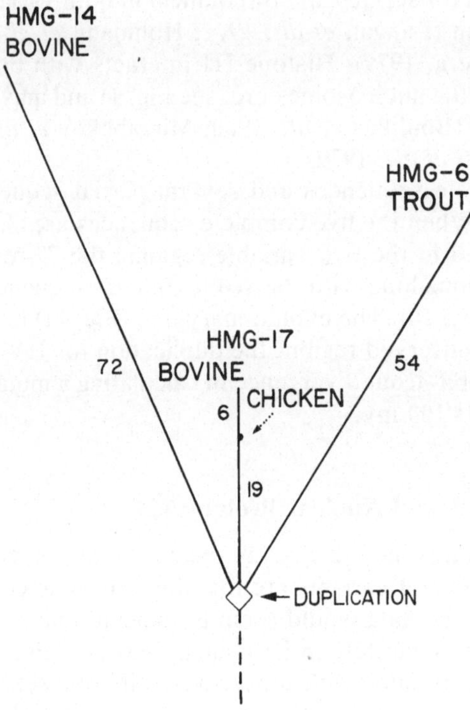

Figure 9. Evolutionary tree of nonhistone chromosomal proteins. The relative lengths of the longer branches are somewhat distorted by the number of gaps required to align the sequences. The trunk (dashed line) is arbitrarily positioned at the duplication.

4.5. Histones H1 and H5

All the histone families except H4 have several variants, either within the same cell type or associated with different cell types and developmental stages of the organism. For example, variants within the same tissue type have been reported for rabbit and calf (Rall and Cole, 1971) and within and among tissues for rat (Seyedin and Kistler, 1980). Histone H5 may be considered as an H1 variant found in a specialized cell, the avian erythrocyte (Yaguchi *et al.*, 1979; von Holt *et al.*, 1979). Histone H1 (with H5) has the longest and most variable sequence of the five types of histones. The protein has three domains: a middle, less basic, somewhat hydrophobic, globular, conserved region; and the amino and carboxyl basic and very variable regions, having some small repeating units and containing mainly Lys, Ala, Ser, and Pro residues (von Holt *et al.*, 1979). Secondary modifications of the protein (see Fig. 3) include acetylation, phosphorylation (of serines and threonines) in both variable regions, and ADP-ribosylation (Langan *et al.*, 1971; Hohmann *et al.*, 1976; Iwasa *et al.*, 1980; Isenberg, 1979). Histone H1 interacts with both histone H2A and the DNA in the nucleosome core (see Fig. 1) and may help to stabilize the nucleosome (Boulikas *et al.*, 1980; Mirzabekov *et al.*, 1980; Klug *et al.*, 1980; Thoma *et al.*, 1979).

Five complete sequences and several partial sequences of H1 and H5 are known. When the five complete sequences are aligned, numerous gaps are required in the two variable regions; the 72-residue conserved region (73 positions long with the sea urchin H1 sequence) has only two gap areas (see Fig. 10). The evolutionary tree (Fig. 11) is based on a PAM matrix for the conserved region; the duplication for H1–H5 is indicated. We used the rabbit–trout divergence in calculating a mutation acceptance rate of 2.4 PAMs/100 my.

4.6. Histone H2A and Nuclear Protein A24

Mammalian histone H2A is 129 residues long; others are a few residues shorter. Several variants are found in the same cell type of higher vertebrates, but the data would seem to indicate that these minor duplications occurred separately in individual lines (see Fig. 12). Variants in sea urchins are correlated with activation of different sets of histone genes during various stages of the life cycle: embryos (blastula, early gastrula, late gastrula), adult, and sperm (Brandt *et al.*, 1979; Hieter *et al.*, 1979; von Holt *et al.*, 1979). Variants have been found for histones H1, H2A, and H2B. There is more difference between variants of a given sea urchin

```
                        4                 5                   6
                        2 3 4 5 6 7 8 9 0 1 2 3 4 5 6 7 8 9 0 1 2 3 4 5 6
HISTONE H1
  1 Rabbit, 38-108      G P P V S E L I T K A V A A S K E R N G L S L A A
  2 Trout, 28-98        G P A V G E L A G K A V A A S K E R S G V S L A A
  3 Sea urchin, 42-113  H P P V L E M V Q A A I T A M K E R K G S S A A K
HISTONE H5
  4 Goose, 26-96        H P T Y S E M I A A A I R A D K S R G G S S R Q S
  5 Chicken, 25-95      H P T Y S E M I A A A I R A E K S R G G S S R Q S

           Conserved    H P   V S E M I   K A I   A   K E R   G S S   A
                          ,       ,         ,     ,   ,   ,   ,   ,


                                7                   8                   9
                        7 8 9 0 1 2 3 4 5 6 7 8 9 0 1 2 3 4 5 6 7 8 9 0
HISTONE H1
  1 Rabbit, 38-108      L K K A L A A G G Y D V E K N - - N S R I K L G
  2 Trout, 28-98        L K K S L A A G G Y D V E K N - - N S R V K I A
  3 Sea urchin, 42-113  I K S Y M A A - N Y R V D M N V L A P H V R R A
HISTONE H5
  4 Goose, 26-96        I Q K Y V K S - H Y K V G Q H A - D L Q I K L A
  5 Chicken, 25-95      I Q K Y I K S - H Y K V G H N A - D L Q I K L S

           Conserved    I K K Y   A A     Y   V     N           I K L A
                                          ,   ,


                                          10                  11
                        1 2 3 4 5 6 7 8 9 0 1 2 3 4 5 6 7 8 9 0 1 2 3 4
HISTONE H1
  1 Rabbit, 38-108      L K S L V S K G T L V E T K G T G A S Q S F K L
  2 Trout, 28-98        V K S L V T K G T L V E T K G T G A S G S F K L
  3 Sea urchin, 42-113  L R N G V A S G A L K Q V T G T G A S Q R F R V
HISTONE H5
  4 Goose, 26-96        I R R L L T T G V L K Q T K G V G A S Q S F R L
  5 Chicken, 25-95      I R R L L A A G V L K Q T K G V G A S Q S F R L

           Conserved      R   L V     G   L K Q T K G T G A S G S F R L
                                      ,   ,         ,   , , , ,   ,
```

Figure 10. Alignment of the central conserved region of histones H1 and H5. This region begins at position 42 in our alignment and includes residues 38–108 from rabbit H1.3 (M. Hsiang, C.R. Largman, and R.D. Cole, unpublished results; cited by Cole, 1977), 28–98 from trout (MacLeod *et al.*, 1977), 42–113 from the sea urchin *Parechinus angulosus* gonad (W.N. Strickland *et al.*, 1976; W.N. Strickland *et al.*, 1978, cited by von Holt *et al.*, 1979), 26–96 from goose (*Anser anser*) H5 (Yaguchi *et al.*, 1979), and 25–95 from chicken H5 (Garel *et al.*, 1975; Sautière *et al.*, 1975; Briand *et al.*, 1980). Conserved residues are those appearing in at least three of the five sequences.

histone at different developmental stages in the same species than between variants from different species at the same stage (Brandt *et al.*, 1979). Histone H2A (residues 31–95) interacts strongly with H2B in the octamer and less strongly with the other core histones. Residues 106–129 interact with the carboxyl half of the conserved region of histone H1 (58–109)

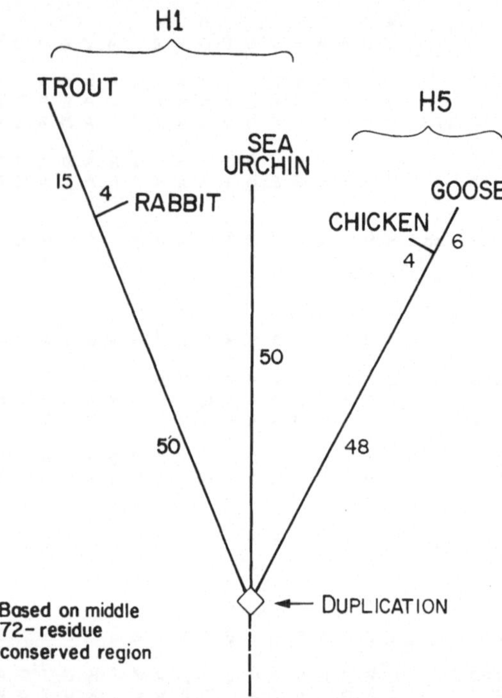

Figure 11. Evolutionary tree of histones H1 and H5, based on the alignment of Fig. 10. The trunk (dashed line) is arbitrarily positioned.

(McGhee and Felsenfeld, 1980). H2A may be phosphorylated and acetylated in the amino-terminal region (Isenberg, 1979; McGhee and Felsenfeld, 1980); Lys-119 may bind to ubiquitin in 5–10% of the nucleosomes (Goldknopf *et al.*, 1980) to form nuclear protein A24 (see Figs. 3 and 5). Bohm *et al.* (1980) have proposed that the highly conserved central region (residues 12–118) is less exposed in the nucleosome than are the terminal, more variable regions, which may then have different functional roles.

Sequence data available for two bovine variants (Yeoman *et al.*, 1972; Sautière *et al.*, 1974; Franklin and Zweidler, 1977), two mouse variants (Blankstein *et al.*, 1977), three rat variants (Laine *et al.*, 1976), two chicken variants (Laine *et al.*, 1978; Urban *et al.*, 1979), rainbow trout (Bailey and Dixon, 1973), and the sea urchins *Psammechinus miliaris* (Wouters

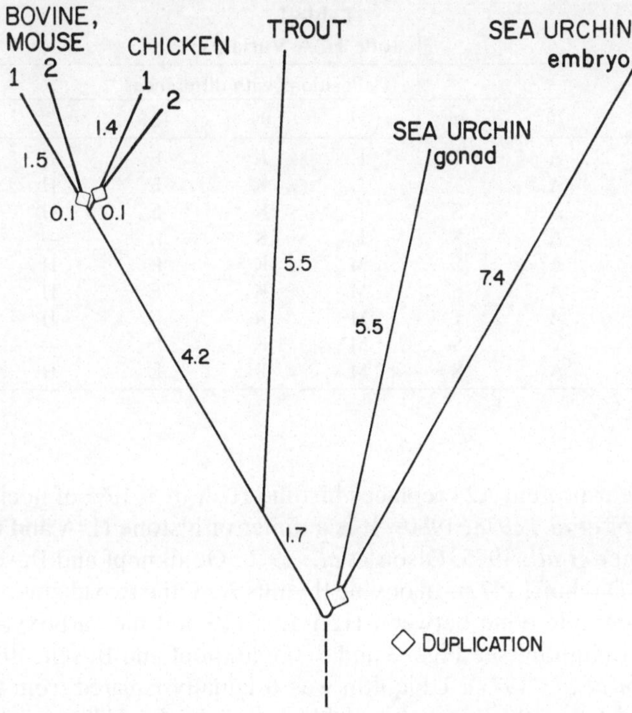

Figure 12. Evolutionary tree of histone H2A. The sequences from *Psammechinus* and *Strongylocentrotus* embryos are identical but very different from the sequence from *Psammechinus* gonad. The trunk (dashed line) is arbitrarily positioned.

et al., 1978; Busslinger *et al.*, 1980) and *Strongylocentrotus purpuratus* (Sures *et al.*, 1978) were used to construct an alignment of 129 positions. Only seven variable positions have been found so far in mammalian and avian sequences, as presented in Table I. The most variable region includes the last nine positions; otherwise, variable positions among all the sequences, including four of the nine positions with gaps, are scattered along the alignment. The tree of Fig. 12 was derived from this alignment. The mammal–bird and the mammal–fish divergence points are approximately 300 and 400 my bp, respectively. The mutation acceptance rate for the mammal–bird branch is 0.54 PAM/100 my.

Table I
Histone H2A Variants

	Positions with differences						
	10	16	51	99	121	124	128
Bovine 1	A	T	L	K	E	H	G
Mouse 1	A	T	L	K	E	H	G
Rat 1	A	S	L	K	E	H	G
Chicken 1	A	S	L	K	D	—	A
Bovine 2	A	S	M	K	E	H	G
Mouse 2	A	S	M	K	E	H	G
Rat 2	A	T	M	R	E	H	G
Chicken 2	V	S	M	K	E	—	S
Rat 3	A	S	M	R	E	H	G

Nuclear protein A24 replaces histone H2A in 5–10% of nucleosomes (Goldknopf *et al.*, 1978, 1980). It is a dimer of histone H2A and ubiquitin (Schlesinger *et al.*, 1975; Olson *et al.*, 1976; Goldknopf and Busch, 1980; Hunt and Dayhoff, 1977). In bovine thymus A24, the two chains are linked by an isopeptide bond between H2A Lys-119 and the carboxyl-terminal Gly-76 in ubiquitin (see Figs. 2 and 5) (Goldknopf and Busch, 1977, 1980; Goldknopf *et al.*, 1977). Ubiquitin was originally isolated from the cytoplasm of bovine thymus cells (Schlesinger *et al.*, 1975) and has been identified in the cytoplasm of all tissues and species (even prokaryotes) examined (Goldstein *et al.*, 1975), just as ubiquitin-containing protein A24 is found in all chromatin examined.

Ubiquitin has also been called thymosin polypeptide β_1 (Low and Goldstein, 1979), protein S in trout testis (Watson *et al.*, 1978), and nonhistone chromosomal protein HMG-20 (Walker *et al.*, 1978b). Complete or partial sequences are now known from several organisms or tissues: bovine thymus (Schlesinger *et al.*, 1975; Olson *et al.*, 1976; Walker *et al.*, 1978b; Low and Goldstein, 1979), bovine brain (Seidah *et al.*, 1978), bovine parathyroid gland (Hamilton and Rouse, 1980), human thymus (Schlesinger and Goldstein, 1975), rat liver (Goldknopf *et al.*, 1976), and trout testis (Watson *et al.*, 1978). We have not yet calculated a mutation acceptance rate for ubiquitin. On the one hand, the human sequence appears to be identical with that from bovine thymus, and the trout sequence may also be identical (except that half of the molecules end with Arg-74 and the other half have only one additional carboxyl-terminal Gly). On the other hand, the bovine parathyroid ubiquitin appears to have four differences in the first 50 residues compared with the bovine thymus sequence.

At any rate it has a highly conserved structure, which implies an important function in the chromatin and/or in the cytoplasm. In protein A24, ubiquitin is neither phosphorylated nor acetylated (Goldknopf *et al.*, 1979), but it may be ADP-ribosylated (Okayama and Hayaishi, 1978). Ubiquitin is synthesized and incorporated into A24 throughout interphase (Goldknopf *et al.*, 1980), but A24 disappears and free ubiquitin appears during mitosis (Matsui *et al.*, 1979). This cleavage of protein A24 (Andersen *et al.*, 1980) has also been correlated with transcriptionally active chromatin (Goldknopf *et al.*, 1978). Matsui *et al.* (1979) have suggested that removal of ubiquitin from A24 would free the positively charged carboxyl end of H2A to interact with phosphorylated H1 and H3 and promote chromatin condensation; that is, the presence of A24 would inhibit condensation.

Other questions are how and when ubiquitin moves into the cytoplasm and what function, if any, it has there. Free ubiquitin could perhaps diffuse into the cytoplasm during mitosis when a certain amount is available and when an intact nuclear membrane is not present, or it may be actively transported there. Although ubiquitin was earlier reported to stimulate T lymphocytes and cyclic AMP, it was later found to be inactive in both respects (Low and Goldstein, 1979) and also was not found to have any structural similarity to thymosin α_1. Incidentally, we have recently noticed a suggestive resemblance between the sequences of thymopoietin II and ubiquitin (see Fig. 13), although the two are out of phase, as is found for the legume lectins (Hemperly *et al.*, 1979; Cunningham *et al.*, 1979); it is possible that the two proteins had a very distant common ancestor, or this could be a case of convergent evolution. Recently, however, it was discovered (Ciechanover *et al.*, 1980; Wilkinson *et al.*, 1980) that the ATP-dependent proteolysis factor I is ubiquitin. This catalytic factor forms a covalent conjugate with intracellular proteins to be degraded, presumably in the same way that protein A24 is formed, and may serve as a recognition site for proteases. As this proteolytic system appears to be as ubiquitous as ubiquitin (Wilkinson *et al.*, 1980), the highly conserved sequence of the latter may be the result of having both a cytoplasmic and a nuclear function.

4.7. Histone H2B

In histone H2B, as in H2A, intraspecies variants are found. The protein has an amino-terminal variable region and a longer carboxyl-terminal conserved region (Isenberg, 1979; von Holt *et al.*, 1979). The length of the protein is variable due to the variability of the amino-terminal domain, which possesses several to many small repeating units and has

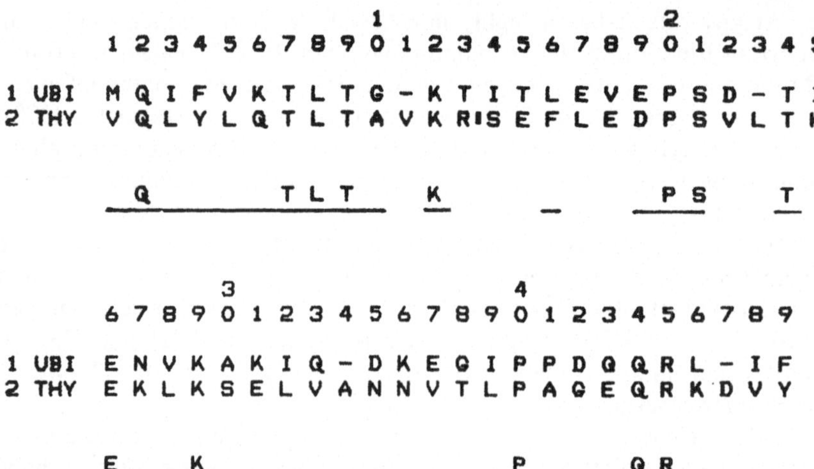

Figure 13. Alignment of ubiquitin and thymopoietin II from bovine thymus. Residues 37–49 followed by residues 1–36 of thymopoietin II (THY) are aligned with residues 1–45 of ubiquitin (UBI). The mark between positions 13 and 14 separates the last and first residues of thymopoietin II. Underlining indicates similar residues; there is 67% chemical similarity between the two sequences aligned in this way.

a high proportion of Lys, Ala, Pro, and Ser (von Holt *et al.*, 1979). This structural feature is particularly evident in the three sea urchin sperm H2B variants (M. Strickland *et al.*, 1977, 1978; W. N. Strickland *et al.*, 1977; W. N. Strickland *et al.*, 1978, cited by von Holt *et al.*, 1979; Brandt *et al.*, 1979; Hieter *et al.*, 1979). H2B is also the core histone with the greatest amount of secondary modification, including ADP-ribosylation (Ogata *et al.*, 1980) as well as phosphorylation and acetylation (Isenberg, 1979; McGhee and Felsenfeld, 1980) (see Figs. 3 and 4). It interacts strongly in the octamer with H2A (31–95) and H4 (38–102) (McGhee and Felsenfeld, 1980; Isenberg, 1979). This region of interaction (residues 37–114 in bovine and human) corresponds to the conserved portion of the chain.

Table II presents the variable positions in the known mammalian and avian variants: three human (Ohe *et al.*, 1979), two bovine (Iwai *et al.*, 1972; Franklin and Zweidler, 1977), three mouse (Franklin and Zweidler, 1977), and two chicken (van Helden *et al.*, 1978; Urban *et al.*, 1979); the trout is added for comparison. The alignment included bovine variant 1, chicken variant 1, the brown trout *Salmo trutta* (Kootstra and Bailey, 1978) the fruit fly *Drosophila melanogaster* (Elgin *et al.*, 1979), the limpet *Patella granatina* (van Helden *et al.*, 1979), embryonic sea urchins *Stron-*

Table II
Histone H2B Variants

	Positions with differences											
	21	25	26	32	38	39	41	60	75	76	77	124
Human 1	A	D	G	S	S	V	V	G	G	E	A	S
Bovine 1	A	D	G	S	S	V	V	G	G	E	A	S
Mouse 1	A	D	G	S	S	V	V	G	G	E	A	S
Chicken 1	T	G	D	S	S	I	V	G	G	E	A	S
Human 2	A	D	G	S	S	I	V	G	G	E	A	S
Bovine 2	A	D	G	S	S	V	V	G	G	Q	A	S
Mouse 2	A	D	G	S	S	V	V	G	G	Q	A	S
Chicken 2	T	G	D	A	S	I	V	S	G	E	A	S
Human 3	A	D	G	S	S	V	V	G	G	E	A	A
Mouse 3	A	D	G	S	S	V	V	G	S	E	A	S
Trout	T	G	G	S	A	I	I	I	G	E	S	S

gylocentrotus purpuratus (Sures *et al.*, 1978) and *Psammechinus miliaris* (Busslinger *et al.*, 1980), and the three variants of the sea urchin *Parechinus angulosus* sperm. The evolutionary tree of Fig. 14 was derived from this alignment, using the 98-residue conserved domain. The mammal–bird, mammal–fish, and mammal–insect divergences are approximately 300, 400, and 700 my bp, respectively. We used these groups to calculate the mutation acceptance rate of 0.39 PAM/100 my.

4.8. Histone H3

Histone H3 is one of the two most highly conserved proteins known. Secondary modifications in the amino-terminal region (see Fig. 4) include phosphorylation, methylation, and acetylation (Isenberg, 1979; McGhee and Felsenfeld, 1980). In the nucleosome, H3 (residues 40–120) interacts strongly with H4 (residues 38–102) (see Fig. 3) (McGhee and Felsenfeld, 1980; Klug *et al.*, 1980; Isenberg, 1979) and also with itself to form H3 dimers in which the single cysteines (at position 110) are close enough to form an interchain disulfide bond (Camerini-Otero and Felsenfeld, 1977). All sequences so far studied are 135 residues long and may be aligned without gaps.

Intraspecies variants are present in bovine (DeLange *et al.*, 1973; Patthy and Smith, 1975; Franklin and Zweidler, 1977) and chicken H3 (Brandt and von Holt, 1974; Urban *et al.*, 1979), as well as heterogeneity at position 127 in bovine (Franklin and Zweidler, 1977). Table III presents the variable positions in H3 for nine species: three variants of bovine,

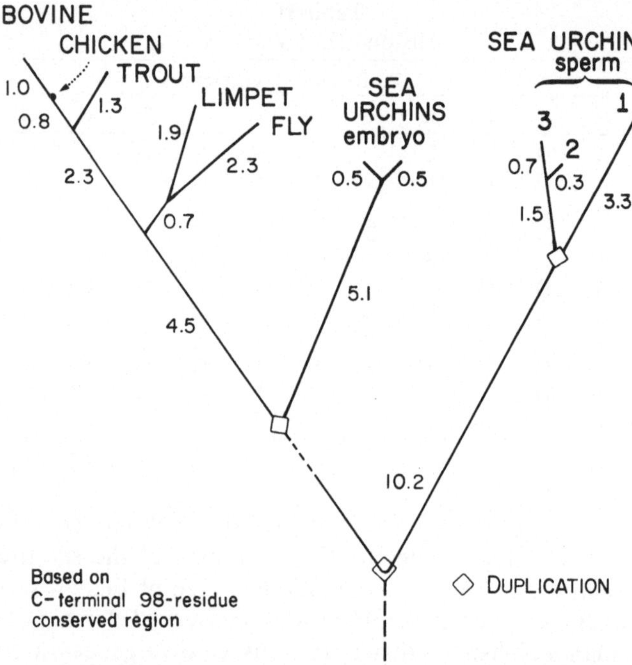

Figure 14. Evolutionary tree of histone H2B, derived from the alignment of the carboxyl-terminal conserved region. The dashed end of the branch for sea urchin sperm histones indicates uncertainty in its attachment point to the branch leading to sea urchin embryonic histones. The trunk (dashed line) is arbitrarily positioned.

two variants of chicken, the smallmouth buffalo fish *Ictiobus bubalus* (Hooper *et al.*, 1973), the shark *Poroderma africanus* (Brandt *et al.*, 1974b), the sea urchins *Psammechinus miliaris* (Schaffner *et al.*, 1978) and *Strongylocentrotus purpuratus* (Sures *et al.*, 1978), the partial sequence of the limpet *Patella granatina* (Brandt *et al.*, 1974a), the garden pea *Pisum sativum* (Patthy *et al.*, 1973), and the partial sequence of the cycad *Encephalartos caffer* (Brandt *et al.*, 1974a). An almost complete sequence from the fruit fly *Drosophila melanogaster* (Goldberg, 1979) is identical with the bovine H3(2) variant. It can be seen that, aside from the bovine heterogeneity, there are only four variable positions among the animals and these are clustered in a 16-residue segment from positions 81 to 96; additionally, two of these four differences are intraspecies var-

Table III
Histone H3

	Positions with differences						
	41	53	81	89	90	96	127
Vertebrates							
Bovine 3	Y	R	D	I	G	S	A/T
Chicken 3	Y	R	D	I	G	S	A
Bovine 1	Y	R	D	V	M	C	A/T
Bovine 2	Y	R	D	V	M	S	A/T
Chicken 2	Y	R	D	V	M	S	A
Buffalo fish	Y	R	D	V	M	S	A
Shark	Y	R	D	V	M	S	A
Invertebrates							
Sea urchins[a]	Y	R	E	V	M	S	A
Limpet (fr.)	Y/						
Plants							
Pea	F	K	D	V	S	A	A
Cycad (fr.)	F/						

[a] *Psammechinus miliaris* and *Strongylocentrotus purpuratus*.

iants produced by two H3 genes and present in both mammal and bird lines. Pea H3 has two other variable sites when compared with animal H3, although it has only four differences from most of the vertebrate sequences.

The evolutionary tree of Fig. 15 shows both interspecies and intraspecies divergences. All three kinds of H3 sequence found so far among vertebrates can be represented by the three bovine variants. The plant–animal divergence (arbitrarily indicated by the dashed line) would have been about one billion years ago. The mutation acceptance rate for histone H3, including chicken, is 0.15 PAM/100 my.

4.9. Histone H4

Histone H4 is the most highly conserved protein known. No intraspecies variants have been found. Secondary modifications in the amino-terminal region (see Figs. 3 and 4) include phosphorylation, acetylation, and methylation (Isenberg, 1979; McGhee and Felsenfeld, 1980). Histone H4 (residues 38–102) interacts primarily with histone H3 (40–120), but also at its carboxyl end (85–102) with histone H2B (65–125) (Isenberg, 1979; McGhee and Felsenfeld, 1980; Klug *et al.*, 1980). The H3–H4 tet-

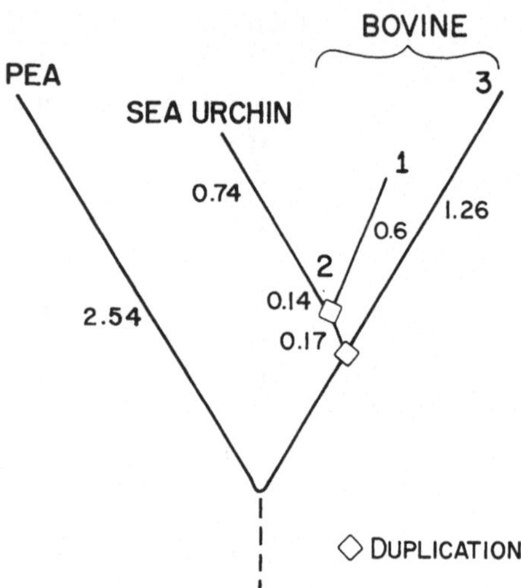

Figure 15. Evolutionary tree of histone H3. The plant–animal divergence is indicated by the arbitrarily positioned trunk (dashed line).

ramer alone is able to organize the DNA of the nucleosome core particle into the central turn (Klug *et al.*, 1980; Mirzabekov *et al.*, 1980; McGhee and Felsenfeld, 1980) and some evidence suggests that the carboxyl-terminal regions of these histones are especially important (Palau *et al.*, 1980; Eshaghpour *et al.*, 1980).

Only a few histone H4 sequences are available. These include complete sequences from bovine (Ogawa *et al.*, 1969), pig (Sautière *et al.*, 1971b), rat (Sautière *et al.*, 1971a), chicken (Urban *et al.*, 1979), the sea urchin *Psammechinus miliaris* (Wouters-Tyrou *et al.*, 1976; Busslinger *et al.*, 1980), and the garden pea *Pisum sativum* (DeLange *et al.*, 1969) and partial sequences from the fruit fly *Drosophila melanogaster* (Goldberg, 1979) and from the macronucleus of the ciliate protozoan *Tetrahymena thermophila* (Glover and Gorovsky, 1979). The four vertebrate sequences are identical; sea urchin and pea differ at one and two positions, respectively (see Tables IV and V). The amino-terminal 72 residues from fruit fly differ from the vertebrate sequences in having Thr instead of Ser at position 1, so it is at least as distant as the sea urchin H4. The evolutionary tree of H4 (Fig. 16) indicates that the sequences from higher eukaryotes have diverged very little during the last one billion years, since the plant–animal divergence. The mutation acceptance rate for the higher eukaryotes, calculated from the number of changes on all branches (from the

Table IV
Histone H4: Positions with Differences

	60	73	77
Bovine	V	T	K
Sea urchin	V	C	K
Pea	I	T	R

PAM matrix) divided by the total time elapsed on all branches, is 0.10 PAM/100 my and is the lowest for any protein known. However, this rate is derived from only six sequences and it may change if more H4 sequences from a wide variety of eukaryotes are determined.

The partial sequence of macronuclear H4 from *Tetrahymena* includes two fragments that correspond to positions 1–53 and 84–96 of the other sequences; however, the *Tetrahymena* sequence introduces two gaps, as it lacks a residue corresponding to Arg-3 and has a Ser between His-18 and Arg-19. Including these two positions, there are 15 differences (22% difference) at least from the higher eukaryote sequences, making a very long branch (at least 25 PAMs) on the tree. Presumably the rate of H4 evolution has been faster in some lower groups. The genetic state of *Tetrahymena* is functionally different from that of higher eukaryotes. In the vegetative stage the ciliate has two nuclei, a micronucleus that will transmit genetic information in sexual reproduction and then generate a new macronucleus, and a macronucleus that divides amitotically, is transcriptionally active, and degenerates during sexual reproduction (Gorovsky *et al.*, 1978). This macronuclear H4 may have accumulated variation during asexual multiplication. The micronuclear histone H4 would probably provide more useful information on evolutionary relationships.

Table V
Histone H4: Matrix of Differences[a]

		1	2	3	4
1	Bovine		1	2	15
2	Sea urchin	**1**		3	15
3	Pea	**2**	**3**		15
4	*Tetrahymena* (53 and 13 residues)	**22**	**22**	**22**	

[a]Number differences are represented in lightface type and percent differences are represented in boldface type.

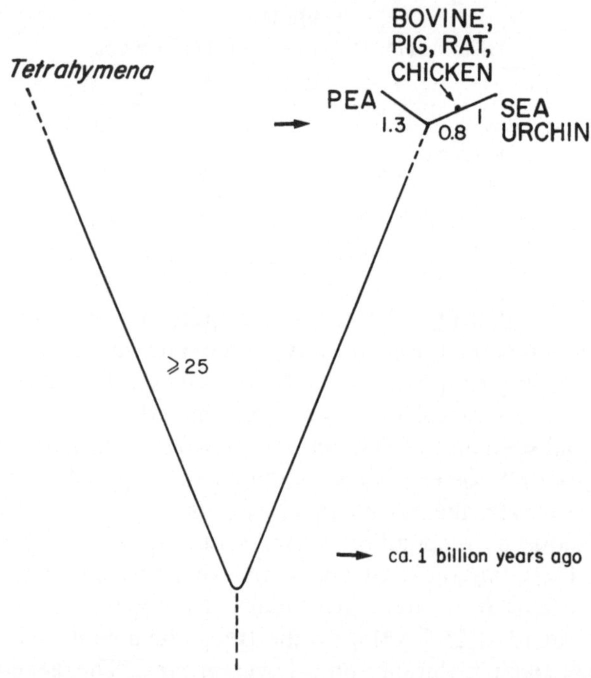

Figure 16. Evolutionary tree of histone H4. The dashed ends of the *Tetrahymena* branch indicate uncertainty about the point where it joins the tree and uncertainty about the amount of difference, as the sequence is incomplete. The trunk (dashed line) is arbitrarily positioned.

4.10. Mutation Acceptance Rates

Table VI summarizes the rates for six of the chromosomal protein families. The four core histones have the most highly conserved sequences known. Nonhistone chromosomal protein HMG-17 does not appear to be as conserved as the core histones; its rate lies with those of glucagon and glutamate dehydrogenase (Dayhoff, 1979). However, the most conserved region of histone H1 has a definitely higher rate than the core histones, even higher than cytochrome c (Dayhoff, 1979). More sequence data from distantly related eukaryotes are needed, especially for histones H4 and H3 and for ubiquitin, to confirm the degree of sequence conservation and to reveal any changes in rates that may have occurred during the evolution of these proteins. The time estimates for the divergence of lines are taken from Dayhoff (1979).

Table VI
Rates of Mutation Acceptance

Divergence[a]	Protein	Rate PAMs/100 my
A,B,D	Histone H4	0.10
A,B,D	Histone H3	0.15
B,C,D	Histone H2B[b]	0.39
D	Histone H2A	0.54
C	Histone H1[b]	2.40
D	NHC protein HMG-17	0.97

[a]A: Animal–plant, 1000 my bp. B: Mammal–arthropod/mollusc, 700 my bp. C: Mammal– bony fish, 400 my bp. D: Mammal–bird/reptile, 300 my bp.
[b]Based on conserved region only.

4.11. The Core Histone Superfamily

4.11.1. Alignments

Lastly we will consider possible relationships among the sequences of the four core histone families. Temussi (1975) presented some paired alignments of all five histones; he found that all were related to one another according to six different criteria for homologies that he selected. More recently Reeck *et al.* (1978) derived an alignment of the carboxyl regions of the four core histones and the amino-terminal region of H2A by pairwise computer analyses of the sequences to detect and evaluate distant relationships. They calculated that the probability of this set of similarities being due to chance is $<10^{-5}$. They found no similarity of H1 to the others.

Previously we found that the four core histones appeared to be related when compared using the unitary matrix in the computer analyses but not so when using the mutation data matrix (Dayhoff, 1976b). We had also found a relationship between the amino-terminal halves of H2A and H4 (Dayhoff, 1973). Since then we have revised the mutation data matrix to include new data. Now, when we compared carboxyl regions of the four histones with program ALIGN, the alignment scores (AS) were suggestively positive, especially an AS > 4.00 for H3 and H4. From these comparisons we derived our own, slightly different [from that of Reeck *et al.* (1978)] alignment of the carboxyl regions of the four bovine histones (see Fig. 17). We also compared again the amino-terminal halves of bovine H2A and H4 and obtained an AS > 4.00; Fig. 18 is an alignment derived from this comparison.

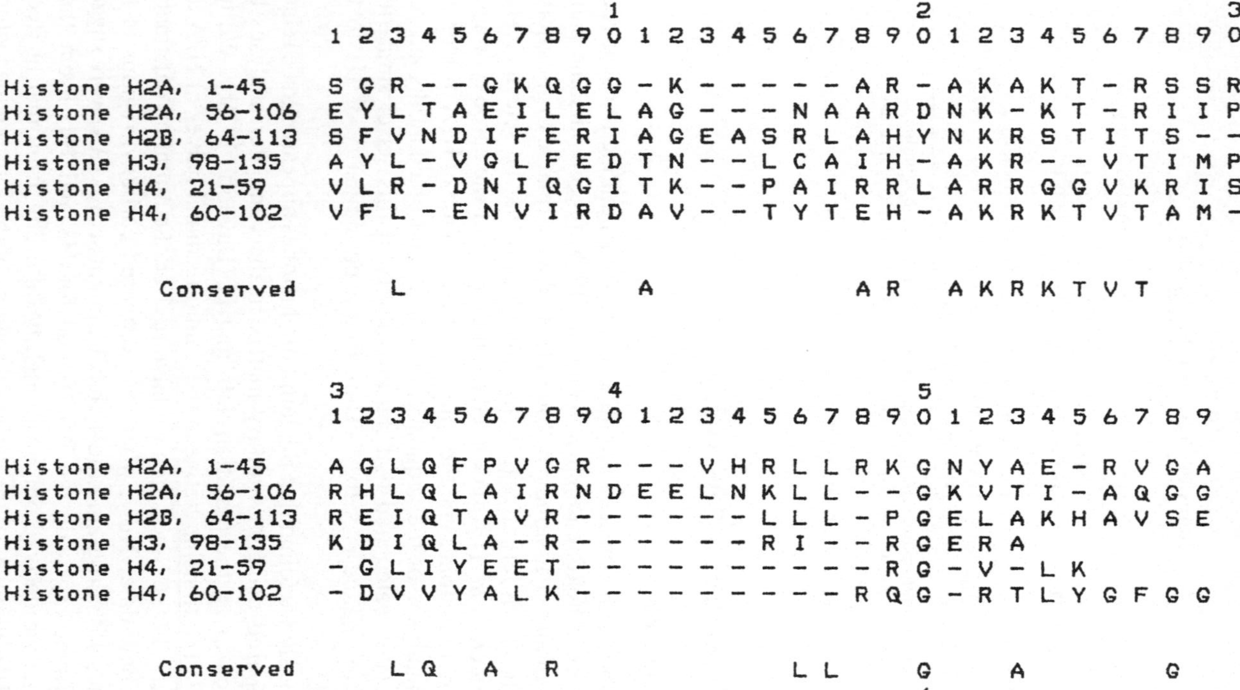

```
                                        1                   2                   3
                     1 2 3 4 5 6 7 8 9 0 1 2 3 4 5 6 7 8 9 0 1 2 3 4 5 6 7 8 9 0

1 Histone H2A,  1-45    S G R - - G K Q G G - K - - - - - A R - A K A K T - R S S R
2 Histone H2A,  56-106  E Y L T A E I L E L A G - - - N A A R D N K - K T - R I I P
3 Histone H2B,  64-113  S F V N D I F E R I A G E A S R L A H Y N K R S T I T S - -
4 Histone H3,   98-135  A Y L - V G L F E D T N - - L C A I H - A K R - - V T I M P
5 Histone H4,   21-59   V L R - D N I Q G I T K - - P A I R R L A R R G G V K R I S
6 Histone H4,   60-102  V F L - E N V I R D A V - - T Y T E H - A K R K T V T A M -

            Conserved       L                 A                   A R   A K R K T V T


                        3                   4                   5
                        1 2 3 4 5 6 7 8 9 0 1 2 3 4 5 6 7 8 9 0 1 2 3 4 5 6 7 8 9

1 Histone H2A,  1-45    A G L Q F P V G R - - - V H R L L R K G N Y A E - R V G A
2 Histone H2A,  56-106  R H L Q L A I R N D E E L N K L L - - G K V T I - A Q G G
3 Histone H2B,  64-113  R E I Q T A V R - - - - - - L L L - P G E L A K H A V S E
4 Histone H3,   98-135  K D I Q L A - R - - - - - - R I - - R G E R A
5 Histone H4,   21-59   - G L I Y E E T - - - - - - - - - - R G - V - L K
6 Histone H4,   60-102  - D V V Y A L K - - - - - - - - - R Q G - R T L Y G F G G

            Conserved       L Q   A   R             L L       G     A         G
                                                              ′
```

Figure 17. Alignment of the bovine nucleosome core histones. Residue numbers are indicated for each sequence segment, the four carboxyl regions and the two amino regions (H2A and H4). Conserved residues are those appearing in 50% or more of the sequences; at position 19 both Arg and His are present in 50% of the sequences.

The nucleotide sequences of the core histones from the sea urchin *Psammechinus miliaris* are now available (Busslinger *et al.*, 1980), so we could compare both the nucleotide and the protein sequences for these regions of H2A (1–58) and H4 (1–62) with program ALIGN. For the nucleotide comparison: (1) with no constraint to compare by codons, counting matches as 1, and with a gap penalty of 2, the AS is 3.45; and (2) with nucleotide comparisons by codons and with a gap penalty of 1, the AS is 3.48. Both scores are lower than that for the protein sequence comparison. This confirms our experience that distant relationships may be more readily detected using protein sequences and the mutation data matrix.

4.11.2. Duplications

In their analysis of histone evolution, Reeck *et al.* (1978) also looked for possible internal homologies that could be evidence of earlier duplication(s) of a smaller gene. The only statistically likely candidate was H2A, when residues 1–60 were matched against 61–121. We get an AS of about 4.00 for this comparison, and as 3.50–4.00 is usually the borderline area, we would consider the duplication as quite likely, but not definitely proven, in agreement with Reeck *et al.* We next compared the two halves of H4 with ALIGN, on the assumption that the most conserved histone sequence might retain evidence of a duplication. The AS was only 1.76, but in the alignment of residues 21–59 and 60–98 there is 67% similarity (ten identities plus 18 single-base exchanges).

Figure 17 and Table VII give the alignment and the matrix of differences for the bovine core histones, including the four carboxyl regions and the amino regions of H2A (1–45) and H4 (21–59). The percent differences between sequences (from 73% to 92%) are all at the superfamily level. However, these percent differences and the various alignment scores must be evaluated in relation to the functional similarities among the core histones, as established by the accumulated evidence from many laboratories, including but not limited to those cited in this paper. The four are able to interact with one another in a precise fashion and with the DNA helix to form the structure of the nucleosome, the basic functional unit of chromatin.

4.11.3. The Evolutionary Tree

The tree (Fig. 19) derived from the core histone alignment is the shortest all-positive topology and shows the expected duplications, which

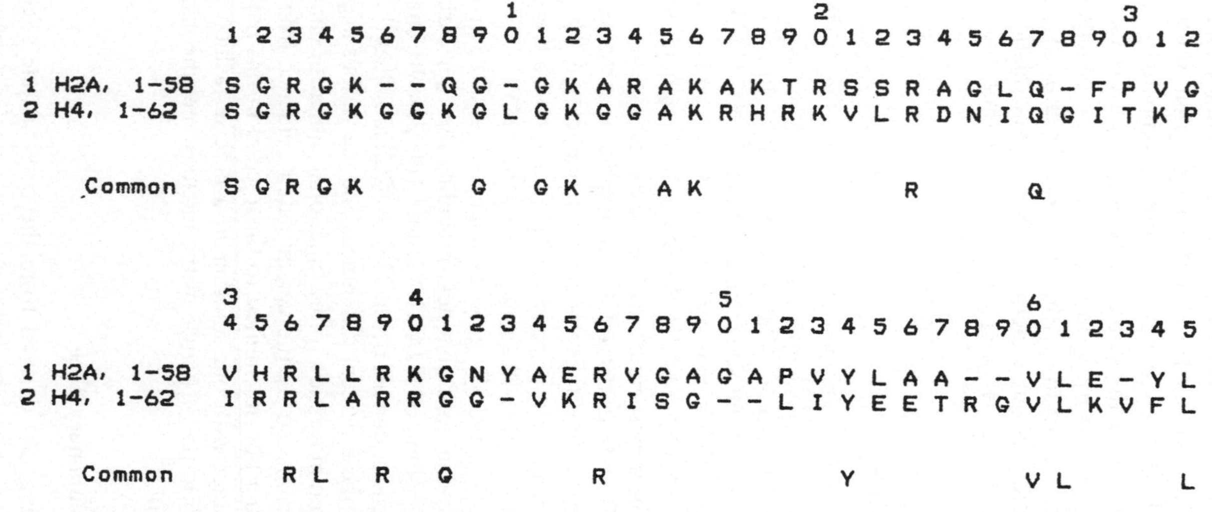

1 2 3
1 2 3 4 5 6 7 8 9 0 1 2 3 4 5 6 7 8 9 0 1 2 3 4 5 6 7 8 9 0 1 2 3
1 H2A, 1-58 S G R G K - - Q G - G K A R A K A K T R S S R A G L Q - F P V G R
2 H4, 1-62 S G R G K G G K G L G K G G A K R H R K V L R D N I Q G I T K P A
Common S G R G K G G K A K R Q
3 4 5 6
4 5 6 7 8 9 0 1 2 3 4 5 6 7 8 9 0 1 2 3 4 5 6 7 8 9 0 1 2 3 4 5
1 H2A, 1-58 V H R L L R K G N Y A E R V G A G A P V Y L A A - - V L E - Y L
2 H4, 1-62 I R R L A R R G G - V K R I S G - - L I Y E E T R G V L K V F L
Common R L R G R Y V L L

Figure 18. Alignment of the amino-terminal halves of bovine histones H2A and H4. Residue numbers are indicated.

Table VII
Bovine Histones of the Nucleosome Core: Matrix of Differences[a]

		1	2	3	4	5	6
1	Histone H2A, 1–45		42	44	44	40	45
2	Histone H2A, 56–106	**78**		44	43	46	47
3	Histone H2B, 64–113	**79**	**76**		41	44	42
4	Histone H3, 98–135	**86**	**78**	**79**		38	35
5	Histone H4, 21–59	**82**	**88**	**92**	**86**		33
6	Histone H4, 60–102	**87**	**82**	**81**	**73**	**80**	

[a]Number differences are represented in lightface type and percent differences are repre- ented in boldface type.

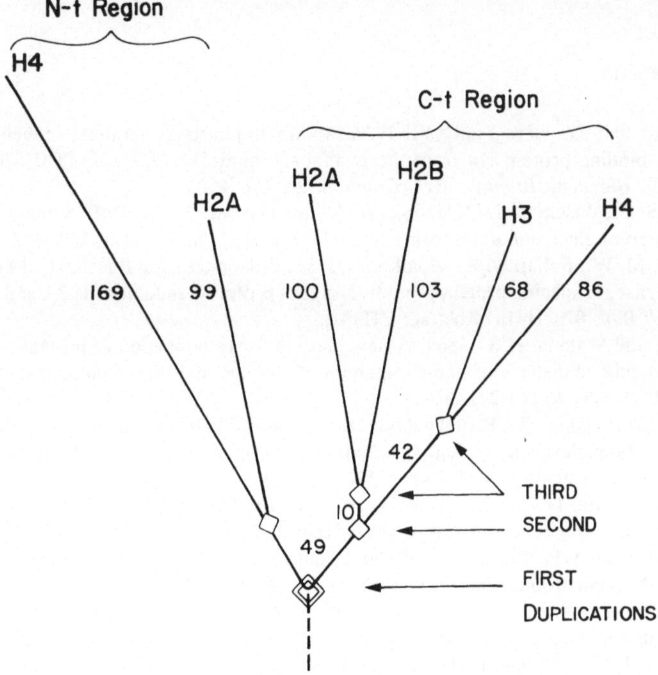

Figure 19. Evolutionary tree of the bovine nucleosome core histones, indicating possible duplications during their very early evolution. The trunk (dashed line) is arbitrarily positioned at the proposed point of the first duplication.

probably occurred more than 1.5 billion years ago (before the eukaryote host–mitochondrial symbiosis). It appears quite possible that, as Reeck *et al.* (1978) proposed, an ancestral histone gene about half the size of H4 first doubled its length, after which two more duplications produced ancestors of the four core histones. More doublings of some or all of these have occurred since to give the various intraspecies variants. Thus it appears that the core histone families may belong in one superfamily of proteins, more than 50% different but sharing common ancestry, similarity of structure, and fundamentally important functions.

ACKNOWLEDGMENTS. We thank Margaret C. Blomquist for expert editorial assistance in the preparation of the manuscript and for technical assistance, Karen C. Lawson for assistance with the illustrations, and K. Lynette Wilz for typing the manuscript. We are grateful to Drs. Robert J. DeLange and Dennis G. Searcy for communicating their data to us before publication. This work was supported by NIH grant GM-08710 from the Institute of General Medical Sciences.

References

Aitken, A., and Rouvière-Yaniv, J., 1979, Amino and carboxy terminal sequences of the DNA-binding protein Hu from the cyanobacterium *Synechocystis* PCC 6701 (ATCC 27170), *Biochem. Biophys. Res. Commun.* **91**:461–467.

Albright, S. C., Wiseman, J. M., Lange, R. A., and Garrard, W. T., 1980, Subunit structures of different electrophoretic forms of nucleosomes, *J. Biol. Chem.* **255**:3673–3684.

Andersen, M. W., Ballal, N. R., Goldknopf, I. L., Wilson, G., and Busch, H., 1980, Protein A24 lyase: A specific protease that hydrolyzes protein A24 to histone 2A and ubiquitin, *J. Cell Biol.* **87**:CH41a (Abstract CH307).

Ando, T., and Watanabe, S., 1969, A new method for fractionation of protamines and the amino acid sequences of one component of salmine and three components of iridine, *Int. J. Protein Res.* **1**:221–224.

Arens, M., Yamashita, T., Padmanabhan, R., Tsuruo, T., and Green, M., 1977, Adenovirus deoxyribonucleic acid replication: Characterization of the enzyme activities of a soluble replication system, *J. Biol. Chem.* **252**:7947–7954.

Bailey, G. S., and Dixon, G. H., 1973, Histone IIb1 from rainbow trout: Comparison in amino acid sequence with calf thymus IIB1, *J. Biol. Chem.* **248**:5463–5472.

Bellard, M., Oudet, P., Germond, J.-E., and Chambon, P., 1976, Subunit structure of simian-virus-40 minichromosome, *Eur. J. Biochem.* **70**:543–553.

Black, J. A., and Dixon, G. H., 1967, Evolution of protamine: A further example of partial gene duplication, *Nature* **216**:152–154.

Blankstein, L. A., Stollar, B. D., Franklin, S. G., Zweidler, A., and Levy, S. B., 1977, Biochemical and immunological characterization of two distinct variants of histone H2A in Friend leukemia, *Biochemistry* **16**:4557–4562.

Boffa, L. C., Sterner, R., Vidali, G., and Allfrey, V. G., 1979, Post-synthetic modifications of nuclear proteins: High mobility group proteins are methylated, *Biochem. Biophys. Res. Commun.* **89**:1322–1327.

Bohm, L., Crane-Robinson, C., and Sautière, P., 1980, Proteolytic digestion studies of chromatin core-histone structure: Identification of a limit peptide of histone H2A, *Eur. J. Biochem.* **106**:525–530.

Boulikas, T., Wiseman, J. M., and Garrard, W. T., 1980, Points of contact between histone H1 and the histone octamer, *Proc. Natl. Acad. Sci. USA* **77**:127–131.

Brandt, W. F., and von Holt, C., 1974, The determination of the primary structure of histone F3 from chicken erythrocytes by automatic Edman degradation. 2. Sequence analysis of histone F3, *Eur. J. Biochem.* **46**:419–429.

Brandt, W. F., Strickland, W. N., Morgan, M., and von Holt, C., 1974a, Comparison of the N-terminal amino acid sequences of histone F3 from a mammal, a bird, a shark, an echinoderm, a mollusc and a plant, *FEBS Lett.* **40**:167–172.

Brandt, W. F., Strickland, W. N., and von Holt, C., 1974b, The primary structure of histone F3 from shark erythrocytes, *FEBS Lett.* **40**:349–352.

Brandt, W. F., Strickland, W. N., Strickland, M., Carlisle, L., Woods, D., and von Holt, C., 1979, A histone programme during the life cycle of the sea urchin, *Eur. J. Biochem.* **94**:1–10.

Bretzel, G., 1972, Über Thynnin, das Protamin des Thunfisches: Die vollständige Aminosäuressequenz von Thynnin Y2, *Hoppe-Seyler's Z. Physiol. Chem.* **353**:933–943.

Bretzel, G., 1973a, Über Thynnin, das Protamin des Thunfisches: Die Aminosäuresequenz von Thynnin Z1, *Hoppe-Seyler's Z. Physiol. Chem.* **354**:312–320.

Bretzel, G., 1973b, Über Thynnin, das Protamin des Thunfisches: Die Aminosäuresequenz von Thynnin Z2, *Hoppe-Seyler's Z. Physiol. Chem.* **354**:543–549.

Briand, G., Kmiecik, D., Sautière, P., Wouters, D., Borie-Loy, O., Biserte, G., Mazen, A., and Champagne, M., 1980, Chicken erythrocyte histone H5. IV. Sequence of the carboxy-terminal half of the molecule (96 residues) and complete sequence, *FEBS Lett.* **112**:147–151.

Busslinger, M., Portmann, R., Irminger, J. C., and Birnstiel, M. L., 1980, Ubiquitous and gene-specific regulatory 5′ sequences in a sea urchin histone DNA clone coding for histone protein variants, *Nucl. Acids Res.* **8**:957–977.

Camerini-Otero, R. D., and Felsenfeld, G., 1977, Histone H3 disulfide dimers and nucleosome structure, *Proc. Natl. Acad. Sci. USA* **74**:5519–5523.

Ciechanover, A., Elias, S., Heller, H., Ferber, S., and Hershko, A., 1980, Characterization of the heat-stable polypeptide of the ATP-dependent proteolytic system from reticulocytes, *J. Biol. Chem.* **255**:7525–7528.

Coelingh, J. P., and Rozijn, T. H., 1975, Comparative studies on the basic nuclear proteins of mammalian and other spermatozoa, *Biol. J. Linn. Soc.* **7**(Suppl. 1):245–256.

Cole, R. D., 1977, Special features of the structures of H1 histones, in: *The Molecular Biology of the Mammalian Genetic Apparatus*, Volume 1 (P. O. P. Ts'o, ed.), North-Holland, Amsterdam, pp. 93–104.

Corden, J., Engelking, H. M., and Pearson, G. D., 1976, Chromatin-like organization of the adenovirus chromosome, *Proc. Natl. Acad. Sci. USA* **73**:401–404.

Cunningham, B. A., Hemperly, J. J., Hopp, T. P., and Edelman, G. M., 1979, Favin versus concanavalin A: Circularly permuted amino acid sequences, *Proc. Natl. Acad. Sci. USA* **76**:3218–3222.

D'Anna, J. A., Jr., and Isenberg, I., 1974, Interactions of histone LAK (f2a2) with histones KAS (f2b) and GRK (f2a1), *Biochemistry* **13**:2098–2104.

Dayhoff, M. O., 1973, Survey of new material, in: *Atlas of Protein Sequence and Structure*, Volume 5, Supplement 1 (M. O. Dayhoff, ed.), National Biomedical Research Foundation, Washington, D.C., pp. S1–S8.

Dayhoff, M. O., 1976a, The origin and evolution of protein superfamilies, *Fed. Proc.* **35**:2132–2138.

Dayhoff, M. O. (ed.), 1976b, *Atlas of Protein Sequence and Structure*, Volume 5, Supplement 2, 1976, National Biomedical Research Foundation, Washington, D.C.

Dayhoff, M. O. (ed.), 1979, *Atlas of Protein Sequence Structure*, Volume 5, Supplement 3, 1978, National Biomedical Research Foundation, Washington, D.C.

Dayhoff, M. O., McLaughlin, P. J., Barker, W. C., and Hunt, L. T., 1975, Evolution of sequences within protein superfamilies, *Naturwissenschaften* **62:**154–161.

DeLange, R. J., Fambrough, D. M., Smith, E. L., and Bonner, J., 1969, Calf and pea histone IV. III. Complete amino acid sequence of pea seedling histone IV; comparison with the homologous calf thymus histone, *J. Biol. Chem.* **244:**5669–5679.

DeLange, R. J., Hooper, J. A., and Smith, E. L., 1973, Histone III. III. Sequence studies on the cyanogen bromide peptides; Complete amino acid sequence of calf thymus histone III, *J. Biol. Chem.* **248:**3261–3274.

DeLange, R. J., Williams, L. C., and Searcy, D. G., 1981, A histone-like protein (HTa) from *Thermoplasma acidophilum*. II. Complete amino acid sequence, *J. Biol. Chem.* **256:**905–911.

Elgin, S. C. R., Schilling, J., and Hood, L. E., 1979, Sequence of histone 2B of *Drosophila melanogaster*, *Biochemistry* **18:**5679–5685.

Eshaghpour, H., Dieterich, A. E., Cantor, C. R., and Crothers, D. M., 1980, Singlet–singlet energy transfer studies of the internal organization of nucleosomes, *Biochemistry* **19:**1797–1805.

Fedor, M. J., and Daniell, E., 1980, Acetylation of histone-like proteins of adenovirus type 5, *J. Virol.* **35:**637–643.

Felsenfeld, G., 1978, Chromatin, *Nature* **271:**115–122.

Finch, J. T., and Klug, A., 1978, X-ray and electron microscope analyses of crystals of nucleosome cores, *Cold Spring Harbor Symp. Quant. Biol.* **42:**1–9.

Finch, J. T., Lutter, L. C., Rhodes, D., Brown, R. S., Rushton, B., Levitt, M., and Klug, A., 1977, Structure of nucleosome core particles of chromatin, *Nature* **269:**29–36.

Fitch, W. M., 1966, An improved method of testing for evolutionary homology, *J. Mol. Biol.* **16:**9–16.

Fitch, W. M., 1971, Evolution of clupeine Z, a probable crossover product, *Nature New Biol.* **229:**245–247; **231:**256.

Fitch, W. M., and Margoliash, E., 1967, Construction of phylogenetic trees, *Science* **155:**279–284.

Franklin, S. G., and Zweidler, A., 1977, Non-allelic variants of histones 2a, 2b, and 3 in mammals, *Nature* **266:**273–275.

Garel, A., Mazen, A., Champagne, M., Sautière, P., Kmiecik, D., Loy, O., and Biserte, G., 1975, Chicken erythrocyte histone H_5; I. Amino terminal sequence (70 residues), *FEBS Lett.* **50:**195–199.

Garrad, W. T., Nobis, P., and Hancock, R., 1977, Histone H3 disulfide reactions in interphase, mitotic, and native chromatin, *J. Biol. Chem.* **252:**4962–4967.

Gazit, B., Panet, A., and Cedar, H., 1980, Reconstitution of a deoxyribonuclease I-sensitive structure on active genes, *Proc. Natl. Acad. Sci. USA* **77:**1787–1790.

Glover, C. V. C., and Gorovsky, M. A., 1979, Amino-acid sequence of *Tetrahymena* histone H4 differs from that of higher eukaryotes, *Proc. Natl. Acad. Sci. USA* **76:**585–589.

Goldberg, M. L., 1979, Sequence analysis of *Drosophila* histone genes, Ph.D. thesis, Stanford University.

Goldknopf, I. L., and Busch, H., 1977, Isopeptide linkage between nonhistone and histone 2A polypeptides of chromosomal conjugate-protein A24, *Proc. Natl. Acad. Sci. USA* **74:**864–868.

Goldknopf, I. L., and Busch, H., 1980, *N*-Bromosuccinimide fragments of protein A24 (uH2A): An implication that ubiquitin is the precursor of conjugation *in vivo*, *Biochem. Biophys. Res. Commun.* **96:**1724–1731.

Goldknopf, I., Olson, M., James, T., Mays, J., and Guetzow, K., 1976, Partial sequence analysis of the protein A24, a conjugate of the 2A histone, *Fed. Proc.* **35:**1722 (Abstract 1854).

Goldknopf, I. L., French, M. F., Musso, R., and Busch, H., 1977, Presence of protein A24 in rat liver nucleosomes, *Proc. Natl. Acad. Sci. USA* **74:**5492–5495.

Goldknopf, I. L., French, M. F., Daskal, Y., and Busch, H., 1978, A reciprocal relationship between contents of free ubiquitin and protein A24, its conjugate with histone 2A, in chromatin fractions obtained by the DNase II, Mg^{++} procedure, *Biochem. Biophys. Res. Commun.* **84:**786–793.

Goldknopf, I. L., Rosenbaum, F., Sterner, R., Vidali, G., Allfrey, V. G., and Busch, H., 1979, Phosphorylation and acetylation of chromatin conjugate protein A24, *Biochem. Biophys. Res. Commun.* **90:**269–277.

Goldknopf, I. L., Sudhakar, S., Rosenbaum, F., and Busch, H., 1980, Timing of ubiquitin synthesis and conjugation into protein A24 during the HeLa cell cycle, *Biochem. Biophys. Res. Commun.* **95:**1253–1260.

Goldstein, G., Scheid, M., Hammerling, U., Boyse, E. A., Schlesinger, D. H., and Niall, H. D., 1975, Isolation of a polypeptide that has lymphocyte-differentiating properties and is probably represented universally in living cells, *Proc. Natl. Acad. Sci. USA* **72:**11–15.

Goodwin, G. H., Woodhead, L., and Johns, E. W., 1977, The presence of high mobility group non-histone chromatin proteins in isolated nucleosomes, *FEBS Lett.* **73:**85–88.

Goodwin, G. H., Walker, J. M., and Johns, E. W., 1978, The high mobility group (HMG) nonhistone chromosomal proteins, in: *The Cell Nucleus*, Volume VI, *Chromatin*, Part C (H. Busch, ed.), Academic Press, New York, pp. 181–219.

Gorovsky, M. A., Glover, C., Johmann, C. A., Keevert, J. B., Mathis, D. J., and Samuelson, M., 1978, Histones and chromatin structure in *Tetrahymena* macro- and micronuclei, *Cold Spring Harbor Symp. Quant. Biol.* **42:**493–503.

Green, M., Wold, W. S. M., Brackmann, K., and Cartas, M. A., 1980, Studies on early proteins and transformation proteins of human adenoviruses, *Cold Spring Harbor Symp. Quant. Biol.* **44:**457–469.

Griffith, J. D., 1975, Chromatin structure: Deduced from a minichromosome, *Science* **187:**1202–1203.

Hamilton, J. W., and Rouse, J. B., 1980, The biosynthesis of ubiquitin by parathyroid gland, *Biochem. Biophys. Res. Commun.* **96:**114–120.

Hemperly, J. J., Hopp, T. P., Becker, J. W., and Cunningham, B. A., 1979, The chemical characterization of favin, a lectin isolated from *Vicia faba*, *J. Biol. Chem.* **254:**6803–6810.

Hieter, P. A., Hendricks, M. B., Hemminki, K., and Weinberg, E. S., 1979, Histone gene switch in the sea urchin embryo. Identification of late embryonic histone messenger ribonucleic acids and the control of their synthesis, *Biochemistry* **18:**2707–2716.

Hohmann, P., Tobey, R. A., and Gurley, L. R., 1976, Phosphorylation of distinct regions of f1 histone: Relationships to the cell cycle, *J. Biol. Chem.* **251:**3685–3692.

Hooper, J. A., Smith, E. L., Sommer, K. R., and Chalkley, R., 1973, Histone III. IV. Amino acid sequence of histone III of the testes of the carp, *Ictiobus bubalus*, *J. Biol. Chem.* **248:**3275–3279.

Horwitz, M. S., 1978, Temperature-sensitive replication of H5ts125 adenovirus DNA *in vitro*, *Proc. Natl. Acad. Sci. USA* **75:**4291–4295.

Hunt, L. T., and Dayhoff, M. O., 1977, Amino-terminal sequence identity of ubiquitin and the nonhistone component of nuclear protein A24, *Biochem. Biophys. Res. Commun.* **74:**650–655.

Huntley, G. H., and Dixon, G. H., 1972, The primary structure of the NH_2-terminal region of histone T, *J. Biol. Chem.* **247:**4916–4919.

Inoue, A., Tei, Y., Hasuma, T., Yukioka, M., and Morisawa, S., 1980, Phosphorylation of HMG17 by protein kinase NII from rat liver cell nuclei, *FEBS Lett.* **117:**68–72.

Isenberg, I., 1979, Histones, *Annu. Rev. Biochem.* **48:**159–191.

Iwai, K., Nakahara, C., and Ando, T., 1971, Studies on protamines. XV. The complete amino acid sequence of the Z component of clupeine. Application of N→O acyl rearrangement and selective hydrolysis in sequence determination, *J. Biochem.* **69:**493–509.

Iwai, K., Hayashi, H., and Ishikawa, K., 1972, Calf thymus lysine- and serine-rich histone, III. Complete amino acid sequence and its implication for interactions of histones with DNA, *J. Biochem.* **72:**357–367.

Iwasa, Y., Takai, Y., Kikkawa, U., and Nishizuka, Y., 1980, Phosphorylation of calf thymus H1 histone by calcium-activated, phospholipid-dependent protein kinase, *Biochem. Biophys. Res. Commun.* **96:**180–187.

Kitamura, N., Adler, C. J., Rothberg, P. G., Martinko, J., Nathenson, S. G., and Wimmer, E., 1980, The genome-linked protein of picornaviruses. VII. Genetic mapping of poliovirus VPg by protein and RNA sequence studies, *Cell* **21:**295–302.

Klug, A., Rhodes, D., Smith, J., Finch, J. T., and Thomas, J. O., 1980, A low resolution structure for the histone core of the nucleosome, *Nature* **287:**509–516.

Kootstra, A., and Bailey, G. S., 1978, Primary structure of histone H2B from trout (*Salmo trutta*) testes, *Biochemistry* **17:**2504–2510.

Kornberg, R., 1977, Structure of chromatin, *Annu. Rev. Biochem.* **46:**931–954.

Kuehl, L., 1979, Synthesis of high mobility group proteins in regenerating rat liver, *J. Biol. Chem.* **254:**7276–7281.

Kuehl, L., Lyness, T., Dixon, G. H., and Levy-Wilson, B., 1980, Distribution of high mobility group proteins among domains of trout testis chromatin differing in their susceptibility to micrococcal nuclease, *J. Biol. Chem.* **255:**1090–1095.

Laine, B., Sautière, P., and Biserte, G., 1976, Primary structure and microheterogeneities of rat chloroleukemia histone H_{2A} (histone ALK, IIb_1, or F_{2a2}), *Biochemistry* **15:**1640–1645.

Laine, B., Kmiecik, D., Sautière, P., and Biserte, G., 1978, Primary structure of chicken erythrocyte histone H2A, *Biochimie* **60:**147–150.

Langan, T. A., Rall, S. C., and Cole, R. D., 1971, Variation in primary structure at a phosphorylation site in lysine-rich histones, *J. Biol. Chem.* **246:**1942–1944.

Lathe, R., Buc, H., Lecocq, J.-P., and Bautz, E. K. F., 1980, Prokaryotic histone-like protein interacting with RNA polynerase, *Proc. Natl. Acad. Sci. USA* **77:**3548–3552.

Levy-Wilson, B., Kuehl, L., and Dixon, G., 1980, The release of high mobility group protein H6 and protamine gene sequences upon selective DNase I degradation of trout testis chromatin, *Nucl. Acids Res.* **8:**2859–2869.

Lilley, D. M. J., and Pardon, J. F., 1979, Structure and function of chromatin, *Annu. Rev. Genet.* **13:**197–233.

Lischwe, M. A., and Sung, M. T., 1977, A histone-like protein from adenovirus chromatin, *Nature* **267:**552–554.

Low, T. L. K., and Goldstein, A. L., 1979, The chemistry and biology of thymosin. II. Amino acid sequence analysis of thymosin α_1 and polypeptide β_1, *J. Biol. Chem.* **254:**987–995.

MacLeod, A. R., Wong, N. C. W., and Dixon, G. H., 1977, The amino-acid sequence of trout-testis histone H1, *Eur. J. Biochem.* **78:**281–291.

Mamrack, M. D., Olson, M. O. J., and Busch, H., 1979, Amino acid sequence and sites of phosphorylation in a highly acidic region of nucleolar nonhistone protein C23, *Biochemistry* **18**:3381–3386.

Mardian, J. K. W., Paton, A. E., Bunick, G. J., and Olins, D. E., 1980, Nucleosome cores have two specific binding sites for nonhistone chromosomal proteins HMG14 and HMG17, *Science* **209**:1534–1536.

Mathew, C. G. P., Goodwin, G. H., and Johns, E. W., 1979, Studies on the association of the high mobility group non-histone chromatin proteins with isolated nucleosomes, *Nucl. Acids Res.* **6**:167–179.

Matsui, S., Seon, B. K., and Sandberg, A. A., 1979, Disappearance of a structural chromatin protein A24 in mitosis: Implications for molecular basis of chromatin condensation, *Proc. Natl. Acad. Sci. USA* **76**:6386–6390.

McGhee, J. D., and Felsenfeld, G., 1980, Nucleosome structure, *Annu. Rev. Biochem.* **49**:1115–1156.

Mende, L., Timm, B., and Subramanian, A. R., 1978, Primary structures of two homologous ribosome-associated DNA-binding proteins of *Escherichia coli*, *FEBS Lett.* **96**:395–398.

Mirzabekov, A. D., Shick, V. V., Belyavsky, A. V., and Bavykin, S. G., 1978, Primary organization of nucleosome core particle of chromatin: Sequence of histone arrangement along DNA, *Proc. Natl. Acad. Sci. USA* **75**:4184–4188.

Mirzabekov, A. D., Belyavsky, A. V., Bavykin, S. G., and Shick, V. V., 1980, Primary organization of nucleosomes and its functional implications, *Biosystems* **12**:265–271.

Needleman, S. B., and Wunsch, C. D., 1970, A general method applicable to the search for similarities in the amino acid sequence of two proteins, *J. Mol. Biol.* **48**:443–453.

Ogata, N., Ueda, K., and Hayaishi, O., 1980, ADP-ribosylation of histone H2B: Identification of glutamic acid residue 2 as the modification site, *J. Biol. Chem.* **255**:7610–7615.

Ogawa, Y., Quagliarotti, G., Jordan, J., Taylor, C. W., Starbuck, W. C., and Busch, H., 1969, Structural analysis of the glycine-rich, arginine-rich histone. III. Sequence of the amino-terminal half of the molecule containing the modified lysine residues and the total sequence, *J. Biol. Chem.* **244**:4387–4392.

Ohe, Y., Hayashi, H., and Iwai, K., 1979, Human spleen histone H2B: Isolation and amino acid sequence, *J. Biochem.* **85**:615–624.

Okayama, H., and Hayaishi, O., 1978, ADP-ribosylation of nuclear protein A24, *Biochem. Biophys. Res. Commun.* **84**:755–762.

Olson, M. O. J., Goldknopf, I. L., Guetzow, K. A., James, G. T., Hawkins, T. C., Mays-Rothberg, C. J., and Busch, H., 1976, The NH_2- and COOH-terminal amino acid sequence of nuclear protein A24, *J. Biol. Chem.* **251**:5901–5903.

Palau, J., Mozo, A., and Querol, E., 1980, On the interactions of histone H4 and H4 peptides with DNA. Electrooptical, hydrodynamic and electron microscopy studies, *Biochimie* **62**:241–249.

Patthy, L., and Smith, E. L., 1975, Histone III. VI. Two forms of calf thymus histone III, *J. Biol. Chem.* **250**:1919–1920.

Patthy, L., Smith, E. L., and Johnson, J., 1973, Histone III. V. The amino acid sequence of pea embryo histone III, *J. Biol. Chem.* **248**:6834–6840.

Rall, S. C., and Cole, R. D., 1971, Amino acid sequence and sequence variability of the amino-terminal regions of lysine-rich histones, *J. Biol. Chem.* **246**:7175–7190.

Reeck, G. R., Swanson, E., and Teller, D. C., 1978, The evolution of histones, *J. Mol. Evol.* **10**:309–317.

Reeves, R., and Candido, E. P. M., 1980, Partial inhibition of histone deacetylase in active chromatin by HMG14 and HMG17, *Nucl. Acids Res.* **8**:1947–1963.

Reudelhuber, T. L., Boulikas, T., and Garrard, W. T., 1980, A nonamer of histones in chromatin, *J. Biol. Chem.* **255**:4511–4515.

Rouvière-Yaniv, J., and Kjeldgaard, N. O., 1979, Native *Escherichia coli* Hu protein is a heterotypic dimer, *FEBS Lett.* **106**:297–300.

Saffer, J. D., and Glazer, R. I., 1980, The phosphorylation of high mobility group proteins 14 and 17 from Ehrlich ascites and L1210 *in vitro*, *Biochem. Biophys. Res. Commun.* **93**:1280–1285.

Sakai, M., Fujii-Kuriyama, Y., and Muramatsu, M., 1978, Number and frequency of protamine genes in rainbow trout testis, *Biochemistry* **17**:5510–5515.

Salas, M., and Viñuela, E., 1980, Proteins covalently linked to viral nucleic acids, *Trends Biochem. Sci.* **5**:191–193.

Sautière, P., Tyrou, D., Moschetto, Y., and Biserte, G., 1971a, Primary structure of the glycine and arginine-rich histone isolated from chloro-leucemic tumor in the rat, *Biochimie* **53**:479–483.

Sautière, P., Lambelin-Breynaert, M.-D., Moschetto, Y., and Biserte, G., 1971b, A glycine and arginine-rich histone from hog thymus: Study of the tryptic peptides and complete sequence, *Biochimie* **53**:711–715.

Sautière, P., Tyrou, D., Laine, B., Mizon, J., Ruffin, P., and Biserte, G., 1974, Covalent structure of calf-thymus ALK-histone, *Eur. J. Biochem.* **41**:563–576.

Sautière, P., Kmiecik, D., Loy, O., Briand, G., Biserte, G., Garel, A., and Champagne, M., 1975, Chicken erythrocyte histone H_5: II. Amino acid sequence adjacent to the phenylalanine residue, *FEBS Lett.* **50**:200–203.

Schaffhausen, B. S., and Benjamin, T. L., 1976, Deficiency in histone acetylation in nontransforming host range mutants of polyoma virus, *Proc. Natl. Acad. Sci. USA* **73**:1092–1096.

Schaffner, W., Kunz, G., Daetwyler, H., Telford, J., Smith, H. O., and Birnstiel, M. L., 1978, Genes and spacers of cloned sea urchin histone DNA analyzed by sequencing, *Cell* **14**:655–671.

Schechter, N. M., Davies, W., and Anderson, C. W., 1980, Adenovirus coded deoxyribonucleic acid binding protein. Isolation, physical properties, and effects of proteolytic digestion, *Biochemistry* **19**:2802–2810.

Schlesinger, D. H., and Goldstein, G., 1975, Molecular conservation of 74 amino acid sequence of ubiquitin between cattle and man, *Nature* **255**:423–424.

Schlesinger, D. H., Goldstein, G., and Niall, H. D., 1975, The complete amino acid sequence of ubiquitin, an adenylate cyclase stimulating polypeptide probably universal in living cells, *Biochemistry* **14**:2214–2218.

Searcy, D. G., and DeLange, R. J., 1980, *Thermoplasma acidophilum* histone-like protein: Partial amino acid sequence suggestive of homology to eukaryote histones, *Biochim. Biophys. Acta* **609**:197–200.

Searcy, D. G., and Stein, D. B., 1980, Nucleoprotein subunit structure in an unusual prokaryotic organism: *Thermoplasma acidophilum*, *Biochem. Biophys. Acta* **609**:180–195.

Seidah, N. G., Crine, P., Benjannet, S., Scherrer, H., and Chrètien, M., 1978, Isolation and partial characterization of a biosynthetic N-terminal methionyl peptide of bovine pars intermedia: Relationship to ubiquitin, *Biochem. Biophys. Res. Commun.* **80**:600–608.

Sergeant, A., Tigges, M. A., and Raskas, H. J., 1979, Nucleosome-like structural subunits of intranuclear parental adenovirus type 2 DNA, *J. Virol.* **29**:888–898.

Seyedin, S. M., and Kistler, W. S., 1980, Isolation and characterization of rat testis H1t: An H1 histone variant associated with spermatogenesis, *J. Biol. Chem.* **255**:5949–5954.

Spiker, S., Mardian, J. K. W., and Isenberg, I., 1978, Chromosomal HMG proteins occur in three eukaryotic kingdoms, *Biochem. Biophys. Res. Commun.* **82**:129–135.

Stein, D. B., and Searcy, D. G., 1978, Physiologically important stabilization of DNA by a prokaryotic histone-like protein, *Science* **202**:219–221.

Strickland, M., Strickland, W. N., Brandt, W. F., and von Holt, C., 1977, The complete amino-acid sequence of histone $H2B_{(1)}$ from sperm of the sea urchin *Parechinus angulosus*, *Eur. J. Biochem.* **77**:263–275.

Strickland, M., Strickland, W. N., Brandt, W. F., von Holt, C., Wittmann-Liebold, B., and Lehmann, A., 1978, The complete amino-acid sequence of histone $H2B_{(3)}$ from sperm of the sea urchin *Parechinus angulosus*, *Eur. J. Biochem.* **89**:443–452.

Strickland, W. N., Schaller, H., Strickland, M., and von Holt, C., 1976, Partial amino acid sequence of histone H1 from sperm of the sea urchin, *Parechinus angulosus*, *FEBS Lett.* **66**:322–327.

Strickland, W. N., Strickland, M., Brandt, W. F., and von Holt, C., 1977, The complete amino-acid sequence of histone $H2B_{(2)}$ from sperm of the sea urchin *Parechinus angulosus*, *Eur. J. Biochem.* **77**:277–286.

Strickland, W. N., Strickland, M., Brandt, W. F., von Holt, C., Lehmann, A., and Wittmann-Liebold, B., unpublished results (1978), cited by von Holt, C., Strickland, W. N., Brandt, W. F., and Strickland, M. S., 1979, More histone structures, *FEBS Lett.* **100**:201–218.

Suau, P., Bradbury, E. M., and Baldwin, J. P., 1979, Higher-order structures of chromatin in solution, *Eur. J. Biochem.* **97**:593–602.

Sung, M. T., Lischwe, M. A., Richards, J. C., and Hosokawa, K., 1977, Adenovirus chromatin. I. Isolation and characterization of the major core protein VII and precursor pro-VII, *J. Biol. Chem.* **252**:4981–4987.

Sures, I., Lowry, J., and Kedes, L. H., 1978, The DNA sequence of sea urchin (*S. purpuratus*) H2A, H2B, and H3 histone coding and spacer regions, *Cell* **15**:1033–1044.

Suzuki, K., and Ando, T., 1972a, Studies on protamines. XVI. The complete amino acid sequence of clupeine YII, *J. Biochem.* **72**:1419–1432.

Suzuki, K., and Ando, T., 1972b, Studies on protamines. XVII. The complete amino acid sequence of clupeine YI, *J. Biochem.* **72**:1433–1445.

Temussi, P. A., 1975, Automatic comparison of the sequences of calf thymus histones, *J. Theoret. Biol.* **50**:25–33.

Thoma, F., Koller, T., and Klug, A., 1979, Involvement of histone H1 in the organization of the nucleosome and of the salt-dependent super structures of chromatin, *J. Cell Biol.* **83**:403–427.

Toniolo, C., 1980, Secondary structure prediction of fish protamines, *Biochim. Biophys. Acta* **624**:420–427.

Trifonov, E., 1978, The helical model of the nucleosome core, *Nucl. Acids Res.* **5**:1371–1380.

Urban, M. K., Franklin, S. G., and Zweidler, A., 1979, Isolation and characterization of the histone variants in chicken erythrocytes, *Biochemistry* **18**:3952–3960.

van der Vliet, P. C., Keegstra, W., and Jansz, H. S., 1978, Complex formation between the adenovirus type 5 DNA-binding protein and single-stranded DNA, *Eur. J. Biochem.* **86**:389–398.

van Helden, P., Strickland, W. N., Brandt, W. F., and von Holt, C., 1978, Histone H2B variants from the erythrocytes of an amphibian, a reptile and a bird, *Biochim. Biophys. Acta* **533**:278–281.

van Helden, P. D., Strickland, W. N., Brandt, W. F., and von Holt, C., 1979, The complete amino-acid sequence of histone H2B from the mollusc *Patella granatina*, *Eur. J. Biochem.* **93**:71–78.

Varshavsky, A. J., Bakayev, V. V., Chumackov, P. M., and Georgiev, G. P., 1976, Minichromosome of simian virus 40: Presence of histone H1, *Nucl. Acids Res.* **3**:2101–2113.

Varshavsky, A. J., Bakayev, V. V., Nedospasov, S. A., and Georgiev, G. P., 1978, On the structure of eukaryotic, prokaryotic, and viral chromatin, *Cold Spring Harbor Symp. Quant. Biol.* **42**:457–473.

Vartapetian, A. B., Drygin, Y. F., Chumakov, K. M., and Bogdanov, A. A., 1980, The structure of the covalent linkage between proteins and RNA in encephalomyocarditis virus, *Nucl. Acids Res.* **8**:3729–3742.

von Holt, C., Strickland, W. N., Brandt, W. F., and Strickland, M. S., 1979, More histone structures, *FEBS Lett.* **100**:201–218.

Walker, J. M., and Johns, E. W., 1980, The isolation, characterization and partial sequences of the chicken erythrocyte non-histone chromosomal proteins HMG14 and HMG17: Comparison with the homologous calf thymus proteins, *Biochem. J.* 185:383–386.

Walker, J. M., Goodwin, G. H., and Johns, E. W., 1976a, The similarity between the primary structures of two non-histone chromosomal proteins, *Eur. J. Biochem.* **62**:461–469.

Walker, J. M., Hastings, J. R. B., Johns, E. W., and Gaastra, W., 1976b, The partial amino acid sequence of a non-histone chromosomal protein, *Biochem. Biophys. Res. Commun.* **73**:72–78.

Walker, J. M., Goodwin, G. H., Johns, E. W., Wietzes, P., and Gaastra, W., 1977a, Comparison of the amino-terminal sequences of two calf-thymus chromatin non-histone proteins, *Int. J. Pept. Protein Res.* **9**:220–223.

Walker, J. M., Hastings, J. R. B., and Johns, E. W., 1977b, The primary structure of a non-histone chromosomal protein, *Eur. J. Biochem.* **76**:461–468.

Walker, J. M., Goodwin, G. H., and Johns, E. W., 1978a, Chromosomal proteins: The amino terminal sequence of high mobility group non-histone chromosomal protein HMG 14, showing sequence homologies with two other chromosomal proteins, *Int. J. Pept. Protein Res.* **11**:301–304.

Walker, J. M., Goodwin, G. H., and Johns, E. W., 1978b, The isolation and identification of ubiquitin from the high mobility (HMG) non-histone protein fraction, *FEBS Lett.* **90**:327–330.

Walker, J. M., Hastings, J. R. B., and Johns, E. W., 1978c, A novel continuous sequence of 41 aspartic and glutamic residues in a non-histone chromosomal protein, *Nature* **271**:281–282.

Walker, J. M., Gooderham, K., and Johns, E. W., 1979a, The isolation, characterization and partial sequence of a peptide rich in glutamic acid and aspartic acid (HGA-2 peptide) from calf thymus non-histone chromosomal protein HMG 2: Comparison with a similar peptide (HGA-1 peptide) from calf thymus non-histone chromosomal protein HMG 1, *Biochem. J.* **179**:253–255.

Walker, J. M., Gooderham, K., and Johns, E. W., 1979b, The isolation and partial sequence of peptides produced by cyanogen bromide cleavage of calf thymus non-histone chromosomal high-mobility-group protein 2: Sequence homology with non-histone chromosomal high-mobility-group protein 1, *Biochem. J.* **181**:659–665.

Walker, J. M., Goodwin, G. H., and Johns, E. W., 1979c, The primary structure of the nucleosome-associated chromosomal protein HMG-14, *FEBS Lett.* **100**:394–398.

Walker, J. M., Stearn, C., and Johns, E. W., 1980, The primary structure of non-histone chromosomal protein HMG17 from chicken erythrocyte nuclei, *FEBS Lett.* **112**:207–210.

Warrant, R. W., and Kim, S.-H., 1978, α-Helix–double helix interaction shown in the structure of a protamine–transfer RNA complex and a nucleoprotamine model, *Nature* **271**:130–135.

Watson, D. C., Peters, E. H., and Dixon, G. H., 1977, The purification, characterization and partial sequence determination of a trout testis non-histone protein, HMG-T, *Eur. J. Biochem.* **74**:53–60.

Watson, D. C., Levy-Wilson, B., and Dixon, G. H., 1978, Free ubiquitin is a non-histone protein of trout testis chromatin, *Nature* **276**:196–198.

Watson, D. C., Wong, N. C. W., and Dixon, G. H., 1979, The complete amino-acid sequence of a trout-testis non-histone protein, H6, localized in a subset of nucleosomes and its similarity to calf-thymus non-histone proteins HMG-14 and HMG-17, *Eur. J. Biochem.* **95**:193–202.

Weber, S., and Isenberg, I., 1980, High mobility group proteins of *Saccharomyces cerevisiae*, *Biochemistry* **19**:2236–2240.

Weisbrod, S., Groudine, M., and Weintraub, H., 1980, Interaction of HMG 14 and 17 with actively transcribed genes, *Cell* **19**:289–301.

Wilkinson, K. D., Urban, M. K., and Haas, A. L., 1980, Ubiquitin is the ATP-dependent proteolysis factor I of rabbit reticulocytes, *J. Biol. Chem.* **255**:7529–7532.

Wong, N. C. W., Poirier, G. C., and Dixon, G. H., 1977, Adenosine diphosphoribosylation of certain basic chromosomal proteins in isolated trout testis nuclei, *Eur. J. Biochem.* **77**:11–21.

Worcel, A., and Benyajati, C., 1977, Higher order coiling of DNA in chromatin, *Cell* **12**:83–100.

Wouters, D., Sautière, P., and Biserte, G., 1978, Primary structure of histone H2A from gonad of the sea urchin *Psammechinus miliaris*, *Eur. J. Biochem.* **90**:231–239.

Wouters-Tyrou, D., Sautière, P., and Biserte, G., 1976, Covalent structure of the sea urchin histone H_4, *FEBS Lett.* **65**:225–228.

Yaguchi, M., Roy, C., and Seligy, V. L., 1979, Complete amino acid sequence of goose erythrocyte H5 histone and the homology between H1 and H5 histones, *Biochem. Biophys. Res. Commun.* **90**:1400–1406.

Yeoman, L. C., Olson, M. O. J., Sugano, N., Jordan, J. J., Taylor, C. W., Starbuck, W. C., and Busch, H., 1972, Amino acid sequence of the center of the arginine-lysine-rich histone from calf thymus: The total sequence, *J. Biol. Chem.* **247**:6018–6023.

Yulikova, E. P., Evseenko, L. K., Baratova, L. A., Belyanova, L. P., Rybin, V. K., and Silaev, A. B., 1976, The primary structure of sturine B, a protamine from Caspian sturgeon, *Bioorg. Khim.* **2**:1613–1617.

Yulikova, E. P., Rybin, V. K., and Silaev, A. B., 1979, The primary structure of stellin A, *Bioorg. Khim.* **5**:5–10.

PART II

Modeling the Process of Sequence Divergence

CHAPTER 6

Simulation of the Evolution of Macromolecular Sequences by Random Fixation of Allowed Codons

MICHAEL COATES and SIMON STONE

1. Introduction

One approach to the study of the evolution of macromolecular sequences (proteins, DNA, RNA) is computer simulation. In this chapter we describe simulation studies on the evolution of five proteins: fibrinopeptides A and B, hemoglobin α and β chains, and cytochromes c. The first requisite of a simulation, if it is to conform to the pattern observed in the real evolution of these proteins, is that it begin with an "ancestral," or starting, sequence and through a series of duplications followed by divergence end with a set of "contemporary," or final, sequences. This process will follow a phylogeny, or genealogy, in which the duplication events are the nodes, or branch points, and divergence occurs along the branches.

Further, it is known that divergence occurs by a process whereby some of the mutations which continually occur in the DNA sequence coding for the protein (the exons) are accepted, or fixed, and thus become characteristic of a population of organisms. The great majority of such accepted mutations in exons are single-nucleotide replacements (point mutations) (Dayhoff, 1972; Vogel and Kopun, 1977; Nishioka and Leder, 1979). Some of these lead to amino acid substitutions, whereas others,

MICHAEL COATES • School of Australian Environmental Studies, Griffith University, Nathan, Queensland, Australia 4111. *SIMON STONE* • Department of Botany, University of Adelaide, Adelaide, South Australia 5001.

because of the degenerate nature of the genetic code, do not. Furthermore, the amount of sequence divergence is not a direct function of the number of fixations, because of: (1) back substitutions, where a nucleotide or amino acid substitution is canceled by a subsequent substitution back to the original nucleotide or amino acid; (2) coincidental identities, where two or more contemporary sequences have identical amino acids or nucleotides at a particular position due not to lack of divergence but to coincidental substitutions to the same nucleotide, or amino acid, after any number of interim changes; and (3) multiple substitutions at the same site, which occur because the number of sites able to accept substitutions in any protein is limited and which results in a slowing of sequence divergence with time as a "saturation point," or limit to total sequence divergence, is approached.

As will be discussed in this chapter, all of the characteristics of real evolution mentioned above are found in the simulation model. The starting protein sequences are arbitrarily arrived at by selecting the most commonly occurring amino acid at each position in the known sequences for each protein studied (Section 2.2). The starting DNA sequences are derived from these by chance selection of one of the synonomous codons for each amino acid in the starting protein. The pattern of duplications for each of the proteins follows the phylogenies of organisms for which sequences of the protein under investigation have been obtained (Section 2.6). These phylogenies are based on fossil evidence and comparative morphology. Point mutations occur in the DNA at random, in terms of both the site of mutation and the three possible replacement nucleotides (Section 2.3). Since the genetic code is used to translate the nucleotide sequences into amino acid sequences, degenerate nucleotide replacements occur in the simulation. The number and type of substitutions which may occur at each site are limited, as will be explained below (Section 2.4). Further, the length of the DNA and corresponding protein sequences conform to the real proteins whose evolution is simulated. These facts, together with the stochastic nature of mutation, and hence fixation, in the simulation model ensure that back substitutions, coincidental identities, and multiple fixations at the same site all occur (Section 2.3).

There are simplifying assumptions in nearly every aspect of the simulation model, which will be discussed in later sections. Although it is important that these assumptions are recognized, they are not all crucial. What is crucial in the simulation model are the suppositions concerning the mechanism controlling the rate of nucleotide fixation and the criteria by which mutations are accepted. In part, the studies discussed in this chapter are a test of these suppositions.

Initially, we employ for the simulation model the extreme case of the neutral evolution hypothesis (Kimura, 1968, 1969, 1979; King and Jukes, 1969). In the extreme case of this hypothesis all fixations involve neutral mutations. Thus, at each position in the protein some amino acids are *allowed,* meaning that substitutions among them will not alter the fitness of the organism. Naturally at each position some amino acids are not allowed, meaning that they are deleterious (reduce fitness). Mutations resulting in amino acids that are not allowed will be eliminated by selection. The existence of deleterious mutations is an integral part of the neutral evolution hypothesis (Kimura and Ohta, 1973, 1974; Kimura, 1979).

It is important to remember that the simulation involves mutation in the DNA and translation into protein by the genetic code so that each allowed amino acid is specified by a set of allowed, synonymous codons. Hence, we call our simulation model "random fixation of allowed codons" (RFAC).

A further aspect of the extreme case of the neutral hypothesis employed initially is that the allowed variation in amino acids for each protein remains constant during simulated evolution. Since in the neutral hypothesis the rate of fixation is the rate of occurrence of neutral mutations (King and Jukes, 1969; Kimura and Ohta, 1974; Kimura, 1979), constant allowed variation will produce a uniform rate of fixation of nucleotide replacements if the intrinsic mutation rate remains constant with time. (The intrinsic mutation rate here equals the rate of DNA replication error/unit time.) We choose this condition initially because an apparent uniform rate of evolution in real proteins[the evolutionary clock (Zuckerkandl and Pauling, 1965)] is considered as evidence for the validity of the neutral hypothesis (Kimura, 1968, 1969, 1979; King and Jukes, 1969; Kimura and Ohta, 1973; Fitch and Langley, 1978).

The model used in simulation of macromolecular evolution by RFAC is similar to the random evolutionary hit (REH) model of Holmquist [Holmquist (1972) and Holmquist *et al.* (1972); see Holmquist (1976, 1978) and references therein for review]. Both are stochastic models constrained by a deterministic aspect representing selection. There are, however, some important differences, which will be discussed in Section 3. RFAC is capable of giving results in essential agreement with REH and with the augmented maximum parsimony, termed augmented distance (AD), method [see Moore *et al.* (1976) and Holmquist *et al.* (1976) for comparison of REH and AD and Coates and Stone (1981) for comparison of RFAC with the others]. Differences between RFAC and the covarion model of Fitch (1971) are also discussed in Section 3.

One use of the simulation is to determine if evolution entirely by random fixation of neutral mutations at a constant rate can account for the observed evolution of particular proteins. We simulate the evolution of fibrinopeptides A and B in mammals, hemoglobin α and β chains in vertebrates, and cytochromes *c* in vertebrates, insects, and plants. As mentioned above, the phylogenies followed are inferred from fossil evidence and comparative morphology. As will be explained in detail, the variation permitted in each case is arrived at by consideration of the known sequences for each protein (Section 2.4). The results of simulation are tabulated as the percentage sequence difference accumulated during the evolutionary time separating each pair of sequences (twice the time since divergence from a common ancestor). These results are compared graphically and statistically with the comparable results of the real evolution of each protein. It is shown that simulation based on the extreme case of the neutral hypothesis fails to provide results that closely fit those of the real evolution of any of the proteins (Coates and Stone, 1981). However, the fit is significantly better in the cases of fibrinopeptides and cytochromes *c* than in the case of the hemoglobin chains (Sections 5.2.1.1–5.2.1.3, 5.2.2.1, and 5.2.2.2).

Possible explanations for these results are given (Sections 5.2.1.1, 5.2.2.1, 5.2.2.2, and 6). Furthermore, it is shown that if the rate of fixation or the amount of allowed variation is altered (increased or decreased) in specific parts of the phylogenies of proteins, better fit to the real results of evolution is obtained (Sections 5.2.2.1 and 5.2.2.2). Restriction of allowed variation is equivalent to Darwinian or directional selection. On the other hand, relaxation of selectional constrains, a simple increase in allowed variation, is accommodated by the neutral hypothesis.

Another use of the simulation is to determine the relative effects of rate of fixation and amount of allowed variation on protein evolution (Sections 5.1 and 5.3). This is done by altering one of these two parameters during simulation while all other factors are held constant. It is worthwhile to point out that the effects of these factors can be assessed even if the simulation model does not conform entirely to reality. For instance, regardless of the mechanism whereby mutations are fixed in populations, the larger the allowed variation for a protein, the greater will be the amount of sequence divergence which can occur. Further, as will be shown (Section 5.3), protein evolution becomes more "efficient" (Holmquist and Pearl, 1980) as allowed variation increases. This stems from the fact that nucleotide replacements are less likely to be degenerate, and the influence of back substitution and multiple fixations at the same site is less, when allowed variation is large. Thus, as allowed variation increases, the rate of amino acid sequence divergence, for any particular rate of fixation of nucleotide replacements, increases *and* the ratio of amino acid to nucleotide sequence differences at any evolutionary time decreases.

Increasing the rate of fixation will shorten the time required to reach the limit of sequence divergence in a particular protein. However, alteration of the rate of fixation will not alter the total amount of sequence divergence which can occur (Section 5.1).

2. The Simulation Model—RFAC

2.1. The Input Data

Essentially the model simulates the descent of a particular protein and its gene from a common ancestor to the final versions found in a related group of living organisms. A large number of events must occur during this simulation of macromolecular evolution, necessitating the use of a large digital computer. The input data for the simulation program are: the starting DNA sequence, the allowed amino acids at each of the encoding sites, the genetic code, the branching pattern of the phylogenetic tree, and the number of fixations which will occur on each branch. The origins of each of these sources of data, as well as the assumptions involved in their derivation, are discussed below.

2.2. The Starting DNA Sequence

First a starting *protein sequence* is derived. For each protein we inspect all known sequences from the taxon encompassing the group of species whose phylogenies will be followed in the simulation, for instance, fibrinopeptides A in mammals, or cytochromes *c* in vertebrates. For each amino acid site in the starting sequence we select the most commonly occurring amino acid. In the case of a tie we choose one of the most common amino acids at random.

The starting DNA sequence is derived by randomly choosing one of the synonomous codons for each of the amino acids in the starting protein sequence. Here we assume that the synonomous codons are used equally in the DNA coding for protein. This assumption may not be true, however; as yet there are insufficient data to allow us to incorporate any consistent bias in codon usage into the simulation model.

The starting sequences so derived will of course bear only accidental similarity to the real ancestral sequences. However, this is not important for the purposes of simulation. The results of simulation are expressed as the percentage sequence difference; the nature of the differences is ignored. Although the number of final sequence differences will depend upon the kinds of codons present in the allowed variation, the actual starting sequence does not have significant influence so long as all sites contain allowed codons (Coates and Stone, unpublished results).

The lengths of the starting DNA sequences are: 20 codons for fibrinopeptides A, 22 codons for fibrinopeptides B, 104 codons for cytochromes *c*, 144 codons for hemoglobin α chains, and 146 codons for hemoglobin β chains.

2.3. Random Point Mutation

Mutation in RFAC is limited to single-nucleotide replacements. Thus no additions or deletions can occur. In the proteins studied this is significant only in the case of the fibrinopeptides in which a number of additions or deletions have occurred during real evolution. In RFAC double- and triple-nucleotide substitutions can occur, but only as a result of successive single-nucleotide substitutions within the same codon.

For each mutation a random number generator (GGUBFS in the International Mathematical and Statistical Library) is employed to pick randomly: first a nucleotide site in the DNA sequence and then one of the three nucleotides different from the one present at the site. Once again the assumption of equal use of codons is employed. It is also assumed that the four nucleotides are exchanged with equal probability. This latter assumption is also unlikely to be always true. Indeed, Vogel and Kopun (1977) report a small bias in favor of transitions. However, the effect of including such a bias is small, causing the simulation to be slightly less efficient in terms of number of fixations required to bring about a particular level of amino acid sequence divergence (Coates and Stone, unpublished results).

The randomly selected nucleotide will be substituted for the old one—become fixed—at the randomly selected site only if the resultant codon codes for one of the allowed amino acids at this site.

2.4. Allowed Amino Acids

The concept that a certain amount of structural variation in proteins is consistent with proper functioning is not new (Anfinsen, 1959; Barnard *et al.*, 1972). Recently, this idea has been enunciated by Matsuda (1978) and called the "structural free degree" of proteins and by Yockey (1977a, b, c), in which the concept of "synonymous amino acids" at each site of a protein is utilized. Although Yockey's term, synonymous amino acids, seems to imply that there is no selective difference between certain amino acids at certain sites, he in fact makes no assumption of selective neutrality. Matsuda clearly states that he believes the structural free degree to consist of sequence differences that have been selected for. In RFAC we assume that a certain amount of variation in sequence is allowed and that the allowed amino acids at each site of the protein sequence are selectively neutral with respect to one another. This enormously simplifies the simulation and allows us to test the neutral hypothesis.

Initially, the allowed variation for each protein consists of all amino acids observed at each site in the known sequences from a particular taxon. The taxa correspond to those used in deriving the starting sequences and are those containing the species whose known phylogenies will be followed in the simulation.

To the allowed variation a certain number of "intermediate" amino acids must be added, because at some sites not all of the observed alternatives can be interchanged by single-nucleotide replacements. Therefore, at these sites we add to the allowed variation one, or sometimes two, amino acids so that such an interchange is possible. When possible the added amino acids are chosen for their chemical similarity (see below) to the observed variants.

The observed alternatives plus the added intermediates we call "observed variation." The observed variation corresponds to the smallest amount of allowed variation used in these studies.

Since known sequences are only a sample of all sequences, the observed variation is undoubtedly an underestimate of the actual allowed variation. Because of this consideration, and in order to explore the effect of increasing allowed variation over observed, we double the number of amino acids allowed at each site in each protein. To accomplish this we classify each amino acid as polar-charged, polar-uncharged, or nonpolar. Amino acids were added to those allowed at each site if they belonged to the same categories as the observed variants. Where the observed variants belonged to more than one category an attempt was made to maintain the proportions of amino acids in each. Again intermediates were added where necessary to ensure complete interchangeability between allowed amino acids by single-nucleotide replacements. Intermediates were added according to the same rules of chemical similarity. We call the allowed variation derived in this way "twice observed variation."

Yockey (1977a) employs a different method for expanding the synonymous amino acids over those observed. Using a formulation by Grantham (1974), he applies a "prescription which predicts functionally equivalent residues at given sites in protein sequences." In this the amino acids are assigned a value which is a composite of composition, polarity, and molecular volume. Amino acids that have values intermediate to the pair of greatest difference among the observed variants at a site are added to the synonymous amino acids at that site.

Unlike our doubling technique, sites that are observed to be invariant remain so after the prescription has been applied. The true situation clearly lies somewhere between the two techniques, because as more proteins are sequenced some presently invariant sites will undoubtedly be revealed as variants. Furthermore, Yockey's prescription is more sensitive to the properties of amino acids observed at each site, whereas our doubling method is more sensitive to the numbers of amino acids at each site.

Interestingly from Yockey (1977a, Table I) we find that the ratio of synonymous amino acids after the prescription has been applied to the observed is 1.87. This is very close to a doubling. The differences in effect on RFAC when the doubling technique and Yockey's prescription are used is investigated in Section 5.2.1.2.

Table I gives the mean numbers (and standard deviations) of amino

Table I

Mean Numbers of Allowed Amino Acids per Site and Number of Variable Sites for the Different Amounts of Allowed Variation Used in RFAC Simulation

Protein	Allowed variation	X number of amino acids allowed per site	Standard deviation	Number of variable sites
Fibrinopeptide A	Observed	3.88	2.15	17
	Twice observed	7.75	4.30	20
Fibrinopeptide B	Observed	4.27	2.80	19
	Twice observed	8.91	5.75	22
Cytochrome c	Observed (vertebrates)	1.80	1.35	37
	Twice observed (vertebrates)	3.60	2.70	104
	Observed (Yockey[a])	3.48	2.48	74
	Twice observed (Yockey[a])	6.49	5.61	104
	Yockey's (1977a) prescription	7.02	4.92	74
Hemoglobin α chain	Observed	3.19	1.63	109
	Twice observed	6.37	3.27	144
Hemoglobin β chain	Observed	3.21	1.83	115
	Twice observed	6.41	3.66	146

[a] Yockey (1977a), Table 1.

acids per site and numbers of variable sites in each protein when allowed variation equals observed and twice observed and from Yockey (1977a, Table I) before and after application of the prescription to cytochrome *c*.

During the initial simulations by RFAC the amount of allowed variation remained constant throughout. Later we attempted to obtain better fit to the results of real evolution by altering the allowed variation, either by expanding or contracting it in certain parts of the phylogeny. The effect of doing this is quite dramatic and the operation draws attention to an aspect of the nature of selection which has been implicit in much of the work in this field but which has not, in our view, been adequately exposed. This is the inherent relative nature of selection coefficients (Sections 5.2.2 and 6).

2.5. Fixation

After a point mutation in the evolving DNA sequence the resulting codon is translated into an amino acid using the table of the genetic code. If mutation results in a terminating codon another mutation is immediately tried. If this amino acid matches with one of those allowed at the site, including the same amino acid in the case of a degenerate mutation, the new nucleotide replaces the old. In this event the fixation is counted and simulation continues. If the mutant codon does not match with one of the allowed codons at the site, no fixation occurs, no count is made, and the cycle, starting with a random point mutation, is repeated.

2.6. Phylogenies

In RFAC simulation, the starting DNA sequence is first duplicated and then a predetermined number of fixations occur in each duplicate before the next duplication event is called for, and so on until a number of final sequences is produced. For each protein whose evolution is simulated the system of duplications and fixations corresponds to a "known" phylogeny of species. Each phylogeny is that of a group of species for which the amino acid sequences of the protein are known. Thus the results of simulation can be compared to those of real evolution.

These phylogenies are based on fossil evidence and comparative morphology. In order to avoid circularity in comparing the simulated results to the real results of evolution, the phylogenies in no way depend upon macromolecular data. For vertebrates the phylogenies of Young (1962) and Romer (1966) are used, with the exception of the times of divergence

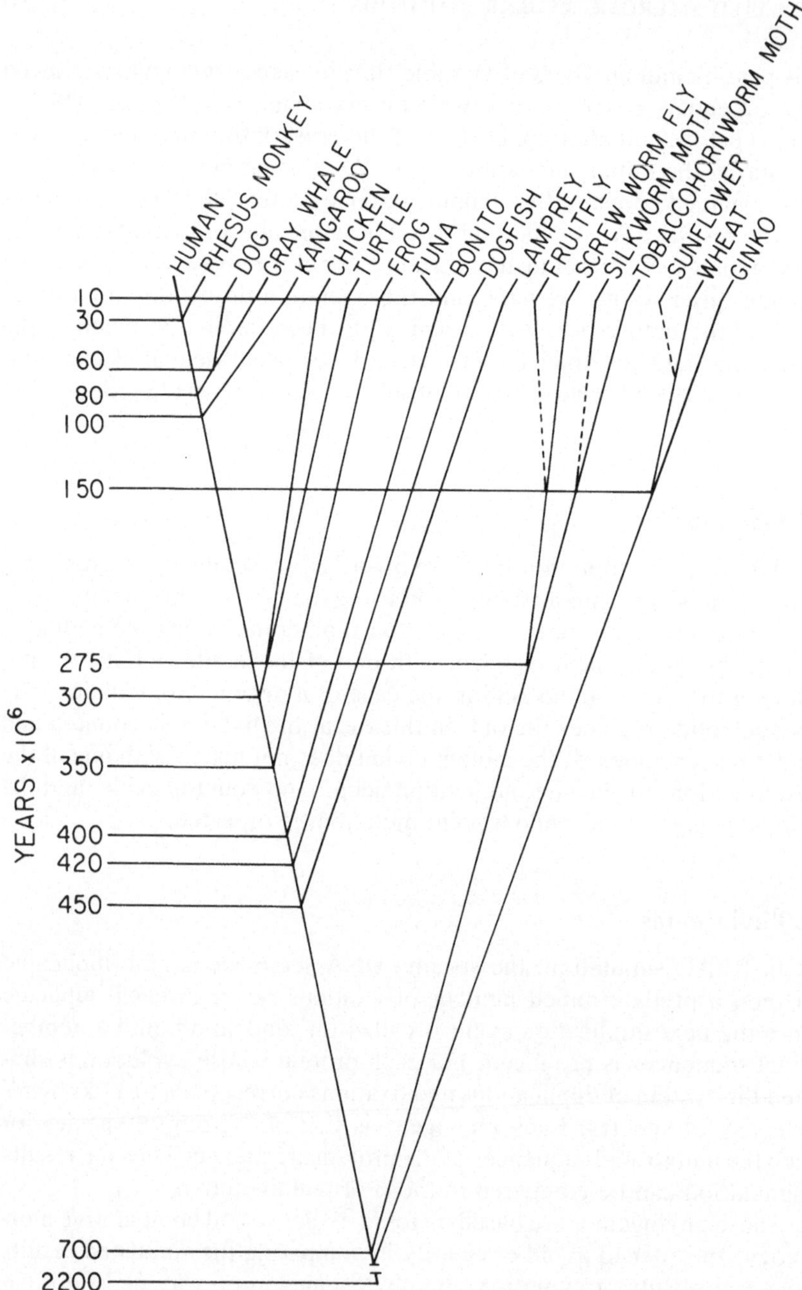

Figure 1. Phylogenetic tree of some vertebrates, insects, and plants and their cytochromes *c*.

of the primates, which are from Spuhler and Jorde (1975). For the insects Hennig (1966) is followed, and for plants Hufford (1978). The absolute times on the branches of the phylogenies are corrected according to Van Eysinga (1975). The phylogeny of vertebrates, insects, and plants used for simulation of cytochrome *c* evolution is given in Fig. 1. The phylogenies of fibrinopeptide A in mammals, hemoglobina α and β chains in vertebrates, and cytochrome *c* in vertebrates are those in Fig. 1 of Coates and Stone (1981). The phylogeny of fibrinopeptides B in mammals is not shown (however, see Table III).

In most simulations by RFAC the number of fixations occurring on each branch of the phylogeny retain a common relationship to the temporal length of the branches throughout. This amounts to a constant rate of fixation during the entire evolution of the protein. In order to determine the effect of rate of fixation overall this was varied during different simulation runs. In certain simulations of cytochrome *c* evolution, allowed variation was expanded in part of the phylogeny and therefore the rate of fixation was increased, in accordance with the neutral hypothesis (Section 5.2.2.1).

It is important to note that our simulation assumes a reasonably accurate knowledge of divergence times, and hence branch lengths, of the phylogenies. In fact no standard deviation can be determined for the values given. The assumption of accurate divergence times, estimated from fossil evidence, has been severely questioned (Wilson *et al.*, 1977). This is a difficulty in all studies addressed to the problem of the rate of macromolecular evolution (Fitch and Langley, 1978; Wilson *et al.*, 1977). However, a certain amount of inaccuracy in the phylogenies is not fatal to the conclusions we wish to draw. We return to this problem in Section 6.

It may also appear that the model assumes that gene duplications occur only at the phylogenetic nodes and thus the only gene duplications that have occurred are exactly correlated with species splits and all genes are orthologous (Fitch, 1970). This is not necessarily the case; the assumption we make is that if duplications *have* occurred within branches of the phylogeny, and there are thus paralogous genes (Fitch, 1970) in the sample, these genes have *not* evolved independently within a species. In support of this is the concept of gene matching [see Maniatis *et al.* (1980) and references therein for review]. An example consists of the duplicate hemoglobin chains in primates. DNA sequence studies of flanking parts of the α-chain genes suggest that α-chain gene duplication occurred prior to the time of primate divergence. Although the α-chain sequences have diverged since that time, the duplicates are much more similar within species than between species. This indicates that the duplicates have been corrected against one another (Maniatis *et al.*, 1980, p. 23). Zuckerkandl (1978) has pointed out the inadequacy, as a generality, of the notion that gene duplication within a species will allow one of the copies to diverge rapidly. Another point is that in every case the proteins sequenced in the various species are the major component, in other words, the protein serving the function of hemoglobin α chain, cytochrome c, or whatever. Hence, it seems valid to assume these proteins will retain the selectional constraints, and hence the allowed variation, concomitant with serving these functional roles.

3. Comparison of RFAC with Other Models

As mentioned in the introduction, the model of macromolecular evolution most closely related to RFAC is the REH model of Holmquist. In both models mutation in the DNA sequence is random, nucleotides are equally interchangeable, and all of the synonomous codons of an amino acid are used equally [however, see Holmquist and Pearl (1980) and Holmquist *et al.*, this volume, Chapter 7]. Degenerate nucleotide replacements, back substitutions, multiple fixations at the same site, and coincidental identities are accounted for in each. In both, selection determines the number of sites in a protein that are variable (termed T_2 by Holmquist) and the variability of these sites. In both, replacements within the selectional constraints are assumed to occur by a stochastic process. One major difference between the two models is the manner in which fixations are distributed among the variable sites. In the earlier version of REH this was approximated by a Poisson process with mean U_2, which is termed the fixation intensity. In a later version (Holmquist and Pearl, 1980; Holmquist *et al.*, this volume, Chapter 7) a better accounting for the "nonrandomicities" in distribution of fixations among the variable sites is achieved with a negative binomial distribution. In RFAC the distribution of nucleotide fixations among and within the variable codon sites is a function of the precisely specified allowed variation at each site. In RFAC the distribution of fixations is *not* a Poisson process; however, it may be approximated by a negative binomial distribution.

Although Holmquist assumes, for the purpose of calculating REH, that replacement with the constraints imposed by selection occurs by a stochastic process during evolution, he does not assume that the process is by chance fixation of neutral mutations. Indeed, in each pairwise comparison of sequences, U_2 and T_2 may change in value and there is no constraint on the rate of fixation.

Another model of macromolecular evolution is the augmented maximum parsimony method [Moore *et al.* (1976) and referenced therein]. As the name implies, this model assumes evolution occurs in the most parsimonious way. Therefore, back substitutions, degenerate substitutions, multiple fixations at the same site, and coincidental identities are not completely accounted for, although some of the last three are detected. This model is completely constrained by the maximum parsimony assumption, but no *a priori* assumption about the effects of selection are made.

Interestingly, REH and AD (augmented distance = augmented maximum parsimony distance) give similar results in terms of the number of nucleotide replacements estimated to have occurred in cytochromes *c* (Moore *et al.*, 1976) and myoglobin and the hemoglobin chains (Holmquist *et al.*, 1976). In RFAC the number of nucleotide fixations required to give a particular amount of amino acid sequence difference is very sensitive to the amount of variation allowed. However, RFAC also gives nucleotide replacement values close to those of the other methods when allowed variation is set at levels that appear reasonable—between observed and twice observed for cytochromes *c* and between twice observed and total variation for hemoglobin chains (Coates and Stone, 1981).

Another model is the covarion model of Fitch (Fitch and Markowitz, 1970; Fitch, 1971) in which at any particular time in the evolutionary history of a protein, certain sites are concomitantly variable and fixations can occur only at these sites, mutations at other sites being eliminated by natural selection. During evolution the variable codons change. In this model, as in REH and RFAC, selection determines the extent of variability, but within that variability, replacement is by chance. Neither REH nor RFAC incorporates the concept of a limited number of concomitantly variable sites that change with time. All sites observed to be variable over a long time span are assumed to be free to fix mutations during that time. Thus the covarion model would appear to be more constrained than the other two. However, Karon (1979) has shown that REH and the covarion model may give consistent results over short evolutionary times.

When protein sequences are examined, REH, AD, the covarion model, and RFAC may all give similar estimates of the number of nucleotide replacements at the level of the DNA. This probably indicates that the differences in their assumptions are not important. However, Holmquist (1980) has pointed out that when the α- and β-hemoglobin messenger RNAs are inspected it is clear that REH and AD, based on protein sequence data, give considerable *underestimates* of the number of nucleotide replacements that have occurred over relatively short evolutionary times. This is true also for RFAC simulation. Both methods give better estimates of numbers of nucleotide replacements when mRNA sequences are considered. The implications of this are discussed in Section 5.3, where REH and RFAC are compared using data from Holmquist (1980).

4. Methods of Comparison of Macromolecular Sequence Divergence in Real and Simulated Evolution

After RFAC simulation, matrices of amino acid differences containing all pairwise comparisons of the final protein sequences are produced. These can be compared with corresponding matrices of amino acid differences between real proteins. Hotelling's T-square test can then be used to determine the probability that the results of real evolution will be found in a population of simulated results. This was done after the simulation of the evolution of fibrinopeptides A in mammals and cytochromes *c* and hemoglobin α and β chains in vertebrates, in which the allowed variation and rate of fixation remained constant during simulation (Coates and Stone, 1981). It was found that some statistical fit could be obtained between simulated and real results for fibrinopeptides A and cytochromes *c* but none for hemoglobin α and β chains. We speculated that this difference might be a result of more constancy in the selective constraints on fibrinopeptide A and cytochrome *c* than in hemoglobin during the evolution of the taxa studied. In Section 5.2.2.2 we explore this possibility further. However, we need first to look at the general effects of altering rate of fixation and allowed variation during evolution as shown by RFAC simulation (Section 5.1).

For visualization of the results of real and simulated evolution we employ a graphic technique. Percentage of amino acid, or nucleotide, sequence difference is plotted against evolutionary time, where evolutionary time is twice the time since the nearest common ancestor (Fig. 2). Thus from Fig. 1 the evolutionary time between dogfish and turtle is 840×10^6 years. It can be seen that a phylogeny, as in Fig. 1, will provide a number of pairwise comparisons for each evolutionary time. For instance, at 200×10^6 years of evolutionary time in Fig. 2 there are four comparisons: man, rhesus monkey, dog, and gray whale with kangaroo. For a set of such comparisons there will be a mean and standard deviation. The evolutionary times, comparisons made, means and standard deviation of percentages of real and simulated sequence difference, and conditions of simulation are given in Tables II–VI for each of the proteins whose evolution is simulated. In practice the RFAC simulation program is repeated ten times to obtain each set of results. Accordingly, in calculating the standard deviation the variance is multiplied by ten in accordance with the central limit theorem. Mean percentage sequence differences only are plotted against evolutionary time in Figs. 2–10.

5. Results

5.1. Effects of Rate of Fixation and Amount of Allowed Variation on Macromolecular Sequence Divergence

Figure 2 shows the general effects of allowed variation and rate of fixation. Because for any gene there is a limit on the number of variable sites and the extent to which they can vary even when all 20 amino acids are allowed at each site, the curve of percentage sequence difference versus evolutionary time will asymptotically approach a saturation value (call it D_{max}). Clearly, if the protein is to retain its functional identity, D_{max} will be less than 100%. Such a curve is shown as curve 1 in Fig. 2.

An increase in the rate of fixation will decrease the evolutionary time required to reach $D_{max}/2$ (call it ET_{50}), while D_{max} remains unchanged. Compare curves 1 and 2 in Fig. 2. If allowed variation is increased both D_{max} and ET_{50} are affected; D_{max} increases while ET_{50} decreases (compare curves 1 and 3 in Fig. 2).

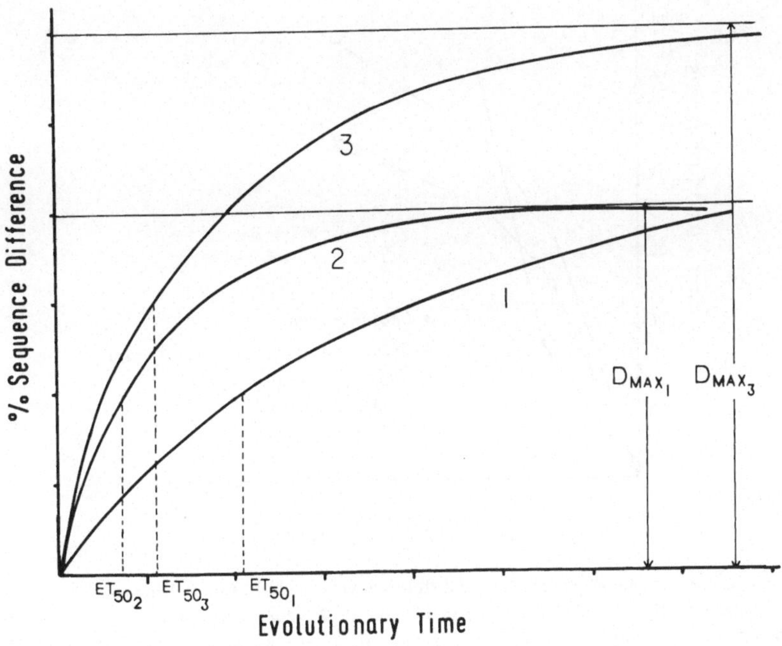

Figure 2. Curves of percentage macromolecular sequence difference versus evolutionary time, showing effects of allowed variation and rate of fixation. See text for further explanation.

In Fig. 2 the allowed variation and rate of fixation remain constant during simulated evolution, and hence the curves are smooth. If either of these parameters is altered during the course of simulation, dips or bumps in the curves can be produced.

5.2. Fitting of Simulated to Real Results of Evolution

Under the neutral hypothesis the amount of allowed variation and rate of fixation are not independent as in Fig. 2. In this hypothesis the rate of fixation is the rate of occurrence of neutral mutations (King and Jukes, 1969; Kimura and Ohta, 1974; Kimura, 1979). If all of the sequences within the set of allowed variants are selectively neutral, then the rate of fixation will be a function of the number of these sequences. Thus when an "observed" uniform rate of macromolecular evolution due to an un-

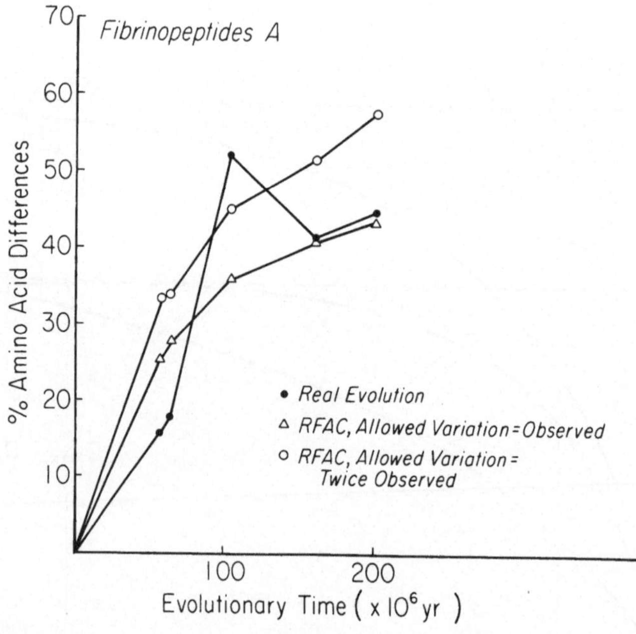

Figure 3. Mean percentage amino acid difference versus evolutionary time for fibrinopeptide A in mammals. Symbols are the same as in Table II: ●, real evolution; △, RFAC simulation, allowed variation equals observed, rate of fixation per codon per year equals 10^{-8}; ○, RFAC simulation, allowed variation equals twice observed, rate of fixation per codon per year equals 10^{-8}.

derlying constant rate of fixation of neutral mutations is invoked as support for the neutral hypothesis (King and Jukes, 1969; Kimura, 1968, 1979), a constant amount of allowed variation is implicit in the invocation. Thus initially we carry out RFAC simulation under constant conditions in order to determine how well an extreme case of the neutral hypothesis can fit the real data.

5.2.1. Amount of Allowed Variation and Rate of Fixation Constant during Simulation

5.2.1.1. Fibrinopeptides. Figures 3 and 4 show mean percentage amino acid differences versus evolutionary time from real and RFAC simulated

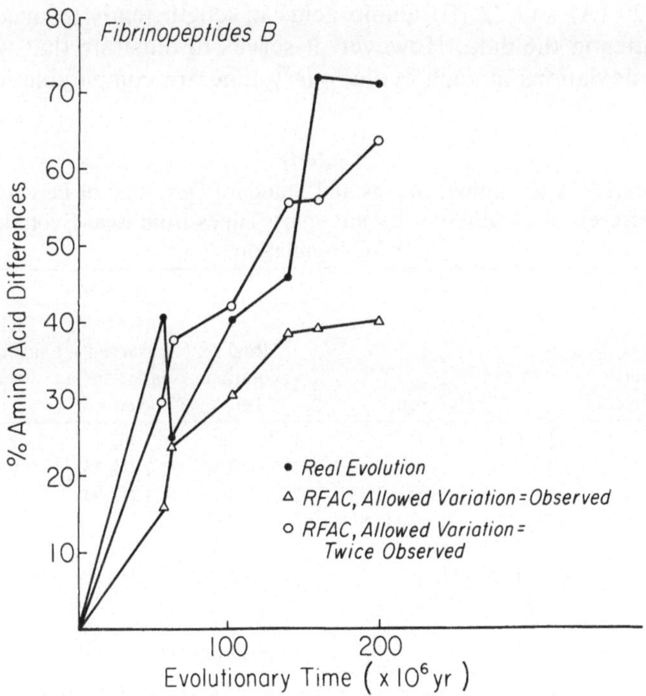

Figure 4. Mean percentage amino acid difference versus evolutionary time for fibrinopeptide B in mammals. Symbols are the same as in Table III: ●, real evolution; △, RFAC simulation, allowed variation equals observed, rate of fixation per codon per year equals 10^{-8}; ○, RFAC simulation, allowed variation equals twice observed, rate of fixation per codon per year equals 10^{-8}.

evolution of fibrinopeptides A and B, respectively, in mammals. The data are from Tables II and III. In each case the amounts of allowed variation used in RFAC simulation are observed and twice observed (see Section 2.4). The rate of fixation is 10^{-8} per codon per year in all cases. It has already been mentioned that the statistical fit between real and simulated results obtained for fibrinopeptides A was better than that for the hemoglobin chains or cytochrome *c* (Coates and Stone, 1981). Simulation of fibrinopeptide B evolution gave statistical fits comparable to those of fibrinopeptide A: *p* value of 0.008 for twice observed variation (0.004 for fibrinopeptide A), *p* value of 0.004 for observed variation (0.008 for fibrinopeptide A). This is perhaps expected because the fibrinopeptides are recognized as proteins with few functional constraints and thus unlikely to have evolved much by selection for selectively advantageous mutations (Dickerson, 1971; Kimura and Ohta, 1973).

Plotting mean *percentage* amino acid differences of fibrinopeptides that are 20 (A) and 22 (B) amino acids in length tends to magnify the irregularities in the data. However, it serves to illustrate that while the standard deviations at each evolutionary time are comparable for simu-

Table II

Fibrinopeptide A Evolution: Means and Standard Deviation of Percent Amino Acid Differences at Different Evolutionary Times from Real Evolution and RFAC Simulation[a]

		X (SD)[c]		
Evolutionary time × 10^6 years	Comparisons[b]	Real evolution (●)	RFAC observed variation[d] (△)	RFAC twice observed variation[d] (○)
56	9/10, 9/11	15.8 (7.44)	25.3 (0.0)	33.2 (4.71)
64	1/3, 2/3	17.9 (20.16)	27.7 (18.26)	33.9 (12.04)
104	6/8, 6/9, 6/10, 6/11, 7/8, 7/9, 7/10, 7/11, 8/9, 8/10, 8/11	25.0 (7.09)	36.1 (18.94)	45.0 (18.31)
160	1/4–11, 2/4–11, 3/4–11, 4/6–11, 5/6–11	41.2 (8.84)	41.1 (10.26)	56.4 (12.05)
200	1–11/12	44.8 (12.04)	43.3 (5.06)	57.5 (4.00)

[a]All real sequences and alignments are from Dayhoff (1972, 1973, 1976, 1978) except where noted.

[b]Species whose fibrinopeptides A are compared are numbered as follows: 1, human; 2, gibbon; 3, rhesus monkey; 4, dog; 5, bear; 6, pig; 7, peccary; 8, camel; 9, red deer; 10, European bison; 11, bovine; 12, kangaroo.

[c]Symbols (●, △, ○) as in Fig. 3.

[d]Rate of fixation equals 10^{-8} per codon per year. See text for further explanation.

Table III
Fibrinopeptide B Evolution: Means and Standard Deviations of Percent Amino Acid Differences at Different Evolutionary Times from Real Evolution and RFAC Simulation.[a]

Evolutionary time × 10^6 years	Comparison[b]	$\overline{X}$(SD)[c] Real evolution (●)	RFAC observed variation[d] (△)	RFAC twice observed variation[d] (○)
56	9/10, 9/11	40.5 (3.37)	16.4 (0.0)	29.5 (2.03)
64	1/3, 2/3	25.0 (15.37)	24.1 (2.03)	37.7 (0.0)
104	7/8, 7/9, 7/10, 7/11, 8/9, 8/10, 8/11	40.0 (6.15)	30.3 (6.53)	41.9 (2.72)
140	6/7–11	45.9 (8.09)	37.6 (7.00)	55.5 (5.02)
160	1/4–11, 2/4–11, 3/4–11, 4/6–11, 5/6–11	72.8 (11.06)	38.7 (8.15)	55.3 (8.81)
200	1–11/12	71.3 (9.65)	40.0 (5.47)	63.4 (8.24)

[a]See Table II, footnote *a*.
[b]Species whose fibrinopeptides B are compared are numbered as follows: 1, human; 2, siamang; 3, green monkey; 4, dog; 5, cat; 6, horse; 7, pig; 8, camel; 9, red deer; 10, European bison; 11, bovine; 12, kangaroo.
[c]Symbols (●, △, ○) as in Fig. 4.
[d]Rate of fixation equals 10^{-8} per codon per year. See text for further explanation.

lated and real results (Tables II and III), the major difference between them is the irregularity of accumulation of sequence differences with time in real evolution compared to simulated evolution under these conditions. Thus we must conclude that our extreme case of the neutral hypothesis, namely, evolution entirely by random fixation of neutral mutations at a constant rate, cannot fully account for fibrinopeptide evolution. However, a greater portion of their evolution has been by this means than is the case for the other proteins studied. Since there is no discernible pattern in the irregularity of the evolution of the fibrinopeptides, we have not attempted to modify the conditions of RFAC simulation to obtain a better fit to the real data.

5.2.1.2. Cytochrome *c*. Simulation by RFAC with constant allowed variation and rate of fixation gave a reasonable graphic fit to the real evolution of cytochrome *c* in the vertebrates (Coates and Stone, 1981). However, the statistical fit was not as good as for the fibrinopeptides: p value of 0.001 when allowed variation equaled observed and rate of fixation equaled 1.9×10^{-9} per codon per year *and* when allowed variation equaled twice observed and rate of fixation equaled 4.8×10^{-10} per codon per year.

Under the conditions of constant allowed variation and rate of fixation it is impossible to fit RFAC results to those of the real evolution of cytochrome *c* over the longer time span represented in Fig. 1. Figure 5 shows the real evolution curve and curves from RFAC simulation under constant conditions, as well as the effect of changing both the allowed variation and rate of fixation during simulation. These data are contained in Table IV. Comparison of the upper RFAC curve in Fig. 5 (see also the curves in Fig. 6) with the real evolution curve illustrates, in part, why no fit is achieved when conditions are constant. In order to fit the first part of the real curve the amount of allowed variation must be increased to twice observed and the rate of fixation set to 9.6×10^{-10} fixations per year per codon; this causes the amount of divergence to be too great at longer evolutionary times. Another approach, not shown, is to *decrease* allowed variation overall while *increasing* rate of fixation. However, when this route is taken a fit to the first part of the real evolution curve occurs

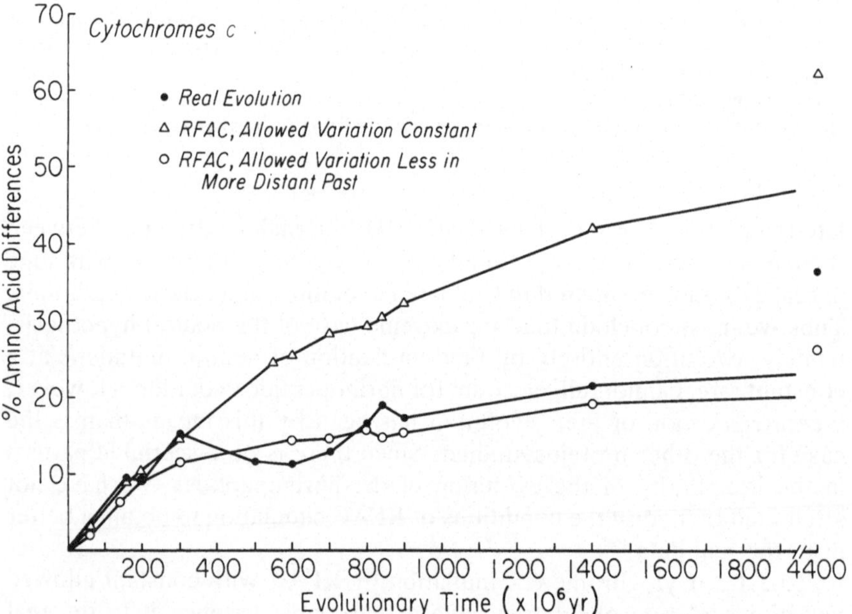

Figure 5. Mean percentage amino acid difference versus evolutionary time for cytochrome *c* in vertebrates, insects, and plants. Symbols are the same as in Table IV: ●, real evolution: △, RFAC simulation, allowed variation equals twice observed, rate of fixation per codon per year equals 9.6×10^{-10}: ○, RFAC simulation, allowed variation up to 300 million years evolutionary time equals twice observed, rate of fixation per codon per year equals 9.6×10^{-10}, for longer evolutionary times allowed variation equals observed, rate of fixation per year per codon equals 4.8×10^{-10}.

Table IV
Cytochrome *c* Evolution: Means and Standard Deviations of Percent Amino Acid Difference at Different Evolutionary Times from Real Evolution and RFAC Simulation.[a]

		$\overline{X}$ (SD)[c]			
				RFAC	
Evolutionary time × 10^6 years	Comparisons[b]	Real evolution (●)	RFAC twice observed variation[d] (△)	Twice observed variation[d] (○)	Observed variation[e] (○)
160	1/3, 1/4, 2/3, 2/4	9.6 (0.76)	8.2 (0.82)	6.4 (0.50)	—
200	1/5, 2/5, 3/5, 4/5	9.1 (2.26)	10.6 (1.40)	9.0 (2.89)	—
300	17/19, 18/19	15.3 (1.37)	13.9 (1.57)	11.5 (0.42)	—
550	6/7, 13/15, 14/15, 13/16, 14/16	11.8 (2.52)	24.4 (1.97)	—	—
600	1/6, 2/6, 3/6, 4/6, 5/6, 1/7, 2/7, 3/7, 4/7, 5/7	11.1 (2.28)	25.6 (3.16)	—	14.5 (1.60)
700	1–7/8	12.8 (3.03)	28.3 (2.08)	—	14.9 (2.93)
800	1–8/9, 1–8/10	16.6 (2.03)	28.9 (2.87)	—	15.6 (2.44)
840	1–10/11	19.0 (2.35)	30.4 (1.95)	—	14.7 (2.66)
900	1–11/12	17.3 (2.03)	32.3 (3.19)	—	15.2 (2.10)
1400	1–12/13, 1–12/14, 1–12/15, 1–12/16	21.8 (3.18)	42.0 (3.36)	—	19.3 (0.46)
4400	1–16/17, 1–16/18	36.6 (2.26)	62.5 (2.49)	—	26.2 (0.91)

[a]See Table II, footnote *a*.

[b]Species whose cytochromes *c* are compared are numbered as follows: 1, human; 2, rhesus monkey; 3, dog; 4, gray whale; 5, kangaroo; 6, chicken; 7, turtle; 8, frog; 9, tuna; 10, bonito; 11, dogfish; 12, lamprey; 13, fruitfly; 14, screw worm fly; 15, silk worm moth; 16, tobacco worm moth; 17, sunflower; 18, wheat; 19, ginko.

[c]Symbols (●, △, ○) are as in Fig. 5.

[d]Rate of fixation equals 9.6×10^{-10} per codon per year.

[e]Rate of fixation equals 4.8×10^{-10} per codon per year. See text for further explanation.

only when the allowed variation is too low to allow a fit to the later part of curve. The conditions under which the lower RFAC curve were produced will be discussed in Section 5.2.2.1.

In Section 2.4 we discussed the difference between two methods for expanding the allowed variation used in RFAC, our doubling technique and the prescription technique of Yockey. In Fig. 6 we compare the two methods when used in RFAC. Only the mean percentage amino acid differences at the different evolutionary times are plotted; the standard deviations are comparable for the two methods. The phylogenetic tree

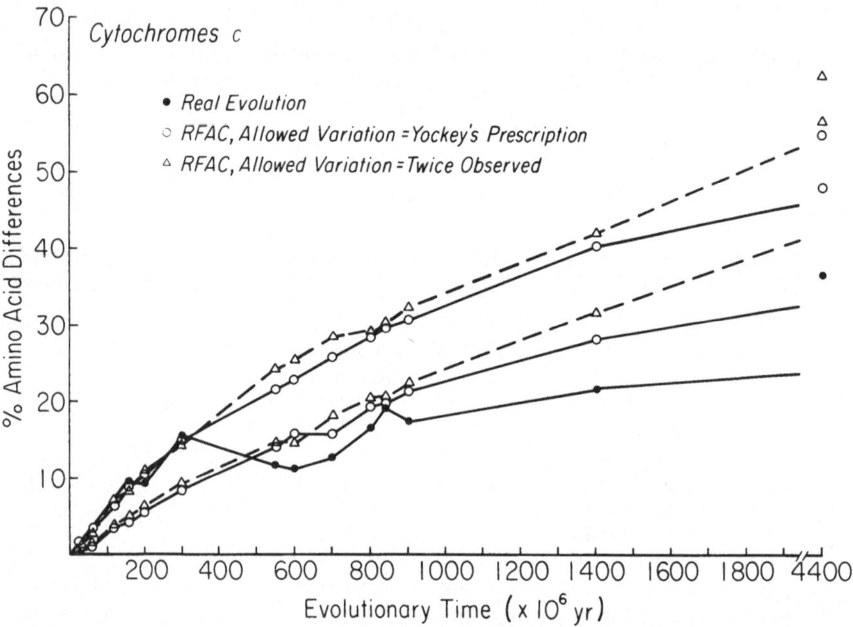

Figure 6. Comparison of RFAC simulation results using Yockey's prescription and our doubling technique to expand allowed variation over that observed. ●, Real evolution; ○, RFAC simulation, allowed variation equals Yockey's prescription; △, RFAC simulation, allowed variation equals twice observed. Lower RFAC curves, rate of fixation per codon per year equals 4.8×10^{-10}; upper RFAC curves, rate of fixation per codon per year equals 9.6×10^{-10}.

for cytochrome *c* shown in Fig. 1 is used for the simulation. To obtain twice the observed variation in this simulation we applied our doubling technique to the observed variation for cytochrome *c* given in Yockey (1977a, Table 1). This is the data from which Yockey derived the prescription used in this simulation. We have run the simulation with twice allowed variation and Yockey's prescription at two rates of fixation. As can be seen from Fig. 6, the two methods give similar results until the longer evolutionary times, after which twice observed variation gives greater divergence. This is due to the greater number of variable sites in twice observed variation. As discussed in Section 2.4 the "true" method of expanding observed variation is probably intermediate between these two methods. However, the experiment depicted in Fig. 6 shows that, at least for evolutionary times less than 1400 million years, corresponding approximately to the divergence of arthropods and vertebrates, either method may be used.

5.2.1.3. Hemoglobin Chains. We noted previously (Coates and Stone, 1981) that the real evolution curves for the hemoglobin chains appear to be composite. Divergence increases until about 360 million years of evolutionary time, whereupon it levels off, increasing again at about 600 million years (see Figs. 7 and 8). This general trend is also detected in the work of Goodman *et al.* (1974, 1975) and Holmquist *et al.* (1976). In Figs. 7 and 8, which are plotted from data in Tables V and VI, the curves obtained with RFAC under constant conditions had allowed variation equal to twice observed and rate of fixation set at 2.4×10^{-9} fixations per codon per year for hemoglobin α and β chains, respectively. Although there is some fit with the first and last parts of the real evolution curves, the "dip" in both real curves cannot be matched by RFAC under constant conditions. In Section 5.2.2.2 we discuss the conditions that gave better fits to the real evolution curves.

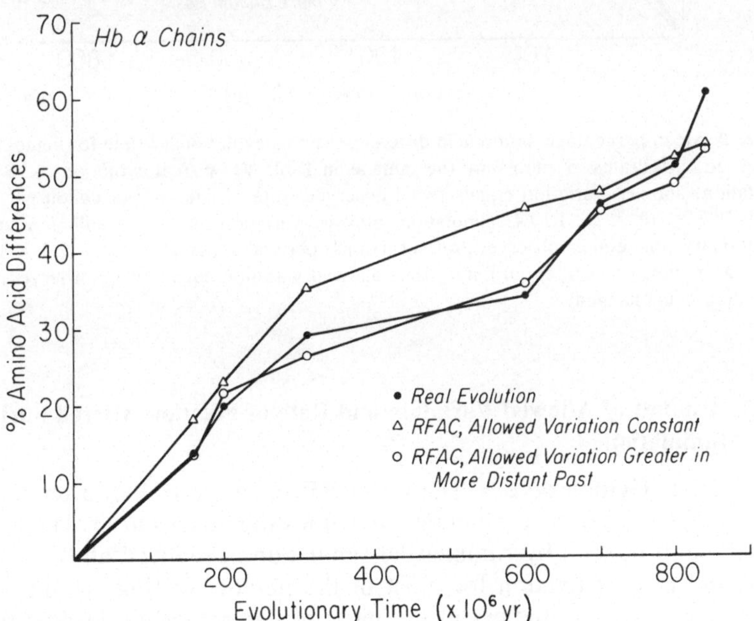

Figure 7. Mean percentage amino acid difference versus evolutionary time for hemoglobin α chain in vertebrates. Symbols are the same as in Table V: ●, real evolution; △, RFAC simulation, allowed variation equals twice observed, rate of fixation per codon per year equals 2.4×10^{-9}; ○, RFAC simulation, allowed variation up to 600 million years of evolutionary time equals observed, rate of fixation per codon per year equals 2.4×10^{-9}; after 600 million years of evolutionary time allowed variation equals twice observed, rate of fixation is unchanged.

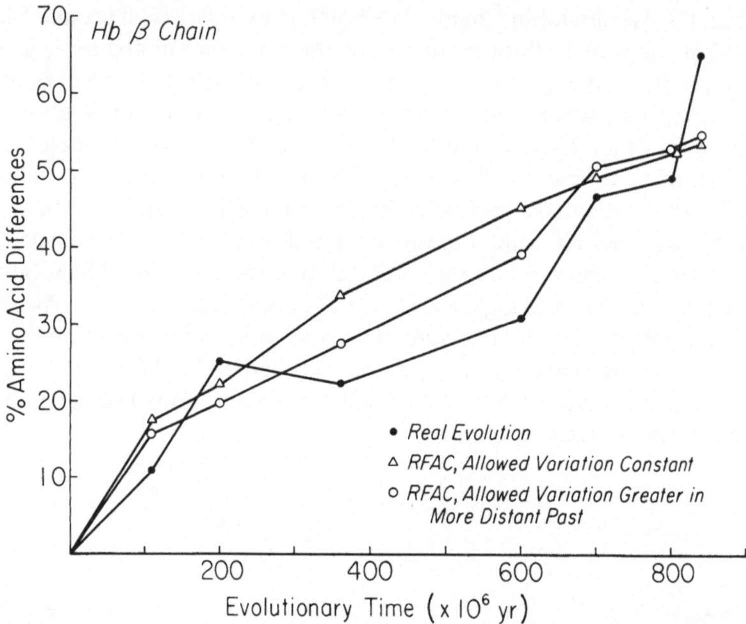

Figure 8. Mean percentage amino acid difference versus evolutionary time for hemoglobin β chain in vertebrates. Symbols are the same as in Table VI: ●, real evolution; △, RFAC simulation, allowed variation equals twice observed, rate of fixation per codon per year equals 2.3×10^{-9}; ○, RFAC simulation, allowed variation up to 600 million years of evolutionary time equals observed, rate of fixation per codon per year equals 2.3×10^{-9}; after 600 million years of evolutionary time allowed variation equals twice observed, rate of fixation is unchanged.

5.2.2. Amount of Allowed Variation and Rate of Fixation Altered during Simulation

5.2.2.1. Cytochrome *c*. The lower RFAC curve in Fig. 5 obviously fits the real evolution curve better than does the upper curve. The lower curve was produced by running the simulation at reduced allowed variation and rate of fixation for most of the phylogeny (Fig. 1) and then increasing these parameters in the parts of the phylogeny leading from the branch point at 300 million years ago to the mammals and from 300 million years ago on the branch leading to the plants. For most of the phylogeny, that is, from the starting sequence leading to all the vertebrates other than mammals and the insects, the allowed variation was that observed in vertebrates and rate of fixation was 4.8×10^{-10} fixations per codon per year. In the branches leading to the mammals and plants,

Table V

Hemoglobin α-Chain Evolution: Means and Standard Deviations of Percent Amino Acid Difference at Different Evolutionary Times for Real Evolution and RFAC Simulation[a]

		$\bar{X}$ (SD)[c]			
				RFAC	
Evolutionary time × 10^6 years	Comparisons[b]	Real evolution (●)	RFAC twice observed variation[d] (△)	Observed variation[d] (○)	Twice observed variation[d] (○)
160	1/3, 1/4, 2/3, 2/4	14.7 (3.14)	18.7 (1.14)	13.7 (1.55)	—
200	1/5, 2/5, 3/5, 4/5	21.3 (2.52)	23.3 (2.21)	16.7 (1.76)	—
360	1/6, 2/6, 3/6, 4/6, 5/6	29.4 (4.15)	35.4 (1.18)	26.2 (2.54)	—
600	1–6/7, 1–6/8	33.9 (6.41)	45.2 (2.22)	35.5 (2.26)	—
700	1–8/9, 1–9/10	46.5 (3.14)	48.2 (1.75)	—	44.7 (5.24)
800	1–10/11, 1–10/12	50.9 (2.53)	52.0 (3.10)	—	48.4 (2.17)
840	1–12/13	60.7 (2.84)	53.8 (2.25)	—	57.6 (2.40)

[a]See Table II, footnote *a*.

[b]Species whose hemoglobin α chains are compared are as follows: 1, human; 2, rhesus monkey; 3, dog; 4, horse; 5, kangaroo; 6, echidna; 7, chicken; 8, viper; 9, tadpole [tadpole hemoglobin α chain from Maruyama *et al.* (1980)]; 10, newt; 11, carp; 12, sucker; 13, shark.

[c]Symbols (●, △, ○) as in Fig. 7.

[d]Rate of fixation equals 2.4×10^{-9} per codon per year. See text for further explanation.

Table VI

Hemoglobin β-Chain Evolution: Means and Standard Deviations of Percent Amino Acid Difference at Different Evolutionary Times for Real Evolution and RFAC Simulation[a]

		$\bar{X}$ (SD)[c]			
				RFAC	
Evolutionary time × 10^6 years	Comparisons[b]	Real evolution (●)	RFAC twice observed variation[d] (△)	Observed variation[d] (○)	Twice observed variation[d] (○)
160	1/3, 2/3, 1/4, 2/4, 3/4	11.1 (1.90)	17.5 (1.11)	15.7 (1.25)	—
200	1/5, 2/5, 3/5, 4/5	25.0 (0.88)	22.4 (2.15)	19.6 (2.21)	—
360	1/6, 2/6, 3/6, 4/6, 5/6	22.3 (2.45)	34.4 (3.32)	27.3 (3.87)	—
600	1–6/7	31.3 (1.71)	45.6 (1.35)	39.2 (3.05)	—
700	1–7/8, 1–8/9	46.7 (3.02)	49.7 (2.69)	—	50.8 (1.15)
800	1–9/10	48.9 (3.82)	53.8 (4.89)	—	53.0 (1.83)
840	1–10/11	65.0 (3.01)	54.4 (0.61)	—	54.6 (3.33)

[a]Table II, footnote *a*.

[b]Species whose hemoglobin β chains are compared are as follows: 1, human; 2, rhesus monkey; 3, dog; 4, rabbit; 5, kangaroo; 6, echidna; 7, chicken; 8, frog; 9, tadpole [tadpole hemoglobin β chain from Watt *et al.* (1980)]; 10, carp; 11, shark.

[c]Symbols (●, △, ○) as in Fig. 8.

[d]Rate of fixation equals 2.3×10^{-9} per codon per year.

beginning 300 million years ago, allowed variation was twice that observed in vertebrates and rate of fixation per codon per year was 9.6×10^{-10}.

It will be noted that this operation is equivalent to a relaxing of selectional constraints in certain parts of a phylogeny. Under the neutral hypothesis this means that the number of selectionally equivalent sequences increases, and hence the number of possible neutral mutations increases. These factors together cause an *increase* in the rate of accumulation of amino acid sequence differences in certain parts of the phylogeny as compared with other parts. Thus, under these conditions the neutral theory does *not* predict a constant rate of evolution. This aspect of the neutral hypothesis has been invoked by Kimura and Ohta (1974) to explain the rapid rate of evolution of guinea pig insulin.

Although the selectional constraints upon cytochrome *c* are relatively great, its evolution nevertheless is reasonably well accounted for by a model based entirely upon random fixation of neutral mutations. This may be because the functional environment of this protein has been stable for a very long time and mutations bestowing greater fitness have all been tried and accepted. The evolution of this protein over the last two billion (2×10^9) years or so may have been largely limited to random drift among a small number of selectively equivalent sequences. We would not care to guess why the selectional constraints may have been relaxed somewhat in modern plants and mammals.

5.2.2.2. Hemoglobin Chains. The RFAC curves giving better fit to the real evolution curves in Figs. 7 and 8 were produced by *reducing* the amount of allowed variation in later parts of the phylogenies. In both cases allowed variation was reduced from twice that observed in vertebrates to that observed at the branch point at 300 million years ago leading to birds, reptiles, and mammals. Rate of fixation was not altered and remained at 2.4×10^{-9} and 2.3×10^{-9} per codon per year for α and β chains, respectively. Thus, in Figs. 7 and 8 and Tables V and VI, for the better fitting RFAC curves the percentage of amino acid differences at evolutionary times up to 600 million years were produced by RFAC with observed variation, while beyond 600 million years they result from RFAC with twice observed variation. This produces a significant dip in both curves.

The success of the above-described operation on RFAC suggests that there has been an increase in the selectional constraints on hemoglobin during its relatively recent evolution in vertabrates. We have previously suggested (Coates and Stone, 1981) that this correlates with the emergence of the homeothermic vertebrates, mammals, and birds, with their higher metabolisms and complete dependence on efficient air-breathing.

In the case of hemoglobin, we did not alter the rate of fixation to conform with allowed variation as was done with cytochrome *c*, for two reasons. First we wanted to show conclusively that the rate of evolution could be rendered nonuniform *simply* by altering the selectional constraints. This shows that Van Valen (1974) is wrong when she says: "It isn't constant selective constraints that give a constant rate of evolution, but something causing a constant rate of substitution within these constraints." But more importantly, we did not alter rate of fixation as the neutral hypothesis demands, because when selective constraints are increased, the neutral hypothesis does not apply.

When selective constraints are increased (allowed variation decreased) amino acids that were previously nondeleterious at certain sites become deleterious. These will be rapidly replaced by mutations resulting in nondeleterious amino acids at these sites. For example, say that previous to the restriction in allowed variation Ala, Leu, Ile, and Val are allowed at a particular site. The great majority of individuals of species A have Ala at this site. After the restriction only Leu and Val are allowed. Now in species A mutations producing Leu or Val at the site will be selectively favored over wild type and their frequency in the species will increase rapidly. This is what is often termed "positive" Darwinian selection, in which a newly selectively advantageous trait replaces a selectively disadvantageous trait. This is not genetic drift and the rate of fixation will not be constant, at least during the transition period during which the new limits on allowed variation are being established.

We note that in this example new mutations resulting in Leu (or Val) are selectively advantageous with respect to Ala and Ile, but selectively neutral with respect to Val (or Leu).

5.3 Effect of Amount of Allowed Variation on "Efficiency" of Evolution

Figures 9 and 10 show plots of percentage amino acid and nucleotide difference versus evolutionary time for hemoglobin α and β chains and the coding parts of their genes (these may also be thought of as messenger RNAs). The curves were produced by RFAC simulation with allowed variation equal to observed and twice observed in vertebrates. The rate of fixation per codon per year was 2.4×10^{-9} for α chain and 2.3×10^{-9} for β chain in all cases.

For both the α and β chains of hemoglobin an increase in allowed variation causes evolution to be more efficient in terms of amount of sequence divergence for a given number of nucleotide replacements (fixations) (Figs. 9 and 10). Furthermore, in the case of hemoglobin α chain, doubling allowed variation causes an increase in efficiency in terms of the

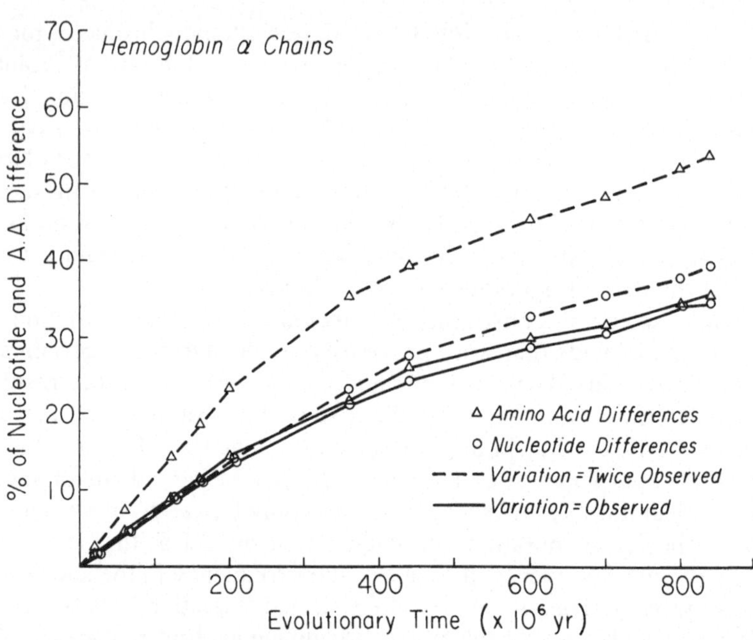

Figure 9. Comparison of means of percentage nucleotide and amino acid differences at different evolutionary times given by RFAC simulation of hemoglobin α-chain evolution. $\triangle$, mean percentage amino acid differences; $\bigcirc$, mean percentage nucleotide differences; dashed lines, allowed variation equals twice observed; solid lines, allowed variation equals observed. Rate of fixation per codon per year equals 2.4×10^{-9} throughout.

rate of amino acid sequence divergence relative to the rate of nucleotide sequence divergence (Fig. 9). This effect was also observed for the fibrinopeptides and cytochrome *c* (Coates and Stone, unpublished results). This second type of efficiency is a consequence of the degeneracy of the genetic code. As an extreme example, consider a site within a protein where only one amino acid, say Ser, is allowed. Six fixations, corresponding to the six synonymous codons for serine, may occur in the gene, whereas no substitutions can occur in the protein at this site. When allowed variation is doubled, allowing, say, Ser and Thr at the site, some fixations in the gene can cause substitutions in the protein at that site.

With allowed variation equal to observed, the efficiency, in terms of the ratio of amino acid sequence to nucleotide sequence differences, is less in hemoglobin α chains partly because there are fewer variable sites than in β chains (Table I). Theoretically, if allowed variation were decreased below that observed in vertebrates for this protein, the rate of increase of percentage nucleotide sequence difference would surpass that

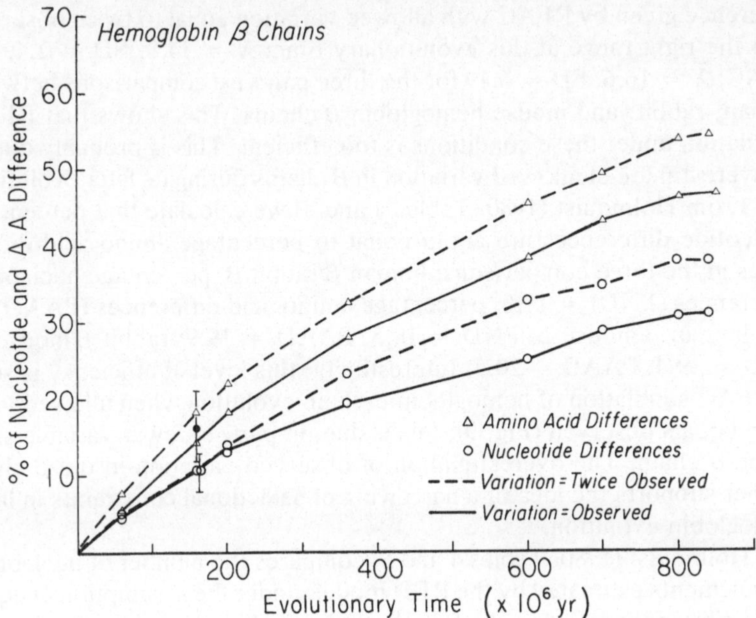

Figure 10. Comparison of means of percentage nucleotide and amino acid differences at different evolutionary times given by RFAC simulation of hemoglobin β-chain evolution. △, mean percentage amino acid differences; ○, mean percentage nucleotide differences; dashed lines, allowed variation equals twice observed; solid lines, allowed variation equals observed. Rate of fixation per codon per year equals 2.3×10^{-9} throughout.

of percentage amino acid sequence difference. However, the difference in observed variation between the α and β chains of hemoglobin is small. Furthermore, an additional increase in allowed variation over observed in hemoglobin β chain does not increase this type of efficiency. These facts indicate that the kinds of amino acids allowed may also influence the efficiency of amino acid to nucleotide sequence divergence. This question deserves further investigation.

The messenger RNA sequences for hemoglobin β chain have been determined for three mammals,—human, rabbit, and mouse. These are compared with each other, as are their corresponding proteins, by Holmquist (1980). The evolutionary time span separating all of these species is about 160 million years. The mean and standard deviation of the percent nucleotide differences from comparisons of these sequences are given in Fig. 10 ($\bar{X}$ = 16.5, SD = 4.27). The percentage nucleotide sequence differences given by RFAC for this evolutionary time are too low ($\bar{X}$ = 11.3, SD = 2.01) (Fig. 10). However, the percentage *amino acid* sequence difference given by RFAC with allowed variation equal to twice observed is in the right range at this evolutionary time; $\bar{X} = 14.8$, SD = 2.49 for RFAC; $\bar{X} = 16.6$, SD = 6.19 for the three pairwise comparisons between human, rabbit, and mouse hemoglobin β chains. This shows that RFAC simulation under these conditions is too efficient. This is probably due to an overestimate of allowed variation in β chains during its later evolution.

From Holmquist (1980, Tables 4 and 5) we calculate that percentage nucleotide differences are about equal to percentage amino acid differences in the three comparisons: human β/rabbit β, percentage nucleotide differences (PND) = 11.6, percentage amino acid differences (PAAD) = 9.5; human β/mouse β, PND = 18.5, PAAD = 19.9; rabbit β/mouse β, PND = 19.4, PAAD = 20.5. Interestingly, this level of efficiency is seen in RFAC simulation of hemoglobin α-chain evolution when allowed variation equals observed (Fig. 9); this is due, in part, to fewer variable sites in the α chain. The overestimation of observed variation in the β chain further supports the idea of a narrowing of selectional constraints in later hemoglobin evolution.

Holmquist (1980, Tables 4 and 5) compares the number of nucleotide replacements estimated by the REH model, under the assumption of equal probability of genetic events, from both the nucleotide and the amino acid sequence data. These estimates can be compared with the number of nucleotide replacements required in RFAC simulation to give the observed nucleotide and amino acid sequence differences between the three species' messenger RNAs and proteins. We use the RFAC curves in Fig. 10, where allowed variation equals observed. For human β/rabbit β there are 51 total nucleotide replacements observed, or 11.6% (100 x 51/3 x 146 codons) nucleotide difference, REH gives 61 nucleotide replacements, and RFAC gives 57; for 9.6% (100 x 14/146 codons) amino acid difference, REH gives 28 nucleotide replacements, and RFAC gives 25. For human β/mouse β there is 18.5% nucleotide difference; REH gives 105 nucleotide replacements, RFAC gives 113; there is 19.9% amino acid difference; REH gives 77 nucleotide replacements, and RFAC gives 58. For rabbit β/mouse β there is 19.4% nucleotide difference; REH gives 102 nucleotide replacements, and RFAC gives 120 there is 20.5% amino acid difference; REH gives 71 nucleotide replacements, and RFAC gives 60. Thus, when nucleotide sequence data are used both methods give better estimates of the actual number of fixations that have occurred. Both underestimate the number of fixations to about the same extent when amino acid sequence data are used. This shows that in *both* methods, when protein data are used, the amount of variation allowed in the protein is overestimated. However, the better results when messenger RNA sequences are considered indicate that the underlying stochastic aspects of the models are adequate. It is hoped that messenger RNA sequences from lower vertebrates will be forthcoming, so that the generality of these conclusions can be assessed.

6. Conclusions and Discussion

The question of the validity of the neutral hypothesis has become very entangled. Perhaps the major progress made in more than a decade since the hypothesis was explicitly stated is that now most people agree the question is no longer, "Does random fixation of neutral mutations occur in macromolecules?," but rather, "How much of the evolution of macromolecules is due to this process?" We think that even this phrasing of the question is inadequate unless it is asked for each particular macromolecule (protein and gene) in turn.

Even put this way, there is still the question of how to address the question. The neutral hypothesis purports to explain the evolutionary clock, and thus the existence of a uniform rate of macromolecular evolution is cited in favor of the hypothesis. Indeed, Fitch and Langley (1978) state: "If 99% of all mutations fixed during evolution of proteins are neutral, then the evolution of these proteins should be proceeding at a uniform rate equal to the neutral mutation rate." As Kimura and Ohta (1974) have pointed out and we have shown (Section 5.2.2.1),this is not strictly true. Accelerated rates of protein evolution may occur when selectional constraints are relaxed, even if 100% of all mutations fixed are neutral. Furthermore, given the slowing down of the rate of accumulation of sequence differences which must occur as the saturation point of divergence is approached (Section 5.2), we wonder what "uniform rate" must mean in this context.

In any case, there is considerable disagreement as to whether an evolutionary clock actually exists. Some workers accept its existence (Zuckerkandl and Pauling, 1965; Kimura, 1968, 1969; Dickerson, 1971; Air and Thompson, 1971; Dayhoff, 1972; Kimura and Ohta, 1974; Van Valen, 1974; Zuckerkandl, 1976; Wilson *et al.*, 1977; Kimura, 1979); others find that nonuniform rather than uniform rates characterize protein evolution (Moore *et al.*, 1976; Holmquist *et al.*, 1976; Holmquist, 1976; Goodman *et al.*, 1975; Goodman, 1978). Fitch and Langley (1978) (see also Langley and Fitch, 1973, 1974) find that: "Rates of amino acid changing nucleotide substitutions are significantly nonuniform in the same protein in different lines of descent and between different proteins in the same line of descent," but that averaged over long time periods, divergence times measured in nucleotide substitutions agree with those measured by the fossil record. In other words, a molecular clock exists but is not very accurate over short time periods due to perhaps selection or perhaps stochastic fluctuations (Wilson *et al.*, 1977). Kimura (1979) suggests that, despite the fluctuations, the existence of a clock over long time spans is still important and vouches for the neutral hypothesis. Fitch and Langley (1978) say that the fluctuations remove support for the neutral hypothesis, but conclude that the possibility is not ruled out that a "sizeable fraction" of all substitutions are neutral.

Finally, models have been proposed which claim to show that evolution occurring entirely by fixation of selectively advantageous mutations in a constant environment can result in uniform rates of macromolecular evolution (Van Valen, 1974; Zuckerkandl, 1976).

A major conclusion of our work is that a model that simulates macromolecular evolution entirely by random fixation of neutral mutations at a constant rate produces a more regular accumulation of protein sequence differences with time than does real evolution, at least for the four proteins studied. This is so despite the fact that RFAC simulation gives variances of sequence divergence between versions of the same protein at each evolutionary time that are comparable to those of real evolution (Coates and Stone, 1981, Tables II–VI).

Thus, macromolecular evolution entirely by random drift would seem to be ruled out. This is in agreement with others who have measured nonuniform rates of protein evolution (Goodman *et al.*, 1975; Moore *et al.*, 1976; Goodman, 1978; Langley and Fitch, 1973, 1974; Fitch and Langley, 1978). However, can it be argued that the observed irregularities in real evolution are the result of stochastic fluctuations in an essentially random process, or are due to errors in measurement of rates of evolution stemming from misinterpretation of the fossil record (Wilson *et al.*, 1977), or a combination of these? Furthermore, even if our conclusion is correct, are we any closer to an answer to the question of how much of evolution is by genetic drift?

This brings us to the second major conclusion of our work. The validity of the neutral hypothesis must be assessed for each protein separately. The comparison of RFAC simulation results for different proteins yields important information. For instance, RFAC simulation under conditions of neutrality gives a better fit to the results of real evolution in the case of the fibrinopeotides and cytochrome *c* than it does for the hemoglobin chains. This coincides with our expectations based upon other considerations. The functional roles served by the different proteins lead us to expect that the fibrinopeptides and cytochromes *c* may have evolved more by random fixations of neutral mutations than has hemoglobin (Sections 5.2.1.1, 5.2.2.1, 5.2.2.2). The nature of the data with which we must work, namely, contemporary protein and gene sequences, may not permit a determination of the importance in any macromolecule of neutral evolution in absolute terms (Langley and Fitch, 1973; Holmquist, 1976). The best we may expect is knowledge of its *relative* importance in different proteins.

Comparison of RFAC simulation results for different proteins also helps with the question of whether fluctuations in real macromolecular evolution are significant. The same fossil evidence (virtually the same phylogeny) was used for the simulation of both cytochrome *c* and the hemoglobin chains in the vertebrates (Coates and Stone, 1981). Both hemoglobin chains show major irregularities in accumulation of sequence differences with time compared with RFAC simulation, whereas those of cytochrome *c* are relatively minor. Those who would invoke stochastic fluctuations or errors in the phylogeny will have to explain why these affect hemoglobin chains and not cytochrome *c*. We again suggest that cytochrome *c* is a more likely candidate for neutral evolution than is hemoglobin.

Another major conclusion is that RFAC simulation can be remarkably predictive in terms of the number of nucleotide replacements required to generate a particular amount of macromolecular sequence divergence. However, its accuracy, at least for protein sequences, is very dependent on the amount of allowed variation used. This demonstrates that selectional constraints, measured as allowed variation, are probably the most important determinating factor in macromolecular evolution. The manner of handling these selectional constraints as allowed variation is the unique aspect of our model. Using RFAC simulation, we can easily measure the effects of allowed variation.

Our analysis has shown that evolution accompanying a *relaxation* of selectional constraints can be viewed as neutral, whereas evolution during a *tightening* of selectional constraints cannot. In the first, evolution will continue through random fixations of selectively neutral mutations, whereas in the second, *some* fixations will occur through selection for newly advantageous mutations. The first will cause an increase in the rate of sequence divergence and the second will cause a decrease (Sections 5.2.2.1 and 5.2.2.2).

The case where selectional constraints are tightened and a new mutation belonging to the new set of allowed variants replaces one that belonged to the old set but is now deleterious points up the relative nature of selection coefficients. Here we have a new mutation that is selectively neutral or advantageous with respect to wild type, depending upon which of the former set of allowed variants occupies the site in the species. Failure to recognize the relative nature of selection coefficients may have led to some confusion about the meaning of selection coefficients and the true nature of selection. Indeed, Zuckerkandl (1976) writes of positive and negative selection coefficients, forgetting that s (selection coefficient) is defined in terms of w (relative Darwinian fitness), where $w = 1 - s$ and the greatest value of w is one (Crow and Kimura, 1970; Dobzhansky, 1970). We have also often heard it said that *negative* selection, *against* deleterious mutations, is unimportant in evolution and only *positive* selection, *for* advantageous mutants, need be considered. But how can there be one without the other? For these reasons we suggest that the phrase "positive Darwinian selection" be dropped and in its place "directional selection" or "selection for newly advantageous mutations" be substituted.

Finally, we wish to briefly mention a modification RFAC simulation which is now under investigation. That is the altering of allowed variation in different parts of the phylogeny not by simple expansion or contraction, but by changing to a different set of allowed amino acids, of perhaps the same size. Clearly this is simulation of directional selection, because newly advantageous mutations will replace newly deleterious ones at the time of transition, but with a neutral component, because between transition points, drift among selectively neutral codons will occur. We feel intuitively that this is a realistic view of macromolecular evolution. This brings RFAC simulation closer to the covarion model of Fitch (1971). It can be seen that this will have the effect of producing irregular rates of sequence divergence during the time span covered by the entire phylogeny (Langley and Fitch, 1973) if only for the reason that when different sets of allowed variants are used in different parts of the phylogeny the possibilities for sequence divergence are automatically increased. However, by permitting smaller sets of allowed variation in any particular part of the phylogeny, say, for mammals, it will bring the percentage nucleotide and amino acid sequence differences closer together, as is required by the data from mammalian hemoglobin β chains and their messenger RNAs (Section 5.3).

Of course, the extreme case of this model occurs when each point in the phylogeny of a protein has only one allowed amino acid sequence. This is the extreme selectionist view of evolution. We predict, on the basis of the known degeneracy of the genetic code, that in this case the amount of nucleotide sequence divergence will far exceed the amount of amino acid sequence divergence, contrary to the known results of real evolution.

References

Air, G. M., and Thompson, E. P. O., 1971, Studies on marsupial proteins. IV. Amino acid sequence of myoglobin from the red kangaroo *Megaleia rufa*, *Aust. J. Biol. Sci.* **24**(1):75–95.

Anfinsen, C. B., 1959, *The Molecular Basis of Evolution*, Wiley, New York.

Barnard, E. A., Cohen, M. S., Gold., M. H., and Kim, J.–K., 1972, Evolution of ribonuclease in relation to polypeptide folding mechanisms, *Nature* **240**(5381):395–398.

Coates, M., and Stone, S., 1981, Simulation of protein evolution by random fixation of allowed codons, *J. Mol. Evol.* **17**(6):311–328.

Crow, F., and Kimura, M., 1970, *An Introduction to Population Genetics Theory*, Harper and Row, New York.

Dayhoff, M. O. (ed.), 1972, *Atlas of Protein Sequence and Structure*, Volume 5, National Biomedical Research Foundation, Washington, D. C.

Dayhoff, M. O. (ed.), 1973, *Atlas of Protein Sequence and Structure*, Volume 5, Supplement 1, National Biomedical Research Foundation, Washington, D. C.

Dayhoff, M. O. (ed.), 1976, *Atlas of Protein Sequence and Structure*, Volume 5, Supplement 2, National Biomedical Research Foundation, Washington, D. C.

Dayhoff, M. O. (ed.), 1978, *Atlas of Protein Sequence and Structure*, Volume 5, Supplement 3, National Biomedical Research Foundation, Washington, D. C.

Dickerson, R. E., 1971, The structure of cytochrome c and rates of molecular evolution, *J. Mol. Evol.* **1**(1):26–45.

Dobzhansky, T., 1970, *Genetics of the Evolutionary Process*, Columbia University Press, New York.

Fitch, W. M., 1970, Distinguishing homologous from analogous proteins, *Syst. Zool.* **19**(2):99–113.

Fitch, W. M., 1971, Rate of change of concomitantly variable codons, *J. Mol. Evol.* **1**(1):84–96.

Fitch, W. M., and Langley, C. H., 1978, Protein evolution and the molecular clock, in: *Evolution of Protein Molecules* (M. Hirosha, and T. Yamanaka, eds.), Japan Scientific Societies Press, Tokyo, pp. 45–60.

Fitch, W. M., and Markowitz, E., 1970, An improved method for determining codon variability in a gene and its application to the rate of fixations of mutations in evolution, *Biochem. Genet.* **4**(4):579–593.

Goodman, M., 1978, Substitutional trends and non-random changes in rates during protein evolution, in: *Evolution of Protein Molecules* (H. Matsubara and T. Yamanako, eds.), Japan Scientific Societies Press, Tokyo, pp. 17–32.

Goodman, M., Moore, G. W., Barnabas, J., and Matsuda, G., 1974, The phylogeny of human globin genes investigated by the maximum parsimony method, *J. Mol. Evol.* **3**(1):1–48.

Goodman, M., Moore, G. W., and Matsuda, G., 1975, Darwinian evolution in the geneology of haemoglobin, *Nature* **253**(5493):603–608.

Grantham, R., 1974, Amino acid difference formula to help explain protein evolution, *Science* **185**(4154):862–864.

Hennig, W., 1966, *Phylogenetic Systematics*, University of Illinois Press, Urbana, Illinois.

Holmquist, R., 1972, Theoretical foundations of paleogenetics, in: *Proceedings of the Sixth Berkeley Symposium on Mathematical Statitistics and Probability: Darwinian, Neo-Darwinian, and Non-Darwinian Evolution*, Volume 5 (L. LeCam, J. Neyman, and E. Scott, eds.), University of California Press, Berkeley, California, pp. 315–350.

Holmquist, R., 1976, Random and nonrandom processes in the molecular evolution of higher organisms, in: *Molecular Anthropology* (M. Goodman, and R. E. Tashian, eds.), Plenum Press, New York, pp. 89–116.

Holmquist, R., 1978, The REH theory of protein and nucleic acid divergence? A retrospective update, *J. Mol. Evol.* **11**(4):361–374.

Holmquist, R., 1980, Evolutionary analysis of α and β hemoglobin genes by REH theory under the assumption of the equiprobability of genetic events, *J. Mol. Evol.* **15**(2):149–159.

Holmquist, R., and Pearl, D., 1980, Theoretical foundations of quantitative paleogenetics Part III: The molecular divergence of nucleic acids and proteins for the case of genetic events of unequal probability, *J. Mol. Evol.* **16**(4):211–267.

Holmquist, R., Cantor, C., and Jukes, T. H., 1972, Improved procedures for comparing homologous sequences in molecules of proteins and nucleic acids, *J. Mol. Biol.* **64**(1):145–161.

Holmquist, R., Jukes, T. H., Moise, H., Goodman, M., and Moore, W. G., 1976, The evolution of the globin family genes: Concordance of stochastic and augmented maximum parsimony genetic distances for α hemoglobin, β hemoglobin and myoglobin phylogenies, *J. Mol. Biol.* **105**(1):39–74.

Hufford, T. L., 1978, *Botany: Basic Concepts in Plant Biology*, Harper & Row, New York.

Karon, J. M., 1979, The covarion model for the evolution of proteins: Parameter estimates and comparison with Holmquist, Cantor and Jukes' stochastic model, *J. Mol. Evol.* **12**(3):197–218.

Kimura, M., 1968, Evolutionary rate at the molecular level, *Nature* **217**(5129):624–626.

Kimura, M., 1969, The rate of molecular evolution considered from the standpoint of population genetics, *Proc. Natl. Acad. Sci. USA* **63**(6):1181–1188.

Kimura, M., 1979, The neutral theory of molecular evolution, *Sci. Am.* **24**(5):98–126.

Kimura, M., and Ohta, T., 1973, Mutation and evolution at the molecular level, *Genet. Supp.* **73**:19–35.

Kimura, M., and Ohta, T., 1974, On some principles governing molecular evolution, *Proc. Natl. Acad. Sci. USA* **71**(7):2848–2852.

King, J. L., and Jukes, T. H., 1969, Non-Darwinian evolution, *Science* **164**(3881):788–798.

Langley, C. H., and Fitch, W. M., 1973, The constancy of evolution: A statistical analysis of the α and β hemoglobins, cytochrome C, and fibrinopeotide A, in: *Genetic Structure of Populations* (N. E. Morton, ed.), University Press of Hawaii, Honolulu, pp. 246–262.

Langley, C. H., and Fitch, W. M., 1974, An examination of the constancy of the rate of molecular evolution, *J. Mol. Evol.* **3**(3):161–177.

Maniatis, T., Fritsch, E. F., Lauer, J., and Lawn, R. M., 1980, The molecular genetics of human hemoglobins, *Annu. Rev. Genet.* **14**:145–178.

Maruyama, T., Watt, K. W. K., and Riggs, A., 1980, Hemoglobins of the tadpole of the bullfrog, *Rana catesbeiana*. Amino acid sequence of the α chain of a major component, *J. Biol. Chem.* **255**(8):3285–3293.

Matsuda, G., 1978, Molecular evolution of hemoglobins, in: *Evolution of Protein Molecules* (H. Matsubara, and T. Yamanaka, eds.), Japan Scientific Societies Press, Tokyo, pp 175–186.

Moore, W. G., Goodman, M., Callahan, C., Holmquist, R., and Moise, H., 1976, Stochastic versus augmented maximum parsimony method for estimating superimposed mutations in the divergent evolution of protein sequences. Methods tested on cytochrome C amino acid sequences, *J. Mol. Biol.* **105**(1):15–37.

Nishioka, Y., and Leder, P., 1979, The complete sequence of a chromosomal mouse α-

globin gene reveals elements conserved throughout vertebrate evolution, *Cell* **18**:875–882.

Romer, A. S., 1966, *Vertebrate Paleontology*, 3rd ed., University of Chicago Press, Chicago.

Spuhler, J. N., and Jorde, L. B., 1975, Primate phylogeny, ecology, and social behavior, *J. Anthropal. Res.* **31**:376–405.

Van Eysinga, F. W. B., 1975, *Geological Time Table*, 3rd ed., Elsevier Scientific Publishing Company, Amsterdam.

Van Valen, L., 1974, Molecular evolution as predicted by natural selection, *J. Mol. Evol.* **3**(2):89–101.

Vogel, F., and Kopun, M., 1977, Higher frequencies of transition among point mutations, *J. Mol. Evol.* **9**(2):159–180.

Watt, K. W. K., Maruyama, T., and Riggs, A., 1980, Hemoglobins of the tadpole of the bull frog, *Rana catesbeiana*. Amino acid sequence of the β chain of a major component, *J. Biol. Chem.* **255**(8):3294–3301.

Wilson, A. C., Carlson, S. S., and White, T. J., 1977, Biochemical evolution, *Annu. Rev. Biochem.* **46**:573–639.

Yockey, H. P., 1977a, A prescription which predicts functionally equivalent residues at given sites in protein sequences, *J. Theor. Biol.* **67**(3):337–343.

Yockey, H. P., 1977b, On the information content of cytochrome C, *J. Theor. Biol.* **67**(3):345–376.

Yockey, H. P., 1977c, A calculation of the probability of spontaneous biogenesis by information theory, *J. Theor. Biol.* **67**(3):377–398.

Young, J. Z., 1962, *The Life of Vertebrates*, 2nd ed., Oxford University Press, Oxford.

Zuckerkandl, E., 1976, Evolutionary processes and evolutionary noise at the molecular level. II. A selectionist model for random fixations in proteins, *J. Mol. Evol.* **7**(4):269–311.

Zuckerkandl, E., 1978, Multilocus enzymes, gene regulation, and genetic sufficiency, *J. Mol. Evol.* **12**(1):57–89.

Zuckerkandl, E., and Pauling, L., 1965, Evolutionary divergence and convergence in proteins, in: *Evolving Genes and Proteins* (V. Bryson, and H. J. Vogel, eds.), Academic Press, New York, pp. 97–166.

CHAPTER 7

Nonuniform Molecular Divergence

The Quantitative Evolutionary Analysis of Genes and Messenger RNAs under Selective Structural Constraints

RICHARD HOLMQUIST, DENNIS PEARL, and THOMAS H. JUKES

1. Introduction

The divergence of species is accompanied by molecular changes in the primary structure of genes and their messenger RNA products. In the present chapter we consider the simplest of such changes, those caused by point mutation: the replacement of one nucleoside—adenosine (A), cytidine (C), guanosine (G), or thymidine (T)—by another and fixation of that replacement by natural selection or by random drift.

Homologous genes are those descendant from a common ancestral gene. As the genes accumulate fixed point mutations their structures differ from that of their common ancestor as well as from each other. To interpret the observed changes between homologous pairs of genes isolated from extant species, one must have a falsifiable theory capable of explaining how the stepwise base replacements lead from the common ancestral gene to the gene structures observed experimentally in present-day organisms.

A theory is falsifiable if any of its predictions can be in principle in disagreement with observation. A theory may explain the observational

RICHARD HOLMQUIST and THOMAS H. JUKES • Space Sciences Laboratory, University of California at Berkeley, Berkeley, California 94720. *DENNIS PEARL* • Department of Statistics, University of California at Berkeley, Berkeley, California 94720.

data perfectly without being falsifiable; in that case its correctness becomes a matter of belief rather than of experimental test. Independent of its falsifiability, a theory may be useful. Although the theory of parsimony (Fitch, 1971) is not falsifiable, in conjunction with a topologically correct branching order for species divergence, it yields approximate lower bounds (Moore, 1977) for the total number of fixed mutations separating two genes. It is a reasonable biologic requirement that any other theory purporting to explain the observations predict a total number of fixed mutations at least as large as the number inferred by parsimony.

The viewpoint to be developed in the present chapter derives from our belief that it is not possible, even in principle, to know accurately every evolutionary detail of the divergence of genes from each other. A case in point is provided by the ancestral gene sequences inferred by the theory of parsimony. In such sequences, correctly inferred sites derive mostly from unvaried nucleotide loci. The invariant loci do not contain any evolutionary information (Ratner, 1977). At varied loci the reconstructed ancestral sequences are prone to err (Holmquist, 1979), the extent of error increasing with the distance of the reconstructed ancestor from the present and with the sparseness of the sequence data. The fewer the contemporary sequences forming branch points for the given distance to the present, the greater the error.

The source of our uncertainty about the detailed pathways of genic change lies in the probabilistic mechanisms of mutation (Holmquist, 1976) as well as in the probabilistic processes by which such mutations are fixed in natural populations of organisms interacting with their environment (Nei, 1975; Matsuda *et al.*, 1978). Another source of uncertainty is our lack of knowledge about the functional constraints on structure during the period of divergence of two genes from their common ancestor to the present.

Faced with these unpleasant realities, we will show in the following sections that there is nonetheless a significant amount of evolutionary information that can be recovered from the statistical patterns present in the primary structure of contemporary genes. The recovery of this information is not dependent on a detailed knowledge of hypothetical ancestral states, but upon our ability to categorize the experimental differences in the primary structure of two homologous genes into simply defined *observational* classes.

2. Constrained Stochastic Theory

Changes in gene structure are the result of genetic drift as well as of a constant testing of those structures for successful biologic function: the trial and error process of Darwinian selection. It is natural, because of the probabilistic nature of these processes, that stochastic theory should be an important resource for studying the dynamics of this selection at the molecular level provided we allow for those nonrandom* structural features imposed by biologic function.

Among stochastic models, theories of gene evolution can be divided into two classes: random models, in which genetic events of various types are assumed to be equiprobable (Jukes and Cantor, 1969; Holmquist, 1972a, 1980; Holmquist *et al.*, 1972); and nonrandom models, in which the various types of genetic events have differing probabilities (Jukes and Holmquist, 1972; Dayhoff and McLaughlin, 1972; Nei and Tateno, 1978).

In both classes of models it has been assumed that all gene or protein families, their different biologic functions and chemical structures notwithstanding, can be analyzed for evolutionary content by some "typical" or "average" mutation probability matrix. The assumption is not tenable in view of the fact—but one of numerous examples—that the base composition of mRNA for several β-hemoglobins (Efstratiadis *et al.*, 1977; Marotta *et al.*, 1977; Konkel *et al.*, 1978) at the third position within codons is around A:C:G:U::6:29:37:28, while that of gene J in ΦX174 (Sanger *et al.*, 1977) is very different, 24:11:18:47. One should try to take such obvious effects of natural selection into account.

Four important constraints on gene structure are:

1. Nonuniform base composition at each of the three positions within codons: at the first two positions in structural genes this nonuniformity arises from selection against the genetic code (Jukes *et al.*, 1975; Holmquist, 1978a).
2. Unequal replacement† frequencies of one base by another.
3. Each variable codon site may have a different probability of fixing a mutation, so that the density of fixed mutations among codons is not always Poissonian.
4. The fact that the three positions within a codon may have different probabilities of fixing a mutation, the third position in particular normally fixing many more mutations than the other two, because of the degeneracy of the genetic code.

*In this chapter we use the words *nonrandom* and *nonuniform* interchangeably. This usage is common in biology; we nevertheless wish to apologize to statisticians who use the word *nonrandom* to mean, specifically, *deterministic*, and who use the word *uniform* for a particular type of statistical distribution.

†Throughout this paper we have used the term *base replacement* rather than *base substitution*. We realize the word "substitution" is widely used for base changes in nucleic acids, and the word "replacement" for amino acid changes in proteins. However, the former word has the connotation of replacement by an approximate equivalent, whereas the replacement of one base by another sometimes has functional consequences nonequivalent to those of the original base.

The first constraint implies unequal usage of degenerate codons for the same amino acid (King and Jukes, 1969). A mathematical formalism for studying the effect of these constraints on gene divergence and the estimation of evolutionary parameters has been published (Holmquist and Pearl, 1980). For brevity we have designated this formalism nonrandom REH theory. In nonrandom REH theory no *a priori* experimentally unjustified boundaries are placed on the magnitude of the above four nonuniformities, and consequently an arbitrarily large extent of nonrandom structure in genes can be accommodated within the formalism. Within the envelope determined by known constraints a great many evolutionary pathways are possible, many, and possibly most of them, compatible with biologic function. Detailed consideration of these pathways does not make an evolutionary analysis mathematically intractable. On the contrary, it permits a more accurate estimation of evolutionary parameters than is possible with random models in which genetic events of various types are assumed to be equiprobable.

In the present chapter we do not intend to detail the calculational methodology. That can be found in Holmquist and Pearl (1980). Rather, we will emphasize the qualitative biologic concepts on which the evolutionary estimations are based. These concepts will be illustrated by concrete numerical results from real gene or messenger RNA sequences for cytochrome *c*, α-hemoglobin, β-hemoglobin, and the two viral capsid proteins VP1 and VP2 from three tumor viruses BKV, SV40, and polyoma. At each stage of this discussion, the close relationship between theoretical concepts and the experimentally observed structures in genes and mRNAs is stressed.

3. Magnitude of Selective Constraints

3.1. Base Composition

SV40, BKV, and polyoma viruses induce tumors in humans, monkeys, and mice, respectively. The nucleotide sequences of the genomes of these viruses have been published (Reddy *et al.*, 1978; Seif *et al.*, 1979; Soeda *et al.*, 1980a) and show significant homology. We have calculated the base composition at 323 codon loci for the gene that codes for the VP1 capsid protein for each of these viruses. The part of the gene that overlaps the VP2/3 gene was excluded. The average base composition is given in Table I. The four bases are not present at 25% each, but range from 14% for thymidine at the first position within the codons to 33% for adenosine at the second position within the codons. Because of the extensive degeneracy of the genetic code at the third position within a codon,

Table I
Average Base Composition at Each Position within the Codon at 323 Compared Codon Loci in the VP1 Gene for Three Tumor Viruses, BKV/SV40, and Polyoma

	A	C	G	T
First position	32.90	20.04	32.82	14.24
Second position	33.27	22.60	17.73	26.40
Third position	27.40	21.52	22.52	28.56

a base replacement at that position does not usually change the amino acid encoded. One might therefore expect that each of the four bases would be present in roughly equal amounts at that position. In the VP1 genes, this is in fact the case (Table I). But it is not true in general, as can be seen from the example for β-hemoglobin in RNA shown in Section 2.

The important concept that we wish to illustrate by Table I is that among genes that code for proteins of a given biologic function, that is, for a particular family of homologous genes, the base composition varies only slightly from one gene to another. By using the average base composition at each of the three codon positions in evolutionary calculations, one will obtain more accurate estimates than by assuming, as is done in random models, that each base is equifrequent in the nucleotide sequence.

3.2. Base Replacement Probabilities

If during gene divergence each of the four bases were equally likely to be replaced by any one of the other three, as in random models, the base composition of the gene would rapidly approach a value of 25% for each of the four bases, and this would be true irrespective of the base composition of the ancestral gene (Holmquist, 1972a, Table II), provided replacements were sufficiently frequent. Thus to maintain, over the long term, the base compositions observed in real genes, the frequencies of each of the 12 possible types of base replacements, A $\rightarrow$ C, for example, cannot be equal. The question then arises of how one determines those base replacement probabilities that do maintain compositional fidelity. Such probabilities cannot be calculated by simply counting the number of each of the 12 types of replacements between two homologous sequenced genes or messenger RNAs. This is because the replacements we want to count have occurred in ancestral sequences about which we have no knowledge. For example, an observed replacement A $\leftrightarrow$ C between two present-day genes could have arisen from the two ancestral replacements G → A in one gene, G → C in the other. A method for calculating base replacement probabilities that maintain the experimentally observed base composition at each position within the codon has been published (Holmquist and Cimino, 1980). The result for the first position within the codons of the VP1 genes is given in Table II. For a random model, the entries in the table would all be equal at 1/12th, that is, 8.33%. In fact, the probabilities range from about 4% for the transversion A → T to about 14% for the transition G → A. Just as importantly, it is not necessary for the replacement probabilities to be symmetric in order to preserve the experimentally observed base composition: in Table II the transversion C → A is almost twice as likely as its reverse A → C. From the right-hand side of the table the conditional base replacement probabilities are not 1/3 as they would be for a random model, but for adenosine range from 0.16 to 0.59. A given base is thus not equally likely to be replaced by any one of the other three.

The biologically important consequence of these unequal base replacement probabilities is to make the passage from one gene structure to another more difficult in the sense of requiring more fixed mutations than would be necessary were the conditional base replacement probabilities all equal at the value of one-third (Holmquist and Pearl, 1980, Fig. 1).

3.3. Density of Fixed Mutations among Variable Codons

Because within a gene different codon loci fix mutations with differing probabilities, a Poisson density is frequently a poor descriptor of the proportion of codons fixing a given number of mutations. To avoid assuming *a priori* that each codon has an equal probability of fixing a mutation, we follow a suggestion of Uzzell and Corbin (1971) that the density of fixed mutations among codons may be more appropriately described

Table II

Composition Faithful Base Replacement Frequencies at the First Position within Codons in the VP1 Gene for Three Tumor Viruses, BKV/SV40, and Polyoma

	New base				$A \to C \mid A$ 0.25 $A \to G \mid A$ 0.59 $A \to T \mid A$ 0.16
Old base	A	C	G	T	.
A	—	5.83	13.95	3.71	.
C	10.52	—	10.47	5.06	.
G	13.96	5.79	—	3.68	$C \to A \mid C$ 0.40
T	9.96	7.15	9.92	—	$T \to G \mid T$ 0.37

by the negative binomial density. The negative binomial density arises naturally when, because of the restrictions of biologic function, different codon sites have different probabilities of fixing a mutation. Only two parameters R and P must be estimated to define this density, so that from a computational point of view it is minimally more complex than the one-parameter Poisson density. It has the further virtue that it includes the Poisson density as a special limiting case ($R \to \infty$, $P \to 0$), so that if the latter density is indeed appropriate to the data, it will not be arbitrarily excluded.

Another limiting case of the negative binomial density occurs when $R = 1$, in which case the density is geometric, a density so far totally ignored in the literature of molecular evolution, but which we shall see as this chapter develops may be the most appropriate statistical descriptor for some gene families.

These relationships are summarized at the top of Fig. 1, and depicted graphically in Fig. 2 for an average fixation intensity of six fixed base replacements per codon. In Fig. 2 the Poisson density is quasisymmetric,

DISTRIBUTION OF FIXED MUTATIONS AMONG CODONS

NEGATIVE BINOMIAL DENSITY

$$PR(N) = \binom{R + N - 1}{N} P^N Q^R$$

LIMITING CASES

POISSON: $PR(N) = e^{-\mu}\mu^N/N!$

$R \to \infty$

$P \to 0$

$\mu \leftarrow RP/Q$

GEOMETRIC: $PR(N) = QP^N$

$R = 1$

DISTRIBUTION OF FIXED MUTATIONS WITHIN CODONS

MULTINOMIAL

$$PR(N; X_1, X_2, X_3) = \frac{N!}{X_1!X_2!X_3!} P_1^{X_1} P_2^{X_2} P_3^{X_3}$$

$$P_1 + P_2 + P_3 = 1$$

$$X_1 + X_2 + X_3 = N$$

Figure 1. Distribution of fixed mutations among and within codons.

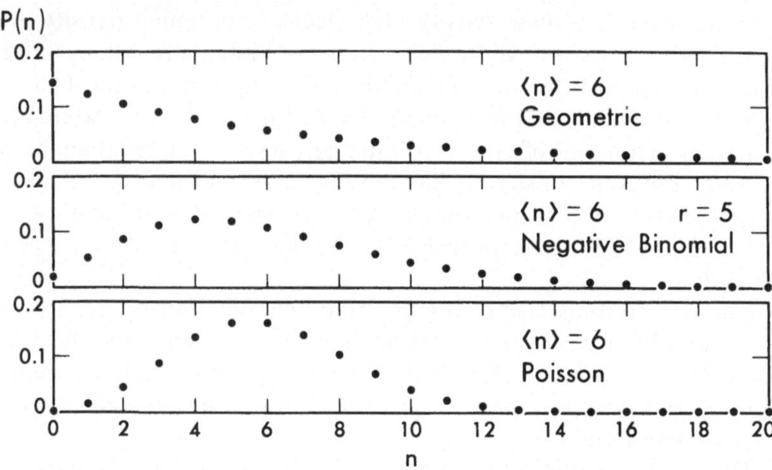

Figure 2. The shape of the negative binomial density and its limiting special cases.

with relatively fewer codons fixing a number of mutations far displaced from either side of the mean value. The geometric density is grossly asymmetric, with a relatively larger number of codon sites fixing very few mutations. For situations between these two limiting cases, the negative binomial density has the intermediate form shown in the center of Fig. 2. For completeness, we note that the parameter R need not be an integer and may assume any nonzero positive value.

In nonrandom REH theory, the experimental distribution of observed base replacements along the gene determines which density is most appropriate.

The evolutionary consequence of a nonuniform distribution of fixed mutations along a gene is to increase the number of base replacements necessary to explain a given observed change between two homologous genes. This is because for a given number of fixed mutations, the number of superimposed replacements within codons is minimal when these replacements can be distributed uniformly among all codons, rather than only among a restricted set of the codons available.

3.4. Distribution of Fixed Mutations within Codons

The three nucleotide positions within codons do not each have an equal probability of fixing a mutation. Nichols and Yanofsky (1979) reported 36, 17, and 146 base replacements between the *trpA* genes from

Salmonella typhimurium and *Escherichia coli*. The distribution of fixed mutations in the sea urchin genes for histones H2A, H2B, H3, and H4 (Grunstein and Grunstein, 1977; Schaffner *et al*., 1978; Sures *et al*., 1978) is even more extreme: 15, 5, and 138 at the three positions, respectively. Although the type of base replacement at a given position within a codon may be somewhat influenced by the nature of the bases at the other two positions (Jukes, 1978), the distribution of base replacements within codons may be approximately described by the multinomial density shown at the bottom of Fig. 1. This density has two parameters P_1 and P_2 ($P_3 = 1 - P_1 - P_2$) and the density gives the probability that for a codon fixing a total of N mutations, X_1 of these will be at the first position within the codon, X_2 at the second, and X_3 at the third.

Accurate values of the parameters P_i cannot be obtained from a simple count of the number of observed base replacements between two sequenced genes at each position within the codon. Thus, in the example of the above paragraph for the tryptophan synthetase A-chain gene, P_1 is poorly estimated by 36/(36 + 17 + 146). Such a calculation does not correct for the distortion caused by superimposed base replacements at the same position within a codon. The magnitude of this distortion is quite severe, as illustrated in Fig. 3. In Fig. 3 the X axis shows the state of a codon from the initial situation in which it has fixed no mutations to the extreme state in which it is saturated with fixed mutations. The parameter μ, designated the fixation intensity (as in Fig. 1, top), is simply the average number of fixed base replacements per codon that the gene has sustained. For simplicity of presentation, we have taken the true values of P_1 and P_2 equal at 0.12, so that P_3 is 0.76. These values are reasonably representative of those we estimate for the divergence of real genes (Table XIX, last three columns). The apparent P_i are shown along the Y axis as a function of the fixation intensity. These apparent values were calculated from

$$P_i = \frac{N_i}{N_1 + N_2 + N_3} \tag{1}$$

the N_i being the expected number of base differences at the ith position within the codons. The N_i were obtained from Eq. (29) in Holmquist and Pearl (1980). To isolate the effect of a nonuniform distribution of fixed mutations within codons, the distribution of fixed mutations among codons was taken to be a Poisson density with parameter μ. To maintain reasonable realism, the base composition at each position within the codon

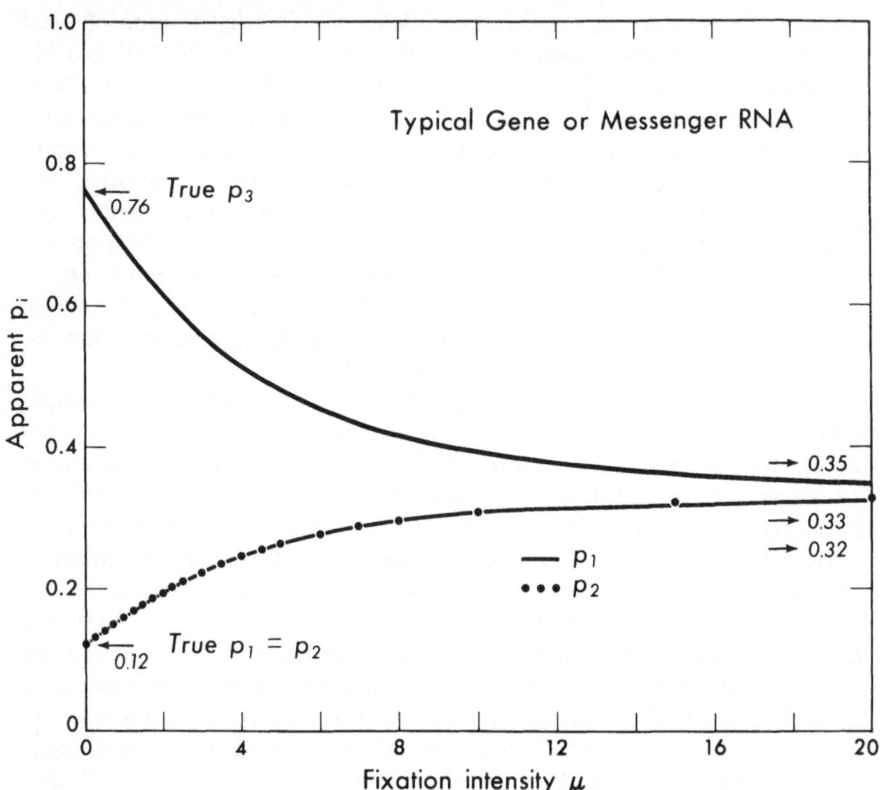

Figure 3. The relationship between the apparent and true distribution of fixed mutations within codons as evolutionary divergence proceeds.

was calculated from the amino acid composition of 189 individual sequences comprising 81 different functional families (Holmquist, 1978a). For *all* values of the fixation intensity μ in Fig. 3, the *true* values of P_1, P_2, and P_3 are 0.12, 0.12, and 0.76, respectively. However, the apparent values of these parameters, that is, those values that would be obtained by a direct count of base differences at each of the three codon positions between two homologous sequenced genes, have means that differ greatly from the true values. As the average number of base replacements per codon increases, P_1, P_2, and P_3 will appear approximately equal (at about 1/3 each) regardless of their true values. This distortion of values—falsely high values of P_1 and P_2, falsely low values of P_3—occurs almost immediately after the gene begins to fix mutations. Figure 3 suggests a simple way to correct for this distortion empirically: because the curves are linear for low values of μ, extrapolation of the apparent P_i values obtained by direct comparison of closely related genes with different known values of μ back to $\mu = 0$ will give Y-axis intercepts approximating the true P_i. The difficulty with this approach, and one that may preclude its use in practice, is that when there have been so few fixed mutations, a valid statistical sampling of the number of them at each position within the codon cannot be obtained. Also, most often μ is not known *a priori*, but must itself be estimated from the data. On the other hand, when the number of replacements has become large enough to give a valid statistical sample, the distortion is so large that linear extrapolation cannot be used.

The importance of the above considerations is that estimates of the total mutations fixed during evolutionary divergence are strong functions of the multinomial parameters. In the next section, we sketch a simple way to estimate these parameters that avoids the above difficulties.

4. Estimation of Parameters

The purpose of the preceding section was twofold: first, to give realistic examples of the magnitude of selective constraints on gene structures; and second, to point out that theories of gene divergence that ignore these constraints will bias the evolutionary information extracted from nucleic acid sequence data.

In the present section, we tackle the problem of assigning numerical values to the parameters of nonrandom REH theory that measure the constraints. There are five parameters to be estimated: the two negative binomial parameters R and P, which describe the distribution of fixed mutations among codons; the parameters P_1 and P_2 of the multinomial distribution which describes the distribution of fixed mutations within codons; and T_2, the average number of codon positions within the homologous gene pair being studied that have been free to fix mutations during the divergence of the two genes from their common ancestor.

4.1. The Observational Principle Underlying Parameter Estimation in Gene Sequences

It is beyond the scope of this chapter to enumerate the details of the estimation procedures. These are given in Holmquist and Pearl (1980). However, the principle behind the estimation procedure is easily stated and illustrated: the five parameters R, P, P_1, P_2, and T_2 are estimated from the fine structure of the gene pair as follows.

Consider the homologous pairs of genes coding for the VP1 capsid protein in BK virus and polyoma virus. The number of base differences between the two genes is classified in two ways: in the first, the numbers of codons having no, one, two, or three observed replacements are tabulated; in the second, the numbers of replacements that occur at each of the three codon positions are tabulated. This tabulation is shown in the first row of numbers of Table III. Because the total number of replacements must be the same by either method of tabulation, we have the relation

$$N(1) + N(2) + N(3) = M(1) + 2M(2) + 3M(3) \tag{2}$$

which makes only five of our observational classes independent. These five independent experimental aspects of the fine structure of the pair of homologous genes were used to estimate the five parameters R, P, P_1, P_2, and T_2.

For some more complex attempt to explain the data, say with a model requiring seven parameters, it would be necessary to consider the fine structure of the pair of homologous genes in more detail by defining at least seven independent observational classes.

Estimates of the five parameters obtained by nonrandom REH theory are shown in Tables IV and V for the VP1 genes of BK virus and polyoma virus. They are: $\hat{R} = 1$; $\hat{P} = 0.84$; $\hat{T}_2 = 308$; $\hat{P}_1 = 0.15$; $\hat{P}_2 = 0.12$; $\hat{P}_3 = 0.73$. From these numerical estimates the expected values of the $M(I)$ and $N(I)$ are calculable from Eqs. (28) and (29) in Holmquist and Pearl (1980), and are in excellent agreement with the observed values (Table III).

Table III
Gene VP1: BKV vs Polyoma[a]

	Number $M(J)$ of codons with J base replacements between MRNAs				Number $N(J)$ of base replacements at 1st, 2nd, and 3rd position within codons		
	$M(0)$	$M(1)$	$M(2)$	$M(3)$	$N(1)$	$N(2)$	$N(3)$
Observed	80	110	93	40	118	105	193
Predicted	80	109	95	39	118	105	193

[a]Total coding triplets compared: 323. Total fixed mutations $= N(1) + N(2) + N(3) = M(1) + 2M(2) + 3M(3)$. Five of these are independent. These five experimentally measureable quantities suffice to determine the evolutionary parameters of the theory: R, P, P_1, P_2, and T_2.

Table IV
Best Fit Evolutionary Parameters for Gene VP1: BKV vs. Polyoma

Negative binomial parameters			
	R 1.0		
	P 0.84		
	$\langle N \rangle = \mu_2 = RP/Q$ 5.4		
Number of varions	T_2 308		
Total fixed mutations	$REH = \mu_2 T_2$ 1663		
Fixed mutations/codon compared:			
$REHC = REH/T_2 = REH/323 = 5.1$			
	Pattern of superimposed mutations at each position within codon:		
Observed replacements	$N(1) = 118$	$N(2) = 105$	$N(3) = 193$
Actual replacements	$P_1 \times REH = 253$	$P_2 \times REH = 196$	$P_3 \times REH = 1214$

Table V further illustrates the point brought out in Section 3.4, that the apparent values of the relative frequency with which the first, second, and third positions within codons fix mutations may be inaccurate estimates of the true relative frequencies. In Table V, the relative frequencies found to explain the experimental observations adequately differ appreciably from the apparent values. The latter, for this example, are greater than the estimated true frequencies by a factor of about two at the first two positions within the codon (0.28 vs. 0.15 for P_1; 0.25 vs. 0.12 for P_2). The apparent frequency of 0.46 at the third position within the codon is very much lower than the estimated true frequency $P_3 = 0.73$.

In making the above estimations we have used the information in five classes of experimental observations to estimate five parameters. It is thus not possible to make a χ^2 test on the observed and expected values

Table V
Relative Rates of Substitution at Three Positions within Codon in the Gene VP1: BKV vs. Polyoma

Observed in structural gene	Estimated actual
$\dfrac{N(1)}{N(1) + N(2) + N(3)} = 0.284$	$P_1 = 0.152$
$\dfrac{N(2)}{N(1) + N(2) + N(3)} = 0.252$	$P_2 = 0.118$
$\dfrac{N(3)}{N(1) + N(2) + N(3)} = 0.464$	$P_3 = 0.730$

shown in Table III. Our model does create real constraints on the set of possible predicted values, so that the agreements found between theory and observation are encouraging and not circular.

A simple extension of the classification in Table III is to consider the eight aspects of the gene fine structure

___   ×__   _×_   __×   ××_   ×_×   _××   ×××

in which a × indicates a nucleotide locus within a codon that has sustained a base replacement. The five-parameter theory is able to predict the expected values of the eight classes, so that a more rigorous test of the adequacy of the methods can be made. In particular, since seven of these classes are independent, a χ^2 goodness-of-fit test with two degrees of freedom may be used.

For a stochastic process of gene divergence, there are many base replacement pathways that lead from an ancestral gene to its descendants found in living species. As we do not, and cannot, know that particular pathway actually taken, any pathway that is consistent with the experimentally sequenced gene structure and any other available experimental information is a scientifically valid explanation of that structure. There is a region of parameter space, rather than a single point in that space, that will provide a satisfactory explanation of the experimental observations. This region can be large for two reasons: (1) the choice of observational classes does not include all the relevant information present in the nucleic acid sequences (see preceding paragraph); and (2) the total information available in the sequences does not suffice to delimit the number of pathways consistent with those sequences to only a few. In geometric terms, the valley about the best-fit estimates of the parameters is broad and shallow rather than narrow and deep. In statistical terms this means that the variance of the estimated parameters may be large. Such a large variance is not necessarily an indication of theoretical inadequacy, because it reflects not only the limits on the precision of information that *has been* (model dependent) extracted from the nucleic acid sequence data, but also the inherent and irreducible limit on the maximal information (model independent) that *can be* extracted from those data (Jaynes, 1979).

4.2. Evolutionary Measures

In Table IV, the parameters calculated in Section 4.1 are related to phenomena of evolutionary interest. For the VP1 genes of BKV virus and polyoma virus:

1. The distribution of fixed mutations among codons is not well explained by the Poisson density. For a Poisson density, $R \rightarrow \infty$ and $P \rightarrow 0$. For these genes, $\hat{R}$ is near unity and $\hat{P}$ is much greater than zero. The distribution of fixed mutations among variable codons is approximately geometric.
2. The fixation intensity μ, the average number of base replacements per variable codon (varion), is given by the product of the negative binomial parameters divided by $Q = 1 - P$ [Holmquist and Pearl, 1980, Eq. (19)]. Its estimated value $\hat{\mu}_2$ was 5.4 replacements per varion.
3. The estimated number of varions $\hat{T}_2$, a measure of the mutable codon sites, is 308, less than the total number of sites compared (323), but greater than the 243 [$= M(1) + M(2) + M(3)$, from Table III] codon sites that show base differences between the two genes.
4. The estimated total mutations fixed $R\hat{E}H$ is simply the product of the fixation intensity by the number of varions, $R\hat{E}H = \hat{\mu}_2\hat{T}_2$, and is about 1700. This striking increase over the 416 [$= N(1) + N(2) + N(3)$, from Table III] observed differences between the two sequenced genes is due to superimposed base replacements at the same nucleotide locus.
5. The pattern of superimposed replacements at each position within the codons can be calculated by multiplying the estimated number of fixed mutations by the proportion that occur at each position. This proportion is taken from the right-hand column of Table V. The calculation is shown in detail at the bottom of Table IV. At the third position within the codon 1021 out of a total of 1214 replacements are superimposed. Thus 84% of the evolutionary history at this position is lost. Even at the first two positions, 50% of the evolutionary history is irretrievable. It is because of situations like this that we express doubts about the accuracy of ancestral sequence reconstructions, however made. For more closely related genes, the loss of evolutionary information would of course be less.

Having outlined the methodology of nonrandom REH theory and its relationship to experimentally observable aspects of gene structures, we now turn to a direct comparison of evolutionary estimates made from the earlier models (Jukes and Holmquist, 1972; Holmquist, 1978b, 1980) in which genetic events of various types were assumed to be equiprobable. For brevity we refer to these earlier models as random REH theory. We will also consider the differences in the estimates made from the amino acid sequence data of proteins, and those made from the nucleic acid sequence data of genes and messenger RNAs.

5. A Comparison of Evolutionary Estimates Made from Protein and Nucleic Acid Sequence Data and from Nonrandom REH Theory

A comparison of the evolutionary estimates made from amino acid and nucleic acid sequence data by theory that does, and does not, consider the selective constraints of Section III will tell us how strongly the information content of the macromolecular sequences used influences those estimates. It will also tell us the quantitative extent to which selective factors are apt to change the estimates made from theories in which the equiprobability of genetic events is assumed. To span a spectrum of possibilities, we will first consider a closely related pair of sequences, the β-hemoglobins or the corresponding mRNAs for rabbit and mouse; and second, we will consider a distantly related pair, the α- and β-hemoglobins or their mRNAs from rabbit.

5.1. A Closely Related Divergence: Mouse and Rabbit β-Hemoglobin

The experimental data (Efstratiadis *et al.*, 1977; Konkel *et al.*, 1978) for rabbit and mouse β-hemoglobin are shown at the top of Table VI, the

Table VI
Experimental Data from Hemoglobin Comparisons

	A closely related pair: rabbit β/mouse β	A distantly related pair: rabbit α/rabbit β
Protein sequence data:		
No. residues with J minimal base replacements		
$J = 0$	116	57
$J = 1$	24	61
$J = 2$	6	21
$J = 3$	0	0
mRNA sequence data:		
No. codons $M(J)$ with J base replacements between mRNAs		
$M(0)$	84	38
$M(1)$	44	42
$M(2)$	14	44
$M(3)$	4	15

corresponding data for the mRNA sequences that code for these proteins being shown in the bottom half of the table.

The 30 amino acid changes between these hemoglobins can be classified into replacements requiring a minimum of no, one, two, and three base replacements in the corresponding codons as introduced by Jukes (1963) and subsequently adopted by Fitch and Margoliash (1967). There are 116 identical amino acids in the two proteins, 24 amino acid replacements requiring a single-base replacement in the codons, six amino acid replacements requiring a two-base replacement, and no amino acid replacements requiring a base replacement at each of the three positions of the codon.

A quite different story is told by the mRNA sequences for mouse and rabbit β-hemoglobin. Not only are the number of codons fixing no base replacements fewer (84 vs. 116), because replacements at the third position within codons are not silent in mRNAs; but the number of codons fixing one or two mutations is markedly larger in the mRNA sequences. There are four codons that have sustained mutations at all three positions of the codon. The number of codons changed is 62, in contrast to the 30 suggested by the amino acid sequence data (Table VII). The 36 minimal base differences are a very poor guide to the 84 actual differences found between the two messenger RNAs. We conclude that amino acid sequence data is of limited value for genetic inference. The fundamental reason for this is that the information content of an amino acid residue is inherently

Table VII

Evolutionary Parameters for a Closely Related Hemoglobin Pair: Rabbit β/Mouse β

Experimental parameter	Estimated from protein sequences	Estimated from mRNA sequences
Amino acid differences	30	—
Total codons changed	—	62
Minimal base differences	36	—
Observed base differences	—	84

Calculated parameter	Protein data, random REH theory (1972)	mRNA data: Random REH theory (1972)	mRNA data: Nonrandom REH theory (1980)
Fixation intensity[a] μ_2	1.45	0.97	1.28
Number of varions T_2	49	104	120
Total replacements[b] *REH*	71	101	154

[a]Average number of base replacements per varion.

[b]Includes superpositions, back mutations, and parallelisms.

less than that of a nucleotide triplet, only $\log_2 20$ or 4.32 bits being required to specify the former, and 3 $\log_2 4$ or 6 bits being required for the latter.

This effect of the redundancy in the genetic code translates into markedly different evolutionary inferences (Table VII, bottom). Even when mRNA sequences are used, random and nonrandom models give substantially different estimates for the evolutionary parameters μ_2, T_2, and *REH*. To explain the observed fine structure of these mRNAs, an estimated total of 154 base replacements are needed. The random model yields a substantially smaller (101 replacements) estimate. The estimated proportion of the structural gene that is able to fix mutations during the mouse/rabbit divergence from their common ancestor increases from 71% (104/146) when structural constraints are ignored to 82% (120/146) when they are considered.

5.2. A Distantly Related Divergence: Rabbit α- and β-Hemoglobin

The experimental data for rabbit α-hemoglobin (Heindell *et al.*, 1978) and β-hemoglobin (Efstratiadis *et al.*, 1977) and their mRNAs are also summarized in Table VI. The ratio of amino acid replacements requiring two base replacements in the codons to those requiring a single base replacement is only 0.34 (21/61) when estimated from amino acid sequence data. The actual ratio of codon replacements in which two base replacements occurred in the codons to those in which a single base replacement occurred is 1.04 (44/42, from lower half of Table VI). Whereas the protein data show that there are no amino acid replacements requiring a base replacement at each of the three positions within codons, of the 82 total amino acid replacements, 15 actually do involve a replacement at each of the three positions.

It is perhaps unanticipated that the number of codons fixing a single base replacement is no higher in the mRNAs for the distant α/β comparison than for the close β/β comparison, being 42 for the former and 45 for the latter (the difference persists even after correction for the slightly different number of codons compared, 139 and 146, in the two cases). An observed decrease for distant divergences in codons fixing a single base replacement is in agreement with the theoretical predictions of REH theory (Holmquist, 1980, Fig. 1).

In Table VIII we again see that amino acid differences are a poor indicator of total codons changed, and minimal base differences are a poor estimate of the actual base differences between mRNAs. The bottom half of Table VIII shows that for this distant divergence the evolutionary estimates made from the protein data differ from those made from the nucleic acid sequence data, but the differences in estimated values are

Table VIII
Evolutionary Parameters for a Distantly Related Hemoglobin Pair: Rabbit α/Rabbitβ

Experimental parameter	Estimated from protein sequences	Estimated from mRNA sequences
Amino acid differences	82	—
Total codons changed	—	101
Minimal base differences	103	—
Observed base differences	—	175

Calculated parameter	Protein data, random REH theory (1972)	mRNA data: Random REH theory (1972)	mRNA data: Nonrandom REH theory (1980)
Fixation intensity μ_2	2.01	2.52	3.72
Number of varions T_2	115	115	114
Total replacements *REH*	231	290	425

relatively smaller than for the closely related mouse/rabbit β-hemoglobin divergence in Table VII. The reason for this is that for such divergent sequences there have been a sufficiently large number of amino acid substitutions of various types to give a better statistical sample of differences. The inclusion of selective constraints, however, is still necessary to obtain accurate estimates, as shown in the last column at the bottom of Table VIII: We estimate that 425 replacements are required to explain the observed fine structure (Table XVIII) of the mRNAs; only 290 would be required to explain the same number of observed differences for mRNAs of random structure and diverging from each other in a random manner.

6. Comparison of Calculation with Experiment

For the VP1 genes of BKV and polyoma virus, we have shown that the evolutionary parameters calculated from nonrandom REH theory and given in Table IV are able to explain the seven categories of gene fine structure tabulated in Table III: observed and calculated values are in good agreement. To be useful a method must have broad applicability, and the purpose of this section is to show that the above agreement is not an accident of the particular genes from particular species chosen for analysis.

In Table IX the experimental data for the iso-1 and iso-2 cytochrome *c* genes from yeast are tabulated (Smith *et al.*, 1979; Montgomery *et al.*,

Table IX
Experimental Data and Evolutionary Parameters Calculated from Nonrandom REH Theory for Yeast Iso-1 vs. Iso-2 Cytochrome *c* Genes

	Number $M(J)$ of codons with J base replacments between messenger mRNAs				Number $N(J)$ of base replacements at lst, 2nd, and 3rd position within codons		
	$M(0)$	$M(1)$	$M(2)$	$M(3)$	$N(1)$	$N(2)$	$N(3)$
Observed	46	50	10	2	11	8	57
Predicted	45	50	12	1	11	8	58

Calculated evolutionary parameters[a]

Negative binominal					Multinomial		
R	P	μ_2	T_2	REH	P_1	P_2	P_3
1.0	0.74	2.78	100	278	0.049	0.034	0.917

[a] $\mu_2 = \langle N \rangle = RP/(1 - P)$; $REH = \mu_2 T_2$.

1980). Evolutionary parameters were estimated for this gene pair and are given at the bottom of the table. For these parameters the expected number of base replacements in each of the seven categories considered were calculated and are given beneath the observed values in the table. The agreement, though not perfect, is quite reasonable. (In all the tables of this section, the expected values have been rounded to the nearest integer. The predicted category totals may, for this reason, differ by a unit or so from the observed totals.)

In Tables X and XI we continue our analysis of the VP1 tumor virus genes at 323 compared loci (969 nucleotide sites). The predicted $M(I)$ and $N(I)$ were calculated from the estimated evolutionary parameters given at the bottom of these tables. For the BKV and SV40 viral gene pair (Table X) the predicted value of $M(2)$ is high. For the polyoma and SV40 viral gene pair (Table XI) the agreement seems reasonable. The former case provides an example where the estimate of the negative binomial parameter R is less than unity.

The gene coding for the VP2 capsid protein in these viruses has a significantly different base composition at each of the three base positions within codons than does the VP1 gene. The two genes do not appear to be homologous. We compared the three pairs of genes at 276 codon positions (828 nucleotide sites). The results are given in Tables XII–XIV. The agreement between expected and observed values is good.

Tables XV and XVI show the comparison of the nucleic acid sequence coding for human β-hemoglobin (Marotta *et al.*, 1977; Lawn *et al.*, 1980)

Table X
Gene VPI: BKV vs. SV40

	Number M(J) of codons with J base replacements between mRNAs				Number $N(J)$ of base replacements at 1st, 2nd, and 3rd position within codons		
	$M(0)$	$M(1)$	$M(2)$	$M(3)$	$N(2)$	nt$N(1)$ $N(3)$	
Observed	143	122	38	20	63	49	146
Predicted	143	111	54	15	64	50	149

Calculated evolutionary parameters[a]

Negative binomial					Multinomial		
R	P	μ_2	T_2	REH	P_1	P_2	P_3
0.5	0.86	2.97	323	959	0.11	0.08	0.81

[a] $<N> = \mu_2 = RP/(1 - P)$; $REH = \mu_2 T_2$.

with that for rabbit (Table XV) and mouse (Table XVI). Calculated and observed values agree. In the context of the limited set of comparisons considered in this chapter, these two are the only ones for which the data are clearly in support of an approximately Poisson density of fixed mutations among the variable codons: $\hat{R}$ is large and $\hat{P}$ is near zero.

Tables XVII and XVIII document the agreement between expected and observed values for the nucleic acid sequence comparisons for the hemoglobins of Section 5.

Table XI
Gene VP1: SV40 vs. Polyoma

	Number $M(J)$ of codons with J base replacements between mRNAs				Number $N(J)$ of base replacements at 1st, 2nd, and 3rd position within codons		
	$M(0)$	$M(1)$	$M(2)$	$M(3)$	$N(1)$	$N(2)$	$N(3)$
Observed	70	125	90	38	111	92	216
Predicted	70	123	95	36	114	94	211

Calculated evolutionary parameters[a]

Negative binomial					Multinomial		
R	P	μ_2	T_2	REH	P_1	P_2	P_3
1	0.92	11.97	303	3626	0.06	0.05	0.89

[a] $<N> = \mu_2 = RP/(1 - P)$ $REH = \mu_2 T_2$.

Table XII
Gene VP2: BKV vs. SV40

	Number $M(J)$ of codons with J base replacements between mRNAs				Number $N(J)$ of base replacements at 1st, 2nd, and 3rd position within codons		
	$M(0)$	$M(1)$	$M(2)$	$M(3)$	$N(1)$	$N(2)$	$N(3)$
Observed	126	106	37	7	50	30	121
Predicted	126	105	39	6	50	30	121
Calculated evolutionary parameters[a]							
Negative binomial					Multinomial		
R	P	μ_2	T_2	REH	P_1	P_2	P_3
1	0.67	2.04	259	527	0.13	0.07	0.80

[a] $\langle N \rangle = \mu_2 = RP/(1 - P)$; $REH = \mu_2 T_2$.

7. Generality of Results

In this chapter we have analyzed gene or messenger RNA structures from five species (yeast, mice, rabbits, monkeys, and humans) and three functional families: the cytochromes c, α- and β-hemoglobin, and two capsid proteins from three tumor viruses. These nucleic acid sequences differ in molecular structure from each other, as do the proteins for which they code. Both closely related and distantly related divergences have been considered.

Table XIII
Gene VP2: BKV vs. Polyoma

	Number $M(J)$ of codons with J base replacements between m RNAs				Number $N(J)$ of base replacements at 1st, 2nd, and 3rd position within codons		
	$M(0)$	$M(1)$	$M(2)$	$M(3)$	$N(1)$	$N(2)$	$N(3)$
Observed	49	81	91	55	149	111	168
Predicted	49	78	98	52	148	111	169
Calculated Evolutionary Parameters[b]							
Negative binomial					Multinomial		
R	P	μ_2	T_2	REH	P_1	P_2	P_3
1	0.92	10.88	262	2850	0.24	0.09	0.67

[a] $\langle N \rangle = \mu_2 = RP/(1 - P)$; $REH = \mu_2 T_2$.

Table XIV
Gene VP2: SV40 vs. Polyoma

	Number $M(J)$ of codons with J base replacements between mRNAs				Number $N(J)$ of base replacements at 1st, 2nd, and 3rd position within codons		
	$M(0)$	$M(1)$	$M(2)$	$M(3)$	$N(1)$	$N(2)$	$N(3)$
Observed	43	88	94	51	152	105	172
Predicted	43	85	100	48	151	105	173

Calculated evolutionary parameters[a]

Negative binomial					Multinomial		
R	P	μ_2	T_2	REH	P_1	P_2	P_3
1	0.90	8.85	275	2434	0.26	0.09	0.65

[a] $<N> = \mu_2 = RP/(1 - P)$; $REH = \mu_2 T_2$.

The estimated evolutionary parameters obtained for the 11 comparisons made are summarized in Table XIX. This is a small sample, but these initial results indicate that nonrandom REH theory can be a useful tool for advancing our understanding of genetic divergence under functional constraints.

The estimated values of the negative binomial parameters R and P

Table XV
Human and Rabbit β-Hemoglobin mRNAs[a]

	Number $M(J)$ of codons with J base replacements between mRNAs				Number $N(J)$ of base replacements at 1st, 2nd, and 3rd position within codons		
	$M(0)$	$M(1)$	$M(2)$	$M(3)$	$N(1)$	$N(2)$	$N(3)$
Observed	106	31	9	0	9	7	33
Predicted	106	31	8	1	9	7	34

Calculated evolutionary parameters[b]

Negative binomial					Multinomial		
R	P	μ_2	T_2	REH	P_1	P_2	P_3
51	0.03	1.45	60	87	0.12	0.09	0.79

[a] In the human messenger RNA, the published nucleotide sequence of codon 53 is GCA (Marotta *et al.*, 1977). In the gene sequence (Lawn *et al.*, 1980) this triplet is GCT. We have used the former.

[b] $<N> = \mu_2 = RP/(1 - P)$; $REH = \mu_2 T_2$.

Table XVI
Human and Mouse β-Hemoglobin mRNAs

	Number $M(J)$ of codons with J base replacements between mRNAs				Number $N(J)$ of base replacements at 1st, 2nd, and 3rd position within codons		
	$M(0)$	$M(1)$	$M(2)$	$M(3)$	$N(1)$	$N(2)$	$N(3)$
Observed	87	40	18	1	22	17	40
Predicted	87	41	17	2	22	17	40

Calculated evolutionary parameters[a]

Negative binomial					Multinomial		
R	P	μ_2	T_2	REH	P_1	P_2	P_3
51	0.03	1.31	88	115	0.25	0.18	0.57

[a] $\langle N \rangle = \mu_2 = RP/(1 - P)$; $REH = \mu_2 T_2$.

indicate that the distribution of fixed mutations among codons is in some cases better described by an approximately geometric density rather than by the Poisson density in current use.

The estimated fixation intensity $\hat{\mu}_2$ ranges from 1.3 base replacements per varion for the close mouse/rabbit β-hemoglobin mRNA comparison, to about 12 for the distant SV40/polyoma viral gene comparison. This means that a given nucleotide locus sustains on the average from less than one to as many as four fixed base replacements.

The values of $\hat{T}_2$ are in most cases less than the total codon sites compared. There thus appears to be a meaningful restriction on the num-

Table XVII
Rabbit and Mouse β-Hemoglobin mRNAs

	Number $M(J)$ of codons with J base replacements between mRNAs				Number $N(J)$ of base replacements at 1st, 2nd, and 3rd position within codons		
	$M(0)$	$M(1)$	$M(2)$	$M(3)$	$N(1)$	$N(2)$	$N(3)$
Observed	84	44	14	4	21	17	46
Predicted	84	43	17	3	21	17	46

Calculated evolutionary parameters[a]

Negative binomial					Multinomial		
R	P	μ_2	T_2	REH	P_1	P_2	P_3
1	0.56	1.28	120	154	0.19	0.14	0.67

[a] $\langle N \rangle = \mu_2 = RP/(1 - P)$; $REH = \mu_2 T_2$.

Table XVIII
Rabbit α- and β-Hemoglobin mRNAs

	Number $M(J)$ of codons with J base replacements between mRNAs				Number $N(J)$ of base replacements at 1st, 2nd, and 3rd position within codons		
	$M(0)$	$M(1)$	$M(2)$	$M(3)$	$N(1)$	$N(2)$	$N(3)$
Observed	38	42	44	15	57	46	72
Predicted	38	42	44	15	57	46	72

Calculated evolutionary parameters[a]

Negative binomial					Multinomial		
R	P	μ_2	T_2	REH	P_1	P_2	P_3
11	0.25	3.72	114	425	0.25	0.16	0.59

[a] $\langle N \rangle = \mu_2 = RP/(1 - P)$; $REH = \mu_2 T_2$.

ber of codon sites that can fix mutations during the divergence of two genes from a common ancestor.

The large estimated number of total replacements fixed $R\hat{E}H$ speaks to the importance of multiple replacements at the same nucleotide locus, revertants, and parallelisms. These processes do not appear to be minor perturbations to evolutionary divergence at the molecular level, but may dominate it (see last two rows of Table IV).

We estimate that the third base position within codons fixes from about 60% of the accepted mutations to about 90% in the yeast iso-1/iso-2 cytochrome *c* comparison. The first base position within codons fixes an estimated 5–26% of all accepted mutations, and the second position within codons from about 3% to 16% of those mutations. In all cases examined, the estimates indicate that the first position fixed a larger proportion of the mutations than did the second, this ratio varying from a low of 1.3 (human and rabbit β-hemoglobin mRNAs) to a high of 3.0 (gene VP2 in the polyoma/SV40 comparison). If we simply average the values in Table XIX, the first, second, and third positions within codons fixed about 16%, 10%, and 74%, respectively, of the total accepted point mutations.

8. Extension to Noncoding Regions

The above methods are also applicable to the noncoding regions of the genome. The homologous pair of nucleotide sequences in the noncoding region is divided into triplets of nucleotides; the base replacements

Table XIX
Evolutionary Parameters for Nucleic Acids Coding for Structural Genes for Cytochrome *c*, Hemoglobins, and Tumor Viruses

	Negative Binomial parameters[a]					Multinomial parameters		
Comparison	R	P	μ_2	T_2	REH	P_1	P_2	P_3
Cytochrome *c*								
Yeast iso-1/iso-2	1	0.74	2.78	100	278	0.05	0.03	0.92
Hemoglobins								
Human β/rabbit β	51	0.03	1.45	60	87	0.12	0.09	0.79
Human β/mouse β	51	0.03	1.31	88	115	0.25	0.18	0.57
Mouse β/rabbit β	1	0.56	1.28	120	154	0.19	0.14	0.67
Rabbit α/rabbit β	11	0.25	3.72	114	425	0.25	0.16	0.59
Tumor proteins								
Gene VP1: BKV/SV40	0.5	0.86	2.97	323	959	0.11	0.08	0.81
Polyoma/BKV	1	0.84	5.40	308	1663	0.15	0.12	0.73
Polyoma/SV40	1	0.92	11.97	303	3626	0.07	0.04	0.89
Gene VP2: BKV/SV40	1	0.67	2.04	259	527	0.13	0.07	0.80
Polyoma/BKV	1	0.92	10.88	262	2850	0.24	0.09	0.67
Polyoma/SV40	1	0.90	8.85	275	2434	0.26	0.09	0.65

[a] $\langle N \rangle = \mu_2 = RP/Q$.

observed in these triplets are classified as in Table III; and the subsequent estimation of evolutionary parameters is the same as in Section 4. The nucleotide triplets to be compared should be free of gaps, but can otherwise be chosen in any convenient manner that does not bias the estimation. The nucleotides within a triplet need not be contiguous in the gene sequences unless ignoring the contiguity would lead to bias.

The presence of numerous genetic deletions or additions in either sequence can make it difficult to choose an alignment of the two sequences reflecting true homology. This is perhaps the factor limiting an adequate analysis of noncoding regions.

9. Importance of Accurate Genetic Distance Estimates to Systematics

The branching order depicting the species genealogy of the hemoglobin genes and the divergence of the viral tumor genes VP1 and VP2 is shown in Fig. 4. The numbers along the branch lengths are in units of estimated total fixed point mutations per 100 codons. The topologies and branch lengths shown are those for which the sum of the branch lengths for a given sequence pair deviates minimally from the REH values given in Table XIX, subject to the metric requirement that all branch lengths be positive (one cannot have a negative number of point mutations). The magnitude of the deviation was measured as the sum of squares of the differences between the reconstructed pairwise distances and the REH values of Table XIX. The more important considerations for inferring phylogenetic topopogies from pairwise distance data between species can

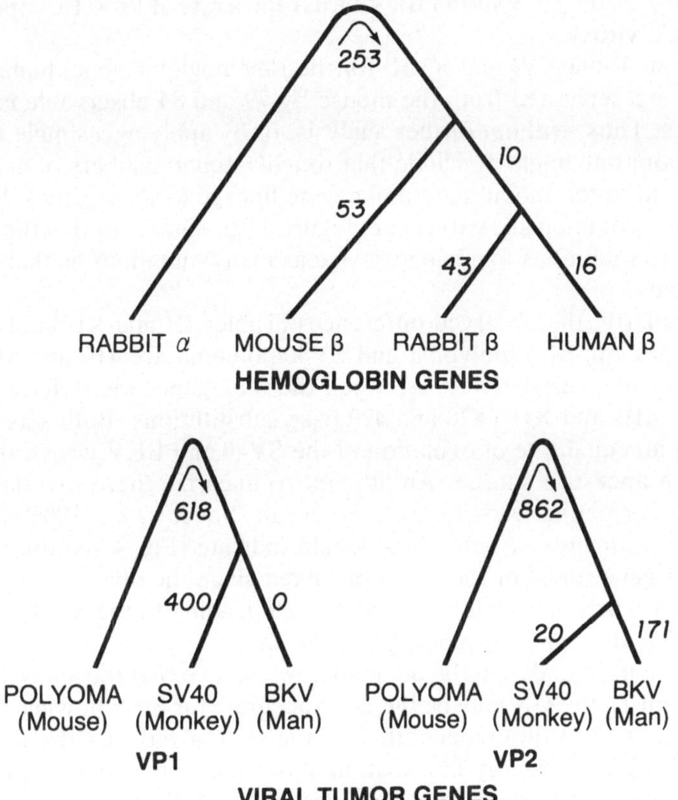

Figure 4. Cladistic relationships between the hemoglobin genes (top) and the tumor virus genes VP1 and VP2 (bottom) as deduced from the *REH* values in Table XIX. The numerical values along the branch lengths are the estimated total number of base substitutions per 100 codons.

be found in Beyer *et al.* (1974) and in Holmquist (1976). Calculational examples are in the Appendix of Holmquist (1972b).

From Fig. 4 (top), the α- and β-hemoglobin genes have diverged from their common ancestor at a rate of about 1.1×10^{-9} fixed mutations per nucleotide site per year if this divergence occurred some 450 million years ago. If the split of rodents, lagomorphs, and primates occurred about 90 million years ago, the β-hemoglobin structural genes have been diverging from their common ancestor at a rate of about 1.6×10^{-9}.

Assuming the tumor virus genes evolved with their host organism (Soeda *et al.*, 1980b), we find that the VP1 genes have diverged from their common ancestor at a rate (Fig. 4, bottom) of about 14×10^{-9} fixed mutations per nucleotide site per year, and the VP2 genes at between a rate of 6.4×10^{-9} (SV40 and BKV virus) and a rate of 19×10^{-9} (polyoma and BKV virus).

From Table XVI and XVII, for the β-hemoglobin gene, humans and rabbits are separated from the mouse by 79 and 84 observable base differences. Thus, without further analysis, or by applying a simple Poisson correction, one might conclude that roughly equal numbers of mutations were fixed in the rabbit and human gene lineage as these genes diverged from their common ancestral gene. Figure 4 (top) indicates that the lineage leading to rabbit has fixed about twice as many mutations as that leading to humans.

Similarly, the observed differences (Tables III and XI) between the VP1 genes for BKV/polyoma and SV40/polyoma are 416 and 419 base replacements, respectively. Between the VP2 genes the differences are (Tables XIII and XIV) 428 and 429 base substitutions. Both sets of data suggest an equal rate of evolution of the SV40 and BKV virus from their common ancestral source. An attempt to linearize these raw data by a Poisson correction leads to the same result (Soeda *et al.*, 1980b).

The estimates in Table XIX would indicate (Fig. 4, bottom) that in the VP1 gene, most of the mutations fixed since the divergence of SV40 and BKV virus occurred along the lineage leading to the SV40 virus. In gene VP2, the opposite seems to be the case.

Without intending to be dogmatic, we believe that the analysis stemming from Table XIX may be the more accurate for the following reasons. A simple total of differences [that is, the sum of the $N(I)$] between two genes does not give any information about the distribution of those differences along the gene or within the codon. A Poisson correction explicitly assumes that each variable nucleotide locus has the same probability of fixing a mutation, and that this probability is very low. From Table XIX, the first assumption is false for all the gene comparisons of this study; the latter assumption appears false for at least eight of the 11 comparisons. Finally, although the Poisson correction was designed to correct for superimposed fixed mutations, in a recent study of this correction involving 247 eukaryotic globins Goodman demonstrated beyond any reasonable doubt that the correction was entirely inadequate: the Poisson formula gave estimates of total fixed mutations (including superpositions) that were as much as a factor of three less than the minimum number of fixed mutations possible (Goodman, 1981a).

10. Discussion

In this chapter our goal has been to outline as realistically as possible the manner in which functional constraints on gene structures can be incorporated into evolutionary theory, and to show how these constraints affect the magnitude of evolutionary estimates and the manner in which these estimates are related to simple problems of interest to the systematist. It should be clear that systematists should be extremely cautious in accepting at face value the accuracy of estimates now in the literature.

Although the accuracy of the estimates given in Table XIX should be more accurate than analogous estimates made from theory in which functional constraints on gene structure and the process by which mutations are fixed are ignored or overidealized, there is a pressing need for further work. Three questions that need additional study are: (1) What are the variances of the parameters estimated in Table XIX?; (2) Have we considered all those functional constraints that are apt to have important quantitative effects on the evolutionary estimates?; and (3) Can the stochastic approach be expanded so that the information from recently diverged sequences is used to refine the estimates made for distantly related sequences in such a manner that the temporal causality is not violated? In the following three paragraphs we consider each of these in turn.

Because the fixation intensity μ_2 and the total mutations fixed REH are each proportional to the ratio P/Q, where $Q = 1 - P$, with P one of the two parameters of the negative binomial density describing the distribution of fixations among codons, a small change in the value of $\hat{P}$ may have a large effect on the values of $\hat{\mu}_2$ and $R\hat{E}H$. For a fixed value of $\hat{R}$, changing $\hat{P}$, from 0.70 to 0.90, a relative increase of 29%, increases estimates of the fixation intensity and total fixed mutations by a factor of about four, a relative increase of 400%. More generally, for a given R and T_2

$$\begin{aligned} \mathrm{Var}(\hat{\mu}_2) &\simeq \left(\frac{R}{Q^2}\right)^2 \sigma^2 \\ \mathrm{Var}(R\hat{E}H) &\simeq \left(\frac{T_2 R}{Q^2}\right)^2 \sigma^2 \end{aligned} \tag{3}$$

where σ^2 is the variance of $\hat{P}$. These variances vary inversely as the fourth power of Q. If Q is small, the variances will be inherently large and there is nothing we can do to reduce them other than estimate P as accurately as possible. The present discussion reinforces the considerations of the final paragraph in Section 4.1. A study of a dozen or so gene sequences coding for homologous proteins with the same biologic function might provide a better estimate of P for that functional family than any single pairwise comparison.

For a functional protein, and for a particular amino acid residue position within that protein, only some of the 20 possible amino acids may be allowed (Yockey, 1977). This in turn restricts the codons possible at the corresponding position within the mRNA or gene coding for the protein. Coates and Stone (1980) have found that such a restriction to functionally equivalent codons significantly increases estimates of total fixed mutations. The $R\hat{E}H$ values reported in Table XIX may thus be underestimates in this respect. At present no method exists, other than realistic computer simulations, to take this effect into account in a general way.

In the present chapter, for a given functional family we have estimated various evolutionary parameters by considering single pairs of gene sequences. A more accurate analysis could be attempted by considering multiple pairs of gene sequences. For example, in Fig. 4 (top), the evolutionary information present in the rabbit/human β-hemoglobin mRNA pair reflects the evolutionary events that occurred between the ancestral gene common to rabbit and human β-hemoglobin and the present-day human β-hemoglobin gene. These same events also occurred between the ancestral β-hemoglobin gene common to the mouse and human and the present-day human gene. By using the information from the rabbit/human pairwise comparison as well as the information present in the mouse/human comparison, it should in principle be possible to estimate the parameters for the latter divergence more accurately than by using the mouse/human comparison alone. Such an expanded analysis, though simple in concept, is technically difficult in execution. Nonetheless, an attempt at such an analysis should probably be made. It would, among other things, permit more meaningful discussion of inequalities in evolutionary rates.

We close the chapter with a comment on the complementarity of stochastic and parsimonious models. Stochastic models consider the properties of the statistical assemblages of nucleotides in gene and mRNA structures. The advantage of such an approach is the analytical ability to take into account quantitatively those nonrandom aspects of genes important for evolutionary inference. This approach also lets us see more clearly the really quite wide range of evolutionary pathways that are consistent with the observational data of such sequences. Such analyses spare us the illusion that the observed sequence had to arise from some preconceived, possibly incorrect *a priori* abiologic theory such as parsimony. The price paid for these advantages is not negligible. By admitting probabilistic notions into our understanding, one loses the ability to say precisely what went on at a particular nucleotide locus or at groups of loci. Though our understanding may be increased, we are accepting that it is forever in principle impossible to give precise values to the evolutionary parameters we are seeking. Though molecular evolution is blessed with an abundance of data that paleontologists can only envy, there is a drawback. The fossil record is given temporal order by the geologic strata in which the fossils are found. But in molecular evolution all the data come from gene sequences of extant organisms. It is as though the entire fossil record were found in a single geologic stratum. We should not allow our desire for simple theories to lull us into believing the accuracy of the estimates made by such theories. Estimates that are well-determined, in the sense of having small variances, in an unreasonable model are not better than estimates with larger variances from a more realistic model.

The drawback of parsimony is that the most parsimonious solution for a given molecular set of gene sequence data is almost certainly wrong in detail because there are infinitely many more nonparsimonious pathways that could have given rise to the same gene sequences. It would be presumptious to believe that the parsimonious solution happened to be close to the pathway actually taken. On the other hand, the analysis of gene and protein sequences by the method of parsimony does have the advantage that an approximate lower bound can be set for the number of point mutations that occur at particular nucleotide or codon loci or the amino acid residues for which the codon codes. The type of residue-by-residue evolutionary analysis of the functionally important residues in hemoglobin made by Goodman *et al.* (1975) (also Goodman, 1981b) using parsimony would be difficult by stochastic techniques.

By the use of both stochastic and parsimonious models we can advance our understanding more rapidly than by dogmatic adherence to one to the exclusion of the other, provided we maintain a clear distinction of the qualitative and quantitative limits of each.

11. Conclusions

Natural selection for biological function and the nature of the genetic code have resulted in experimentally measurable constraints on gene and messenger RNA structures. These include nonuniform base compositions, unequal replacement frequencies of one base by another during evolutionary divergence, codon sites of different mutabilities, and unequal probabilities for fixing a mutation at the three nucleotide loci within codons. These four types of evolutionary nonrandomness, rather than making a quantitative analysis of gene divergence mathematically intractable, in fact permit us to obtain more accurate estimates of evolutionary parameters. The consequence of nonuniform nucleic acid structures and nonrandom mechanisms of evolutionary divergence is to introduce a molecular inefficiency into the stochastic passage from a given gene structure to some other: it requires more fixed mutations to effect this passage for a nonrandom pathway than for a random pathway in which genetic events of various types are equiprobable. The conceptual relationships between nonrandom REH theory and the observed structures of genes and messenger RNAs isolated from extant organisms have been examined. Representative applications of these relationships to extract evolutionary information from such nucleic acid structures have been given for the genes of cytochrome *c*, alpha and beta hemoglobin, and for the genes that code for the VP1 and VP2 capsid proteins of the murine polyoma virus, simian SV40 virus, and human BK virus.

ACKNOWLEDGMENTS. This work was supported by a National Science Foundation Grant PCM 76 18627, Analytical Methods for Examining Protein and Nucleic Acid Evolution, and by a National Aeronautics and Space Administration award NGR–05–003–460, The Chemistry of Living Systems. We thank the Department of Energy for permitting us to use their high-speed CDC 7600 computer at Lawrence Berkeley Laboratory.

References

Beyer, W., Stein, M., Smith, T., and Ulam, S., 1974, A molecular sequence metric and evolutionary trees, *Math. Biosci.* **19**:9–25.

Coates, M., and Stone, S., 1980, Simulation of protein evolution by random fixation of allowed codons, *J. Mol. Evol.* **17**:311–328.

Dayhoff, M. O., and McLaughlin, P. J., 1972, Early evolution: Transfer RNA, in: *Atlas of Protein Sequence and Structure*, Volume 5 (M. O. Dayhoff, ed.), National Biomedical Research Foundation, Washington, D. C., pp. 111–118.

Efstratiadis, A., Kafatos, F. C., and Maniatis, T., 1977, The primary structure of rabbit β-globin mRNA as determined from cloned DNA, *Cell* **10**:571–585.

Fitch, W., 1971, Toward defining the course of evolution: Minimum change for a specific tree topology, *Syst. Zool.* **20**:406–416.

Fitch, W., and Margoliash, E., 1967, Construction of phylogenetic trees, *Science* **155**:279–284.

Goodman, M., 1981a, Globin evolution was apparently very rapid in early vertebrates: A reasonable case against the rate-constancy hypothesis, *J. Mol. Evol.* **17**:114–120.

Goodman, M., 1981b, Decoding the pattern of protein evolution, *Progr. Biophys. Mol. Biol.* **37**:105–164.

Goodman, M., Moore, G. W., and Matsuda, G., 1975, Darwinian evolution in the genealogy of haemoglobin, *Nature* **253**:603–608.

Grunstein, M., and Grunstein, J. E., 1977, The histone H4 gene of *Stronglyocentrotus purpuratus:* DNA and mRNA sequences at the 5_{prime} end, *Cold Spring Harbor Symp. Quant. Biol.* **42**:1083–1092.

Heindell, H. C., Liu, A., Paddock, G. V., Studnicka, G. M., and Salser, W. A., 1978, The primary sequence of rabbit α–globin mRNA, *Cell* **15**:43–54.

Holmquist, R., 1972a, Theoretical foundations of paleogenetics, in: *Sixth Berkeley Symposium on Mathematical Statistics and Probability* (L. LeCam, J. Neyman, and E. L. Scott, eds.), Volume 5, University of California Press, Berkeley, California, pp. 315–350.

Holmquist, R., 1972b, Empirical support for a stochastic model of evolution. *J. Mol. Evol.* **1**:211–222.

Holmquist, R., 1976, Random and nonrandom processes in the molecular evolution of higher organisms, in: *Molecular Anthropology* (M. Goodman, E. E. Tashian, and J. H. Tashian, eds.), Plenum Press, New York, pp. 89–116.

Holmquist, R., 1978a, Evaluation of compositional nonrandomness in proteins, *J. Mol. Evol.* **11**:349–360.

Holmquist, R., 1978b, The REH theory of protein and nucleic acid divergence: A retrospective update, *J. Mol. Evol.* **11**:361–374.

Holmquist, R., 1979, The method of parsimony: An experimental test and theoretical analysis of the adequacy of molecular restoration studies, *J. Mol. Biol.* **135**:939–958.

Holmquist, R., 1980, Evolutionary analysis of α and β hemoglobin genes by REH theory under the assumption of the equiprobability of genetics events, *J. Mol. Evol.* **15**:149–159.

Holmquist, R., and Cimino, J. B., 1980, A general method for biological inference: Illustrated by the estimation of nonrandom gene nucleotide transition probabilities, *Biosystems* **12**:1–22.

Holmquist, R., and Pearl, D., 1980, Theoretical foundations for paleogenetics—Part III: The molecular divergence of nucleic acids and proteins for the case of genetic events of unequal probability, *J. Mol. Evol.* **16**:211–267.

Holmquist, R., Cantor, C., and Jukes, T. H., 1972, Improved procedures for comparing homologous sequences in molecules of proteins and nucleic acids, *J. Mol. Biol.* **64**:145–161.

Jaynes, E. T., 1979, Where do we stand on maximum entropy?, in: *The Maximum Entropy Formalism* (R. D. Levine, and M. Tribus, eds.), MIT Press, Cambridge, Massachusetts, pp. 15–118.

Jukes, T. H., 1963, Some recent advances in studies of the transcription of the genetic message, in: *Advances in Biological and Medical Physics*, Volume 9, Academic Press, New York, pp. 1–41.

Jukes, T. H., 1978, Codons and nearest-neighbor nucleotide pairs in mammalian messenger RNA, *J. Mol. Evol.* **11**:121–127.

Jukes, T. H., and Cantor, C. R., 1969, Evolution of protein molecules, in: *Mammalian Protein Metabolism*, Volume III, Academic Press, New York, pp. 21–132.

Jukes, T. H., and Holmquist, R., 1972, Estimation of evolutionary changes in certain homologous polypeptide chains, *J. Mol. Biol.* **64**:163–179.

Jukes, T. H., Holmquist, R., and Moise, H., 1975, Amino acid composition of proteins: Selection against the genetic code, *Science* **189**:50–51.

King, J. L., and Jukes, T. H., 1969, NonDarwinian evolution, *Science* **164**:788–798.

Konkel, D. A., Tilghman, S. M., and Leder, P., 1978, The sequence of the chromosomal mouse β-globin major gene: Homologies in capping, splicing and poly(A) sites, *Cell* **15**:1125–1132.

Lawn, R. M., Efstratiadis, A., O'Connell, C., and Maniatis, T., 1980 The nucleotide sequence of the human β-globin gene, *Cell* **21**:647–651.

Marotta, C., Wilson, J. T., Forget, B., and Weissman, S. M., 1977, Human β-globin messenger RNA, *J. Biol. Chem.* **252**:5040–5053.

Matsuda, H., Gojobori, T., and Takahata, N., 1978, Theoretical study on protein polymorphism and its bearing on the evolution of protein molecules, in: *Evolution of Protein Molecules* (H. Matsubara, and T. Yamanako, eds.), Japan Scientific Societies Press, Tokyo, pp. 89–100.

Montgomery, D. C., Leung, D. W., Smith, M., Shalit, P., Faye, G., and Hall, B., 1980, Isolation and sequence of the gene for iso-2–cytochrome c in *Saccharomyces cerevisiae*, *Proc. Natl. Acad. Sci. USA* **77**:541–545.

Moore, G. W., 1977, Proof of the populous path algorithm for missing mutations in parsimony trees, *J. Theor. Biol.* **66**:95–106.

Nei, M., 1975, *Molecular Population Genetics and Evolution*, North–Holland/Elsevier, Amsterdam.

Nei, M., and Tateno, Y., 1978, Nonrandom amino acid substitution and estimation of the number of nucleotide substitutions in evolution, *J. Mol. Evol.* **11**:333–347.

Nichols, B., and Yanofsky, C., 1979, Nucleotide sequences of *trpA* of *Salmonella typhimurium* and *Escherichia coli:* An evolutionary comparison, *Proc. Natl. Acad. Sci. USA* **76**:5244–5248.

Ratner, V. A., 1977, An algorithm for constructing phylogenetic trees from amino acid sequence data, in: *Mathematical Models of Evolution and Selection* (V. A. Ratner, ed.), Academy of Sciences of the USSR, Siberian Department, Institute of Cytology and Genetics, Novosibirsk, pp. 5–52.

Reddy, V. B., Thimmappaya, B., Dhar, R., Subramanian, K. N., Zain, B. S., Pan, J., Ghosh, P. K., Celma, C., and Weissman, S. M., 1978, The genome of simian virus SV40, *Science* **200**:459–502.

Sanger, F., Air, G. M., Barrell, B. G., Brown, N. L., Coulson, A. R., Fiddes, J. C., Hutchison III, C. A., Slocombe, P. M., and Smith, M., 1977, Nucleotide sequence of bacteriophage ΦX174 DNA, *Nature* **265**:687–695.

Schaffner, W., Kunz, G., Daetwyler, H., Telford, J., Smith, H. O., and Birnstiel, M. L., 1978, Genes and spacers of cloned sea urchin histone DNA analyzed by sequencing, *Cell* **14**:655–671.

Seif, I., Khoury, G., and Dhar, R. 1979, The genome of human papovavirus BKV, *Cell* **18**:963–977.

Smith, M., Leung, D. W., Gillam, S., and Astell, C. R., 1979, Sequence of the gene for iso–1-cytochrome c in *Saccharomyces cerevisiae*, *Cell* **16**:753–761.

Soeda, E., Arrand, J. R., Smolar, N., Walsh, J. E., and Griffin, B. E., 1980a, Coding potential and regulatory signals of the polyoma virus genome, *Nature* **283:**445–453.

Soeda, E., Maruyama, T., Arrand, J., and Griffin, B. E., 1980b, Host-dependent evolution of three papova viruses, *Nature* **285:**165–167.

Sures, I., Lowry, and Laurence, H. K., 1978, The DNA sequence of sea urchin (*S. purpuratus*) H2A, H2B and H3 histone coding and spacer regions, *Cell* **15:**1033–1044.

Uzzell, T., and Corbin, K. W., 1971, Fitting discrete probability distributions to evolutionary events, *Science* **172:**1089–1096 .

Yockey, H. P., 1977, A prescription which predicts functionally equivalent residues at given sites in protein sequences, *J. Theor. Biol.* **67:**337–343.

PART III

Prospects for Investigating Evolution through Genomic DNA

CHAPTER 8

Genomic DNA:
New Approaches to Evolutionary Problems

ALAN F. SCOTT and KIRBY D. SMITH

1. Introduction

The genome of a haploid cell from most eukaryotes contains over three billion base pairs of DNA. The organization of this DNA is complex. Much of it consists of various families of repeated sequences, some of which have changed rapidly in time, while others have been conserved evolutionarily. A relatively small portion of the genome contains the few thousand expressed genes that code for proteins. Regulatory sequences, important in differentiation and adaptation, are also present. Further, there may be sequences that migrate within and between chromosomes, perhaps altering the phenotype, and sequences that might direct the rearrangement of chromosomes, which also could profoundly affect the development of organisms and their morphology. In short, all of the information that constitutes an individual as a member of a species, genus, or higher taxonomic group is encoded in the genome, and it is here that evolutionary change is recorded. In the past, in order to investigate the interrelatedness of species and the genetic processes by which they evolve, systematists and evolutionary biologists have, by necessity, been forced to study the consequences of gene expression. But the revolution in DNA techniques that has emerged from molecular biology now makes it possible to study and compare gene structure and organization directly. The in-

ALAN F. SCOTT and KIRBY D. SMITH • The Johns Hopkins University School of Medicine and Howard Hughes Medical Institute, Laboratory for Human Biochemical Genetics, Baltimore, Maryland 21205.

sights that are emerging from these new approaches for studying the genome have already drastically altered our concept of the gene. In the future they may also alter our view of evolution. In this chapter we hope to survey some of the principal methods that are being used in evolutionary studies of genomic DNA, to describe a few results of their application, and to discuss their prospects for the future. Much of what we present will be speculative or preliminary, but it may indicate some of the directions of evolutionary biology in the years ahead.

2. Overview of Methods

A variety of techniques are available to study the genome, many of which have been applied to questions addressed by evolutionary biologists. Although the newer methods using recombinant DNA have attracted recent attention, much has been learned by other procedures. Two early techniques, base composition and nearest neighbor analysis (where the order of adjacent nucleotides is determined), have been used to distinguish highly divergent groups. For instance, eukaryotic and prokaryotic DNAs were shown to differ in that the former have a relative scarcity of CG doublets (e.g., Russell *et al.*, 1976; Nussinov, 1980). Although the selective forces responsible for these differences are only beginning to be understood (Razin and Riggs, 1980), it is clear that methods such as these have limited systematic utility because of their low resolution.

Another approach used for comparative studies of more closely related groups involves the isolation of distinct subsets of genomic DNA by density equilibrium ultracentrifugation. The use of this technique has shown that one or more components of the DNA from higher organisms are often observed as secondary or "satellite" bands on CsCl gradients because these DNAs have unusual base compositions which alter their densities. Some satellite sequences have subsequently been shown to consist of families of highly repeated simple sequence DNAs that are often associated with particular chromosomal regions, especially telomeres and centromeres. Their relatively fast rate of evolutionary change has made them attractive for comparative studies, such as those made by Rice (1972) for the rodents and Ryder (1980) for the Equidae. The possible significance of these sequences will be discussed below.

Closely related taxa can also be compared by techniques which measure sequence homology, such as the analysis of DNA reassociation or melting characteristics (Britten *et al.*, 1974). These two methods have provided a means to study the relative complexity of the different classes of genomic DNA and a way to measure sequence homology between species. Both methods rely on the fact that the hydrogen bonding that stabilizes the double helix can be weakened by lowering the salt concentration or raising the temperature and that once dissociated, single-stranded molecules, under appropriate conditions, will anneal with their complementary sequences to reform double-stranded DNA. The rate at which reannealing occurs is a function of the relative concentration of a given sequence, and thus highly repeated tracks find their complementary strands more rapidly than do single-copy sequences. When single-stranded DNAs from different species are mixed and allowed to reassociate, the stability of the hybrid molecules is a function of how many bases (hence, hydrogen bonds) they share in common. If the hybrid molecules are then heated, the poorer the sequence similarity, the lower will be the temperature at which they again dissociate. In practice such experiments are often carried out by labeling one of the DNAs with radioisotope and hybridizing it with a large excess of the other DNA, which is not labeled. The samples are mixed, dissociated by boiling, and allowed to hybridize for appropriate lengths of time. The reason that small amounts of labeled sequences are used is so that the majority of the radioactive fragments will be at too low a concentration to hybridize appreciably with themselves. In kinetic experiments the DNA mixture is analyzed at given time intervals and the proportion of single- and double-stranded (reannealed) DNA determined by fractionation on hydroxylapatite (HAP) columns or by digestion of single-stranded DNA molecules with *S*1 nuclease (Britten *et al.*, 1974). In melting experiments the reassociated DNAs are heated to increasing temperatures, and the proportion of double-stranded molecules rendered single-stranded at each point is again measured by HAP chromatography or *S*1 nuclease digestion. These methods can provide valuable information about the nature of particular sequences and may still offer one of the best single measurements for comparing the DNA between species [see Kohne *et al.* (1972) and Gillespie (1977) for examples of, respectively, the use of single-copy and repeated DNAs for the study of primate relatedness, and Sibley and Ahlquist (1980) for studies of avian taxonomy].

A major limitation of the methods described above is that rare DNA sequences cannot be easily studied. For these, high-resolution methods, such as restriction enzyme mapping, cloning, and DNA sequencing, are needed. The discovery and availability of DNA restriction and modification enzymes in the 1970s [reviewed by H.O. Smith (1979)], coupled with the ability to clone pure sequences, has made it possible to study and compare specific genes as well as their nonexpressed flanking regions. Restriction endonucleases are a class of enzymes, isolated primarily from bacteria, that recognize specific short nucleotide sequences in DNA and then cleave the molecules within or near these sites. Nearly four dozen enzymes with different specificities are commercially available as of this writing and many others have been described (Roberts, 1980). The abundance of restriction enzymes is particularly useful for comparative studies because it allows experimenters to estimate the amount of sequence similarity among species by sampling their DNAs for the presence or absence

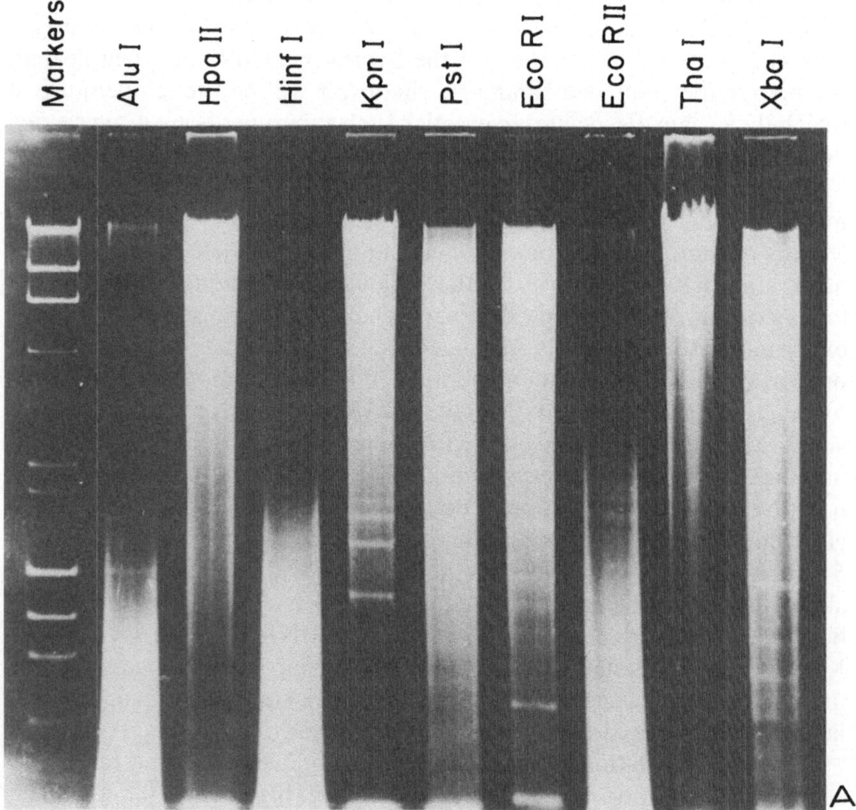

Figure 1A. Ethidium bromide-stained agarose gel of human DNA digested with various restriction edonucleases. Note that most enzymes with 4- or 5-bp recognition sites, such as *Alu* I (5′AGCT3′), *Hinf* I (GANTC), and *Eco* RII [CC(A/T)GG], cleave the DNA into small fragments. Most of the enzymes with 6-bp recognition sequences, such as *Eco* RI (GAATTC), *Xba* I (TCTAGA), *Pst* I (CTGCAG), and *Kpn* I (GGTACC), generate large fragments. *Tha* I (CGCG) produces large fragments because of the relative scarcity of CG doublets in eukaryotic DNA (Russle *et al.*, 1976), while *Hpa* II (CCGG) leaves large fragments not only because a CG doublet occurs in the recognition site, but because it will not cut at this site if the internal C is methylated, which is the case for much eukaryotic DNA from a variety of tissues (Razin and Riggs, 1980).

of a particular enzyme site at a number of homologous positions. Because restriction enzymes recognize specific sites, they digest the DNA molecules from a given organism at those sites (with the exception of those affected by methylation; see Fig 1A) into discrete fragments which can then be fractionated by electrophoresis in agarose or acrylamide gels. Because genomic DNA is so complex, restriction enzyme digests generally produce a smear of fragments extending throughout the gel. Digests with enzymes that recognize common sequences (e.g., *Hae* III, which recognizes 5'GGCC3') tend to generate small fragments, while digestion with enzymes having rare recognition sites (e.g., *Pst* I, which recognizes 5'CTGGAG3') leave mostly large molecules (see Fig. 1A). Discrete bands sometimes seen in ethidium bromide-stained restriction enzyme digests (e.g., *Kpn* I in Fig. 1A) result from families of repeated DNAs that have the same equally spaced restriction sites and occur hundreds or thousands of times in the genome (Manuelidis, 1976; Wu and Manuelidis, 1980).

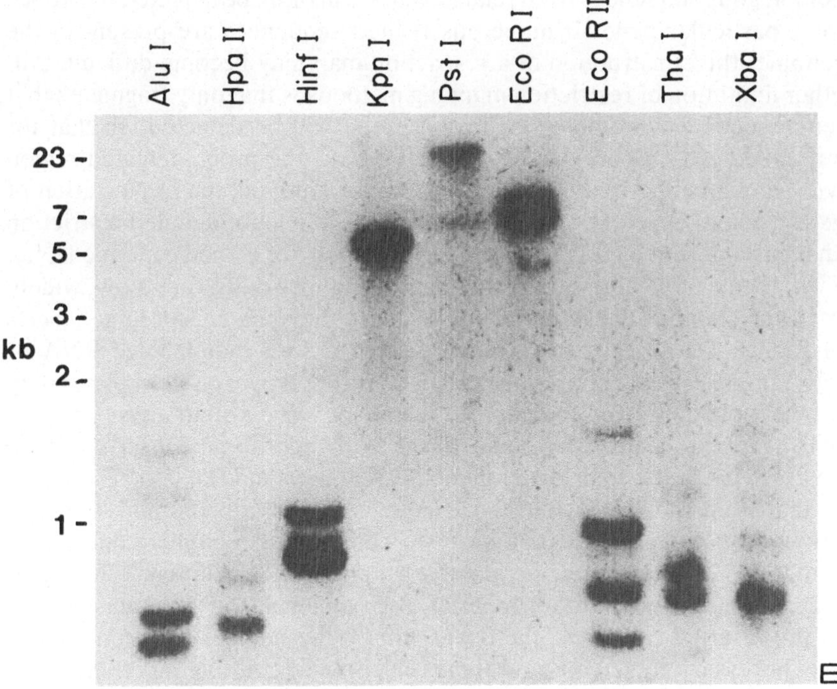

Figure 1B. DNA from Fig. 1A transferred to nitrocellulose sheets and hybridized with I^{125}-radiolabeled 18S rRNA.

Digestion products that occur only once or a few times cannot be observed by staining but can be detected by hybridization of the digested DNA with a radiolabeled complementary sequence followed by autoradiography. In this procedure digested DNA, fractionated in an agarose gel, is denatured by soaking the gel in alkali and is then transferred to a nitrocellulose sheet by blotting (Southern, 1975). Once the DNA is immobilized on the nitrocellulose, the sheets are incubated in a solution containing a specific radiolabeled RNA or DNA probe which hybridizes to complementary DNA fragments on the sheet. Figure 1B illustrates the use of this method with various enzymatic digests of human DNA which were transferred and hybridized with iodinated 18S ribosomal RNA. By using this technique with appropriate digests it is possible to construct maps of the restriction enzyme sites in and around genes. Such "Southern blotting" has enabled us to learn a great deal about the organization of genes both within and between species and has also proven to be a powerful method for population studies (e.g., Kan and Dozy, 1980), chromosomal mapping (e.g., Scott *et al.*, 1979), and a number of other related problems.

A difficulty with restriction mapping studies of genomic DNA is that homologous sequences from related genes will often be detected by probes for a particular gene. If numerous related sequences are present in the genome, the construction of a restriction map may become difficult. Another limitation of restriction mapping methods is that only fragments that retain sequences homologous to the probe will be detected, so that for any given enzyme only those sites closest to the probe-detected region will be mapped. An alternative strategy for studying the organization of gene clusters which also provides material suitable for detailed restriction analysis and sequencing is to screen "libraries" of cloned genomic DNA. Two types of vectors, bacteriophages and plasmids, are now widely used for cloning. Phage are useful because they can accept large inserts of foreign DNA and can be easily screened with radiolabeled RNA or DNA probes. When phage-infected bacteria are grown on Petri plates they produce plaques of lysed cells. All the phage DNA within a given plaque will have the same insert, and phage with appropriate inserts can be detected by a filter hybridization method (Benton and Davis, 1977) similar to that used for gels. A positive plaque can be recovered and its phage grown in large quantities. In order to construct as complete and representative a collection of the genome as possible, a "library" should be made from a random assortment of large molecules. The larger the size of the inserted fragment, the fewer will be the number of recombinants it is necessary to screen, and the easier it will be to overlap cloned segments into larger maps. Many of the bacteriophage cloning vectors can accept inserted fragments of 15,000–20,000 base pairs (15–20 kb), which makes it possible to create "libraries" of genomic DNAs with a reasonably small number of recombinants that include, statistically, all of the genetic material. Also, the more random the process is for creating fragments to be cloned, the more likely one is to find the gene of interest. Lawn *et al.* (1978) created a human genomic "library" by partially digesting human DNA with the restriction enzymes *Hae* III and *Alu* I. Because each of these enzymes recognizes a four-base-pair (bp) sequence that would occur, on the average, once every 256 nucleotides, the resulting collection of fragments represents a highly random sample of the genome. Moreover, because of partial digestion, a portion of the random sample would be in the size range of 15–20 kb. Another method (Fig. 2) involves partially digesting DNA with *Eco* RI and then inserting those fragments directly into the *Eco* RI sites of a certain phage, such as Charon 4A (Williams and Blattner, 1979). The disadvantage of the latter method is that *Eco* RI cuts, on the average, only once in 4096 bases, and a certain percentage of the genome will not be represented because the fragments will not fall in the size range appropriate for insertion into the phage (usually they will be too large). Plasmids, which are small circular DNAs found in bacteria and yeast, are best suited for cloning short fragments, such as cDNAs copied from messenger RNAs or sequences resulting from digests of recombinant phage. Their chief advantage is that they are easily grown and can readily provide the amounts of DNA needed for detailed characterization and sequencing.

Once a fragment has been cloned and grown in moderate quantities, it is possible to obtain its sequence relatively easily. Two sequencing methods have become available which are widely used. Both methods rely on the generation of a collection of fragments of different lengths terminating at a given designated base. The chemical method of Maxam and Gilbert (1977, 1980) involves end-labeling a purified DNA fragment with radioisotope and subjecting the DNA to a variety of partial base-specific modification and cleavage reactions. The enzymatic method (Sanger *et al.*, 1977) uses DNA polymerase to synthesize copies of the DNA fragment of interest which are terminated prematurely by the addition of dideoxy analogs of the four nucleotides. Once a collection of fragments has been produced by either method, these can be separated on acrylamide gels and the sequence read directly. Typically 200 or more nucleotides can be determined per fragment. The application of sequencing to evolutionary studies is already well advanced (e.g., Jukes, 1980) and promises to be even more important in the future.

Given the present repertoire of techniques, it is possible to study a variety of types of genomic DNA in great detail. A few examples will be illustrated below.

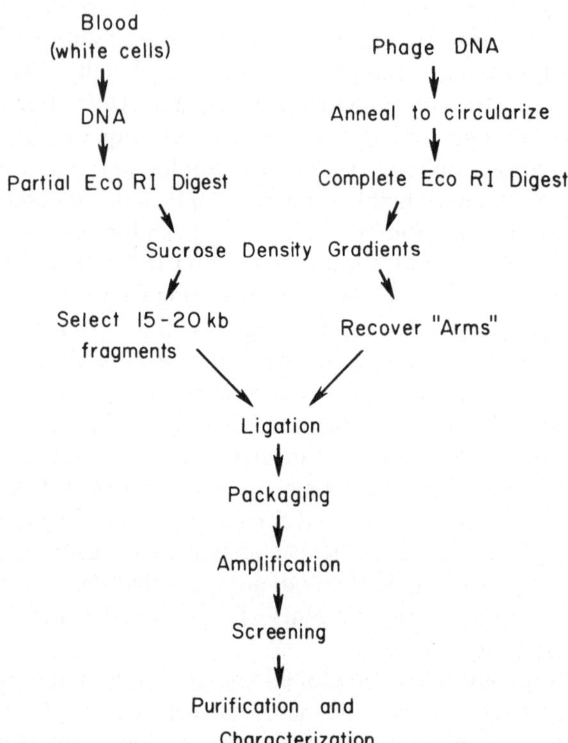

Figure 2. One procedure for constructing genomic "libraries." DNA made from white cells or other tissues is partially digested with *Eco* RI and molecules of 15–20 kb are selected from sucrose gradients (Maniatis *et al.*, 1978; Robbins *et al.*, 1979). DNA from the bacteriophage Charon 4A (Williams and Blattner, 1979) is annealed to form circular molecules and digested completely with *Eco* RI, and the phage "arms" and internal fragments, which are to be replaced with genomic fragments, are separated by sucrose density gradient centrifugation. The large phage "arms" and genomic fragments each have complementary "sticky" ends due to the fact that *Eco* RI produces four bases of single-stranded DNA at the end of each double-stranded molecule. Under appropriate conditions the *Eco* RI ends can hybridize to each other. Once hybridized, the "arms" and DNA fragments are convalently linked with DNA ligase and the recombinant phages are "packaged" by mixing with extracts containing bacteriophage "head" and "tail" proteins (Enquist and Sternberg, 1979). The then viable recombinants are mixed with appropriate *E. coli* cells and grown as plaques on Petri plates. The plaques can then be screened by blotting and hybridization with radiolabeled probes.

3. Comparative Studies of Genomic DNA

Reassociation kinetics have shown that the genome of higher organisms consists of sequences that occur from once to several thousand times. On the basis of their relative abundance or reiteration frequency, the genome can be categorized into highly repeated, middle-order repeated, low-order repeated, and single-copy DNA (Britten and Davidson, 1971). Needless to say, these are not discrete classes, but are somewhat arbitrary subdivisions of a continuum. Superimposed on this classification is another that distinguishes between "silent" DNA and transcriptionally active sequences, which in turn might be subdivided into those that code for proteins and those that do not. Each of these classes of DNA has been studied in a comparative context, and in this chapter we will present examples of a moderately repeated DNA fragment that may not be transcribed, a low-order repeated gene that is transcribed, and a single-copy gene that is both transcribed and translated. We will also discuss some of the other recently described and potentially important sequences that occur in genomic DNA which may have evolutionary importance.

3.1. Y-Chromosome DNA

Digestion of DNA from human males with the restriction enzyme *Hae* III produces two discrete bands that stand out amid a smear of heterogeneous fragments. These bands are approximately 3.4 and 2.0 kb in size and are contained within the long arm of the Y chromosome (Kunkel *et al.*, 1977). The 3.4-kb fragment may constitute as much as 40% of the DNA from the Y chromosome (Cooke, 1976; Kunkel *et al.*, 1979) and the 2.0-kb fragment up to 20% (Cooke, 1976). The two fragments are unrelated, since they fail to hybridize with one another. The 3.4-kb fragment has been best studied and has been shown to be a heterogeneous collection of related sequences. Although the 3.4-kb fragment that is physically isolated from gels is from the Y chromosome, there are sequences within it that hybridize to female DNA (Kunkel *et al.*, 1979) and human autosomal DNA in somatic cell hybrids (Cooke and McKay, 1978). Male-specific reiterated DNA, isolated by hybridization of radiolabeled male DNA with an excess of unlabeled female DNA (Kunkel *et al.*, 1976), has been shown to occur within the 3.4-kb fragments. Reassociation kinetics estimates that there may be as many as 40 different families of male-specific DNAs included within the set of 3.4-kb sequences. Yet within any particular 3.4-kb fragment only three of the possible 40 sets of reiterated sequences appear to be present. (Fig. 3).

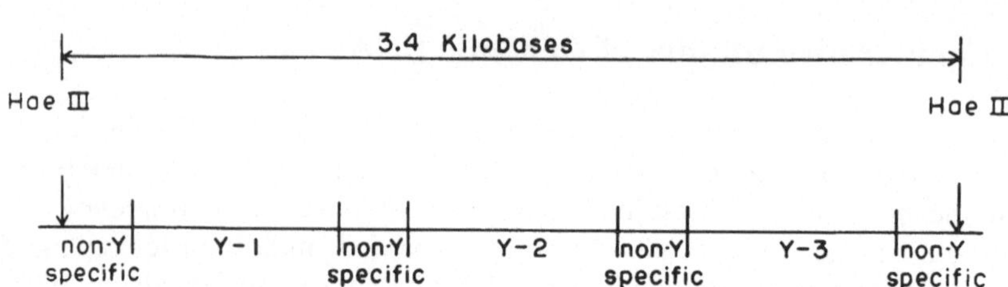

Figure 3. Schematic representation of the human male 3.4-kb family generated by digestion with the restriction enzyme *Hae* III. Each fragment has sequences that are specific to the Y chromosome, but may differ from one another (Y-1, Y-2, Y-3), as well as sequences that occur on other chromosomes (non-Y-specific). The Y-specific components are also human-specific, whereas the non-Y-specific components have sequences that are both human-specific and sequences that occur in other primates. The actual arrangement of these components is not entirely known.

Digestion of various primate DNAs with a large number of restriction endonucleases has failed to identify specific fragments associated with only their Y chromosomes. Southern blot hybridization, using radiolabeled human 3.4-kb fragments as probe, also did not reveal male-specific fragments in various primate DNAs digested with *Hae* III (Cooke and McKay, 1978; Kunkel and Smith, in press). However, as shown in Fig. 4A, when radiolabeled 3.4-kb fragments were hybridized in solution to various primate DNAs, it was found that about half of the radioactivity reassociated. At the same criteria of hybridization, human female DNA also bound about 50% of the radioactivity. The reduction in reassociation rate seen among the primates suggests that the number of sequences with 3.4-kb homology decreases as a function of the evolutionary distance from humans. The similarity in reassociation rates for male and female DNA from each of the nonhuman primates suggests that the homologous sequences are missing from their Y chromosomes. This has been confirmed by *in situ* hybridization, where the only detectable regions of homology to the 3.4-kb fragment were on primate autosomes. When DNAs from chimpanzees and gorillas were studied with only human male-specific DNA at standard criteria (which allows about 25% sequence mismatching), no homology was observed (Fig. 4B). At lower criteria, where the number of mismatched base pairs is allowed to increase to about 40%, there was hybridization to primate DNA (Kunkel and Smith, in press). But, as with the non-Y-specific portions of the 3.4-kb fragments, the similarity in reassociation rates of male and female DNA and results of *in situ* hybridization indicate the homology is with autosomal rather than Y-chromosome sequences. Instead, the homologous sequences in the primates were distributed over many autosomes (Szabo *et al.*, in preparation). When the extent of divergence between the non-Y-specific 3.4-kb sequences and their autosomal homologs in nonhuman primates was examined by the thermal stability of hybrid DNA molecules, a subset of the sequences was found that form stable hybrids with gorilla, chimpanzee, and orangutan DNA, indicating little, if any, sequence divergence. A second subset was found which showed no sequence divergence from human female DNA but was very different from all of the primate DNAs (Kunkel and Smith, in press).

Thus a significant portion of the human Y chromosome is composed of a repeating unit which is defined by the regular occurrence of *Hae* III restriction sites at 3.4-kb intervals yet is heterogeneous, with some components that are specific to the human Y chromosome and others that have homology with sequences on human and primate autosomes. Therefore, much of the human Y chromosome must have arisen from autosomal sequences. Curiously, the gorilla Y chromosome is nearly identical structurally to the human Y, yet, by this analysis, the sequences of which it is comprised must be very different from those in humans. It has been shown that there is no correlation between the presence of the 3.4 and 2.0-kb fragments and male determination (Kunkel *et al.*, 1977). No obvious function has yet been ascribed to these sequences. Their role, if any, in determining fitness remains to be demonstrated.

3.2. Other Middle-Order Reiterated DNAs

Another middle-order reiterated family of DNA that is analogous to the 3.4-kb fragment has been detected in the human genome and is noteworthy because of its association with the β-globin gene (Fig. 8) (Kaufman *et al.*, 1980). Homologs of this family of sequences have also been detected in the gorilla (Scott *et al.*, in preparation) and other primates (Adams *et al.*, submitted). In humans these sequences are about 6.4 kb long and occur about 3000 times. A sequence from a similar family also occurs 3′ of the rabbit adult β gene and has been shown by *in situ* hybridization to be distributed on several chromosomes with a reiteration frequency of about 4000–5000 (Hoeijmakers-van Dommelen *et al.*, 1980). Thus, in each species this repeat constitutes approximately 1% of the genome. Middle-order reiterated families have also been studied in *Drosophila*, where they have been described as "fluid" because of the way they appear to "migrate" between chromosomes when they are mapped in different strains of flies (Young, 1979).

It remains to be seen what benefit most repeated DNAs provide organisms. The way in which many of these sequences multiply and spread has been described as "parasitic" or "selfish" (Doolittle and Sapienza, 1980; Orgel and Crick, 1980), implying that they evolve without editing by natural selection. However, it is difficult to prove that a structure is

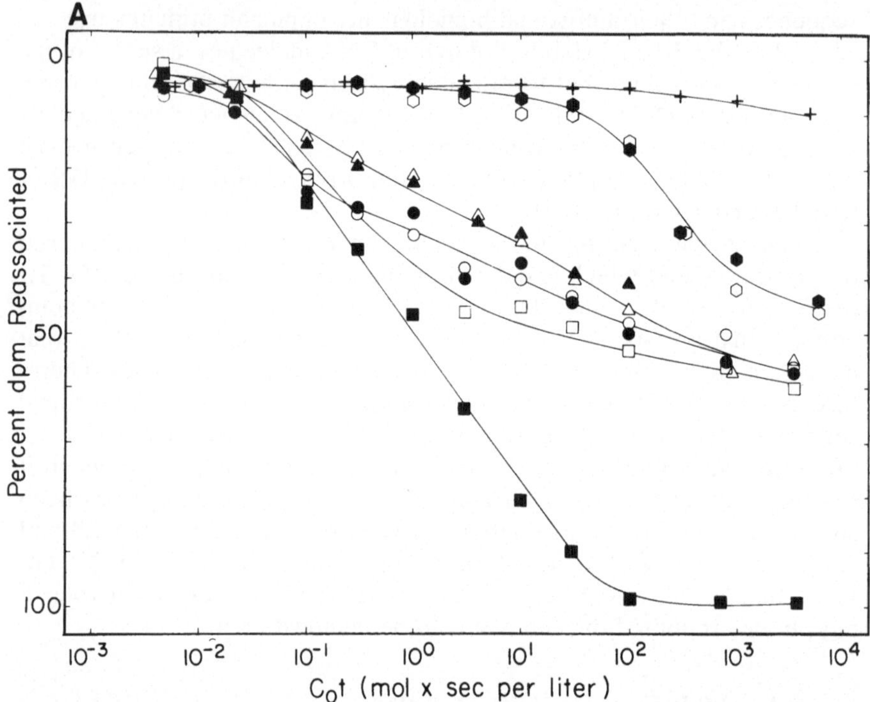

Figure 4A. Reassocation (C_0t) curve of radiolabeled human male 3.4-kb *Hae* III fragments hybridized with an excess of various DNAs at standard criteria (see Kunkel and Smith, in press). Human male ■; human female □; male chimpanzee ▲; female chimpanzee △; male gorilla ●; female gorilla ○; male orangutan ⬢; female orangutan ⬡; and *E. coli* +. The abscissa represents the product of the concentration of DNA in the reaction (C_0) and the length of time of the incubation (t). The percent DPM (disintegrations per minute) reassociated is calculated as the ratio of radiolabeled double-stranded DNA recovered from hydroxylapatite columns over the total of all labeled DNA present (see text).

worthless in an evolutionary sense. Many investigators, in fact, believe that reiterated sequences may function in the regulation of gene expression (e.g., Davidson and Britten, 1979) and thus are of major evolutionary importance. It is possible that these DNAs evolve without selective constraints on the actual sequence *per se*, but still may be regulatory if regulation is a consequence of positioning homologous tracks in appropriate locations. Thus, selection would require that such blocks of DNA be present, although their actual base composition might vary. Comparative studies are likely to have a significant impact on our understanding of these problems.

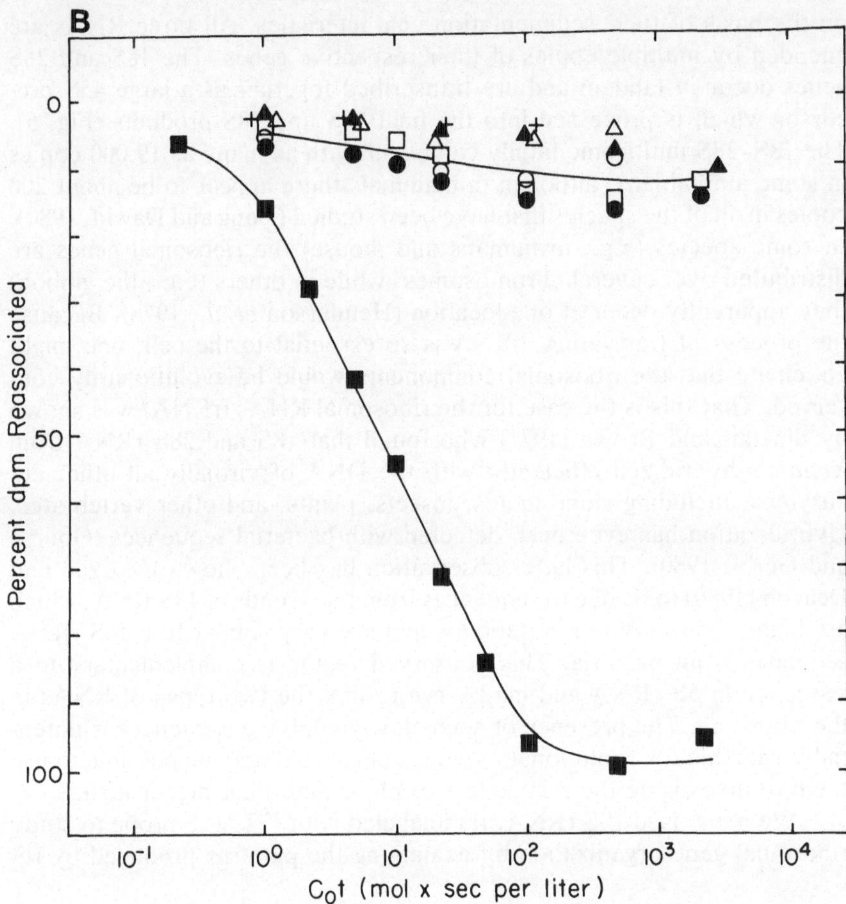

Figure 4B. C_0t curve of radiolabeled human male-specific DNA hybridized to the DNAs of various nonhuman primates. This probe was prepared by hybridizing labeled male sequences to a large excess of female DNA (from Kunkel and Smith, in press). DNA that did not bind to female DNA after two rounds of hybridization was considered male-specific. Symbols are as in Fig. 4A.

3.3. Studies of the Ribosomal Genes

The ribosomes are an essential part of the cellular machinery for converting the information stored in nucleic acid into protein. They are complex structures which, in eukaryotes, consist of over 50 proteins and three types of RNA. These RNAs have been designated 5S, 18S, and 28S on the basis of their sedimentation characteristics. All three RNAs are encoded by multiple copies of their respective genes. The 18S and 28S genes occur in tandem and are transcribed together as a large 45S precursor which is processed into the final 18S and 28S products (Fig. 5). The 18S–28S multigenic family can occur with as many as 19,000 copies in some amphibians, although in mammals there appear to be about 200 copies in all of the species that have been studied (Long and Dawid, 1980). In some species (e.g., in humans and mouse) the ribosomal genes are distributed over several chromosomes, while in others (e.g., the gibbon) they apparently occur at one location (Henderson *et al.*, 1976). Because the process of translating mRNA is so essential to the cell, one might anticipate that the ribosomal components would be evolutionarily conserved. That this is the case for the ribosomal RNAs (rRNA) was shown by Sinclair and Brown (1971) who found that 18S and 28S rRNA from *Xenopus* hybridized efficiently with the DNA of virtually all other eukaryotes, including slime molds, insects, plants, and other vertebrates. Hybridization has even been detected with bacterial sequences (Gourse and Gerbi, 1980). This latter observation has been shown by Azad and Deacon (1980) to be due to sequences from the 3′ ends of 18S RNA, which are highly conserved in vertebrates and are very similar to a 16S rRNA sequence from bacteria. This conserved region is complementary to a sequence in 5S rRNA and may serve to link the two types of RNAs in the ribosome. The presence of such slowly evolving sequences is potentially valuable for evolutionary studies, because it may be possible to use them to investigate the relatedness of phyla and other higher taxa.

We have used 18S rRNA, radiolabeled with ^{125}I, as a probe to study ribosomal gene organization by examining the patterns produced by re-

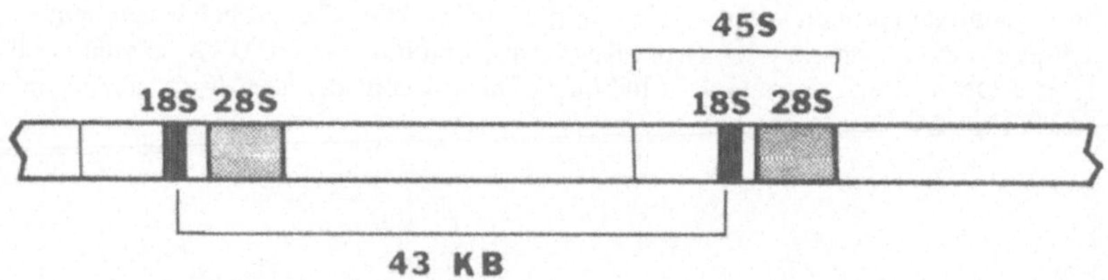

Figure 5. Map of the 18S and 28S ribosomal RNA gene family. One repeat is shown. These genes are transcribed as a large 45S precursor RNA, which is processed to give the final 18S and 28S molecules, indicated by the solid and hatched regions, respectively. In humans and the mouse the repeat length is about 43 kb and there are about 200 copies (Arnheim and Southern, 1977; Cory and Adams, 1977; Long and Dawid, 1980) occurring on several chromosomes.

striction enzyme digestion of various mammalian DNAs. As described above, the digested DNA was separated on agarose gels, denatured, neutralized, and transferred to nitrocellulose sheets (Southern, 1975). These were then hybridized with the probe. Because the ribosomal genes are repeated, the labeled rRNA can detect complementary sequences in small amounts of genomic DNA. Figure 6 illustrates typical results obtained with digests of various mammalian DNAs. A number of points can be made about these data. First, most of the enzymes give simple patterns, which might not be expected, since we are looking at the composite of approximately 200 copies of this gene occurring at several chromosomal locations. Second, most species can be distinguished from each other on the basis of at least one enzyme. Only the patterns from closely related species, such as the Chinese and Syrian hamsters (not illustrated), were identical with all of the enzymes tested (Scott *et al.*, in preparation). Interestingly, although these rodents are assigned to different genera by

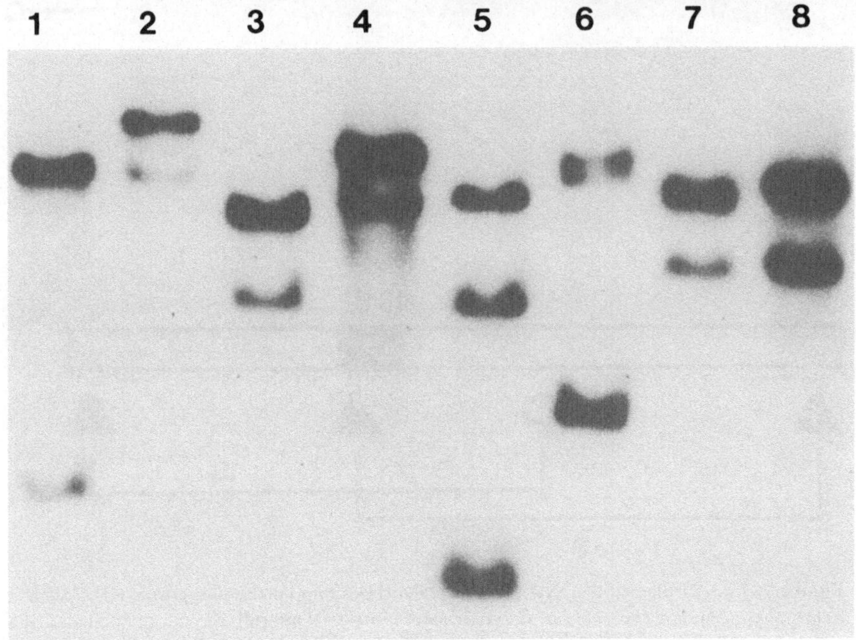

Figure 6. Various mammalian DNAs digested with *Pst* I and hybridized with ^{125}I-radiolabeled 18S rRNA. Lane 1, pig-tailed macaque (*Macaca nemestrina*); lane 2, baboon (*Papio cynocephalus*); lane 3, green monkey (*Cercopithecus aethiops*); lane 4, orangutan (*Pongo pygmaeus*); lane 5, mink (*Mustella vison*); lane 6, muntjac (*Muntiacus muntjak*); lane 7, buffalo (*Bison bison*); lane 8, sheep (*Ovis aries*).

classical methods, they are much more alike by this analysis than are other congeneric species that were examined. These observations suggest that all, or nearly all, of the 18S rRNA genes and flanking regions within a given animal are very similar and that a mechanism must exist whereby a new restriction site occurring in one copy can spread relatively quickly

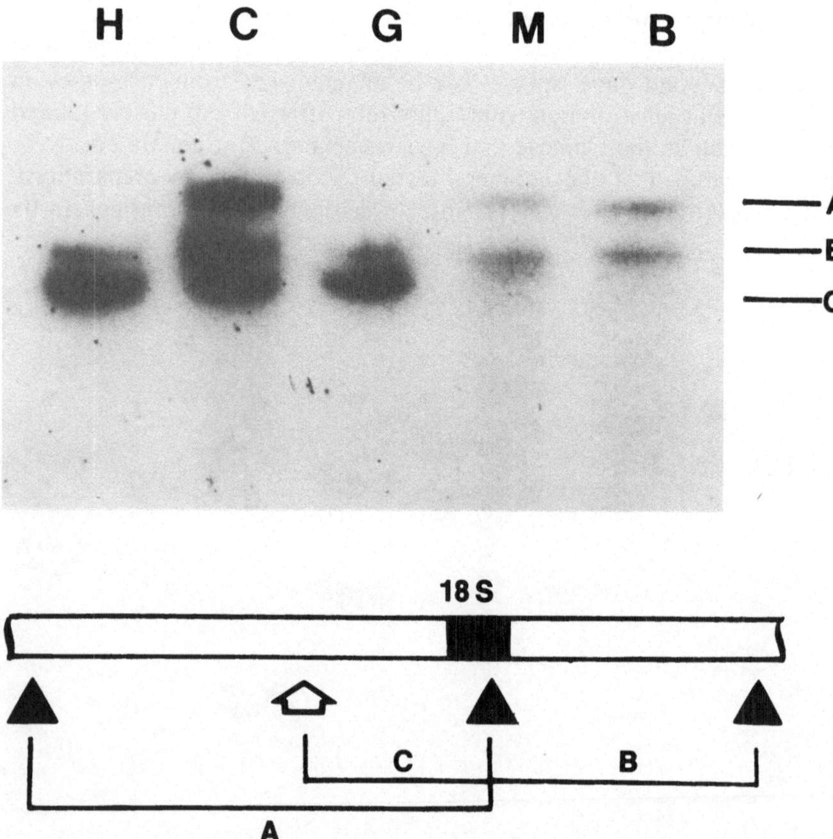

Figure 7. *Eco* RI digests of DNA from humans (H), chimpanzee (C), gorilla (G), *Macaca nemestrina* (M), and the baboon, *Papio cynocephalus* (B), hybridized with radiolabeled 18S rRNA probe, and a schematic interpretation of the results. The labeled bands correspond to the regions indicated in the map. A variable site (open arrow) results in intensity differences of bands A, B, and C between the species. In the macaque and baboon the site is apparently absent in all copies of the gene. In the gorilla and human the site is present in most of the copies, giving rise to prominent B and C bands. In the chimpanzee the site is present in about half of the copies, so that all three bands are observed.

and give rise to species-specific band patterns. Theoretical studies of repeated genes predict that they undergo such changes by a process of unequal crossing-over (G.P. Smith, 1976). If this spreading process is also coupled with a mechanism to discard copies, then we might expect a given copy to increase or decrease in abundance. We might also expect to see clusters of ribosomal genes on separate chromosomes evolving independently of one another, as Arnheim (personal communication) has shown. This last observation may, in part, account for the heterogeneity observed among copies of these genes (Arnheim and Southern, 1977). Heterogeneity for a particular enzyme site accounts for the differences seen in *Eco* RI digests of various primate DNAs (Fig. 7). Three *Eco* RI sites are present adjacent to the 18S rRNA gene in these primates, with the middle site occurring at variable frequencies in the different species. In humans and gorilla one *Eco* RI site located to the 5′ side of the 18S gene is present in the majority of copies, giving rise to a 6-kb band as well as a 7-kb band from the 3′ side. In the chimpanzee the variable site is less frequent, so a 14-kb band extending from another *Eco* RI site further 5′ becomes more prominent. In the baboon and macaque the middle site is entirely absent, so only the 14-kb and 7-kb bands are seen. Although there may be heterogeneity within copies and differences between chromosomes, there also must be a process of "correction" between chromosomal clusters or else restriction patterns would become increasingly complex as copies on separate chromosomes diverged, and we would not see the simple species-specific patterns that we do. A similar observation for the α-globin genes has been made by Zimmer *et al.* (1980), who have termed the tendency for a family of genes to evolve in unison "concerted" evolution.

As a tool for systematic comparisons the ribosomal genes appear to be more useful for higher taxonomic levels than are the mitochondrial DNAs. Table I compares the Nei similarity indices for identical species pairs using both mitochondrial and ribosomal data (Brown *et al.*, 1979;

Table I
Comparison of Similarity Indices for Three Primates Calculated by the Method of Nei (1979) for the 18S rRNA Genes and Mitochondrial DNAs (Mt DNA)

Species compared	Mt DNA[a]	18S rRNA
Human vs. rhesus	0.33	0.47
Human vs. baboon	0.30	0.53
Rhesus vs. baboon	0.38	0.67

[a] Brown *et al.* (1979).

Scott *et al.*, in preparation). These data imply that the restriction enzyme sites in and adjacent to the 18S ribosomal RNA genes evolve at a slower rate than do the mitochondrial sequences. It should be emphasized that most of the enzyme site differences that we observe occur in the non-transcribed flanking or spacer sequences. Unlike the 18S and 28S sequences, the spacer regions are not as constrained by selection. Species comparisons within groups vary considerably. Most of the rodents examined were very different from one another and had few bands in common, particularly in comparison with the other groups. A relatively faster evolutionary rate for rodents (in comparison to a taxonomically equivalent group of birds) has also been shown by Avise *et al.* (1980) from protein studies.

One distinct advantage to restriction enzyme analyses for species comparisons is that the resolving power of the comparison can be increased by simply doing additional restriction digests. Arnehim and Southern (1977) were able to distinguish between strains of mice when their DNAs were examined with the appropriate enzymes. In the few cases in our study where none of the tested enzymes could distinguish between two samples, almost certainly the examination of more sites with additional enzymes would have provided a difference.

3.4. Studies of Single-Copy DNA

The difficulty of studying sequences that occur once or a few times in the genome has been a major obstacle in molecular genetics, but the new methods in recombinant DNA [reviewed by Wetzel (1980)] have provided the means to examine this material. With the development of plasmid cloning vectors it has been possible to prepare large amounts of totally pure cDNAs made from the messenger RNAs that code for many proteins. The sequencing of these cDNAs has provided a great deal of information about how genes evolve (e.g., Jukes, 1980). And, as with the rRNA genes, the use of such sequences as probes of genomic DNAs has made possible the mapping and comparison of restriction enzyme sites in and around the genes from which they are transcribed (e.g., Jeffreys and Flavell, 1977a). These probes have also been used to detect genomic sequences from "libraries" of recombinant phage, as described below.

Among the best characterized gene regions from vertebrates are the α-globin and non-α globin gene clusters. The α cluster in humans consists of two embryonic genes, two adult genes, and one α-like sequence that is not translated and has been described as a pseudogene (Proudfoot and Maniatis, 1980). The non-α gene cluster (Fig. 8) occurs on chromosome

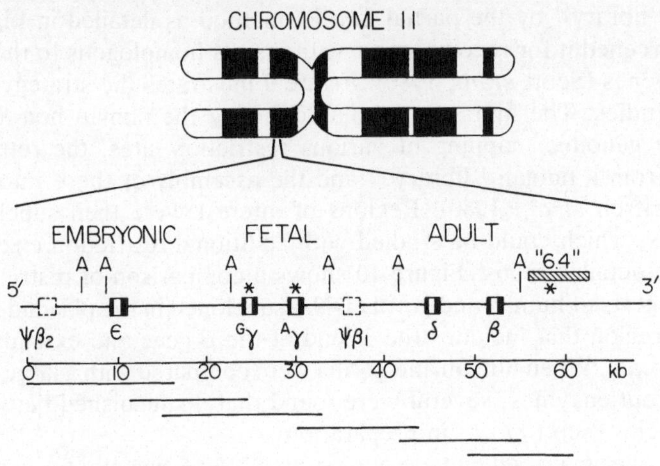

Figure 8. The human non-α globin gene cluster form chromosome 11. The approximate location of the cluster on the chromosome is indicated (Gusella *et al.*, 1979), although the 5′ to 3′ orientation of the cluster with respect to the centromere is not known. The cluster consists of two pseudogenes ($\psi\beta1$ and $\psi\beta2$), an embryonic ϵ gene, two fetal genes ($^{G}\gamma$ and $^{A}\gamma$), and two adult genes (δ and β). Regions marked by an A are "Alu" family sequences (Fritsch *et al.*, 1980; Duncan *et al.*, 1979). The "6.4" indicates a repeat of 6.4 kb occurring about 3000 times in the genome (Kaufman *et al.*, 1980). The asterisks indicate the location of common polymorphisms in human populations (Kan and Dozy, 1978; Jeffreys, 1979). Lines below the map indicate cloned regions recovered from human genomic "libraries" (Fritsch *et al.*, 1980; Kaufman *et al.*, 1980).

11 in humans, and hence is not syntenic with the α cluster on chromosome 16 (Deisseroth *et al.*, 1977). The non-α cluster consists of: two pseudogenes; an embryonic ϵ gene; two fetal genes, "G" γ and "A" γ (so designated because they differ by a glycine and an alanine at position 136); and two adult genes, δ and β. This overall arrangement is conserved in gorillas, chimpanzees, baboons, and two species of macaques (Martin *et al.*, 1980; Barrie, *et al.*, 1981). In primates, the rabbit, the mouse, and probably many other mammals the order of developmental expression for the genes in the non-α gene cluster is the same as their 5′ to 3′ physical linkage (Barrie *et al.*, 1981; Fritsch *et al.*, 1980; Lacy *et al.*, 1979; Jahn *et al.*, 1980). Whether this is coincidental, historical, or a reflection of some fundamental principle is not yet known.

The utility of cloned genomic fragments in comparative studies is illustrated by the analysis of DNA regions from the human non-α cluster and homologous sequences from a gorilla "library." We constructed the gorilla "library" by the partial *Eco* RI method as detailed in Fig. 2 and have screened it for recombinants with inserts homologous to the human globin genes (Scott *et al.*, 1980). Figure 9 illustrates the strategy used in these studies. The first step in characterizing the human non-α cluster was the genomic mapping of various restriction sites, the retrieval of clones from a human "library," and the assembly of these into a large map (Fritsch *et al.*, 1980). Regions of interest were then subcloned in plasmids, which could be studied with additional restriction enzymes to produce detailed maps. Figure 10 shows a comparison of restriction enzyme digests of human and gorilla DNA subcloned in the plasmid pBR322 from a region that includes the 3′ end of the β gene and extends 3.5 kb downstream. When human and gorilla were compared with a large number of different enzymes, several were found that distinguished between the two species (Scott *et al.*, in preparation).

Another major advantage of cloning DNA, especially for systematic comparisons, is that regions of particular interest can be recovered and the DNA sequenced. Although DNA sequencing is only a few years old, the amount of information already available rivals the collection of known protein sequences and is accumulating at approximately 100 kb per year. All of the expressed genes from the non-α globin cluster in humans have now been sequenced, as have some of their homologs from other species. Once such sequences are available, it is possible to construct phylogenetic trees, as illustrated by Hewett-Emmett *et al.* (this volume, Chapter 9). Gene regions can also be examined for homology by another method, shown in Fig. 11. In this figure the sequences of the two adult β genes from the mouse, including flanking regions and intervening sequences, have been compared to one another. A computer was programmed to search for stretches of eight bases in common and then plot their positions with a line of appropriate length. This method visually displays regions of homology. Differences between the sequences due to the relative insertion or deletion of DNA also can be easily observed as displacements in the diagonal of maximum homology.

A major surprise from recombinant DNA studies was that many structural genes were not contiguous (Jeffreys and Flavell, 1977b), and it has been suggested that the expressed sequences or exons may represent ancestral functional domains. This seems to be true for the four coding regions of immunoglobin γ-heavy chain (Gilbert, 1978) and may also apply to the globins. Eaton (1980) has argued that the amino acid sequences of the three exons of the β-globin gene are each associated with different functional properties. Experimental studies of isolated peptides from both the α- and β-globin chains have confirmed that in each chain the region representing the middle exon binds heme (Craik *et al.*, 1980). Thus the

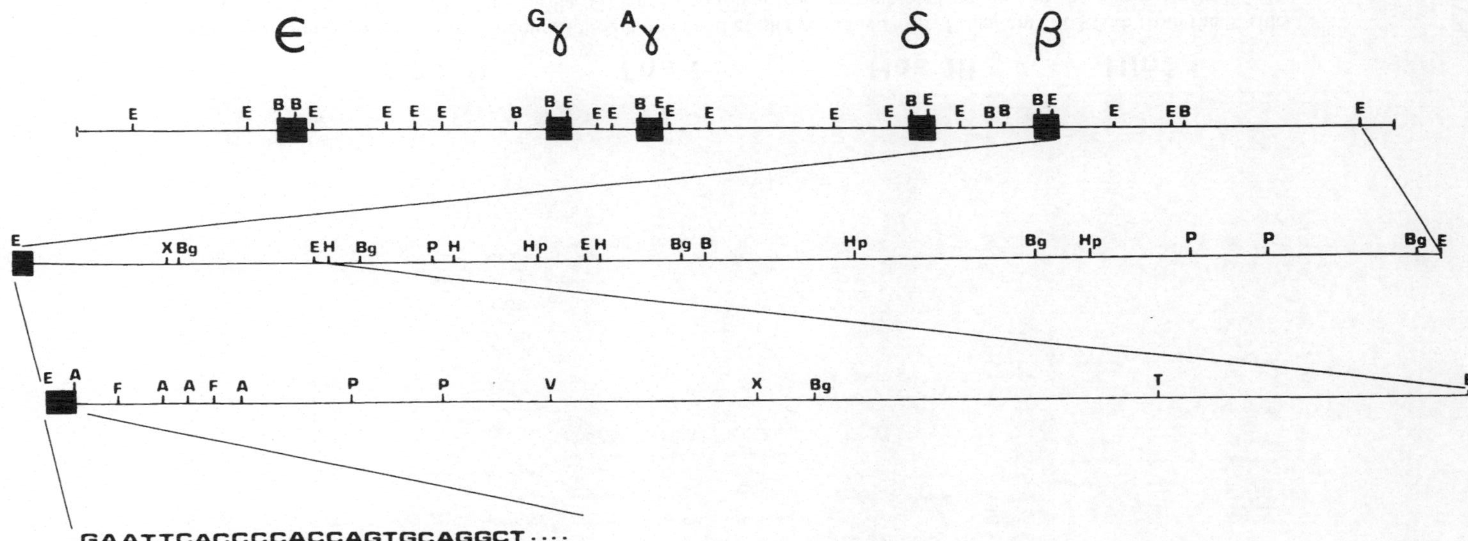

Figure 9. Strategy for the analysis of the non-α gene cluster. The ability to study the genetic material at different levels of resolution is illustrated with the human non-α genes. Restriction enzyme maps were first prepared by restriction mapping of genomic DNA and from analysis of large sequences detected in a human genomic "library" cloned in bacteriophage (Fritsch *et al.*, 1980). In this example, a region of 3.6 kb which includes the 3′ end of the β gene was subcloned in pBR322 (Kaufman *et al.*, 1980). This subclone was then digested with additional restriction enzymes and used for comparison with a homologous clone from a gorilla "library" as illustrated Fig. 10. The highest level of resolution is provided by actual determination of the DNA sequence.

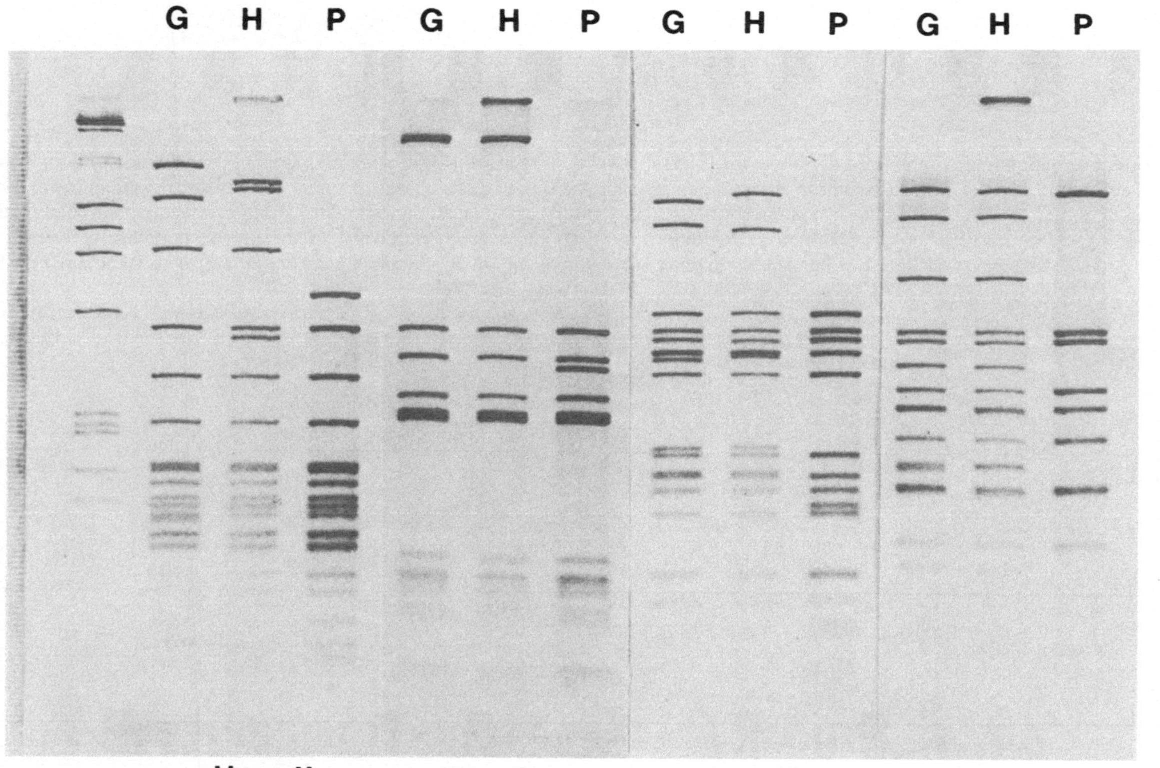

Figure 10. Comparisons of homologous cloned human and gorilla genomic DNA. Fragments of 3.6 kb from the 3′ side of the β gene from human (H) and gorilla (G) were subcloned from recombinant phage into the plasmid pBR322 (P) and digested with a variety of restriction enzymes. Digests of the two recombinants and native pBR322 with four enzymes are shown. Differences within the genomic inserts are observed with *Hpa* II and *Hae* III. *Hinf* I cuts both DNAs in the same positions. *Tha* I does not cut either plasmid within the inserts. Results of the *Tha* I digest again demonstrate the relative rarity of CG doublets in eukaryotic DNA.

possibility exists that many proteins may be the product of smaller amino acid sequences that have become linked by the removal of intervening regions during mRNA processing. These intervening sequences (introns or IVSs), which separate the exons, have also been studied in detail from several genes and have been found to vary significantly in length among homologous genes. Likewise, their sequences have been shown to evolve much more rapidly than do the exon sequences. However, the positions at which they interrupt the coding regions and the sequences at the junctions between coding and intervening regions are strictly maintained. Although the ubiquity of intervening sequences in a number of eukaryotic genes implies that they serve a function, that function remains unclear. Many of the ideas about functional domains and the possibility that such blocks have been linked in the construction of proteins may be confirmed by extending the collection of homologous genes, such as the globins, to other species, especially those that are evolutionarily distant.

One consequence of the study of gene sequences within and between species has been to identify what may be important genetic mechanisms in evolution. For example, studies of the α genes in primates (Zimmer *et al.*, 1980) and of α-thalassemia in humans (Embury *et al.*, 1979; Phillips *et al.*, 1979) have led to the hypothesis that the genes coding for these loci undergo relatively frequent mispairing which results in either loss or amplification. Chromosomes have been identified in human populations with from none to three α loci (Goosens *et al.*, 1980), all apparently the result of unequal crossing-over. Periodic proteins, such as keratin, collagen, metallothionein, and silk fibroin, with simple repeating amino acid sequences that are in turn coded by repeating DNA sequences may also have evolved by a similar process of repeated unequal crossing-over.

When the paired γ loci in the human non-α cluster (Fig. 8) were sequenced, Slightom *et al.* (1980) observed that the $^{\text{"G"}}\gamma$ and $^{\text{"A"}}\gamma$ loci on one chromosome were more alike than the $^{\text{"A"}}\gamma$ alleles from different chromosomes. The authors concluded that a process of "correction" had occurred and sequences from one locus appeared to have been transferred into the tandem locus. Further, the 3′ end of the "corrected" region was characterized by a sequence of repeating TGs which they suggested might have served as a hotspot in the recombination event that produced the "correction." Efstratiadis *et al.* (1980) have argued that many of the deletion mutants of human β-globin, where one or two amino acids have been lost, could be the result of mispairing between small nearby repeats in the DNA. According to their model, which explains both "correction" and deletion, as DNA is replicated the single-stranded region associated with the replication fork can slip backward or forward and mispair with

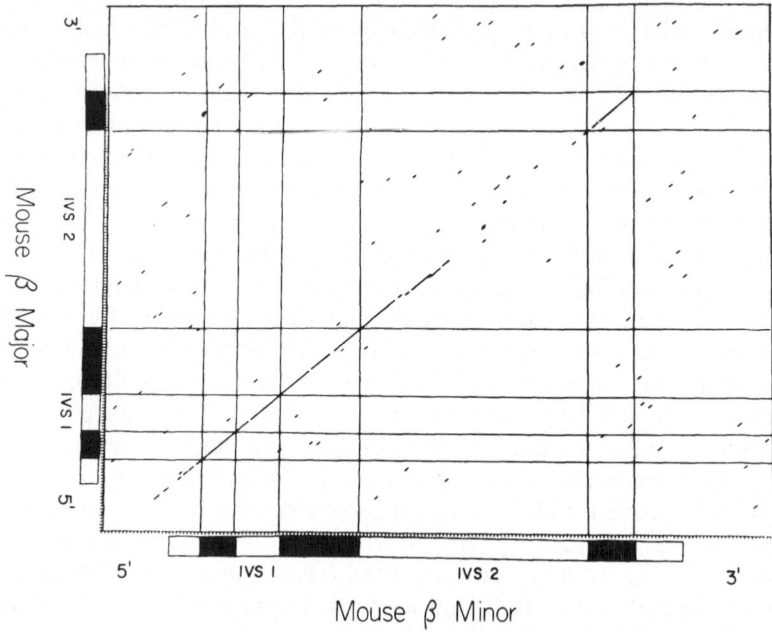

Figure 11. Graphical representation of sequence homology between two mouse genes. The β^{major} and β^{minor} genes were compared by a computer program designed to match regions of homology. These data are reanalyzed from Konkel *et al.* (1979). The match length for this comparison was eight, so that wherever eight nucleotides were found in common between the two sequences a line of appropriate length was drawn. The figure shows that the regions of maximum homology lie within the translated regions (indicated by dark blocks along the axes), within the 5′ nontranslated region, and in the first intervening sequence. Comparisons of these genes with the adult β gene of the rabbit showed little homology within the first IVS, suggesting that, as with the human γ genes (see text), perhaps "gene" correction has occurred between the 5′ regions of these two linked genes.

the wrong homologous sequences. This results in a single-stranded loop of material which cannot participate in base pairing and is then susceptible to excision by various DNA processing enzymes. This process of "correction" may contribute significantly to the way the genome evolves and may be as important as point mutation in reshaping the genetic material.

As indicated above, both the α-globin and non-α-globin clusters in humans contain nonexpressed gene regions called pseudogenes (Fritsch *et al.*, 1980). Such sequences have also been observed in other primates (Barrie *et al.*, 1981), the mouse (Vanin *et al.*, 1980; Nishioka *et al.*, 1980), rabbits (Lacy and Maniatis, 1980), and goats (Cleary *et al.*, 1980). Where pseudogenes have been sequenced, they have generally been found to be abnormal in one or more locations. It is not certain how these sequences arise, although unequal crossing-over may account for at least some of them. It has been proposed that once these sequences are no longer expressed, they might degenerate relatively quickly as new point mutations or other changes no longer alter the fitness of the organism. Some pseudogenes are especially noteworthy, because they lack intervening sequences (Vanin *et al.*, 1980; Nishioka *et al.*, 1980). Since the positions of introns in structural genes appear to be highly conserved even though their sequences change relatively rapidly, the precise loss of such regions in some pseudogenes is difficult to understand. One explanation is that fully processed mRNA is copied into DNA by a reverse transcriptase and then somehow becomes integrated into the genome, although it is difficult to understand how this could happen in germ line cells (Vanin *et al.*, 1980). Regardless of the actual explanation, it is clear that there must exist genetic mechanisms that we only barely understand.

What significance do these curious properties of genomic DNA have for molecular systematics? The processes described above could, theoretically, have a profound effect on protein sequences, which in turn could confuse the reconstruction of phylogenies based on such data. A case of such a potential problem has been proposed by Mahoney and Nute (1980) for the baboon α chain. They argue that the baboon α is less like those of humans and rhesus than would be expected on the basis of the amino acid sequences of other proteins. Most of the substitutions in the baboon chain are found at the amino end (up to position 57). Another peculiar α chain, described by Boyer *et al.* (1973) in certain gorillas and chimpanzees, designated ${}^{3}\alpha$, is most different from "normal" α beyond position 63. Mahoney and Nute (1980) speculate that the baboon α and the ${}^{3}\alpha$ genes may each be crossover products with an as yet unidentified gene in the α cluster. Since the human α cluster is known to consist of two pseudogenes as well as an embryonic chain, it is possible that one of these sequences participated in each event. Other crossover products have been described in humans, such as the Lepore and anti-Lepore hemoglobins, which involve β and δ genes, and hemoglobin Kenya, a crossover product between a γ locus and the β locus. Although these genes have not become fixed in populations, such an eventuality is possible if the crossover product is not clearly deleterious. Along with confounding the construction of phylogenetic trees, events of this nature would also falsely affect estimates of the mutation rate calculated on the basis of evolutionary comparisons of protein sequence.

Another instance of a new gene product resulting from a genetic "accident" may be human chorionic gonadotropin (hCG). The 30 amino acids at the C terminus of this hormone are not homologous to any of the other related glycoprotein hormones, but based on the nucleic acid sequence, Fiddes and Goodman (1980) suggest that this may be due to the loss by mutation of a terminator codon which produces an elongated mRNA. Other readthrough products are known from the globins, but none of these has become fixed in populations as has hCG. While the vast collection of protein sequence data that has been gathered over the last 20 years for use in comparative studies is certainly valuable and most of the conclusions drawn from it probably correct, we must recognize that the model of the genetic material changing simply by the accumulation of point mutations is not necessarily complete.

3.5. DNA Polymorphisms

The use of restriction enzyme digests to detect heterogeneity in the genomic DNA is likely to become as valuable a tool for the study of population structure as are protein methods. To date, the best studied DNA polymorphisms in human populations involve the non-globin genes. The special interest in the globins is, in part, due to the variety of hemoglobinopathies that occur in humans. Kan and Dozy (1978) described a variant *Hpa* I restriction site located about 3 kb to the 3′ side of the β-globin gene. In individuals where the site is present *Hpa* I produces a 7.6-kb band when hybridized with radiolabeled β-cDNA. When the site is absent a 13-kb band appears (Fig. 12). Upon screening patients with sickle-cell (HbS) disease, where the molecular lesion is caused by a single base substitution in the β coding sequence, Kan and Dozy (1978) found an unusually high correlation between the presence of the 13-kb band and the HbS locus. Studies of black populations in North America show that about 60% of the HbS genes occur on chromosomes with the 13-kb fragment, wherease only 9% of normal HbA genes occur on these chromosomes (Feldenzer *et al.*, 1979; Panny *et al.*, 1981). However, surveys of the sickle-cell gene in Saudi Arabian and Indian populations indicated a very different association with the *Hpa* I site from that seen in blacks and have led to some tentative conclusions on the origins and migration of the HbS gene (Kan and Dozy, 1980; Boyer *et al.*, 1981).

Other polymorphic sites involving the enzyme *Hind* III have been observed associated with the fetal γ-globin genes (Jeffreys, 1979; Tuan *et al.*, 1979), and these, too, can be used as markers to follow HbS-bearing chromosomes (Phillips *et al.*, 1980). A practical consequence of these

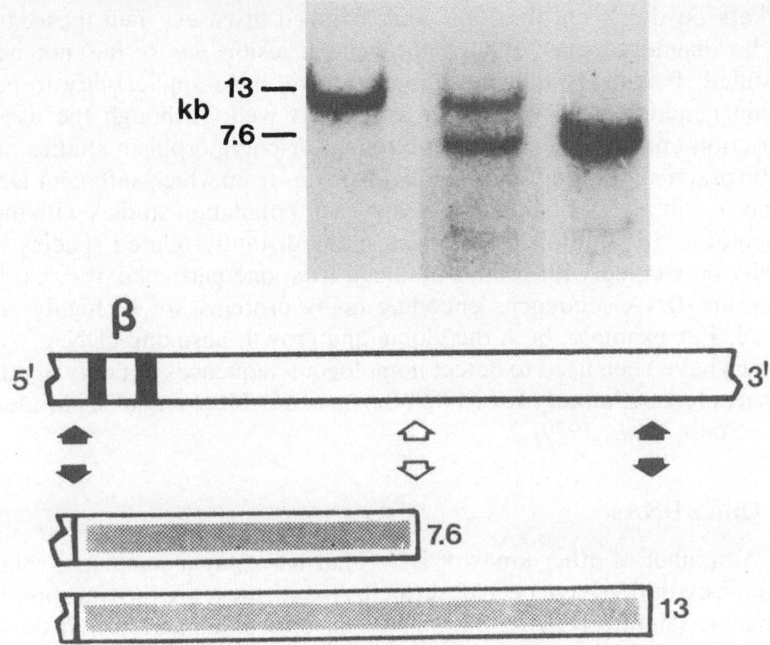

Figure 12. Polymorphic *Hpa* I restriction enzyme site. Digestion of human DNAs with the enzyme *Hpa* I followed by hybridization with a β-globin cDNA probe are illustrated at the top. Two homozygotes and a heterozygote are shown. The enzyme *Hpa* I cleaves DNA adjacent to the human β gene at the positions indicated on the accompanying map (Kan and Dozy, 1978). A polymorphic site (open arrow) for this enzyme generates distinct fragments that can be used to detect chromosomes that carry abnormal β genes. The 13-kb fragments have been shown to be in linkage disequilibrium with the HbS allele of the β gene in a high percentage of the black population with the sickle-cell gene.

polymorphisms is that they frequently permit the prenatal diagnosis of sickle-cell disease and other abnormalities of the β genes from amniocyte DNA (Little *et al.*, 1980; Kazazian *et al.*, 1980). Additional polymorphic restriction sites have been identified adjacent to known structural genes as well as from single-copy sequences randomly chosen from a human "library" (Wyman and White, 1980). Given a sufficient number of polymorphisms from representative sites distributed throughout the genome, it should be possible to construct linkage maps for genetic disorders or traits where the underlying biochemistry is unknown (Botstein *et al.*, 1980). For example, if linkage relationships can be found with various markers on the X chromosome and X-linked diseases, then these, too, can be diagnosed even though the genetic lesion *per se* has not been identified. It is likely that these methods will have applicability to population genetics studies of other species as well. Although the use of restriction enzyme and Southern blotting for polymorphism studies may not be practical for individual small organisms from which sufficient DNA cannot be obtained, it should be suitable for population studies with most vertebrates. In addition, genes from many distantly related species can usually be studied with probes obtained from one particular species, because the DNA sequences encoding many proteins are so highly conserved. For example, both the globin and growth hormone cDNAs from humans have been used to detect homologous sequences not only in other primates (e.g., Zimmer *et al.*, 1980; Barrie *et al.*, 1981), but also in rodents (e.g., Scott *et al.*, 1979).

3.6. Other DNAs

A number of other kinds of DNA that may play a significant role in genomic evolution have been described. Among these are the transposable elements, satellite DNAs, and small transcribed sequences that do not code for proteins.

3.6.1. Transposable Elements

The first of these types of DNA that may have profound physiological and evolutionary consequences are the so-called "jumping genes," transposable elements, or transposons, which appear to move to new locations in the genome and in the process may alter the phenotype (e.g., Roeder and Fink, 1980). Such sequences have been postulated in eukaryotes on classical genetic bases (McClintock, 1956) but have only recently been studied by recombinant DNA methods. These DNAs have been observed in the genomes of different strains of bacteria, yeast, and *Drosophila*. One such element in yeast accounts for 2% of the genome, while in *Drosophila* there are at least three families of transposons, and each may occur on the order of 20–50 times. In the diverse set of species in which these elements have been studied, including a virus-related sequence from the chicken, these DNAs are characterized by being fairly large (about 5 kb) and having identical regions at either end (Dunsmuir *et al.*, 1980; Levis *et al.*, 1980). In *Drosophila* it has been found that although different transposons belonging to the same class have different terminal sequences, each individual transposon has an identical sequence of 300–500 bp at either end. Even among related transposons, whose sequences differ by only a few nucleotides, the ends of any given one are identical (Levis *et al.*, 1980). Thus some mechanism, as with the rRNA genes and the γ-globin genes, must "correct" the ends with respect to one another. In the fruit fly, transposon DNA is transcribed and as much as 3% of the total RNA may hybridize to it. In addition, at least some of this RNA is translated into protein (Ashburner, 1980). These sequences may be important evolutionarily, if they serve to modify gene action by insertion into structural or regulatory sequences, as may be the case for the white eye color locus in *Drosophila* (Ashburner, 1980). Comparative studies have only begun on these DNAs, but it appears that they may share conserved regions adjacent to the insertion point, which suggests a common mechanism for controlling insertion. Whether such sequences are widespread in vertebrates and whether they play any significant role in evolution is not yet known, although it is well documented that genes within the immunoglobin clusters are rearranged during development (Sakano *et al.*, 1980), and the concept of movable DNA sequences may prove to be a fundamental process for many types of genes in a large variety of organisms not only developmentally but evolutionarily.

3.6.2. Chromosome Organization and Satellite DNAs

It has been argued that chromosomal rearrangements may be largely responsible for the phenotypic differences observed between species and that quickly evolving groups undergo faster chromosomal change (Wilson *et al.*, 1974). When various species of mammals are compared, most observable chromosomal differences apparently involve movements of large blocks of DNA. However, there is ample evidence that a number of linkage groups are preserved during this process. Comparisons of mouse and human gene maps have identified about a dozen such conserved clusters, some of which are quite large (Lalley *et al.*, 1978; Ford, 1978). The occurrence of linked groups of genes in very different species may be the result of maintenance by natural selection, or it may be a consequence of the way repeated and single-copy DNAs are distributed on chromosomes. Whether the rate of chromosomal change and morphologic evolution are causally related will be difficult to demonstrate, but we may at least be able to understand the factors determining chromosomal rearrangement.

Chromosomal change may, to some extent, be mediated by satellite DNAs. These DNAs are characterized as simple, highly repeated sequences and G. P. Smith (1976) has argued from computer modeling studies that they may arise by a process of sister-chromatid exchange and unequal crossing-over in the absence of selection. It has also been shown that satellite DNAs are often both species-specific and widely distributed chromosomally. For example, in the family Bovidae, the satellite DNAs from each species that has been studied produce specific patterns after restriction enzyme digestion, but within any given species the particular satellite is distributed over several chromosomes (Maio *et al.*, 1977). The fact that the sequences are both unique to a given species and distributed among many of its chromosomes means that there must either be a saltatory process by which they "jump" between chromosomes and then disperse, or an interchromosomal "correction" mechanism.

It has been suggested that the occurrence of these species-specific DNAs at particular chromosomal locations may direct the pairing of chromosomes, although this has not been experimentally verified. Another possibility is that they may serve as a mechanism by which chromosomal regions can be joined together (Holmquist and Dancis, 1979). If satellite DNAs occur adjacent to linkage groups of structural genes, then the maintenance of these clusters is easier to understand, although again there is little convincing evidence that this is so. A third suggestion is that the presence of satellite DNAs adjacent to the euchromatic regions of chromosomes may influence the rate of recombination in these regions (Bostock, 1980). Although it is possible that satellite DNAs may have significant effects on chromosomal rearrangements, we still know virtually nothing about how such rearrangements might alter phenotypic expression. Here, too, comparative studies are likely to be informative.

3.6.3. "Alu" Family DNAs

Another recently described group of interesting DNA sequences has been identified that may also be important in gene regulation and evolution. In humans these repeats are approximately 300 bp long and have been termed the "Alu" family, because a majority of them, when digested with *Alu* I, produce fragments of 120 and 170 bp (Rubin *et al.*, 1980). The "Alu" sequences may occur up to 300,000 times, constituting about 3% of the human genome, are dispersed throughout the genome, and have been mapped to several locations in the non-α globin cluster (Fig. 8) (Jelinek *et al.*, 1980; Fritsch *et al.*, 1980). At least some of these sequences are transcribed *in vitro* by RNA polymerase III and are also apparently transcribed *in vivo*, because they show sequence homology to small double-stranded RNAs found within the heterogeneous nuclear RNA. The "Alu" repeats have a sequence of 30 bp that is highly conserved in mammalian evolution based on comparisons with similar sequences from other species that are also transcribed by RNA polymerase III. They also show sequence homology to the site of origin of DNA replication in certain viruses, which has led to the idea that they may serve as sites for the initiation of DNA replication in eukaryotes (Jelinek *et al.*, 1980). Why these sequences occur adjacent to structural genes and what their role in gene expression and regulation is, if any, has yet to be clarified, but comparative studies of these sequences should be useful in addressing these questions.

4. Conclusions

DNA studies are likely to become increasingly important in systematics and evolutionary biology. Methods of appropriate resolution are now available to answer questions about both closely and distantly related taxa. For example, base composition assays may be adequate for comparing broad taxonomic groups, melting curve or solution hybridization analyses useful for comparing families or genera, and restriction enzyme mapping and sequencing suitable for species or population comparisons. However, in the future it seems certain that the study of cloned sequences will provide the greatest insights not only into how species are related but into the mechanisms by which they evolved.

The new DNA techniques have become so powerful that the amino acid sequences of several proteins are now known principally or entirely from their DNA sequences. Among these are the interferons, about which only partial protein data were available until the cDNAs were cloned and sequenced (Allen and Fantes, 1980). An even more startling example is the work of Sutcliffe *et al.* (1980), who, in studying cloned viral DNA, discovered a potential gene product on the basis of the DNA sequence. They then synthesized a peptide predicted by that sequence and raised antibodies against it which were then used to detect the presence of the authentic protein in infected cells. Further, sequences are now being isolated from *Drosophila* "libraries" for genes associated with morphologic characters that have been chromosomally mapped by traditional genetics but for which no specific gene product is known. It is entirely likely that genes important in development and behavior will also be retrieved from such "libraries" in the near future, and we will then be able to study genes responsible for the characters we classically think of as distinguishing species. Although it will be considerably harder to find such genes in mammalian "libraries," eventually this may be possible.

The amount of nucleic acid sequence information is bound to become much greater than the accumulated data on proteins. Even for the few genes that are expressed at high levels, cloning the DNA is preferable to protein methods, because such small amounts of blood or tissue are needed.

A few grams of fresh liver or 100 or 200 ml of blood from a mammal (much less from vertebrates with nucleated red cells) should yield more than enough DNA for the construction of a genomic "library." Few proteins can be studied so economically. Further, unlike protein studies, a recombinant "library" can be repeatedly screened with new probes and serves as a resource indefinitely. This obviates the need to obtain additional specimens, which may be a problem with rare species. Also, the techniques for synthesizing artificial polynucleotides have been greatly improved, so that it is now often possible to create a DNA sequence that will code for a particular protein (Itakura, 1980). Such synthetic oligomers are likely to find increasing applications not only in the production of proteins by bacteria [reviewed by Wetzel (1980)], but also as probes to detect the authentic gene. The cytochrome *b* gene of yeast was found by such a procedure (M. Smith *et al.,* 1979). Thus, it may be possible to study the genes which code for many proteins whose amino acid sequences are known, but for which mRNAs have not been isolated.

We would like to close this review by proposing that a concerted effort be made to establish a taxonomically diverse bank of genomic "libraries." Today there are several dozen genes that can be retrieved from such "libraries." An ever-increasing number of probes will be available in the future and it is not fantasy to predict that by the end of the decade several hundred types of genes will be accessible. The most interesting problems about the genome and its evolution have yet to be solved. The new DNA techniques that have become available in the last 5 years have raised more questions than they have answered, but they provide the means to explore questions about the genome and its evolution that could only be guessed about before. As natural habitats are destroyed and many species become increasingly rare, we should, at the very least, make every effort to preserve their genetic information in a form that can be studied well into the future. Although museum collections will continue to be valuable, it would be sad if we continued to save skins and bones but discarded the information that directed their construction.

References

Adams, J. W., Kaufman, R. E., Kretschmer, P. J.,. Harrison, M., and Nienhuis, A. W., 1980, A family of long reiterated DNA sequences, one copy of which is next to the human beta globin gene, *Nucl. Acids Res.* **8**:6113–6128.

Adams J. W., Barton, P. J. R., Cline, A., Malcolm, S., Davis-Moulton, A., Scott, A. F., Smith, K. D., Nienhuis, A. W., 1982, Further characterization of the 6.4 Kb repeated sequence of humans. Nucleotide sequence of the 5′ end, chromosomal distribution, and homology to other primate DNAs, in press.

Allen, G., and Fantes, K. H., 1980, A family of structural genes for human lymphoblastoid (leukocyte-type) interferon, *Nature* **287**:408–411.

Arnheim, N., and Southern, E. M., 1977, Heterogeneity of the ribosomal genes in mice and men, *Cell* **11**:363–370.

Ashburner, M., 1980, *Drosophila* at Kolymbari, *Nature* **288**:538–540.

Avise, J. C., Patton, J. C., and Aquadro, C. F., 1980, Evolutionary genetics of birds, *J. Hered.* **71**:303–310.

Azad, A. A., and Deacon, N. J., 1980, The 3′-terminal primary structure of five eukaryotic 18S rRNAs determined by the direct chemical method of sequencing. The highly conserved sequences include an invariant region complementary to eukaryotic 5S rRNA, *Nucl. Acids Res.* **8**:4365–4376.

Barrie, P. A., Jeffreys, A. J., and Scott, A. F., 1981, Evolution of the β-globin gene cluster in man and the primates, *J. Mol. Biol.* **149**:319–336.

Benton, W. D., and Davis, R. W., 1977, Screening λgt recombinant clones by hybridization to single plaques *in situ*, *Science* **196**:180–182.

Bostock, C., 1980, A function for satellite DNA?, *Trends Biol. Sci.* **5**:117–119.

Botstein, D., White, R. L., Skolnick, M., and Davis, R. W., 1980, Construction of a genetic linkage map in man using restriction fragment length polymorphisms, *Am. J. Hum. Genet.* **32**:314–331.

Boyer, S. H., Noyes, A. N., Boyer, M. L., and Marr, K., 1973, Hemoglobin $^3\alpha$ chains in apes. Primary structures and the presumptive nature of back mutation in a normally silent gene, *J. Biol. Chem.* **248**:992–1003.

Boyer, S. H., Panny, S. R., Smith, K. D., and Dover, G. J., 1981, How many ancestral mutations have led to the hemoglobin A–S polymorphism: Approaches to an answer in: *Birth Defects Symposium XI, Human Mutation: Population and Biological Aspects* (E. Hock, and H. Porter, eds.), Academic Press, New York, pp. 35–47.

Britten, R. J., and Davidson, E. H., 1971, Repetitive and non-repetitive DNA sequences and a speculation on the origins of evolutionary novelty, *Q. Rev. Biol.* **46**:111–133.

Britten, R. J., Graham, D. E., and Neufeld, B. E., 1974, Analysis of repeating DNA sequences by reassociation, in: *Methods in Enzymology* Volume 29 (L. Grossman and K. Moldave, eds.), Academic Press, New York, pp. 363–418.

Brown, W. M., George, M., and Wilson, A. C., 1979, Rapid evolution of animal mitochondrial DNA, *Proc. Natl. Acad. Sci. USA* **76**:1967–1971.

Cleary, M. L., Haynes, J. R., Schon, E. A., and Lingrel, J. B., 1980, Identification by nucleotide sequence analysis of a goat pseudoglobin gene, *Nucl. Acids Res.* **8**:4791–4802.

Cooke, H., 1976, Repeated sequence specific to human males, *Nature* **262**:182–186.

Cooke, H. J., and McKay, R. D. G., 1978, Evolution of a human Y-chromosome specific repeated sequence, *Cell* **13**:453–460.

Cory, S., and Adams, J. M., 1977, A very large repeating unit of mouse DNA containing the 18S, 28S and 5.8S rRNA genes, *Cell* **11**:795–805.

Craik, C. S., Buchman, S. R., and Beychok, S., 1980, Characterization of globin domains: Heme binding to the central exon product, *Proc. Natl. Acad. Sci. USA* **77**:1384–1388.

Davidson, E. H., and Britten, R. J., 1979, Regulation of gene expression: Possible role of repetitive sequences, *Science* **204**:1052–1059.

Deisseroth, A., Nienhuis, A., Turner, P., Velez, R., Anderson, W. F., Ruddle, F., Lawrence, J., Creagen, R., and Kucherlapati, R., 1977, Localization of the human α-globin structural gene to chromosome 16 in somatic cell hybrids by molecular hybridization assay, *Cell* **12**:205–218.

Doolittle, W. F., and Sapienza, C., 1980, Selfish genes, the phenotype paradigm and genome evolution, *Nature* **284**:601–603.

Duncan, C., Biro, P. A., Choudary, P. V., Elder, J. T., Wang, R. R. C., Forget, B. G., DeRiel, J. K., Weissman, S. M., 1979, RNA polymerase III transcriptional units are interspersed among human non-α-globin genes, *Proc. Natl. Acad. Sci. USA* **76:**5095–5099.

Dunsmuir, P., Brorein, W. J., Simon, M. A., and Rubin, G. M., 1980, Insertion of the *Drosophila* transposable element *copia* generates a 5 base pair duplication, *Cell* **21:**575–579.

Eaton, W. A., 1980, The relationship between coding sequences and function in haemoglobin, *Nature* **284:**183–185.

Efstratiadis, A., Posakony, J. W., Maniatis, T., Lawn, R. M., O'Connell, C., Spritz, R. A., DeRiel, J. K., Forget, B. G., Weissman, S. M., Slightom, J. L., Blechl, A. E., Shoulders, C. C., Proudfoot, N. J., 1980, The structure and evolution of the human β-globin gene family, *Cell* **21:**653–668.

Embury, S. E., Lebo, R. V., Dozy, A. M., and Kan, Y. W., 1979, Organization of the α-globin genes in the Chinese α-thalassemia syndromes, *J. Clin. Invest.* **63:**1307–1310.

Enquist, L., Sternberg, N., 1979, *In vitro* packaging of lamda *Dam* vectors and their use in cloning DNA fragments, in: *Methods in Enzymology*, vol. 68, (R. Wu, ed.) Academic Press, New York, pp. 281–298.

Feldenzer, J., Mears, J. G., Burns, A. L., Natta, C., and Bank, A., 1979, Heterogeneity of DNA fragments associated with the sickle globin gene, *J. Clin. Invest.* **64:**751–755.

Fiddes, J. C., and Goodman, H. M., 1980, The cDNA for the β-subunit of human chorionic gonadotropin suggests evolution of a gene by readthrough into the 3′-untranslated region, *Nature* **286:**684–687.

Ford, E. H. B., 1978, Evolutionary conservation of gene linkage, *Nature* **274:**112.

Fritsch, E. F., Lawn, R. M., and Maniatis, T., 1980, Molecular cloning and characterization of the human β-like globin gene cluster, *Cell* **19:**959–972.

Gilbert, W., 1978, Why genes in pieces?, *Nature* **271:**501.

Gillespie, D., 1977, Newly evolved repeated DNA sequences in primates, *Science* **196:**889–891.

Goosens, M., Dozy, A. N., Embury, S. H., Zachariades, Z., Hadjiminas, M. G., Stamatoyannopoulos, G., and Kan, Y. W., 1980, Triplicated α-globin loci in humans, *Proc. Natl. Acad. Sci. USA* **77:**518–521.

Gourse, R. L., and Gerbi, S. A., 1980, Fine structure of ribosomal RNA. III. Location of evolutionarily conserved regions within ribosomal DNA, *J. Mol. Biol.* **140:**321–339.

Gusella, J., Varsanyi-Breiner, A., Kao, F. -T., Jones, C., Puck, T. T., Keys, C., Orkin, S., and Housman, D., 1979, Precise localization of human β-globin gene complex on chromosome 11, *Proc. Natl. Acad. Sci. USA* **76:**5239–5243.

Henderson, A. S., Atwood, K. C., and Warburton, D., 1976, Chromosomal distribution of rDNA in *Pan paniscus, Gorilla gorilla beringei*, and *Symphalangus syndactylus:* Comparison of related primates, *Chromosoma* **59:**147–155.

Hoeijmakers-van Dommelen, H. A. M., Grosveld, G. C., de Boer, E., Flavell, R. A., Varley, J. M., and Jeffreys, A. J., 1980, Localization of repetitive and unique DNA sequences neighbouring the rabbit β-globin gene, *J. Mol. Biol.* **140:**531–547.

Holmquist, G. P., and Dancis, B., 1979, Telomere replication, kinetochore organizers, and satellite DNA evolution, *Proc. Natl. Acad. Sci. USA* **76:**4566–4570.

Itakura, K., 1980, Synthesis of genes, *Trends Biol. Sci.* **5:**114–116.

Jahn, C. L., Hutchison, C. A., Phillips, S. J., Weaver, S., Haigwood, N. L., Voliva, C. F., and Edgell, M. H., 1980, DNA sequence organization of the β-globin complex in the BALB/c mouse, *Cell* **21:**159–168.

Jeffreys, A. J., 1979, DNA sequence variants in $^{G}\gamma$-, $^{A}\gamma$-, δ- and β-globin genes in man, *Cell* **18:**1–10.

Jeffreys, A. J., and R. A., Flavell, 1977a, A physical map of the DNA regions flanking the rabbit β-globin gene, *Cell* **12**:429–439.

Jeffreys, A. J., and Flavell, R. A., 1977b, The rabbit β-globin gene contains a large insert in the coding sequence, *Cell* **12**:1097–1108.

Jelinek, W. R., Toomey, T. P., Leinwand, L., Duncan, C. H., Biro, P. A., Choudary, P. V., Weissman, S. M., Rubin, C. M., Houck, C. M., Deininger, P. L., and Schmid, C. W., 1980, Ubiquitous, interspersed sequences repeated in mammalian genomes, *Proc. Natl. Acad. Sci. USA* **77**:1398–1402.

Jukes, T. H., 1980, Silent nucleotide substitutions and the molecular evolutionary clock, *Science* **210**:973–978.

Kan, Y. W., and Dozy, A. M., 1978, Polymorphism of DNA sequence adjacent to human β globin structural gene: Relationship to sickle mutation, *Proc. Natl. Acad. Sci. USA* **75**:5631–5635.

Kan, Y. W., and Dozy, A. M., 1980, Evolution of the hemoglobin S and C genes in world populations, *Science* **209**:388–391.

Kaufman, R. E., Kretchmer, P. J., Adams, J. W., Coon, H. C., Anderson, W. F., and Nienhuis, A., 1980, Cloning and characterization of DNA sequences surrounding the human β-, δ-, and γ-globin genes, *Proc. Natl. Acad. Sci. USA* **77**:4229–4233.

Kazazian, H. H., Phillips, J. A., Boehm, C. D., Vik, T. A., Mahoney, M. J., and Ritchey, A. K., 1980, Prenatal diagnosis of β-thalassemias by amniocentesis: Linkage analysis using multiple polymorphic restriction endonuclease sites, *Blood* **56**:926–930.

Kohne, D. E., Chiscon, J. A., and Hoyer, B. H., 1972, Evolution of primate DNA sequences, *J. Hum. Evol.* **1**:627–644.

Konkel, D. A., Maizel, J. V., and Leder, P., 1979, The evolution and sequence comparison of two recently diverged mouse chromosomal β-globin genes, *Cell* **18**:865–873.

Kunkel, L. M., and Smith, K. D., 1982, Evolution of human Y-chromosome DNA, *Chromosoma*, in press.

Kunkel, L. M., Smith, K. D., and Boyer, S. H., 1976, Human Y-chromosome-specific reiterated DNA, *Science* **191**:1189–1190.

Kunkel, L. M., Smith, K. D., Boyer, S. H., Borgaonkar, D. S., Wachtel, S. S., Miller, O. J., Breg, W. R., Jones, H. W., and Rary, J. M., 1977, Analysis of human Y-chromosome-specific reiterated DNA in chromosome variants, *Proc. Natl. Acad. Sci. USA* **74**:1245–1249.

Kunkel, L. M., Smith, K. D., and Boyer, S. H., 1979, Organization and heterogeneity of sequences within a repeating unit of human Y chromosome deoxyribonucleic acid, *Biochemistry* **18**:3343–3353.

Lacy, E., and Maniatis, T., 1980, The nucleotide sequence of a rabbit β-globin pseudogene, *Cell* **21**:545–553.

Lacy, E., Hardison, R. C., Quon, D., Maniatis, T., 1979, The linkage arrangement of four rabbit β-like globin genes, *Cell* **18**:1273–1283.

Lalley, P. A., Minna, J. D., and Francke, U., 1978, Conservation of autosomal gene synteny groups in mouse and man, *Nature* **278**:160–163.

Lawn, R. M., Fritisch, E. F., Parker, R. C., Blake, G., Maniatis, T., 1978, The isolation and characterization of linked δ-β-globin genes from a cloned library of human DNA, *Cell* **15**:1157–1174.

Lawn, R. M., Efstratiadis, A., O'Connell, C., and Maniatis, T., 1980, The nucleotide sequence of the human β-globin gene, *Cell* **21**:647–651.

Levis, R., Dunsmuir, P., and Rubin, G. M., 1980, Terminal repeats of the *Drosophila* transposable element *copia:* Nucleotide sequence and genomic organization, *Cell* **21**:581–588.

Little, P. F. R., Annison, G., Darling, S., Williamson, R., Camba, L., and Modell, B.,

1980, Model for antenatal diagnosis of β-thalassaemia and other monogenic disorders by molecular analysis of linked DNA polymorphism, *Nature* **285:**144–147.

Long, E. O., and Dawid, I. B., 1980, Repeated genes in eukaryotes, *Annu. Rev. Biochem.* **49:**727–764.

Mahoney, W. C., and Nute, P. E., 1980, Amino acid sequence of the hemoglobin α chain from a baboon *(Papio cynocephalus):* A product of gene fusion?, *Biochemistry* **19:**1529–1534.

Maio, J. J., Brown, F. L., and Musich, P. R., 1977, Subunit structure of chromatin and the organization of eukaryotic highly repetitive DNA: Recurrent periodicities and models for the evolutionary origins of repetitive DNA, *J. Mol. Biol.* **117:**637–655.

Maniatis, T., Hardison, R. C., Lacy, E., Lauer, J., O'Connell, C., Quon, D., Sim, G. K., and Efstratiadis, A., 1978, The isolation of structural genes from libraries of eucaryotic DNA, *Cell* **15:**687–701.

Manuelidis, L., 1976, Repeating restriction fragments of human DNA, *Nucl. Acids Res.* **3:**3063–3076.

Martin, S. L., Zimmer, E. A., Kan, Y. W., and Wilson, A. C., 1980, Silent δ-globin gene in Old World monkeys, *Proc. Natl. Acad. Sci. USA* **77:**3563–3566.

Maxam, A. M., and Gilbert, W., 1977, A new method for sequencing DNA, *Proc. Natl. Acad. Sci. USA* **74:**560–564.

Maxam, A. M., and Gilbert, W., 1980, Sequencing end-labeled DNA with base-specific chemical cleavages, in: *Methods in Enzymology* Volume 65 (L. Grossman and K. Moldave, Eds.), Academic Press, New York, pp. 499–560.

McClintock, B., 1956, Controlling elements and the gene, *Cold Spring Harbor Symp. Quant. Biol.* **21:**197–216.

Nei, M., and Wen-Hsiung, L., 1979, Mathematical model for studying genetic variation in terms of restriction endonucleases, *Proc. Natl. Acad. Sci. USA* **76:**5269–5273.

Nishioka, Y., Leder, A., and Leder, P., 1980, Unusual α-globin gene that has cleanly lost both globin intervening sequences, *Proc. Natl. Acad. Sci. USA* **77:**2806–2809.

Nussinov, R., 1980, Some rules in the ordering of nucleotides in the DNA, *Nucl. Acids Res.* **8:**4545–4562.

Orgel, L. E., and Crick, F. H. C., 1980, Selfish DNA: The ultimate parasite, *Nature* **284:**604–607.

Panny, S. R., Scott, A. F., Smith, K. D., Phillips, III, J. A., Kazazian, Jr., H. H., Talbot, Jr., C. C., and Boehm, C. D., 1981, Population heterogeneity of the *Hpa I* restriction site associated with the β-globin gene: Implications for prenatal diagnosis, *Am. J. Hum. Genet.* **33:**25–35.

Phillips, J. A., Scott, A. F., Smith, K. D., Young, K. E., Lightbody, K. L., Jiji, R. M., and Kazazian, H. H., 1979, A molecular basis for various α-thalassemia states in American Blacks, *Blood* **54:**1439–1445.

Phillips, J. A., Kazazian, H. H., Boehm, D. D., Panny, S. R., Scott, A. F., and Smith, K. D., 1980, Prenatal diagnosis of sickle cell anemia by restriction endonuclease analysis: *Hind* III polymorphisms in γ-globin genes extend test applicability, *Proc. Natl. Acad. Sci. USA* **77:**2853–2856.

Proudfoot, N. J., and Maniatis, T., 1980, The structure of a human α-globin pseudogene and its relationship to α-globin gene duplication, *Cell* **21:**537–544.

Razin, A., and Riggs, A. D., 1980, DNA methylation and gene function, *Science* **210:**604–610.

Rice, N. R., 1972, Changes in repeated DNA in evolution, *Brookhaven Symp. Biol.* **23:**44–79.

Robbins, J., Rosteck, P., Haynes, J. R., Freyer, G., Cleary, M. L., Kalter, H. D., Smith, K., and Lingrel, J. B., 1979, The isolation and partial characterization of recombinant

DNA containing genomic globin sequences from the goat, *J. Biol. Chem.* **254:**6187–6195.

Roberts, R. J., 1980. Restriction and modification enzymes and their recognition sequences, *Nucl. Acids Res.* **8:**r63–r80.

Roeder, G. S., and Fink, G. R., 1980. DNA rearrangements associated with a transposable element in yeast, *Cell* **21:**239–249.

Rubin, C. M., Houck, C. M., Deininger, P. L., Friedman, T., and Schmid, C. W., 1980. Partial nucleotide sequence of the 300-nucleotide interspersed repeated human DNA sequences, *Nature* **284:**372–374.

Russell, G. J., Walker, P. M. B., Elton, R. A., and Subak-Sharpe, J. H., 1976. Doublet frequency analysis of fractionated vertebrate nuclear DNA, *J. Mol. Biol.* **108:**1–23.

Ryder, O. A., 1980. Satellite DNAs of the Equidae. Abstract. Second International Congress of Systematic and Evolutionary Biology.

Sakano, H., Maki, R., Kurosawa, Y., Roeder, W., and Tonegawa, S. 1980. Two types of somatic recombination are necessary for the generation of complete immunoglobin heavy-chain genes, *Nature* **286:**676–683.

Sanger, F., Nicklen, S., and Coulson, A. R., 1977. DNA sequencing with chain-terminating inhibitors, *Proc. Natl. Acad. Sci. USA* **74:**5463–5467.

Scott, A. F., Phillips, J. A., and Migeon, B. R., 1979. DNA restriction endonuclease analysis for localization of human β- and δ-globin genes on chromosome 11, *Proc. Natl. Acad. Sci. USA* **76:**4563–4565.

Scott, A. F., Panny, S. R., Smith, K. D., Boyer, S. H., and Bush, R. M., 1980. Comparative restriction mapping of the primate non-alpha globin gene region and the construction of a gorilla genomic library. Abstract. Second International Congress of Systematic and Evolutionary Biology.

Sibley, C. G., and Ahlquist, J. E., 1980. The phlogeny of the ratite birds by DNA × DNA hybridization. Abstract. Second International Congress of Systematic and Evolutionary Biology.

Sinclair, J. H., and Brown, D. D., 1971. Retention of common nucleotide sequences in the ribosomal deoxyribonucleic acid of eukaryotes and some of their physical characteristics, *Biochemistry* **10:**2761–2769.

Slightom, J. L., Blechl, A. E., and Smithies, O., 1980. Human fetal $^{G}\gamma$- and $^{A}\gamma$-globin genes: Complete nucleotide sequences suggest that DNA can be exchanged between these duplicated genes, *Cell* **21:**627–638.

Smith, G. P., 1976. Evolution of repeated DNA sequences by unequal crossover, *Science* **191:**528–535.

Smith, H. O., 1979. Nucleotide sequence specificity of restriction endonucleases, *Science* **205:**455–462.

Smith, M., Leung, D. W., Gillam, S., Astell, C. R., Montgomery, D. L., and Hall, B. D., 1979. Sequence of the gene for iso-1-cytochrome c in *Saccharomyces cerevisiae*, *Cell* **16:**753–761.

Southern, E. M., 1975. Detection of specific sequences among DNA fragments separated by gel electrophoresis, *J. Mol. Biol.* **98:**503–517.

Sutcliffe, J. G., Shinnick, T. M., Green, N., Liu, F. -T., Niman, H. L., and Lerner, R. A., 1980. Chemical synthesis of a polypeptide predicted from nucleotide sequence allows detection of a new retroviral gene product, *Nature* **287:**801–805.

Tuan, D., Biro, P. A., DeRiel, J. K., Lazarus, H., and Forget, B. G., 1979. Restriction endonuclease mapping of the human γ globin gene loci, *Nucl. Acids Res.* **6:**2519–2544.

Vanin, E. F., Goldberg, G. I., Tucker, P. W., and Smithies, O., 1980. A mouse α-globin-related pseudogene lacking intervening sequences, *Nature* **286:**222–226.

Wetzel, R., 1980. Applications of recombinant DNA technology, *Am. Sci.* **68:**664–675.

Williams, B. G., and Blattner, F. R., 1979, Construction and characterization of the hybrid bacteriophage lambda Charon vectors for DNA cloning, *J. Virol.* **29**:555–575.

Wilson, A. C., Sarich, V. M., and Maxson, L. R., 1974, The importance of gene rearrangement in evolution: Evidence from studies on rates of chromosomal, protein and anatomical evolution, *Proc. Natl. Acad. Sci. USA* **71**:3028–3030.

Wu, J. C., and Manuelidis, L., 1980, Sequence definition and organization of a human repeated DNA, *J. Mol. Biol.* **142**:363–386.

Wyman, A. R., and White, R., 1980, A highly polymorphic locus in human DNA, *Proc. Natl. Acad. Sci. USA* **77**:6754–6758.

Young, M. W., 1979, Middle repetitive DNA: A fluid component of the *Drosophila* genome, *Proc. Natl. Acad. Sci. USA* **76**:6274–6278.

Zimmer, E. A., Martin, S. L., Beverley, S. M., Kan, Y. W., and Wilson, A. C., 1980, Rapid duplication and loss of genes coding for the α chains of hemoglobin, *Proc. Natl. Acad. Sci. USA* **77**:2158–2162.

CHAPTER 9

Features of Gene Structure, Organization, and Expression That Are Providing Unique Insights into Molecular Evolution and Systematics

DAVID HEWETT-EMMETT, PATRICK J. VENTA, and RICHARD E. TASHIAN

1. Introduction

Over the last 20 or so years, the study of amino acid sequences of proteins derived from different biologic species has greatly enriched our knowledge of both the mechanisms of molecular evolution and the phylogenetic relationships of the species. Increasingly sophisticated computer techniques have been employed in an effort to extract all of the phylogenetic information contained in such data sets. Several problems have arisen along the way, among the most persistent being the uncertainty of whether two sequences are truly orthologous and the species divergence is being examined, or whether they are paralogous and the gene divergences are being examined. Indeed, where gross discrepancies between gene phylogeny and species phylogeny occur, the probability is that paralogous genes are involved (Goodman, 1981; Goodman *et al.*, this volume, Chapter 4).

In contrast to the slow, steady progress in the amino acid sequence studies, the recent developments in nucleic acid sequencing have been

DAVID HEWETT-EMMETT, PATRICK J. VENTA, and RICHARD E. TASHIAN • Department of Human Genetics, University of Michigan Medical School, Ann Arbor, Michigan 48109.

rapid and dramatic. Among these, the following have been particularly relevant to evolutionary studies:

(a) The ability to detect and quantitate silent nucleotide substitutions, i.e., base changes in codons that do not result in amino acid changes.

(b) The discovery that the coding regions of genes are interrupted by intervening sequences, or introns, which are usually less conserved than are the nucleotide sequences of the coding regions.

(c) The discovery that some adjacent, and structurally closely related. genes may frequently "correct" each other, and this correcting, or concerted, evolution *may* indeed be of widespread occurrence.

(d) The discovery of unexpressed genes, or pseudogenes, which are related (via gene duplication) to expressed genes.

In this chapter, we will review briefly some of the above developments and, since a reasonable amount of nucleotide sequences are now available for the coding regions of globin genes, show how these nucleotide sequences can be used to construct phylogenetic trees of the hemoglobin genes. The following two sections of this chapter are intended to provide the reader with background information for the final section, which deals with the construction and analysis of phylogenetic trees based on nucleotide sequences.

2. Recombinant DNA Technology

2.1. Cloning

The basic concepts of recombinant DNA technology, as well as more specific techniques, have been well reviewed (Wu, 1979; Grossman, 1980; Chakrabarty, 1978). The general method of gene cloning will be reviewed here, although the reader may wish to consult the above references for more detail. Segments of complementary DNA (cDNA) or genomic DNA are inserted into replicating vector systems (plasmids, bacteriophage, or a hybrid of the two, called "cosmids") to form recombinant DNA. This recombinant DNA is then used to transform bacterial hosts. To select for bacteria that contain recombinants when using plasmids, drug resistance (e.g., tetracycline resistance) markers are employed. One marker is used to select those bacteria that have been transformed (i.e., have taken up plasmid DNA) and a second marker is used to select those transformants that have inserted DNA. When foreign DNA is inserted into the middle of the drug-resistant gene, function of the gene is lost. Colonies that die when plated on the second antibiotic are saved on a replica plate that contains only the first antibiotic.

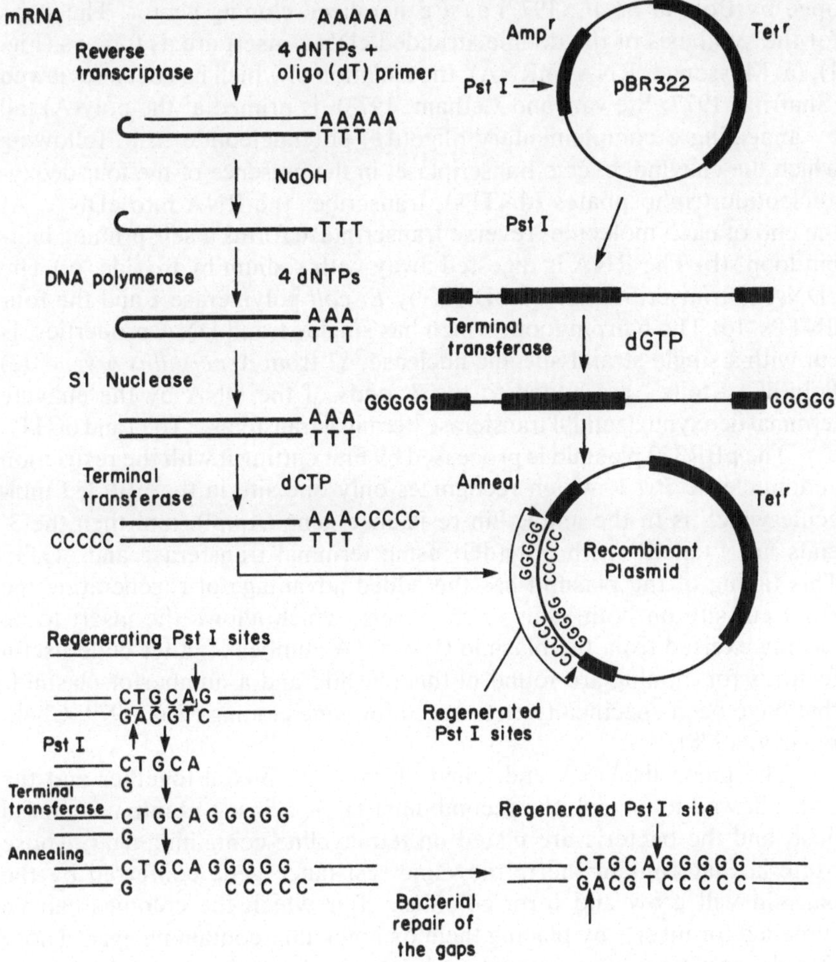

Figure 1. Steps to produce a recombinant cDNA plasmid. The left of the figure shows the steps required to produce the double-stranded cDNA insert. The right side shows the preparation of the plasmid to receive the insert. The regeneration of the *Pst* I is at the bottom of the figure. Individual steps are discussed in the text.

A specific example will serve to further explain the method. Figure 1 shows a strategy in common use for producing cDNA clones. To develop the recombined DNA, it is necessary to synthesize the double-stranded complementary DNA (dscDNA) insert and prepare the plasmid vector to receive the insert. In this case, the plasmid is pBR322 which was developed by Bolivar *et al.* (1977) as a generalized cloning vector. The steps for the synthesis of the double-stranded cDNA insert are as follows (Fig. 1): (a) Messenger RNA (mRNA), the isolation of which has been reviewed (Shafritz, 1977; Stewart and Letham, 1973), is primed at the poly(A) tail by annealing a complementary oligo(dT) polynucleotide to it, following which the enzyme reverse transcriptase, in the presence of the four deoxynucleotide triphosphates (dNTPs), transcribes the RNA into cDNA. At the end of each molecule, reverse transcriptase forms a self-priming hairpin loop. (b) The RNA is digested away with sodium hydroxide. (c) The cDNA is transcribed into dscDNA by *E. coli* polymerase I and the four dNTPs. (d) The hairpin loop, which has single-strand DNA properties, is cut with a single-strand-specific nuclease, *S*1 from *Aspergillus oryzae*. (e) Poly(dC) "tails" are added to the 3′ ends of the DNA by the enzyme terminal deoxynucleotidyl transferase (terminal transferase, TdT) and dCTP.

The pBR322 plasmid is processed by first cutting it with the restriction endonuclease *Pst* I, which recognizes only one site in the pBR322 molecule which is in the ampicillin resistance gene (*Amp*^{r}), and then the 3′ ends are "tailed" with poly(dG) using terminal transferase and dGTP. This tailing of the plasmid has the added advantage of regenerating the *Pst* I cut site on both sides of the insert, which allows the insert to be cleanly excised from the plasmid (Fig. 1). A number of other quite useful features for cloning are found in this plasmid and a number of plasmids that have been specifically engineered for gene cloning (Wu, 1979; Chakrabarty, 1978).

The tailed dscDNA and tailed plasmid are mixed together and the ends allowed to anneal, the recombinant DNA is inserted into a bacterial host, and the bacteria are plated on tetracycline-containing agar. Those bacteria that contain the tetracycline resistance gene conferred by the plasmid will grow and form colonies, after which the colonies can be screened for inserts by placing them on ampicillin-containing agar. Those that do not grow have inserts, and colonies from the same clones that have been saved on tetracycline agar can be used to screen for particular cDNA clones.

Developing full-length cDNA clones has proved to be quite difficult. Those described in the literature as full length usually contain the full coding region but lack varying amounts of the 3′ and 5′ nontranslated regions.

Loss of full-length cDNA inserts can occur during any of the steps and it is probable that losses occur during all the steps. Degradation of the RNA by contaminating RNase in the reverse transcriptase preparation or premature termination of synthesis causes loss of the 5′ end of the mRNA sequence. Premature termination during double-stranding causes loss of the 3′ end of the messenger RNA sequence. Contaminating DNases during any of the steps produce losses of either the 3′ or 5′ end. Less than full-length material has been shown to be produced both during reverse transcription (Buell *et al.*, 1978) and double-stranding (Wickens *et al.*, 1978). Less than full-length cDNA clones have been obtained even after full-length double-stranded inserts have been isolated (Küpper *et al.*, 1981), indicating that losses have occurred during *S*1 nuclease digestion, tailing, transformation, and/or propagation of the dscDNA. Other protocols which differ from the above example in one or more steps (e.g., using synthetic "linkers" and DNA ligase rather than using terminal transferase) have not been more efficient in producing full-length clones. Fortunately, for most purposes full-length cDNA clones, though highly desirable, are not indispensable. Overlapping cDNA clones can be used to obtain necessary structural information (e.g., splice junctions; see below). Most cDNA clones are eventually used as probes to find genomic sequences from which missing sequence information can be obtained. In addition, to sequence 5′ missing portions of the cDNA clone a part of the clone can be used to prime mRNA for reverse transcription sequencing (Wilson *et al.*, 1980). A very recent cloning strategy gives cDNA inserts which are complete from the cap site to the poly(A) addition site (Okayama and Berg, 1982).

Unfortunately, reverse transcriptase is not completely faithful in copying mRNA and sequences within cDNA clones sometimes contain errors which can cause problems in systematic analyses. Estimates of how often reverse transcriptase incorporates an incorrect base have been made. Using synthetic homopolymers, Springgate *et al.* (1973) found the level of infidelity to be as high as 1/530 for a poly(A) poly(dT) template. Using ϕX174 bacteriophage DNA as a template and a biologic assay of back mutation, Loeb *et al.* (1979) found the level of infidelity to be 1/300 to 1/800 [*E. coli* polymerase I and T4 DNA polymerase have much higher fidelities (Loeb *et al.*, 1979), on the order of 1/12,000 or better]. These levels of infidelity should be kept in mind when using such sequence data for developing evolutionary trees, and although the levels may not affect the position of the nodes, they may affect the branch length. Two examples of errors shown in globin cDNA clones are a transition which occurred in a human α-globin cDNA clone [Wilson *et al.* (1980); further discussed below] and a single base deletion which occurred in a rabbit β-globin cDNA clone (Browne *et al.*, 1977). It is also difficult to distinguish real sequence polymorphism from cloning artifacts. For example, in the human globin case, Wilson *et al.* (1977) have shown from analysis of variant globin protein sequences that underlying silent nucleotide sequence polymorphism is expected. In systems other than globin, where an abundance of amino acid sequence data is not available, true nucleotide sequence polymorphism could be mistaken for reverse transcriptase errors or vice versa.

Genomic DNA cloning is more straightforward than cDNA cloning since it is only necessary to cut the DNA with a restriction enzyme or shear it and then ligate the fragments into a vector. Nevertheless, difficulties with genomic cloning can also occur, such as the occasional prediction of incorrect sequence data resulting from rearrangements. This seems to occur most often in clones containing repetitive DNA (Arnheim and Kuehn, 1979). To guard against this artifact, the restriction pattern as seen on gel electrophoresis of uncloned genomic DNA can be hybridized with a radioactive probe and compared with the restriction pattern of the cloned DNA. Differences in the patterns indicate rearrangements are present in the clones. Deletions of sequences are sometimes detected on propagation of a clone and no analysis of genomic DNA is required here.

2.2. DNA and RNA Sequencing

Nucleic acid sequencing was for a long time a very tedious task and only a few special nucleic acids (e.g., tRNAs) could be sequenced at all. There are now a number of rapid DNA sequencing methods and most depend on either chemical degradation or enzymatic synthesis of DNA. Those most commonly employed are the chemical degradation method of Maxam and Gilbert (1980) and the dideoxy method pioneered in the laboratory of Frederick Sanger. These and other methods of DNA sequencing have been amply reviewed (Sanger, 1981; Weissman, 1979; Air, 1979); these reviews should be consulted if more technical information is desired than is given here. The new rapid techniques are based upon four principles:

(a) Reasonable amounts of specific homogeneous polynucleotides can be isolated by cloning and the use of restriction enzymes. Since the restriction enzymes recognize and cut specific sequences (four to six bases in length) within the DNA, each homologous DNA molecule is cut at exactly the same locations, producing molecules that have the same ends.

(b) Nucleic acids can be separated on acrylamide gels when differing in length by only one nucleotide.

(c) It is possible to produce partial lengths of the nucleic acid which end specifically at one of the four base types within a length of nucleic acid.

(d) Reaction conditions can be adjusted so that there is one break, on the average, per DNA strand.

To explain the basic strategy of DNA sequencing we will discuss the chemical degradation method of Maxam and Gilbert. The method relies on the chemical cleavage of the DNA with base-specific procedures. By first processing the DNA so that only one end is radioactively labeled, the number of nucleotides from any base in a sequence can be found by running the reaction on an acrylamide gel and reading the "rungs of the ladder" produced by placing the gel against x-ray film. By running four parallel reactions corresponding to the four bases on the same gel it is possible to read the sequence of the DNA directly from the x-ray film. Figure 2 shows a schematic example of an x-ray film from a sequencing gel.

The original enzymatic method (called the "plus–minus" method for reasons beyond the scope of this discussion) was developed by Sanger

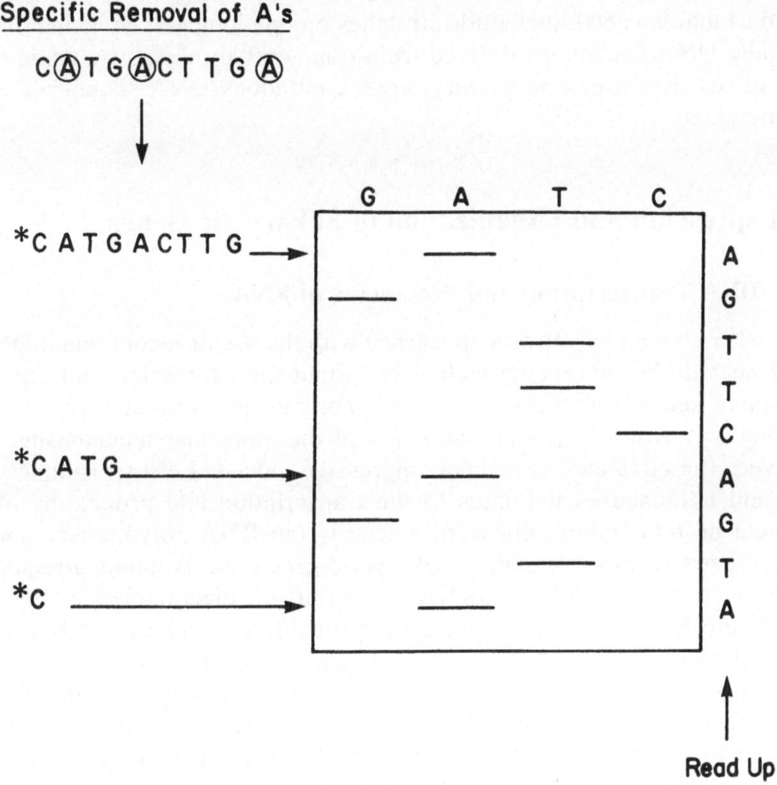

Figure 2. Schematic representation of a DNA sequencing gel. The figure demonstrates how the bands in the A lane are generated. The adenine nucleotides are destroyed by the chemical procedure and, given an average of only one nucleotide destroyed per DNA molecule, the resulting products that are labeled are shown beneath the starting molecule. The asterisk indicates the labeled end of the DNA. By using procedures specific for the other three nucleotides, the four DNA "ladders" can be displayed on one gel. Determining the sequence is then a simple matter of reading the bands in order from the bottom to the top of the gel.

and Coulson (1975). This technique has now been largely replaced by another enzymatic method developed in Sanger's laboratory which makes use of chain-terminating deoxynucleotide analogs and is commonly called the "dideoxy method" (Sanger *et al.*, 1977). This name comes from the dideoxynucleotides which are used for chain termination. By adding small amounts of analogs which are incorporated into the growing polynucleotide chain but which do not have 3′ hydroxyl groups for the attachment of the next nucleotide, the chain is stopped at those nucleotide positions corresponding to that deoxynucleotide analog. This method has two advantages over their earlier enzymatic method: it is simpler to use, and the bands on the gel are cleaner and easier to read. By using ultrathin gels (Sanger and Coulson, 1978) and current technology, it is possible to read as many as 600 nucleotide stretches on one long gel. By using overlapping DNA sequences derived from fragments produced by using different restriction enzymes, very large, continuous DNA sequences can be obtained.

3. Expression and Organization of Eukaryotic Genes

3.1. DNA Transcription and Processing of RNA

A great deal has now been learned with the use of recombinant DNA and nucleotide sequencing techniques about the expression and organization of genes from higher (eukaryotic) organisms. Although we are still a long way from understanding many of the molecular mechanisms involved, the advances have been impressive indeed. Let us examine the present evidence as it relates to the transcription and processing of a typical protein-coding, eukaryotic gene by an RNA polymerase. Some generalized features of such a eukaryotic gene (and its flanking regions) and the processing of its mRNA transcript are diagrammed in Fig. 3. First, the RNA polymerase molecule probably recognizes and binds to certain regions of DNA that are "upstream" (i.e., toward the left or 5′ end) from the specific initiation site where transcription of DNA into RNA will begin. Certain consensus DNA sequences have now been found which appear to serve as promoter regions for RNA polymerase binding. One of these sites is centered 25–30 bases upstream from the initiation site, and has the consensus sequence TATA. It is referred to as the TATA box, or Goldberg–Hogness box, and appears to correspond to the Pribnow box found in similar regions of bacterial DNA. Another region has the consensus sequence CCAAT and is located 70–80 nucleotides upstream from the initiation site. For reviews see Abelson (1979), Proudfoot *et al.*

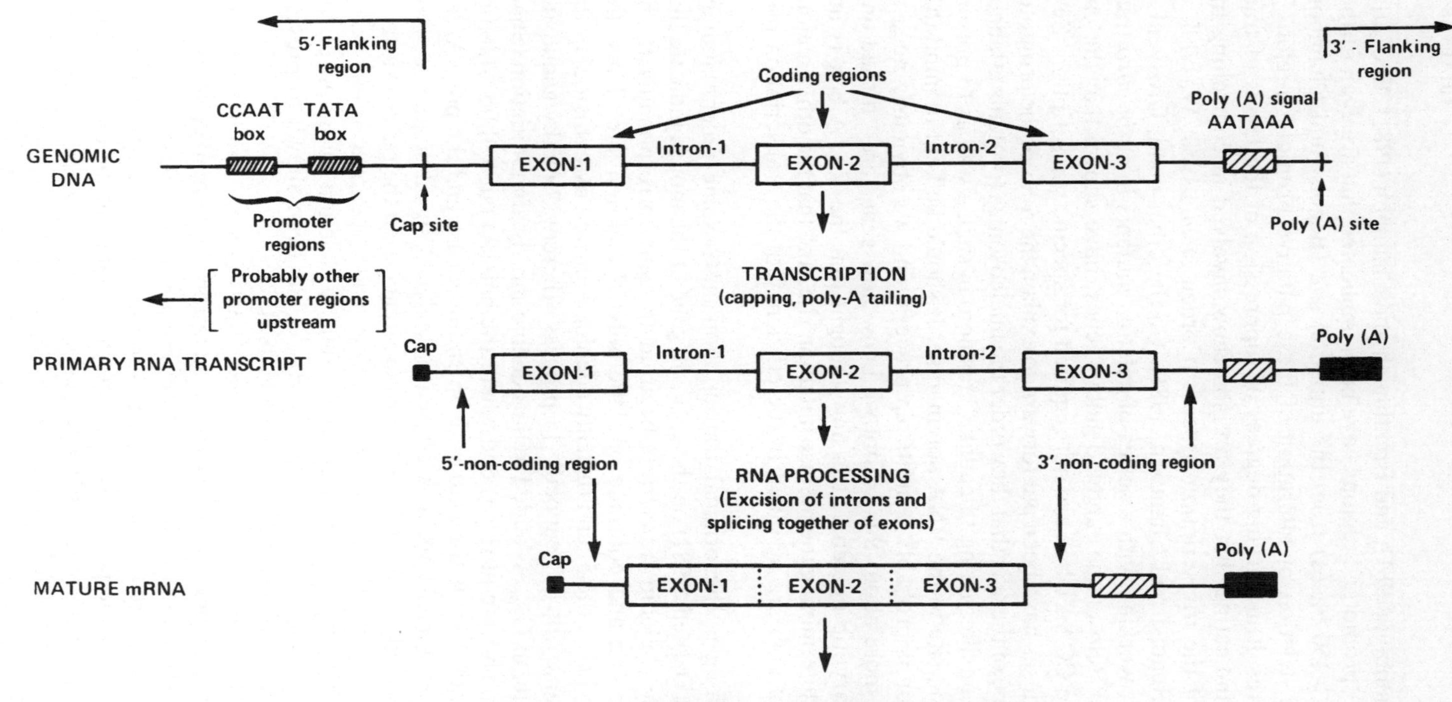

Figure 3. Generalized scheme for the processing and expression of a typical eukaryotic gene coding for a protein.

(1980), Chambon (1981), and Breathnach and Chambon (1981). Recently, additional "promoter" regions have been implicated that are even further upstream (>150 bases) from the initiation site (Benoist and Chambon, 1981). The deletion or alteration of these putative promoter regions is known to result in varying degrees of suppression of the levels of transcription, indicating that they are somehow involved in controlling the initiation of the transcription process (Corden *et al.*, 1980; Wasylyk *et al.*, 1980; Benoist and Chambon, 1981). On the other hand, Grosveld *et al.* (1981a), working with a series of deletion mutants in an *in vitro* transcription system of the rabbit β-globin genes, have shown that the cap site and the CCAAT box are not required for specific transcription. Several hypotheses have been put forward to explain the roles of the promotor regions. One of these is that they order the nucleosomes (proteins attached to DNA in an alternating "beadlike" fashion) in such a way as to present the initiation site to the RNA polymerase. It should be kept in mind that transcription is probably a multistep process, and we obviously need to learn a lot more about this before we know how genes are "turned on" in the eukaryotic nucleus. It is now thought that the TATA box is important in determining the precise initiation site, and that the other regions further upstream are involved in the efficiency of the transcription process.

How long is the original transcription unit? Some recent findings (Hofer and Darnell, 1981) on the β-globin gene of the mouse indicate that most of the β-globin transcripts begin at the site corresponding to the mRNA cap site and continue past the poly(A) addition site. Thus, the first nucleotide at the 5′ end (usually adenine followed by cytosine) of the primary transcript is "capped" [a process whereby 7-methylguanosine triphosphate, m^7G_{ppp}, is added to the adenine shortly after initiation (Shatkin, 1976)]. The site at the 3′ end where the poly(A) tail will be added to the RNA transcript is then cut with an endonuclease and the poly(A) tail added. What appears to be a consensus poly(A) signal sequence, AATAAA, is usually located 11–30 bases upstream from the poly(A) attachment site (Proudfoot and Brownlee, 1974), and appears to direct some aspect of the poly(A) tailing process, since a deletion mutation of the AATAAA sequence in late SV 40 mRNAs has been shown to prevent polyadenylation (Fitzgerald and Shenk, 1981).

The RNA transcript is now "capped" and "tailed," and if it contains intervening sequences (commonly termed introns) not found in the mature mRNA, then such noncoding sequences are normally removed before the mRNA is transported from the nucleus to the cytoplasm. It is the amino acid sequence of the polypeptide chain translated from the coding region of *this* spliced mRNA that has been used by molecular evolutionists to study the patterns and rates of evolution of homologous proteins from different species (see several chapters in this volume).

It should be pointed out that the DNA and RNA sequences that appear to be important for the transcription of RNA from DNA and the processing of the RNA transcript are characteristic of most, but not all, genes which are transcribed in eukaryotic nuclei. Similarly, the capping and polyadenylation processes are not absolute requirements for the production of mature mRNAs (Abelson, 1979; Breathnach and Chambon, 1981).

3.2. The Intervening Sequences of Genes

Because the phenomenon of genes being interrupted, or split, by intervening sequences, or introns, is such an interesting one, we shall discuss it separately. Although these introns are not normally present in the mature mRNA, they nevertheless may prove quite useful to the molecular evolutionist in studying the phylogenetic relationships between closely related species and/or genes. As mentioned in Section 3.1, certain regions of the primary RNA transcript [between the cap site and poly(A) site] are deleted before the RNA chain passes through the nuclear membrane. These introns have now been found in most eukaryotic genes or viral genes expressed in eukaryotic nuclei. However, several eukaryotic genes, such as the interferon and histone genes, do *not* contain introns (Lawn *et al.*, 1981; Kedes, 1979), and it is possible that the genes studied to date may not be entirely representative. Of interest in this respect is the finding that the two genes (*iso*-1 and *iso*-2) coding for cytochrome *c* in yeast do not have introns, whereas the presumably homologous genes in the rat contain a single intervening sequence (Scarpulla *et al.*, 1981). The greatest number of introns reported so far are those of the chicken α-2 collagen gene, with at least 51 introns (Vogeli *et al.*, 1980). Introns are not always found within coding regions (i.e., between the start codon and the stop codons), but may be located in a noncoding region, as is the case of the first intron of the rat, chicken, and human preproinsulin genes, which is located in the 5′ leader sequence between the cap site and the start codon (Lomedico *et al.*, 1979; Perler *et al.*, 1980; Bell *et al.*, 1980; Ullrich *et al.*, 1980).

Inspection of the base sequences of the introns of protein-coding genes has revealed what appears to be an almost universal feature in that the first two bases and last two bases at the 5′ and 3′ ends of the intron are GT and AG, respectively. This GT–AG rule was first recognized by Pierre Chambon and his colleagues (Breathnach *et al.*, 1978) and has been referred to as the Breathnach–Chambon rule. Additional short consensus sequences have been identified for the RNA splicing mechanism (Sharp, 1981; Lewin, 1980a):

G C CCC
5′ (exon) AG/**GT**AAGT. . .(intron). . .TXTTTXC**AG**/G (exon) 3′

The intriguing possibility that some of the small nuclear RNAs (U1 RNAs) possess sequences which can complement the bases of these bordering consensus sequences and initiate some phase of the splicing mechanism has been suggested by several investigators (Lerner *et al.*, 1980; Rogers and Wall, 1980; Murray and Holliday, 1979). The recent findings by Solnick (1981) of an adenovirus mutant with a splicing defect in which the normal 5′ splice site sequence /**GT**GAGG had mutated to /**GT**GAAT suggests that the /**GT**GAGG sequence may play a role in the splicing of adenovirus mRNAs. Further support for the role of small nuclear U1 RNAs in the splicing of pre-mRNAs comes from the studies of Yang *et al.* (1981), who demonstrated that the splicing of adenovirus mRNA was specifically inhibited by antibodies that recognize ribonucleoprotein–U1 RNA complexes.

It now appears that the intron sequences are probably cut out in a stepwise fashion. Grosveld *et al.* (1981b) found that some of the pre-mRNAs of rabbit β-globin transcripts had lost either 40 or all of the 126 bases of the first intron (IVS 1) while retaining all of the second intron (IVS 2). Other intermediate mRNAs lacked all of IVS 1 and 90 bases of IVS 2. It thus appears that removal of IVS 1 precedes that of IVS 2. Whether this stepwise process is generally true must await further studies.

Several interesting examples have recently come to light in which variations in the splicing out of introns can result in: (1) an amino acid deletion "polymorphism" (Stein *et al.*, 1980), (2) α-thalassemia (Orkin *et al.*, 1981), and (3) β-thalassemia (Busslinger, *et al.*, 1981; Fukumaki *et al.*, 1982). In the deletion polymorphism, it appears that the presence of a second /**GT** splice site immediately 5′ to the "normal" splice site of intron F of chicken ovomucoid (i.e., . . .T/**GT**GAGT/**GT**GAG. . .) results in the recognition at times by the splicing enzyme of the 5′ /**GT**GAG sequence, resulting in a Val(GTG)–Ser(AGT) deletion about 20% of the time. In the α-thalassemia, there is a deletion of five bases beginning with the T of the /**GT** in IVS 1 of the α-globin transcript (i.e., . . .AG/**GT**AGCCT. . . → AG/G(- - - - -)CT. . .), resulting in the production of an abnormal α-globin mRNA. In the β-thalassemia, a new **AG**/ site is generated toward the 3′ end of IVS 1 of the β-globin transcript (i.e., . . .TTGG. . . .TT**AG**/. . . → TT**AG**/. . . .TT**AG**/. . .), which results in the differential splicing of IVS 1 and a reduction of normal β-globin mRNA.

The discovery that different sized mRNAs can be produced from the same gene by differential splicing of pre-mRNAs clearly adds a new dimension to our understanding of the mechanisms that can control gene expression. An example of this type of alternative splicing comes from the studies of the human growth hormone gene (Fiddes *et al.*, 1979; Wallis, 1980; DeNoto *et al.*, 1981). One of the introns in the primary transcript appears to be processed in two ways. One of the mRNAs produces a hormone (mol. wt. 20,000) which stimulates growth, while the other mRNA synthesizes a hormone (mol. wt. 22,000) which both has growth-stimulating properties and influences carbohydrate metabolism. Thus, two mRNAs are translated into two somewhat functionally different hormones. Another example is that the single α-amylase gene *Amy*-1^A is expressed in both the liver and salivary gland of the mouse; however, the mRNAs of the liver and salivary gland differ in that the 5′ leader, noncoding sequence of the salivary mRNA (161 nucleotides long) differs completely from the 5′ leader sequence of the liver mRNA (50 nucleotides long). Examination of the genomic DNA revealed that these two sequences have resulted from differential splicing in the 5′ leader regions of their primary transcripts (Young *et al.*, 1981). Thus, the transcripts of the *Amy*-1^A gene are processed differently, depending on the tissues in which they are expressed. Yet another example of possible differential splicing comes from studies on a pyruvate kinase gene in the rat. The pyruvate kinase polypeptide chains of the red cells and liver differ in that their molecular weights are 63,000 and 60,000, respectively. Recent *in vitro* translation experiments by Marie *et al.* (1981) on mRNAs extracted from rat red cells and liver suggest that the differences in the sizes of the two pyruvate kinases are due to different-sized tissue-specific mRNAs. Since these two pyruvate kinases appear to be coded for by a single gene, it seems that differential splicing is responsible for the different mRNAs.

3.3. The Role of Intervening Sequences

What roles might introns play in the function, expression, and evolution of genes? As mentioned in Section 3.2, certain genes which normally do not contain introns can be expressed. A striking example of intron excision comes from an analysis of the two tandemly arranged preproinsulin genes of the rat (preproinsulins I and II). The preproinsulin II gene contains two introns, whereas the other gene contains only one intron; nevertheless, the two insulins produced by these genes are present in approximately equal amounts in the blood (Lomedico *et al.*, 1979). Sequence examination of these two genes showed that the second intron (located in the connecting C peptide of the rat preproinsulin II gene) has been precisely excised from the preproinsulin I gene. The fact that the single preproinsulin genes of humans (Ullrich *et al.*, 1980; Bell *et al.*, 1980) and chickens (Perler *et al.*, 1980) have two introns located in positions exactly homologous to those of the rat preproinsulin II gene suggests that the possession of two introns represents the ancestral state, and that the second intron was probably deleted during or after gene duplication.

The mechanism whereby introns are precisely removed from the gene is baffling indeed [for discussion see Leder *et al.* (1980)]. From an evolutionary standpoint, several examples of the complete deletion of one or more introns have now been reported. The first of these, involving one intron of the preproinsulin II gene, has already been discussed. A second example in which both introns have been exactly excised in a nonfunctioning mouse α-globin pseudogene ($\psi\alpha3$) will be discussed in Section 3.5 on pseudogenes (Nishioka *et al.*, 1980; Vanin *et al.*, 1980). Another very interesting example of what appears to be intron excision comes from a recent study of the soybean leghemoglobin gene (Jensen *et al.*, 1981). This gene contains three introns, one of which is located in the middle of a region which appears to be homologous to the second exon of all vertebrate hemoglobin genes examined to date. This finding suggests that the ancestral globin gene contained three introns. Two of the introns appear to be homologous to the two introns (IVS 1 and IVS 2) of the vertebrate hemoglobin genes in that they are located at homologous positions, as shown in Table I. However, only when we are able to find a homologous third, central intron (IVS 3) in another primitive globin gene such as a myoglobin or an invertebrate hemoglobin will we be able to tell whether having an extra intron is the ancestral condition or whether it has resulted from the unique insertion of an intron in the central exon of leghemoglobin. So far, the evidence seems to favor deletion rather than insertion of introns as the more frequent evolutionary event; however, many more homologous genes must be examined before we can tell whether this is the general trend.

If introns are not always essential for the proper expression of eukaryotic genes, why have most introns been retained at homologous positions, in some cases for over one billion years of evolution? One attractive hypothesis suggested by Gilbert (1978) is that perhaps introns facilitate or speed up gene evolution by permitting the construction of new genes from portions of existing ones (rearrangement of exons) by recombinational events in the introns of genes. If this is true, then perhaps exons might correspond to functional units (domains) of proteins. With regard to the exons of hemoglobin genes, it has been noted that the central exon

Table I

Nucleotide Sequences of Exons Bordering the RNA Splice Points for the Intervening Sequences (IVS) of Various Hemoglobin Chains

	α-Globins		β-Globins	
Mammalian[a]	31 Arg Met Phe AGG ATG TCC ↑ T IVS 1 (113,122)	99 Lys Leu Leu AAG CTC CTG ↑ A IVS 2 (134)	30 Arg Leu Leu AGG CTG CTG ↑ T A IVS 1 (100–130)	104 Lys Arg Leu Leu AGG CTC CTG A ↑ T A IVS 2 (570–950)
Chicken	31 Arg Met Phe AGG ATG TTC ↑ IVS 1[b]	99 Lys Leu Leu AAA CTC CTG ↑ IVS 2[b]	30 Arg Leu Leu AGG CTG CTG ↑ IVS 1	104 Arg Leu Leu AGG CTC CTG ↑ T T IVS 2[b]
Xenopus[c]	31 Arg Met Phe AGG ATG TTC ↑ IVS 1 (~150)	99 Pro Leu Leu CCA TTG CTG ↑ IVS 2 (~310)	29 Arg Leu Leu CGT CTG CTG ↑ IVS 1 (~185)	103 Lys Arg Phe AAG CGC TTC ↑ IVS 2 (~930)

	Leghemoglobin		
Soybean	31 Ser Ile Leu TCG ATT CTG ↑ IVS 1 (~160)	68 Leu Val Arg TTG CTG CGT ↑ IVS 3 (~225)	103 Val Val Val GTG GTG GTT ↑ IVS 2 (~300)

[a]Mammalian globins include: human α, β, δ, γ, and ϵ; rabbit α and β; and mouse α, β^{major}, and β^{minor}. For mammalian and chicken references see Table III. Other references: *Xenopus* (Patient *et al.*, 1980); soybean (Jensen *et al.*, 1981). Numbers above residues indicate position from N terminus. Numbers in parentheses are the lengths (number of bases) or approximate lengths of the introns.

[b]Probable splice sites; genomic DNA not sequenced.

[c]The α- and β-globins are on the same chromosome.

(corresponding to amino acid residue positions 31–104 in the β-globins, and 32–99 in the α-globins) contains the proximal heme-binding histidine residue as well as most of the residues involved in heme contacts (Gilbert, 1978; Blake, 1981; Argos and Rossman, 1979). More recently, Craik *et al.* (1980) confirmed this by isolating a proteolytic fragment corresponding to this region with full heme-binding capacity. It has also been pointed out by Eaton (1980) that almost all of the $\alpha_1\beta_2$ and $\alpha_1\beta_1$ contacts could be assigned to the central exon and third exon, respectively. These observations suggest that the dimer contacts might have evolved in the heme-binding domain (central exon) and that the tetramer contacts evolved later. The data of Goodman *et al.* (1975; also this volume, Chapter 4) indicate that homotetramers had already evolved in the early stem of the ancestral jawed vertebrates. A burst of adaptive substitutions at the $\alpha_1\beta_2$ contacts from the time of the duplication that produced the α and β-globin genes to these ancestral gnathostomes was then associated with the evolution of cooperativity in heterotetrameric hemoglobin.

Further elucidation of the globin domains comes from a recent analysis of the folded β-globin chains by Gō (1981). This study revealed the presence of four well-defined subdomains (compact structures): one is present in each of the two regions corresponding to the first and third exons, and two are present in the central exon, which contains most of the heme-binding residues. This finding of two subdomains in the central exon suggested that perhaps the earliest globin genes contained an intron separating the two central subdomains. This prediction is now supported by the recent finding of Jensen *et al.* (1981) discussed earlier in this section, who found that the soybean leghemoglobin gene contains three introns, the central intron falling between the two central subdomains as determined by Gō (1981).

Other proteins in which functional domains correspond to exons that are interrupted by introns in the genes coding for them are the chicken ovomucoids (Stein *et al.*, 1980), alcohol dehydrogenase of *Drosophila melanogaster* (Benyajati *et al.*, 1981), and chicken lysozyme (Jung *et al.*, 1980). Were we soothsayers, we could predict with some confidence that introns will interrupt the five Kringle loop structures of plasminogen and the two Kringle loop structures of the related proenzyme prothrombin (Hewett-Emmett *et al.*, 1981a). One of the best examples of domains being separated by introns is seen in the immunoglobulin heavy chain, in which four exons correspond to the four functional units of the immunoglobulin molecule (Gilbert, 1978; Sakano *et al.*, 1979; Tonegawa *et al.*, 1978; Brack and Tonegawa, 1977; Calame *et al.*, 1980). Also, β_2-microglobulin, which represents the light, invariant chain of the HL-A (tissue histocompatibility) antigens, corresponds precisely in size and is homologous to a single domain/exon of the immunoglobin heavy or light chain (Smithies and Poulik, 1972; Cunningham *et al.*, 1973).

A unique example of an intron which actually codes for an enzyme ("RNA maturase") that is involved in the excision of the intron that codes for it comes from the studies on the cytochrome *b* gene of yeast (Lazowska *et al.*, 1980; Lewin, 1980b).

3.4. Gene Duplication

As has been evident in the material presented in the preceding sections, the creation of new genes by gene duplication is one of the principal means by which genes evolve. An excellent account of the possible mechanisms and consequences of evolution by gene duplication was published by Ohno (1970) and should be consulted for background information. With the advent of recombinant DNA and associated techniques, several new insights into the fate of duplicated genes have emerged which have already proven extremely useful to students of molecular evolution.

Because more is now known about the molecular aspects of the hemoglobin gene family than other gene families, except perhaps the somewhat unique immunoglobulins, we shall draw heavily on recent findings which have emerged from studies on the organization and structure of the mammalian α- and β-globin genes as well as limited information on globin genes from nonmammalian sources.

When did the gene duplication take place that gave rise to the vertebrate α- and β-globin genes? An analysis of genealogic or phylogenetic trees (constructed on the basis of amino acid sequence data) of the vertebrate and invertebrate globins suggests that this duplication occurred in the ancestors of the jawed vertebrates (gnathostomes) about 425–500 million years ago (Goodman *et al.*, this volume, Chapter 4).

Although sharks (elasmobranchs) and boney fish (teleosts) possess both α- and β-globins, their chromosomal assignments have not been determined. Recent studies on the organization of the α- and β-globin genes of the African clawed toad (*Xenopus laevis*) have demonstrated that these genes are closely linked (Patient *et al.*, 1980). The fact that the linked α and β genes are found on two different chromosomes ($\alpha^1\beta^1$ and $\alpha^2\beta^2$) in *Xenopus laevis*, but are found on only one chromosome in a related toad (*Xenopus tropicalis*), suggests that this type of duplication resulted from chromosome duplication by tetraploidization (Jeffreys *et al.*, 1980).

Since the α- and β-globin genes are found on different chromosomes in birds and mammals, it can be assumed that the dispersal of these genes took place between 300 and 350 million years ago. As seen in Fig. 4, a

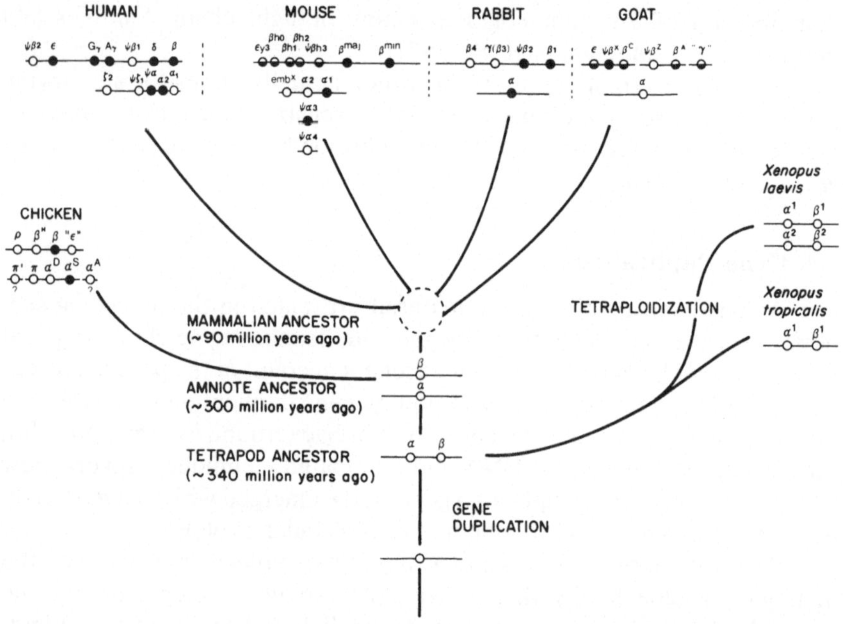

Figure 4. Distribution and linkage of hemoglobin genes in the vertebrates. Linkage data are from the following sources: human (Efstratiadis *et al.*, 1980; Weatherall and Clegg, 1979; Proudfoot and Maniatis, 1980; Proudfoot, personal communication); mouse (Jahn *et al.*, 1980; Leder *et al.*, 1980, 1981; Popp *et al.*, 1981); rabbit (Lacy and Maniatis, 1980; Hardison, 1981); goat (Haynes *et al.*, 1980a,b; Cleary *et al.*, 1980, 1981); *Xenopus laevis* (Patient *et al.*, 1980; Jeffreys *et al.*, 1980); *Xenopus tropicalis* (Jeffrey *et al.*, 1980); chicken, which is still controversial (are α^S and α^A identical?) (Donlan *et al.*, 1981; Chapman *et al.*, 1980; Knöchel *et al.*, 1982). Note that for some species the linkage is not completely known. Where only a single α gene is shown, it is not intended to imply that there is necessarily only one locus, but rather that very little is known. There is evidence in mouse (Popp *et al.*, 1981; Leder *et al.*, 1981) that the pseudoalpha genes are *not* linked to the expressed α genes. We speculate that α was still linked to β in the tetrapod ancestor. Note that in the chicken an embryonic β-like gene ("ϵ") is located 3′ to the hatching (β^H) and adult (β) beta genes (Dolan *et al.*, 1981). This violates what was seeming to be a trend; temporally later expressed genes are 3′ to the early genes. ● Genes (or cDNA) fully sequenced and included in the tree (Fig. 7 and 8). ◐ Partially sequenced genes included in the tree. Since the tree was constructed, other genes have been sequenced, notably human $\psi\zeta1$ (Proudfoot, personal communication), mouse βh2 (Edgell and Hutchinson, personal communication), and *Xenopus* α^1 (Partington and Baralle, 1981), and β^1 (Williams *et al.*, 1980), goat $\psi\beta^Z$ and $\psi\beta^X$ (Cleary *et al.*, 1981), mouse ϵy3 (Hansen *et al.*, 1982) and rabbit "γ"($\beta3$) (Hardison, 1981). The human adult α-globin used in the tree was based on the cDNA sequence of Wilson *et al.* (1980). Subsequent sequence of the $\alpha1$ and $\alpha2$ genes shows that the cDNA sequence corresponds to $\alpha2$; however, there are some minor differences discussed in the footnote to Table III.

number of subsequent duplications have occurred to produce clusters of α-like and β-like globin genes in mammals, and a β-like cluster has recently been reported for the chicken (Dolan *et al.*, 1981). For a more detailed discussion of the evolution of globin genes based on the amino acid sequences of a considerable number of vertebrate and invertebrate globins and plant leghemoglobins see Goodman *et al.* (this volume, Chapter 4).

The structures of all of the hemoglobin genes of animals so far examined are similar to the generalized gene structure shown in Fig. 3, in that they contain two intervening sequences or introns. As shown in Table I, one intron (IVS 2) tends to be longer (especially within the β-like globins) than the other intron (IVS 1). In all hemoglobin genes examined to date from mammalian, avian (chicken), and amphibian (*Xenopus*) sources, these two introns (although varying, sometimes considerably, in length) have remained at exactly the same homologous positions (Table I). The first splice site is found within the homologous arginine codons at position 29, 30, or 31, and the second splice site is located between the Lys, Arg, or Pro codons at position 99, 103, or 104 and the Leu or Arg residues at position 100, 104, or 105.

The recent discovery (Jensen *et al.*, 1981) that the soybean leghemoglobin gene contains three, rather than two, introns (discussed in Section 3.3) suggests that the ancestral hemoglobin genes may have characteristically contained three introns, and that the central intron was subsequently deleted in an ancestral hemoglobin gene prior to duplication. It is noteworthy that the first and third introns of the leghemoglobin gene are homologous to introns 1 and 2 (IVS 1 and IVS 2) of vertebrate hemoglobin genes and are located at the same homologous positions found in the vertebrate genes. As shown in Table I, even the splice site of the first intron (IVS 1) is located between the second and third base of the codon for residue 31, just as is the case for the homologous codons in the α- and β-globin genes of the vertebrates. It is indeed remarkable that homologous splice sites have been maintained for about one billion years of evolution.

Comparative studies of the nucleotide sequences of the hemoglobin genes, as expected, show that the coding regions (exons) have remained much more conservative than the noncoding regions (i.e., the 5′ and 3′ leader and terminal regions and introns). As mentioned previously, certain nucleotide sequences have been preserved in the noncoding regions, such as the GT/AG and consensus sequences bordering the intron splice sites and the consensus signal sequence for poly(A) initiation. However, except for their approximate lengths and positions in the gene, the nucleotide sequences of the noncoding regions have for the most part diverged to a much greater extent than those of the coding regions. There are exceptions, however, such as the high degree of similarity between the first introns of the (recently duplicated?) mouse β^{major} and β^{minor} genes (Konkel *et al.*, 1979), and the 5′ leader sequences of the β^{A}-, β^{C}-, and "γ"-globin genes of the goat (Haynes *et al.*, 1980a). The most dramatic preservation of the sequences of the noncoding regions of duplicated genes, however, is to be found in the human $^{G}\gamma$- and $^{A}\gamma$-globin genes. Slightom *et al.* (1980) have sequenced the human $^{G}\gamma$- and $^{A}\gamma$-globin genes on the same chromosome and most of the $^{A}\gamma$-globin gene on the other chromosome of the same individual. The nucleotide sequences of the first intron in all three genes show only one base difference between the allelic $^{A}\gamma$ genes and no differences between the nonallelic $^{G}\gamma$ and $^{A}\gamma$ genes. Only one base difference is seen among the three genes in the 5′ noncoding leader sequences, and only one difference in the first introns of all three genes. However, in the second intron, a simple (TG-rich) nucleotide sequence is found which seems to be a "hot spot" for recombination. On the 5′ side of this "hot spot" in intron 2 (the second intron), the nonallelic $^{G}\gamma$ and $^{A}\gamma$ genes do not differ in their base sequences, whereas the allelic $^{A}\gamma$ genes differ by 12 bases and two deletions. However, on the 3′ side of this "hot spot" the nonallelic $^{G}\gamma$ and $^{A}\gamma$ genes differ in their sequences at 13 positions and the allelic $^{A}\gamma$ genes at only two positions. A schematic representation of these sequence comparisons of the three γ-globin genes is presented in Fig. 5. The fact that $^{G}\gamma$ is more like $^{A}\gamma$ on the same chromosome (I) than $^{A}\gamma$ on chromsome I is like $^{A}\gamma$ on chromosome II suggests that the 5′ two-thirds of the $^{G}\gamma$ and $^{A}\gamma$ genes on the same chromosome have maintained their high degree of similarity by a mechanism of intergenic exchange (Slightom *et al.*, 1980).

The two human α-globin genes, $\alpha 2$ and $\alpha 1$, which code for identical polypeptide chains, have also been sequenced; however, since these were derived from different individuals, it is difficult to draw such precise conclusions as with the $^{G}\gamma$ and $^{A}\gamma$ genes (Liebhaber *et al.*, 1980; Michelson and Orkin, 1980). Nevertheless, it is interesting to note that the 5′ noncoding leader regions, the first introns, and all exons of both $\alpha 2$ and $\alpha 1$ genes have identical sequences. Several differences are seen in the second introns (four base differences and two deletions or insertions); however, the 3′ noncoding terminal regions differ considerably (17%) between the $\alpha 2$ and $\alpha 1$ genes. Interestingly, the 3′ noncoding region of the adjacent human $\psi\alpha 1$ gene (see Fig. 4) is more similar to the homologous region of $\alpha 2$ than $\alpha 1$ is. It seems, therefore, that the recombination unit comprises the 3′ flanking region of the 5′ gene ($\alpha 1$ or $\alpha 2$), the 5′ flanking region, and the coding region of the 3′ gene ($\alpha 2$ or $\alpha 1$). Support for this comes from the analysis of a Welsh family in which a chromosome coding three expressed α-globin genes occurs (Higgs *et al.*, 1980). From restriction enzyme mapping, it seems as though the $\alpha 1$ gene is repeated (i.e., $\alpha 2$–$\alpha 1$–$\alpha 1$), which is consistent with a crossover point 3′ to the normal $\alpha 1$ gene. The crossover point or "hot spot" in this case is not located in an intron (as in the $^{G}\gamma/^{A}\gamma$ case), but in a 3′ noncoding region. No generalization about such recombination occurring specifically in introns can therefore be drawn, particularly as only three of the six different hybrid Lepore-type hemoglobins (δ–β, β–δ, γ–β) which have been characterized (Weatherall and Clegg, 1979) could be the result of an unequal crossover occurring in an intron. If the two (or possibly three) homologous chicken α genes are indeed orthologous to the two α genes of mammals (Chapman *et al.*, 1980), then the two mammalian α genes would have originated by duplication over 300 million years ago.

Since the α-globin genes in a particular higher primate are more similar to each other than they are to the α-globin genes of related species, it has been hypothesized that their similar sequences are maintained by a process of concerted or corrective evolution [first termed coincidental evolution by Hood *et al.* (1975)] which occurs through a relatively rapid (evolutionarily speaking) mechanism of gene duplication and gene deletion by intergenic recombination (Zimmer *et al.*, 1980). Just how common this process of concerted evolution is remains to be determined. Probably other closely linked genes, such as the primate δ- and β-globin genes, undergo a similar (but less pronounced) type of corrective evolution as discussed in Section 3.5.

3.5 Pseudogenes

After the process of gene duplication, several fates may befall the duplicated gene. It may remain as a fairly faithful copy of the original gene, as in the cases of the human $^{G}\gamma$- and $^{A}\gamma$-globin or the $\alpha 2$- and $\alpha 1$-globin genes discussed above (Section 3.4). Second, the duplicated gene may diverge from the parent gene, but still produce a product similar to that of the original gene, as is found in the α-like or β-like globin genes. Third, the duplicated gene can diverge to such an extent that its protein product has acquired a function completely different from that of the

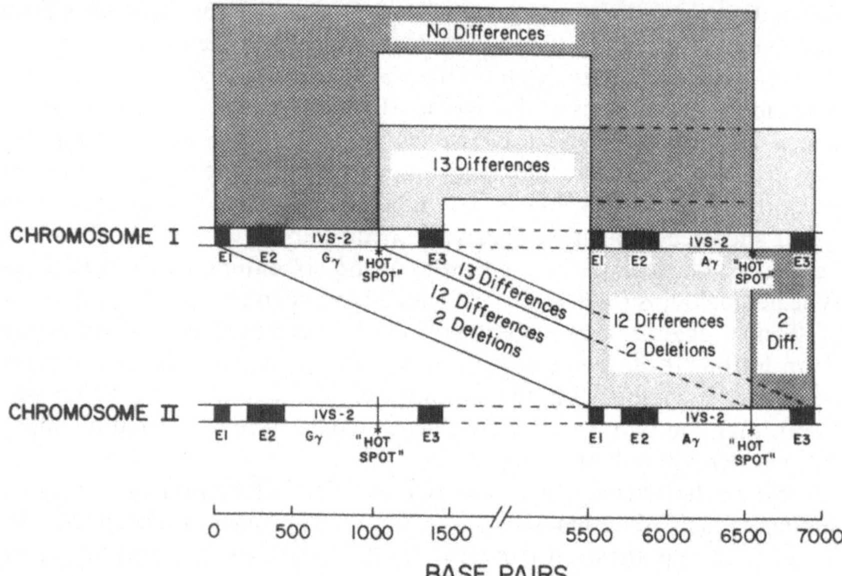

Figure 5. Concerted evolution in the human $^{G}\gamma$ and $^{A}\gamma$-globin gene cluster [based on data of Slightom *et al.* (1980)]. Cloned fragments were isolated from each chromosome of a single individual. Note that the "front" parts of $^{G}\gamma$ and $^{A}\gamma$ (i.e., exon 1, intron 1, exon 2, and part of intron 2) on chromosome I are indistinguishable, whereas the "back" parts (remainder of intron 2 and exon 3) show 13 base substitutions. By contrast, when comparing the same gene ($^{A}\gamma$-globin) on different chromosomes, the "front" part shows 12 substitutions and two small deletions, whereas the "back" part shows only two substitutions. Slightom *et al.* suggest that there is a "hot spot" of recombination located in IVS-2 (intron 2), and that concerted evolution maintains the 5′ regions identical. Note that $^{G}\gamma$ and $^{A}\gamma$ differ at only one residue in their amino acid sequence—γ 136. This is located in the region *not* "corrected".

original gene product. It is only through a careful comparative examination of the amino acid or nucleotide sequences of such genes or gene products that the homologous relationships to the original genes can be ascertained. Examples of these types of duplicated genes are the haptoglobins and the chymotrypsin family of serine proteases (Kurosky *et al.*, 1980; Hewett-Emmett *et al.*, 1981a) or insulin and the insulin-like growth factors: growth factors I and II, nerve growth factor, and relaxin (Bradshaw, 1980; Marquardt *et al.*, 1981). Finally, the gene may be "turned off" (become nonfunctional) because a mutation within the gene prevents its proper expression. These seemingly nonfunctioning genes have been referred to as pseudogenes,* and since their discovery by Jacq *et al.* (1977) in the 5S ribosomal RNA gene family, an increasing number of pseudogenes have been described, particularly in the α- and β-globin gene family (Proudfoot, 1980). Sequence analysis of these pseudogenes can determine their relatedness to the functional gene from which they have been derived. Thus, the α and β pseudogenes ($\psi\alpha$-3 and $\psi\beta$h3)† of the mouse are homologous to the functioning α- and β-globin genes. The extent to which pseudogenes have changed from the parent genes varies considerably. For example, the human ζ pseudogene ($\psi\zeta$1) appears to have been rendered nonfunctional because of a single base change in the sixth codon of the first exon, which produced a stop codon, whereas other pseudogenes show extensive changes in critical regions. This may reflect a recent origin of the gene duplication and the extent of concerted evolution. Table II lists the changes which have occurred in various pseudogenes. As is evident, a variety of interruptions (i.e., base changes, introduced stop codons, deletions) can close down a gene if they occur in such critical or presumably critical regions as promoter boxes, cap sites, normal start and stop codons, and GT/AG intron junctions. One of the most dramatic changes is found in the α-globin pseudogenes of different mouse strains, where the two introns have been precisely deleted from the genes (Nishioka *et al.*, 1980; Vanin *et al.*, 1980).

Just how common are pseudogenes? If the globin gene family is any indication, where about 16% of the genes are pseudogenes, their occurrence is probably not uncommon.

We might also inquire whether we can identify any future pseudogenes. There is a possible candidate in the globin family—the δ-globin gene, which is expressed in man, apes, New World monkeys, and probably the tarsier (see Fig. 6), but *not*, as noted many years ago, in Old

*This term is used to designate genes that through some defect are not able to produce a protein product. It is not intended to imply they have no function at all. As Vanin *et al.* (1980) point out, they may be important in controlling the expression of adjacent genes.

†Jahn *et al.* (1980) use the designation βh3 to delineate beta homologous gene 3, a β-like gene that is unexpressed and by our criteria would be termed a pseudogene. Henceforth we use $\psi\beta$h3 to accentuate the unexpressed nature of the gene while retaining the original nomenclature.

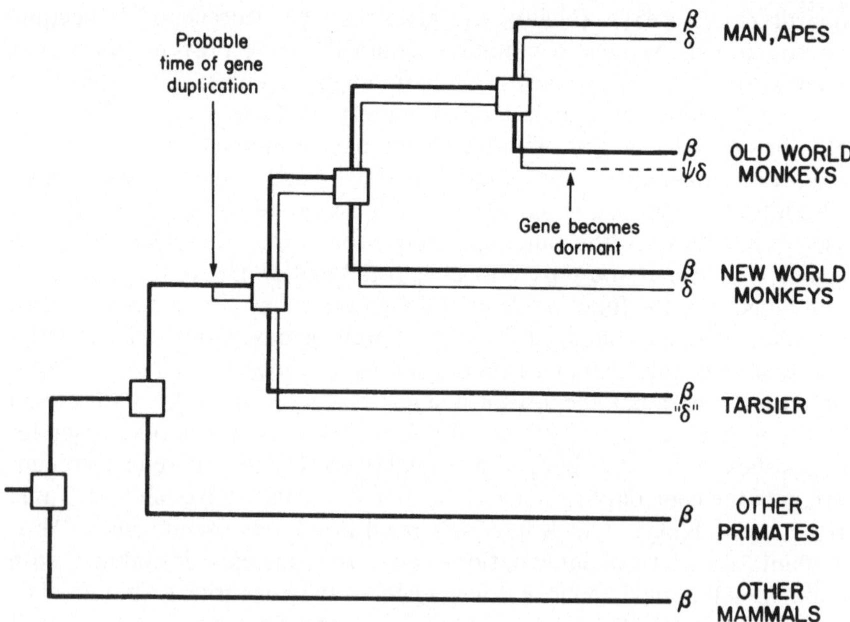

Figure 6. The origin and distribution of the δ-globin gene. The δ-gene product is present in humans, apes, New World monkeys (Boyer *et al.*, 1972), and probably tarsier (Beard *et al.*, 1976), but not Old World monkeys (Barnicot and Wade, 1970). From restriction mapping, Barrie *et al.* (1980) and Martin *et al.* (1980) showed that the δ gene *is* present in Old World monkeys, but not the lemur (a lower primate)., □ Divergence of major primate lineages.

World monkeys (Boyer *et al.*, 1972; Beard *et al.*, 1976; Barnicot and Wade, 1970). The δ chain is expressed concurrently with, but at lower levels than, the β chain and they compete for nascent α chains in forming heterodimers (αδ, αβ) and tetramers ($\alpha_2\beta_2$, $\alpha_2\delta_2$). Martin *et al.* (1980) and Barrie *et al.* (1981) showed that the δ gene, which is located 5′ to the β gene and possesses a characteristic *Pvu* I restriction site, is present in Old World monkeys and thus must qualify, by our criteria, as a pseudogene. It may prove similar to the mouse ψα-3 or the human ψζ1 in possessing few defects, and might therefore, by suitable restoration of the defects (e.g., back mutation), be expressed in some future Old World monkey lineage.

Table II

Changes in Various Regions of Mammalian Hemoglobin Pseudogenes[a]

	Promoter boxes		Start codon	Intron splice sites		Stop codons	Poly(A) signal sequence	Defects in coding regions
				IVS 1	IVS 2			
Normal or consensus sequences	CCAAT	TATAA C	ATG	GT/AG	GT/AG	TGA/TAA/TAG	AATAAA	—
Rabbit $\psi\beta 2$	Missing	CATAA	ATG	GT/CC	GT/CA	TGA	AATAAA	One frameshift and one stop codon in exons 1 and 3
Goat $\psi\beta^z$	CTAAT	CTCAC	ATG	GC/AG	GT/AG	(deletion after seventh codon of exon 3)		Early frameshift in exon 1
Goat $\psi\beta^x$	TTAAT	CTCAC	ATG	GC/AG	GT/AG	TGG[b]	AATTAA	Several frameshifts and stop codons
Mouse $\psi\beta$h3	Missing	Missing	AGG	GT/—[c]	GT/AG	TAA	AATAAA	Numerous frameshifts and stop codons
Mouse $\psi\beta$h2	GTAAC	TAAAG	ATG	GT/AG	GT/AG	TGA	AATAAA	Frameshift in exon 1
Mouse $\psi\alpha 3$	CCAAT	TAGAA	ATG	Deleted	Deleted	TAA	AATAAA	Numerous frameshifts and stop codons
Human $\psi\alpha 1$	CCAAT	CATAA	GTG	GC/AG	GA/AG	TAA	AATGAA	—
Human $\psi\zeta 1$	CCAAT	TATAA	ATG	GT/AG	GT/AG	TGA	AATAAA	Stop codon (TAG) after fifth codon of exon 1

[a]Bases differing from codon or consensus sequence are in bold type. For references, see Table III and legend to Fig. 4.
[b]Seven additional codons until new stop codon.
[c]Deleted.

An interesting aspect of δ/β globin evolution is that, unlike the ${}^{G}\gamma/{}^{A}\gamma$ and $\alpha1/\alpha2$ gene pairs mentioned earlier, there is no strong evidence for concerted evolution. Apart from the putative δ chain in tarsier, δ chains are characterized by residues 116-Arg and 117-Asn (β chains have 116-His and 117-His); interestingly, $\delta5$ is always identical to $\beta5$ in the same species. These three positions are the only sites bearing upon the origins and evolution of δ and β chains, other substitutions being unique to species groups. Delta 5 can be easily explained if concerted evolution kept the 5′ region of the genes identical (Pro in δ and β of man and apes, Gly in δ and β of New World monkeys, and Ala in β and "δ" of tarsier). Since residues 9 and 12 differ in δ and β chains in the same species, concerted evolution (if it is occurring at all) must result only in "correction" of residues 1–8.

Boyer *et al.* (1973) described an unusual minor α-globin in a few adult gorillas and chimpanzees. While residues 1–63 are identical to the respective normal adult α-globin sequence, residues 64–141 show eight differences common to both apes. It now seems likely that this "new" gene is a hybrid produced by a crossover between a normal α-globin gene and a defective pseudogene.

Critics of Ohno's (1970) treatise on evolution (of new functions) by gene duplication found it hard to explain how a duplicated gene relieved of negative selection could retain enough of its partner gene's character and not incorporate a large number of irrevocably destructive mutations. Studies of nucleic acid sequences of gene families have amply supported Ohno's original ideas, and concerted evolution provides a means of keeping two adjacent genes quite similar.

4. Use of Nucleotide Sequences in Evolutionary Tree-Building

4.1. Why Nucleotide Sequences?

So far we have reviewed what studies of mRNA, cDNA, and genomic DNA have told us about gene structure, organization, and expression that laborious analysis of protein structure over 25 years had, at best, only hinted at. Now we will look at what advantages nucleic acid sequences might have over protein sequences in constructing evolutionary trees that depict both species relationships and gene duplications. As with past sections, our review will rely heavily on data gleaned from the globin gene family.

The first analyses of nucleic acid sequences coding for proteins have, on the whole, employed pairwise comparisons (Kafatos *et al.*, 1977; Jukes, 1980; Perler *et al.*, 1980; Efstratiadis *et al.*, 1980; Li *et al.*, 1981; Miyata and Yasunaga, 1981; Kimura, 1981), just as was done with the early studies of protein sequences (Zuckerkandl and Pauling, 1965). These studies have generally supported the concept that "silent" substitutions (i.e., base substitutions not causing an amino acid substitution) occur proportionately more often than amino acid-changing substitutions, taking into account the "opportunity factor," notably that random mutation of DNA should result in three amino acid-changing substitutions for each silent substitution. In addition, these authors have generally found that, making assumptions about the time of speciation based on the fossil record, the silent substitutions provide an adequate molecular clock. These studies, however, have failed to put the sequences into a tree context, with all the attendant back mutations, parallel mutations, etc., that of necessity occur when this is done. Nevertheless, some of the findings are of general interest:

(a) Kafatos *et al.* (1977) suggested (based on their analysis of rabbit and human β-globin mRNA) that, surprisingly, silent substitutions occur less frequently at functional globin sites than at "nonfunctional" sites. We describe our findings regarding this observation in Section 4.3.3.

(b) Perler *et al.* (1980) suggested that the polymorphic rabbit β-globin alleles, which differ by four amino acid-changing nucleotide substitutions, must have undergone Darwinian selection, as no silent substitutions have occurred.

(c) Efstratiadis *et al.* (1980) constructed a tree based on rate constancy criteria that gave plausible β-globin relationships and suggested dates at which duplications occurred that fit quite well with the distribution of these genes in extant species. We discuss this in Section 4.3.1.

(d) Li *et al.* (1981) developed the concept that pseudogenes may have been expressed genes under selective constraints for a period of their existence and, by assuming a molecular clock is operable, they calculated the various dates at which gene duplications occurred and genes subquently became pseudogenes. We comment on this further in Sections 4.3.2 and 4.3.3.

In collaboration with John Czelusniak and Morris Goodman, we have built phylogenetic trees from the available globin nucleotide sequences

(Hewett-Emmett *et al.*, 1981b; Czelusniak *et al.*, 1982) and it is these data that we will review and interpret further in this section.

4.2. Tree-Building Strategy

We have employed the maximum parsimony strategy, which has been extensively used for protein sequence data [for reviews see Goodman (1981) and Goodman *et al.*, this volume, Chapter 4]. Put simply, this approach fits the sequence data to a particular tree in the most economical fashion, providing a nucleotide replacement (NR) length for this tree. Alternative trees (created by a branch-swapping algorithm) are then tested until no tree with a lower NR value can be found. Sometimes this tree represents a "valley" and not the "true" minimum, so further subjective trees are used as inputs for the branch-swapping algorithm. When satisfied that no further improvement can be made, one analyzes the tree. Even with this optimal tree, there are many alternative ways of distributing the nucleotide substitutions. In fact there are two extremes. In the *A solution*, nucleotide replacements are "pushed" to the terminal branches, so that they are shared by as *few* descendant species as possible; in the *B solution*, they are shared by as *many* descendent species as possible. It is well recognized that branches that represent long periods of evolution contain an underestimate of the true number of substitutions, due to a failure to detect multiple substitutions at the same site. The denser the network, the greater the likelihood of picking up these multiple "hits," since speciation or gene duplication events may occur between "hits." Thus, use of the A solution will tend to correct for this inequality as substitutions are more often placed on these external branches of the tree [for a fuller discussion see Goodman (1981)]. From such a tree, direct comparison of substitution rates can be made *without* recourse to the fossil record, by calculating *relative* rates subsequent to a bifurcation in the tree (see Section 4.3.3).

4.3. A Hemoglobin Nucleotide Tree

In Table III, we list the globin nucleotide sequences used in constructing the tree. Since this analysis was undertaken, additional se-

Table III
Globin Nucleotide Sequence Data Used in Constructing Phylogenetic Trees[a]

Globin	Source of DNA sequenced	Percent complete	Reference
1. Human α	cDNA	100	Wilson *et al.*. (1980)
2. Rabbit α	cDNA	100	Heindell *et al.* (1979)
3. Human $\psi\alpha 1$	Genomic	100	Proudfoot and Maniatis (1980)
4. Mouse α	Genomic	100	Nishioka and Leder (1979)
5. Mouse (CD 1 Swiss) $\psi\alpha 3$	Genomic	100	Vanin *et al.* (1980)
6. Mouse (BALB/c) $\psi\alpha 3$	Genomic	100	Nishioka *et al.* (1980)
7. Chicken α	cDNA	100	Richards and Wells (1980)
8. Human β	Genomic	100	Lawn *et al.* (1980)
9. Human δ	Genomic	100	Spritz *et al.* (1980)
10. Rabbit $\beta 1$ (allele 1)	Genomic	100	Van Ooyen *et al.* (1979)
11. Rabbit $\beta 1$ (allele 2)	Genomic	100	Hardison *et al.* (1979)
12. Rabbit $\psi\beta 2$	Genomic	100	Lacy and Maniatis (1980)
13. Goat β^{A}	Genomic	51	Haynes *et al.* (1980a,b)
14. Goat β^{C}	Genomic	65	Haynes *et al.* (1980a,b)
15. Goat "γ"	Genomic	66	Haynes *et al.* (1980a,b)
16. Goat $\psi\beta^{X}$	Genomic	100	Cleary *et al.* (1980)
17. Mouse β^{major}	Genomic	100	Konkel *et al.* (1979)
18. Mouse β^{minor}	Genomic	100	Konkel *et al.* (1979)
19. Mouse $\psi\beta h3$	Genomic	63	Jahn *et al.* (1980)
20. Mouse $\beta h2$[b]	Genomic	36	Jahn *et al.* (1980)
21. Mouse $\beta h1$	Genomic	32	Jahn *et al.* (1980)
22. Mouse $\beta h0$	Genomic	33	Jahn *et al.* (1980)
23. Mouse ϵ	Genomic	51	Jahn *et al.* (1980)
24. Goat ϵ	Genomic	37	Haynes *et al.* (1980a,b)
25. Human ϵ	Genomic	100	Baralle *et al.* (1980)
26. Human ${}^{G}\gamma$	Genomic	100	Slightom *et al.* (1980)
27. Human ${}^{A}\gamma$	Genomic	100	Slightom *et al.* (1980)
28. Chicken β	cDNA	100	Salser *et al.* (1979)[c]

[a]Subsequent sequence analysis of human genomic $\alpha 1$ and $\alpha 2$ globins demonstrates that the cDNA clone was in fact derived from an $\alpha 2$ mRNA. It is the 3′ noncoding region that identifies the cDNA sequence as $\alpha 2$. The three coding sequences differ at five positions: *α15 (Gly)* is GGT in both genomic ($\alpha 1$, $\alpha 2$) sequences but GGC in the cDNA; *α17(Val)* is GTC in both genomic sequences but GTT in the cDNA; *α54 (Gln)* is CAG in $\alpha 1$ (genomic) and the cDNA, but CAA in $\alpha 2$ (genomic); α123(Ala) is GCC in $\alpha 1$ (genomic) and the cDNA, but GCT in $\alpha 2$ (genomic); finally, *α60(Lys)* is AAG in both genomic sequences, but GAG (which would code Glu) in the cDNA. Wilson *et al.* (1980) admit that this change at α60 *must* be the result of a base substitution during or after the cloning process; other differences may represent polymorphisms. Other sequences published too late to include in the tree-building analysis are *Xenopus* α chain (Partington and Barelle, 1981), *Xenopus* β chain (Williams *et al.*, 1980), duck, α chain (Hampe *et al.*, 1981), sheep β^{A},β^{C}, "γ"-chain partial sequences (Kretschmer *et al.*, 1981), chicken β-like globin partial sequences (Dolan *et al.*, 1981), and human $\psi\zeta 1$ (N. J. Proudfoot, personal communication). Complete sequences are now available for goat $\psi\beta^{Z}$ and $\psi\beta^{X}$ (Cleary *et al.*, 1981), mouse ϵY3 (Hansen *et al.*, 1982), and rabbit embryonic $\beta 3$ which is related to primate γ genes (Hardison, 1981).

[b]$\beta h2$ has since been completed in the laboratory of M. H. Edgell and C. A. Hutchinson III (personal communication) and contains a frameshift mutation, which would qualify it as a pseudogene by our criteria (see Section 4.3.2).

[c]An alternative chicken β-globin sequence is available (Richards *et al.*, 1979); however, we used the sequence of Salser *et al.* (1979). We anticipate that the genomic β sequence may show differences from both (Dolan *et al.*, 1981).

quences have become available and they are listed in the footnote to Table III.

The tree with lowest NR value is shown in Figs. 7 (A solution) and 8 (B solution). All calculations and interpretations are based on the A solution for reasons noted above.

4.3.1. Relationships of the Globin Genes

The main features revealed by examination of the tree in Fig. 7 are the relationships of the different globin genes to one another. At this stage, too few different species are represented to draw any conclusions of relevance to systematists, but clearly the potential is there. What emerges is that our analysis must account for gene organization and temporal expression as well as just the nucleotide or amino acid sequences. The main points are:

(a) The α- and β-like globin genes form distinct groups, as previously recognized by use of amino acid sequence data (Goodman *et al.*; 1975, also this volume, Chapter 4). Within the β-like cluster, the δ and β genes are closely related, and the gene duplication is placed after the radiation of the mammals, as previously inferred (see Fig. 6).

(b) The human, and other primate, γ chains show a closer relationship to the ϵ chains than to the δ/β groups. This was also suggested by the analysis of Efstratiadis *et al.*, (1980). Interestingly, other partial ϵ chain sequences (goat and mouse) cluster with the human ϵ chain. This means that the ϵ/γ duplication preceded the divergence of the placental mammals. From protein sequence data, however, there was no evidence addressing this question, due partly, no doubt, to the difficulty in purifying sufficient quantities of nonadult β-like globins. The further implication of this is that the γ-globin gene we recognize only in the higher primates was present in the placental mammal ancestor and thus in other mammalian lineages. What has happened to it? If we use the human cluster as a model (see Fig. 4), we expect the γ gene to be located 3′ to the ϵ gene but 5′ to the adult β gene. In mouse, therefore, the β-like genes βh0, βh1, βh2, and βh3 are candidates. Although not fully sequenced, our tree hints that they might be β-like*. However more complete sequence data used in the tree

**Editor's note:* A new maximum parsimony analysis of 40 globin nucleotide sequences, including the now completed βh0 and βh2 sequences, has revealed two changes for the positioning of the mouse β-like sequences. The βh2 and βh3 sequences group together first before joining the branch of mouse β-major and mouse β-minor, and βh0 and βh1 sequences, which still group together, now joining the human γ branch.

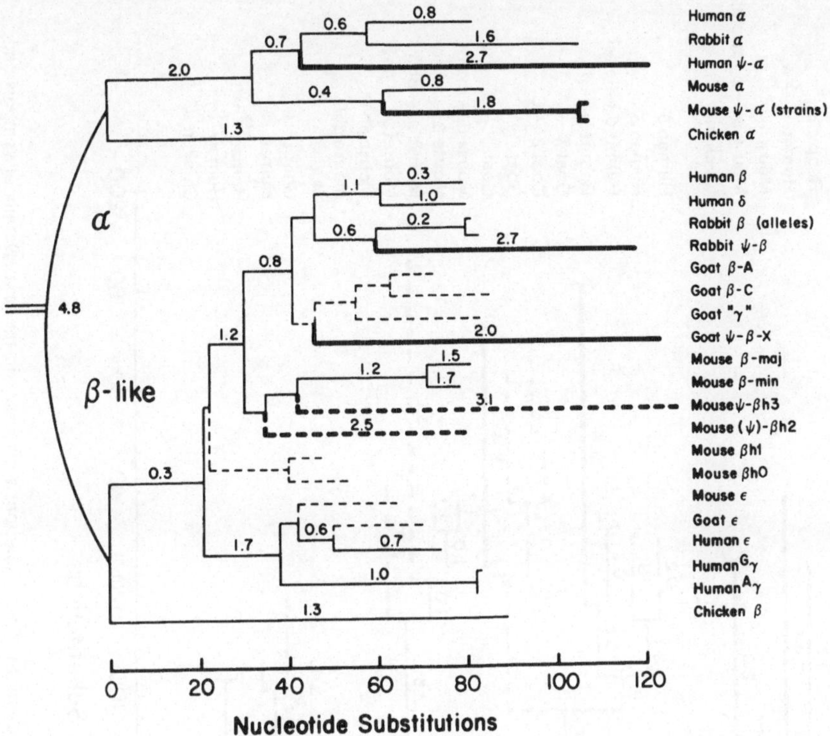

Figure 7. The most parsimonious globin nucleotide tree (A solution). The tree was built from the sequences listed in Table III; only the coding regions were used after optimally aligning the sequences. The alignment contained 449 nucleotide positions. Branches and links on the tree are drawn to scale and the numbers on the longer branches (>8 nucleotide substitution) represent the R_{CS} values (ratio of amino acid-changing to silent substitutions). Bold lines represent pseudogenes and dashed lines represent incomplete sequences. This is the A solution (see text). A more recent tree is described in Czelusniak *et al.* (1982).

of Czelusniak *et al.* (1982) suggest βh0 and βh1 are related to primate γ chains as is rabbit β3. Neither the mice βh0/βh1 nor the rabbit β3 are expressed in fetal life, so the pattern of expression of this locus may have changed in the higher primates. The goat poses an interesting problem. Unlike the primates, it has at least four temporally expressed β-like genes: embryonic (ε), fetal ("γ"), juvenile (β-C) and adult (β-A). From our tree,

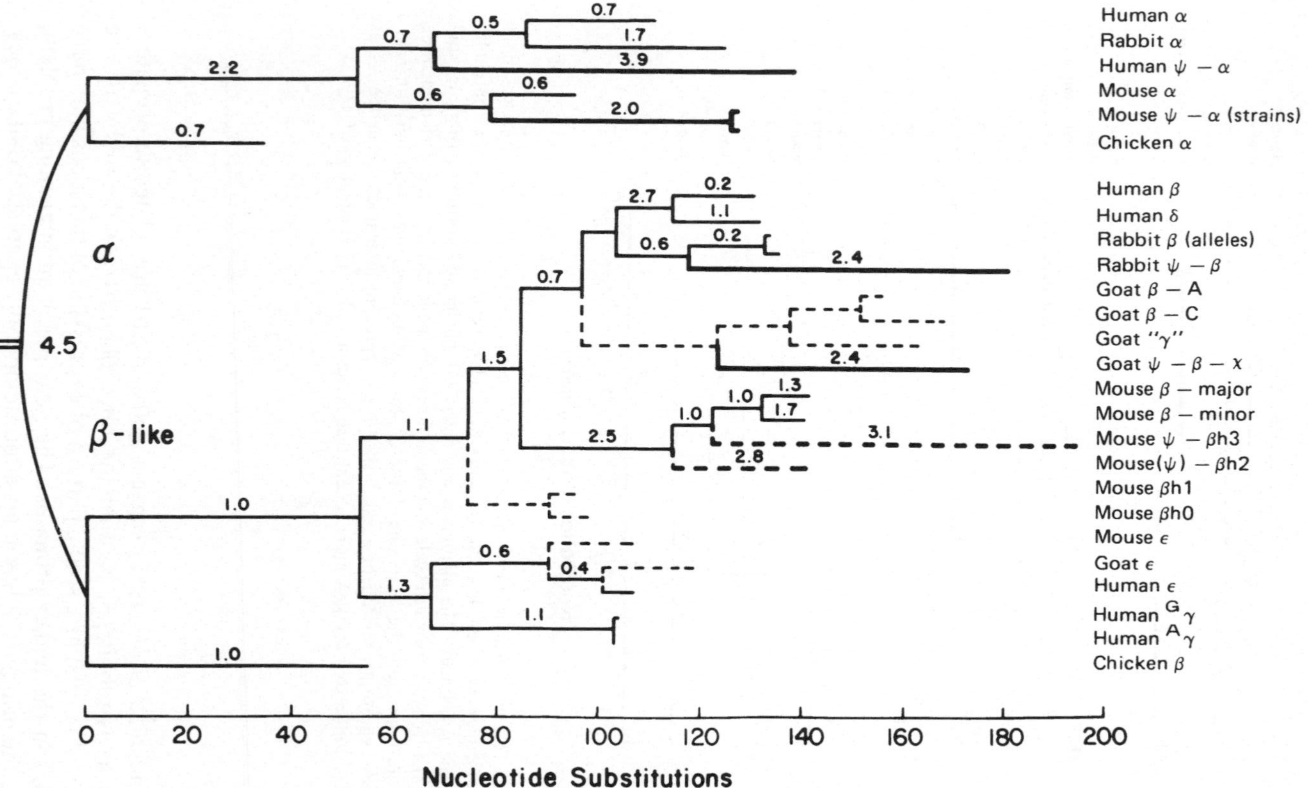

Figure 8. The most parsimonious globin nucleotide tree (B solution). The B solution is discussed in the text. Details of symbols and other information are in Fig. 7.

it is clear that the goat "γ" is related to the adult and juvenile β genes. The possibility remains that what was once the goat γ locus has been "corrected" by concerted evolution.

The lower primates also provide an interesting puzzle. Some years ago, Buettner-Janusch *et al.* (1972) showed that lemur fetal hemoglobins were indistinguishable electrophoretically from their adult hemoglobin counterparts. Recently, however, Barrie *et al.* (1981) have shown that the brown lemur β-gene cluster does possess a gene which hybridizes preferentially to a human γ-globin probe. The brown lemur possesses a simple cluster ϵ–γ–$\psi\beta$–β, which is in fact reminiscent of the rabbit. Clearly, more sequence work is needed to decipher the history and fate of the γ-globin gene locus.

(c)Five pseudogenes are included in the tree; of these, four are most closely related to the corresponding adult globins from the same species. The fifth is human $\psi\alpha$, which *if* the tree is correct, resulted from a gene duplication occurring at about the time of the placental mammal radiation. There are some grounds for treating this with scepticism; the rabbit α-globin sequence is based on cDNA and it may be clustering with human $\psi\alpha$ on the basis of some "silent" site identities. It will be interesting to compare this cDNA sequence with that derived from rabbit genomic DNA.

4.3.2. Pattern of Nucleotide Substitution

Apart from the genealogic aspects of the tree shown in Fig. 7, we can analyze the pattern of nucleotide substitution on the links and branches. By comparing the reconstructed amino acid sequences at adjacent nodes, we can determine whether a particular nucleotide substitution is amino acid-changing or silent. Complications arise if two or three nucleotide substitutions within a codon occur on a single link or branch. For example, conversion of Met (AUG) to Asp (GAU) involves three nucleotide substitutions. This can take six routes:

AUG (Met)	$\xrightarrow{1}$	A*A*G (Lys)	$\xrightarrow{1}$	AA*U* (Asn)	$\xrightarrow{1}$	*G*AU (Asp)
AUG (Met)	$\xrightarrow{1}$	A*A*G (Lys)	$\xrightarrow{1}$	*G*AG (Glu)	$\xrightarrow{1}$	GA*U* (Asp)
AUG (Met)	$\xrightarrow{1}$	*G*UG (Val)	$\xrightarrow{1}$	G*A*G (Glu)	$\xrightarrow{1}$	GA*U* (Asp)
AUG (Met)	$\xrightarrow{1}$	*G*UG (Val)	$\xrightarrow{0}$	GU*U* (Val)	$\xrightarrow{1}$	G*A*U (Asp)
AUG (Met)	$\xrightarrow{1}$	AU*U* (Ile)	$\xrightarrow{1}$	*G*UU (Val)	$\xrightarrow{1}$	G*A*U (Asp)
AUG (Met)	$\xrightarrow{1}$	AU*U* (Ile)	$\xrightarrow{1}$	A*A*U (Asn)	$\xrightarrow{1}$	*G*AU (Asp)

By averaging, we score this interconversion as 2.83 amino acid-changing and 0.17 silent substitutions. These values are calculated for each branch and link of the tree and are documented in Table IV. Note that we also include the number of nucleotide substitutions occurring at the third position of the codon for each branch/link. As might be expected, these third-position substitution values are similar to those for silent substitutions, and since they are more easily obtained (and require no assumptions), they might be used in place of the silent substitution values in some instances, e.g., rate calculations. On each branch of the tree that has at least eight nucleotide substitutions we have indicated the *ratio* of amino acid-*changing* to *silent* substitutions, which we designate R_{cs}. It will be noted that the R_{cs} values for pseudogenes range from 1.8 to 3.1, while, with two exceptions, the range for functional genes is 0.2 to 1.7. The two exceptions are links 34 (leading to mammalian α-globins) and 35 (α–β link) as defined in Table IV; link 34 has an R_{cs} value of 2.0 and link 35 a value of 4.8. Randomly mutating DNA has an R_{cs} value of 3.0; thus, the pseudogenes approach but do not reach this value, indicating that stabilizing selection has become almost inoperative on amino acid-changing substitutions. Li *et al.* (1981) have noted that some pseudogenes seem to have been expressed for a period following the gene duplication (or

Table IV
Distribution and Type of Nucleotide Substitutions for Different Branches and Links of the Globin Phylogeny

Species no.[a]	Globin	Category	Total substitutions	Third base	Amino acid-changing	Silent
1	Human α	α	24	13	10.5	13.5
2	Rabbit α	α	46	21	28	18
3	Human $\psi\alpha1$	$\psi\alpha$	77	27	56.2	20.8
4	Mouse α	α	22	11	10	12
5	Mouse $\psi\alpha3$ (CD1 Swiss)	$\psi\alpha$	1	0	1	0
6	Mouse $\psi\alpha3$ (BALB/C)	$\psi\alpha$	1	0	0	1
7	Chicken α	α	59	30	33.5	25.5
8	Human β	β	16	12	4	12
9	Human δ	β	18	6	9	9
10	Rabbit β (allele 1)	β	3	0	3	0
11	Rabbit β (allele 2)	β	1	0	1	0
12	Rabbit $\psi\beta$	$\psi\beta$	60	20	43.5	16.5
13	Goat β^A	Partial	8	3	6	2

Table IV (*continued*)

Species no.[a]	Globin	Category	Total substitutions	Third base	Amino acid-changing	Silent
14	Goat β^{C}	Partial	22	8	16	6
15	Goat "γ"	Partial	27	11	18	9
16	Goat $\psi\beta^{X}$	$\psi\beta$	71	30	47.2	23.8
17	Mouse β^{major}	β	10	4	6	4
18	Mouse β^{minor}	β	8	3	5	3
19	Mouse $\psi\beta$h3	Partial	78	25	59.2	18.8
20	Mouse (ψ)βh2	Partial	32	12	22.8	9.2
21	Mouse βh1	Partial	7	4	2	5
22	Mouse βh0	Partial	7	2	3.5	3.5
23	Mouse ϵ	Partial	21	15	7	14
24	Goat ϵ	Partial	18	9	8.5	9.5
25	Human ϵ	ϵ/γ	24	14	10	14
26	Human $^{G}\gamma$	ϵ/γ	1	0	1	0
27	Human $^{A}\gamma$	ϵ/γ	0	0	0	0
28	Chicken β	β	90	41	51	39

Link no.	Category	Leads to species	Total substitutions	Third base	Amino acid-changing	Silent
30	α	1,2	14	9	5	9
31	α	1–3	10	7	4	6
32	$\psi\alpha$	5,6	45	18	29.2	15.8
33	α	4–6	29	21	8.5	20.5
34	α	1–6	33	11	22.2	10.8
35	α–β	1–7 and 8–28	75	21	62	13
36	β	8,9	15	8	8	7
37	β	10,11	19	16	3	16
38	β	10–12	14	10	5	9
39	β	8–12	5	1	5	0
40	—	13,14	8	4	6	2
41	—	13–15	9	4	4	5
42	—	13–16	5	3	2	3
43	β	8–16	11	7	5	6
44	β	17,18	29	16	16	13
45	β	17–19	7	3	5	2
46	β	17–20	5	3	3	2
47	β	8–20	11	5	6	5
48	—	21,22	18	10	10.5	7.5
49	β	8–22	1	1	1	0
50	ϵ/γ	24,25	8	5	3	5
51	ϵ/γ	23–25	4	2	3	1
52	ϵ/γ	26,27	44	26	22.5	21.5
53	ϵ/γ	23–27	16	7	10	6
54	β	8–27	20	16	5	15

[a]See Table III for numbering of species.

when concerted evolution ceased). This correlates well with our pseudogene R_{cs} values; those with the higher R_{cs} values (human $\psi\alpha 1$, rabbit $\psi\beta 2$) are thought to have become pseudogenes immediately, while that with a lower R_{cs} value (mouse $\psi\alpha 3$) was probably an expressed gene for a longer period of time. Note that concerted evolution will not affect the R_{cs} value, as it will simply remove mutations of both types from the analysis (see also Section 4.3.3).

In Table V we summarize the composite R_{cs} values for various regions of the tree. Only links leading to branches representing complete sequences are included (see category listings in Table IV). Coding links and branches (α, β, ϵ, γ) have an average R_{cs} value of 1.0, whereas the pseudogenes have an average R_{cs} value of 2.3. How do we explain the R_{cs} value of 4.8 for the α–β link? At first we thought that perhaps the A solution (Fig. 7) in some way distorted this ratio, because (as mentioned in Section 4.2) the substitutions gravitate toward the terminal branch regions of the tree. Therefore, we looked at the B solution (Fig. 8), with its opposite bias. To our satisfaction, the R_{cs} value for the α–β link in this tree is 4.5. Our first inclination is to believe that Darwinian selection has resulted in an elevated rate of fixation of amino acid-changing substitutions, thus increasing the R_{cs} value for this branch approximately fivefold. Some support for this is provided by the fact that proportionately more of the amino acid-changing substitutions fixed were at strong functional sites (i.e., cooperative sites, heme contacts). However, alternative explanations are possible. Could the silent *rate* have been lowered? Kafatos *et al.* (1977) noted that at functional sites, silent substitutions occur less frequently. We notice a similar trend (see Section 4.3.3).

One feature of the R_{cs} values is that it provides a method for identifying possible pseudogenes. Although only 36% of the nucleotide sequence has been completed (Jahn *et al.*, 1980), mouse βh2 has accumu-

Table V
Ratio R_{cs} of Amino Acid-Changing to Silent Substitutions for Various Regions of the Gene Phylogeny

Branches and links[a]	β	α	γ/ϵ	$\psi\beta$	$\psi\alpha$	α–β link
R_{cs}	0.97	1.1	1.0	2.2	2.3	4.8
Number of substitutions	290	237	97	131	124	75
Number of codons	146	141	146	146	141	149[b]

[a]Links leading to both β and $\psi\beta$ are regarded as β, those leading to both α and $\psi\alpha$ as α, and those leading to β and γ/ϵ as β. Only complete sequences or interior links leading to complete sequences are included (see category listings in Table IV).
[b]In complete alignment there are 477 nucleotides (149 codons).

lated as many substitutions as the expressed adult β genes (β^{major} and β^{minor}). Its R_{cs} value, however, is 2.5. This, taken with an amino acid sequence substitution (93-Cys → His), which seems incompatible with a functional globin (Jahn *et al.*, 1980), suggested to us that βh2 was probably a pseudogene (Hewett-Emmett *et al.*, 1981b). This has been confirmed by the finding of a frameshift early in the coding region (M. H. Edgell and C. A. Hutchinson III, personal communication).

So far, we have included in the tree only pseudogenes for which the "parent" expressed gene (actually a sibling expressed gene representing the parent gene) is also sequenced. Recently, Proudfoot and co-workers (personal communication) have completed the nucleotide sequence of human $\psi\zeta 1$ (see Fig. 4), which is an embryonic α-like globin gene. Were we to include this in our tree, however, we would not find an R_{cs} value in the pseudogene range (1.8–3.1). This is because the expressed embryonic α-like gene ($\zeta 2$) has not been sequenced, and so our $\psi\zeta 1$ branch would contain a large number of the substitutions shared by both $\psi\zeta 1$ and $\zeta 2$ (which would be under selective constraint and have a low R_{cs} value) and a fewer number of substitutions which occurred subsequent to $\psi\zeta 1$ becoming a separate lineage and a pseudogene (high R_{cs} value). Therefore, the branch would have an R_{cs} value in the expressed gene range. Once again it is obvious that a knowledge of gene organization and expression is essential before making simplistic interpretations of the globin tree data.

4.3.3. Functional Regions of the Globins

Until now, we have discussed genealogic relationships and nucleotide substitution patterns. We have treated the globins as whole molecules without regard to subdividing them into their known functional regions. By the time of the α/β gene duplication, the ancestral globin presumably bound heme, and was able to form oligomers (Goodman *et al.*, 1975). Since cooperativity between subunits must have developed subsequent to this gene duplication, we might expect to see evidence for adaptive selection on the α–β link (link number 35 in Table IV). We subdivided sites in the α- and β-like globins into cooperative ($\alpha_1\beta_2$ contacts, Bohr effect sites, and, in the case of β-like globins, 2,3,-DPG-binding sites) and "other" sites according to the listing of Goodman (1981). The results of subdividing the globins into categories in this fashion are shown in Table VI.

These data can be analyzed more fully by further subdividing the "other sites" into heme contacts, weak functional (i.e., $\alpha_1\beta_1$ contacts and

Table VI
Analysis of the Number of Nucleotide Replacements in Codons Specifying Cooperative Functions in Various Regions of the Gene Phylogeny

	β	α	γ/ϵ	$\psi\beta$	$\psi\alpha$	α–β link
Cooperative sites[a]						
Amino acid-changing substitutions[b]	4	2	4	16	9.5	17.5
Total substitutions	20	16	9	22	12	22
Number of codons	19	16	19	19	16	25
Other Sites						
Amino acid-changing substitutions[b]	139	120	45.5	74.7	76.8	44.5
Total substitutions	270	222	88	109	112	53
Number of codons	127	125	127	127	125	124

[a]$\alpha_1\beta_2$ contacts, Bohr effect sites, and 2,3-diphosphoglycerate (DPG) binding sites [defined in Goodman (1981)].
[b]Values are not always integral, for reasons described in text (Section 4.3.2).

interior positions), and exterior sites of no apparent function, again using the listing in Goodman (1981). From this, R_{cs} values and relative rates of nucleotide substitution can be calculated and these are shown in Tables VII and VIII. This process of looking at regions of the globins, while illuminating certain features of globin evolution, results in R_{cs} values and relative rates based upon rather small numbers. A denser globin nucleotide

Table VII
Ratio R_{cs} of Amino Acid-Changing to Silent Substitutions for Different Links and Branches of the Tree with Nucleotide Codons Subdivided into Various Functional Categories

	Strong functional		Weak		
	Cooperative	Heme contact	functional[a]	Exterior[b]	Total
β	0.25	0.41	0.83	1.3	0.97
α	0.14	0.09	1.3	1.3	1.1
γ/ϵ	0.44	0.75	1.4	1.0	1.0
$\psi\beta$	2.7	0.88	4.3	2.2	2.2
$\psi\alpha$	3.8	1.9	2.9	2.0	2.3
α–β link	3.9	7.0	4.1	6.3	4.8

[a]These include $\alpha_1\beta_1$ contact sites and interior positions as defined in Goodman (1981).
[b]These are the remaining sites (neither the strong nor weak functional sites), but it should be noted that these exterior sites may well contain sites that are conserved because of at present undefined interactions with other molecules (e.g., haptoglobin).

tree (i.e., more sequences) will therefore be required before some of the general conclusions made here can be verified:

(a) Strong functional sites of adult expressed globin genes have R_{cs} values in the range 0.09–0.41, whereas weak functional sites and exterior sites are in the range 0.83–1.3.

(b) The high R_{cs} value for the α–β link is not based upon any one category of sites (Table VII); however, the highest relative rate of substitution is in the cooperative sites, although *both* the amino acid-changing substitution rate and the silent substitution rate are high (Table VIII).

(c) There is some support for the original observation of Kafatos *et al.* (1977) that the silent substitution rate tends to be lower at strongly functional (i.e., cooperative, heme contact) than other sites (see also Hewett-Emmett *et al.*, 1981b).

(d) The pseudogenes behave as if stabilizing (negative) selection is not operative (high R_{cs} values, relative rates of substitution similar for strong functional and other categories of sites), but if a trend is there, it is that the heme contact sites have a lower R_{cs} value. Further data will be necessary to explore this. An important point is that regions of DNA sequence may be conserved because of their importance in transcriptional or splicing events rather than the product for which they code. Thus, the coding sequence close to an intron splicing junction, for example, may be conserved relative to other regions. As Gō (1981) and others have pointed out, these regions are also the connecting segments between domains.

What possibility is there that the silent substitutions can be used as a molecular clock? Already certain investigators, by use of pairwise comparisons, have attempted to estimate when gene duplications occured (e.g., Efstratiadis *et al.*, 1980; Li *et al.*, 1981). From the data in Tables IV and V, we can chose examples to show that this may indeed be possible (see Fig. 9).

Clearly the pseudogene–parent gene pairs show that while the silent substitutions incorporated have been similar in number, the amino acid-changing substitutions are greatly elevated in the pseudogenes. Estimates of the proportion of the pseudogene lineage that was expressed versus the proportion that was unexpressed can be determined as Li *et al.* (1981) and Miyata and Yasunaga (1981) have done. However, we believe that the values for silent and amino acid-changing substitutions obtained from a large tree are preferable to those obtained by comparing two or three sequences.

Comparisons can be made of duplicated expressed genes. Thus, since the ϵ/γ duplication, both genes have evolved in a similar conservative fashion, although, if anything, the γ locus appears to have evolved, at least

Table VIII

Analysis of Relative Nucleotide Replacement Rates[a] for Different Globin Gene Categories Subdivided into Cooperative, Heme Contact, Weak Functional, and "Nonfunctional" Sites of Hemoglobin

	Amino acid-changing				Silent			
	Cooperative[b]	Heme contact	Weak functional[c]	Exterior, unknown function	Cooperative[b]	Heme contact	Weak functional[c]	Exterior, unknown function
β	0.15	0.22	0.71	1.0	0.80	0.70	1.1	1.0
α	0.11	0.05	0.95	1.0	1.0	0.66	0.96	1.0
γ/ϵ	0.52	0.32	1.0	1.0	0.65	0.43	0.72	1.0
$\psi\beta$	1.3	0.52	1.1	1.0	1.1	1.3	0.56	1.0
$\psi\alpha$	0.86	0.55	0.87	1.0	0.46	0.59	0.65	1.0
α–β link	2.2	0.95	1.4	1.0	3.5	0.86	2.1	1.0

[a]Normalized, so that substitution rate of exterior sites is 1.0.

[b]Cooperative sites are $\alpha_1\beta_2$ contact, DPG-binding, and Bohr effect sites.

[c]Weak functional sites are $\alpha_1\beta_1$ contacts and interior positions.

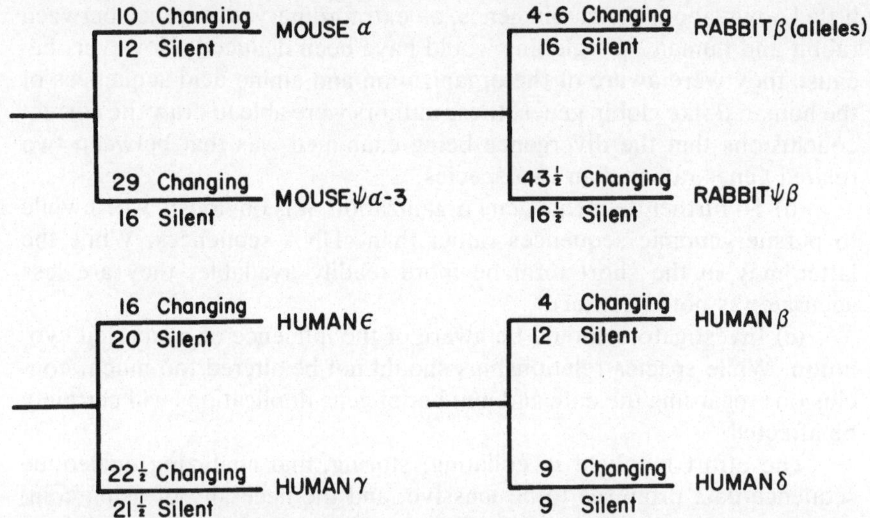

Figure 9. Examples of the constancy of the silent substitution rate. Each of the pairs is taken from the tree in Fig. 7. Note that the number of silent substitutions for each pair is similar; in the case of the pseudogenes, the amino acid-changing substitution rate ranges from three times (mouse $\psi\alpha-3$) to ten times (rabbit $\psi\beta$) the rate in the corresponding expressed gene.

for a period, more rapidly (less negative selection or fixing advantageous substitution?).

In the case of δ and β, there is no real indication that the δ gene is becoming a pseudogene in the human lineage, although we know that its fate might be precisely that in Old World monkeys (see Section 3.5 and Fig. 6).

5. Concluding Remarks

It is clear that we are in the middle of a revolution regarding our understanding of gene organization and expression. In the end it can only aid our understanding of systematic relationships, although at present we may be overwhelmed by this genetic complexity. Several important points should be reemphasized:

(a) To build evolutionary trees from nucleotide sequence data, it is important to have information on gene organization and temporal expression of the genes. Proudfoot and Baralle (1979), using a rabbit β-globin probe, were able to isolate the human embryonic ϵ-globin gene. Were little known about the globin genes, an extraordinary divergence between rabbit and human "β" globins would have been deduced. However, because they were aware of the organization and amino acid sequences of the human β-like globin genes, these authors were able to draw the correct conclusions that the divergence being examined was that between two related genes rather than two species.

(b) To further delineate gene organization, it is obviously worthwhile to pursue genomic sequences rather than cDNA sequences. While the latter may in the short term be more readily available, they are less accurate (as noted earlier).

(c) Investigators should be aware of the influence of concerted evolution. While species relationships should not be altered too much, conclusions regarding the date and number of gene duplications will certainly be affected.

The effort involved in collating, storing, and analyzing nucleotide sequence data promises to be massive, and the necessity of using computers even greater, but the likelihood is that a richer understanding of our evolutionary history and that of all living organisms will gradually emerge. That further surprises are in store is certain. As Leder *et al.* (1980) stated in an issue of *Science* devoted entirely to recombinant DNA, and as discussed in Sections 3.3 and 3.5, ". . . there is a special irony in the uninterrupted structure of the [mouseψ]α_3 gene.. . . Had it been discovered in 1977, we would have been left with the comfortable feeling that evolution created no surprises when dealing with fundamental structures like genes. Just 3 years later, such a finding creates exactly the opposite impression."

ACKNOWLEDGMENTS. We thank Drs. Jerry B. Lingrel, Marshall H. Edgell, Clyde Hutchinson III, and Nicholas J. Proudfoot for providing us with globin nucleotide sequence data prior to publication. We are particularly grateful to John Czelusniak and Dr. Morris Goodman for their collaborative efforts in building the globin nucleotide tree and encouragement to analyze it further. Without Lisa Campeau, who assembled the manuscript, it is safe to say it would never have been completed.

This work was supported in part by NIH grant GM24681 and NSF grant DEB 7810717 to M. Goodman.

References

Abelson, J., 1979, RNA processing and the intervening sequence problem. *Annu. Rev. Biochem.* **48:**1035–1069.

Air, G. M., 1979, Rapid DNA sequence analysis, *CRC Crit. Rev. Biochem.* **6**:1–33.

Argos, P., and Rossman, M. G., 1979, Structural comparisons of heme binding-proteins, *Biochemistry* **22**:4951–4960.

Arnheim, N., and Kuehn, M., 1979, The genetic behavior of a cloned mouse ribosomal segment mimics mouse ribosomal gene evolution, *J. Mol. Biol.* **134**:743–765.

Baralle, F. E., Shoulders, C. C., and Proudfoot, N. J., 1980, The primary structure of the human epsilon-globin gene, *Cell* **21**:621–626.

Barnicot, N. A., and Wade, P. T., 1970, Protein structure and the systematics of Old World monkeys, in: *Old World Monkeys* (J. R. Napier and P. H. Napier, eds.), Academic Press, New York, pp. 227–260.

Barrie, P. A., Jeffreys, A. J., and Scott, A. F., 1981, Evolution of the β-globin gene cluster in man and the primates, *J. Mol. Biol.* **149**:319–336.

Beard, J. M., Barnicot, N. A., and Hewett-Emmett, D., 1976, α and β chains of the major haemoglobin and a note on the minor component of *Tarsius*, *Nature* **259**:333–341.

Bell, G. I., Pictet, R. L., Rutter, W. J., Cordell, B., Tischer, E., and Goodman, H. M., 1980, Sequence of the human insulin gene, *Nature* **284**:26–32.

Benoist, C., and Chambon, P., 1981, *In vivo* sequence requirements of the SV40 early promoter region, *Nature* **290**:304–310.

Benyajati, C., Place, A. R., Powers, D. A., and Sofer, W., 1981, Alcohol dehydrogenase gene of *Drosophila melanogaster:* Relationship of intervening sequences to functional domains in the protein, *Proc. Natl. Acad. Sci. USA* **78**:2717–2721.

Blake, C. C. F., 1981, Exons and the structure, function and evolution of haemoglobin, *Nature* **291**:616.

Bolivar, F., Rodriguez, R. L., Green, P. J., Betlach, M. C., Heynecker, H. L., and Boyer, H. W., 1977, Construction and characterization of new cloning vehicles. II. A multipurpose cloning system, *Gene* **2**:95–113.

Boyer, S. H., Noyes, A. N., Timmons, C. F., and Young, R. A., 1972, Primate hemoglobins: Polymorphisms and evolutionary patterns, *J. Hum. Evol.* **1**:515–543.

Boyer, S. H., Noyes, A. N., Boyer, M. L., and Marr, K., 1973, Hemoglobin 3alpha chains in apes—Primary structures and presumptive nature of back mutation in a normally silent gene, *J. Biol. Chem.* **248**:992–1003.

Brack, C., and Tonegawa, S., 1977, Variable and constant parts of the immunoglobulin light chain gene of a mouse myeloma cell are 1250 nontranslated bases apart, *Proc. Natl. Acad. Sci. USA* **74**:5652–5656.

Bradshaw, R. A., 1980, Insulin-related growth factors, *Protides Biol. Fluids* **28**:165–168.

Breathnach, R., and Chambon, P., 1981, Organization and expression of eucaryotic split genes coding for proteins, *Annu. Rev. Biochem.* **50**:349–383.

Breathnach, R., Benoist, C., O'Hare, K., Gannon, F., and Chambon, P., 1978, Ovalbumin gene—Evidence for a leader sequence in messenger-RNA and DNA sequences at exon–intron boundaries, *Proc. Natl. Acad. Sci. USA* **75**:4853–4857.

Browne, J. K., Paddock, G. V., Liu, A., Clarke, P., Heindell, H. C., and Salser, W., 1977, Nucleotide sequences from the rabbit beta globin gene inserted into *Escherichia coli* plasmids, *Science* **195**:389–391.

Buell, G. N., Wickens, M. P., Payvar, F., and Shimke, R. T., 1978, Synthesis of full length cDNAs from four partially purified oviduct mRNAs, *J. Biol. Chem.* **253**:2471–2482.

Buettner-Janusch, J., Buettner-Janusch, V., and Coppenhaver, D., 1972, Properties of the hemoglobins of newborn and adult prosimians (Prosimii: lemuriformes and lorisiformes), *Folia Primatol.* **17**:177–192.

Busslinger, M., Moschonas, N., and Flavell, R. A., 1981, β^+ Thalassemia: aberrant splicing results from a single point mutation in an intron, *Cell* **27**:289–298.

Calame, K., Rogers, J., Early, P., Davis, M., Livant, D., Wall, R., and Hood, L., 1980, Mouse Cμ heavy chain immunoglobulin gene segment contains three intervening sequences separating domains, *Nature* **284**:452–455.

Chakrabarty, A. M. (ed.), 1978, *Genetic Engineering*, CRC Press, Boca Raton, Florida.

Chambon, P., 1981, Split genes, *Sci Am.* **244**:60–71.

Chapman, B. S., Tobin, A. J., and Hood, L. E., 1980, Complete amino acid sequences of the major early embryonic α-like globins of the chicken, *J. Biol. Chem.* **255**:9051–9059.

Cleary, M. L., Haynes, J. R., Schon, E. A., and Lingrel, J. B., 1980, Identification by nucleotide sequence analysis of a goat pseudoglobin gene, *Nucl. Acids Res.* **8**:4791–4802.

Cleary, M. L., Schon, E. A., and Lingrel, J. B., 1981, Two related pseudogenes are the result of a gene duplication in the goat β-globin locus, *Cell* **26**:181–190.

Corden, J., Wasylyk, B., Buchwalder, A., Sassone-Corsi, P., Kedinger, C., and Chambon, P., 1980, Promoter sequences of eukaryotic protein-coding genes, *Science* **209**:1406–1414.

Craik, C. S., Buchman, S. R., and Beychok, S., 1980, Characterization of globin domains: Heme binding to the central exon product, *Proc. Natl. Acad. Sci. USA* **77**:1384–1388.

Cunningham, B. A., Wang, J. L., Berggard, I., and Peterson, P. A., 1973, Complete amino-acid sequence of beta2-microglobulin, *Biochemistry* **12**:4811–4822.

Czelusniak, J., Goodman, M., Hewett-Emmett, D., Weiss, M. L., Venta, P. J., and Tashian, R. E., 1982, Phylogenetic origins and adaptive evolution of avian and mammalian haemoglobin genes which are expressed differentially during ontogeny, *Nature*, in press.

DeNoto, F. M., Moore, D. D., and Goodman, M., 1981, Human growth hormone DNA sequence and mRNA structure: possible alternative splicing, *Nucl. Acids Res.* **9**:3719–3730.

Dolan, M., Sugarman, B. J., Dodgson, J. B., and Engel, J. D., 1981, Chromosomal arrangement of the chicken β-type globin genes, *Cell* **24**:669–677.

Eaton, W. A., 1980, The relationship between coding sequences and function in haemoglobin, *Nature* **284**:183–185.

Efstratiadis, A., Posakony, J. W., Maniatis, T., Lawn, R., O'Connell, C. O., Spritz, R. A., DeRiel, J. K., Forget, B. G., Weissman, S. M., Slightom, J. L., Blechl, A. E., Smithies, O., Baralle, F. E., Shoulders, C. C., and Proudfoot, N. J., 1980, The structure and evolution of the human β-globin gene family, *Cell* **21**:653–668.

Fiddes, J. C., Seeburg, P. H., DeNoto, F. M., Hallewell, R. A., Baxter, J. D., and Goodman, H. M., 1979, Structure of genes for human growth hormone and chorionic somatomammotropin, *Proc. Natl. Acad. Sci. USA* **76**:4294–4298.

Fitzgerald, M., and Shenk, T., 1981, The sequence 5′-AAUAAA-3′ forms part of the recognition site for polyadenylation of late SV40 mRNAs, *Cell* **24**:251–260.

Fukumaki, Y., Ghosh, P. K., Benz, Jr., E. J., Reddy, V. B., Lebowitz, B., Forget, B. G., and Weissman, S. M., 1982, Abnormally spliced messenger RNA in erythroid cells from patients with β^+-thalassemia and monkey cells expressing a cloned β^+-thalassemic gene, *Cell* **28**:535–593.

Gilbert, W., 1978, Why genes in pieces? *Nature* **271**:501.

Gō, M., 1981, Correlation of DNA exonic regions with protein structural units in haemoglobin, *Nature* **291**:90–92.

Goodman, M., 1981, Decoding the pattern of protein evolution, *Prog. Biophys. Mol. Biol.* **37**:105–164.

Goodman, M., Moore, G. W., and Matsuda, G., 1975, Darwinian evolution in the genealogy of haemoglobin, *Nature* **253**:603–608.

Grossman, L. (ed.), 1980, *Methods in Enzymology*, Volume 65, *Nucleic Acids*, Part I, Academic Press, New York.

Grosveld, G. C., Shewmaker, C. K., Jat, P., and Flavell, R. A., 1981a, Localization of DNA sequences necessary for transcription of the rabbit β-globin gene *in vitro*, *Cell* **25**:215–226.

Grosveld, G. C., Koster, A., and Flavell, R. A., 1981b, A transcription map for the rabbit β-globin gene, *Cell* **23**:573–584.

Hampe, A., Therwath, A., Sonario, P., and Galibert, F., 1981, Nucleotide sequence of a cloned duck β-globin cDNA, *Gene* **14**:11–21.

Hansen, J. N., Konkel, D. A., and Leder, P., 1982, The sequence of a mouse embryonic β-globin gene, *J. Biol. Chem.* **257**:1048–1052.

Hardison, R. C., 1981, The nucleotide sequence of rabbit embryonic globin gene β3, *J. Biol. Chem.* **256**:11780–11786.

Hardison, R. C., Butler, E. T., III, Lacy, E., and Maniatis, T., 1979, The structure and transcription of four linked rabbit β-like globin genes, *Cell* **18**:1285–1297.

Haynes J. R., Rosteck, P., Jr., and Lingrel, J. B., 1980a, Unusual sequence homology at the 5′ ends of the developmentally regulated β^A-, β^C-, and γ-globin genes of the goat, *Proc. Natl. Acad. Sci. USA* **77**:7127–7131.

Haynes, J. R., Rosteck, P., Jr., Schon, E. A., Gallagher, P. M., Burke, D. J., Smith, K., and Lingrel, J. B., 1980b, The isolation of the β^A-, β^C-, and γ-globin genes and a presumptive embryonic globin gene from a goat DNA recombinant library, *J. Biol. Chem.* **255**:6355–6367.

Heindell, H. C., Liu, A., Paddock, G. V., Studnicka, G. M., and Salser, W. A., 1978, The primary sequence of rabbit α-globin mRNA, *Cell* **15**:43–54.

Hewett-Emmett, D., Czelusniak, J., and Goodman, M., 1981a, The evolutionary relationships of the enzymes involved in blood coagulation and hemostasis, *Ann. N.Y. Acad. Sci.* **370**:511–527.

Hewett-Emmett, D., Czelusniak, J., Goodman, M., Venta, P. J., and Tashian, R. E., 1981b, Evolution of nucleotide sequences coding for hemoglobin chains, *Fed. Proc.* **40**:1591.

Higgs, D. R., Old, J. M., Pressley, L., Clegg, J. B., and Weatherall, D. J., 1980, A novel α-globin gene arrangement in man, *Nature* **284**:632–635.

Hofer, E., and Darnell, J. E., 1981, The primary transcription unit of the mouse β-major globin gene, *Cell* **23**:585–593.

Hood, L., Campbell, J. H., and Elgin, S. C. R., 1975, The organization, expression, and evolution of antibody genes and other multigene families, *Annu. Rev. Genet.* **9**:305–353.

Jacq, C., Miller, J. R., and Brownlee, G. G., 1977, A pseudogene structure in 5S DNA of *Xenopus laevis*, *Cell* **12**:109–120.

Jahn, C. L., Hutchinson, C. A., Phillips, S. J., Weaver, S., Haigwood, N. L., Voliva, C. F., and Edgell, M. H., 1980, DNA sequence organization of the β-globin complex in the BALB/c mouse, *Cell* **21**:159–168.

Jeffreys, A. J., Wilson, V., Wood, D., and Simons, J. P., 1980, Linkage of adult α- and β-globin genes in *X. laevis* and gene duplication by tetraploidization, *Cell* **21**:555–564.

Jensen, E. Ø., Paludan, K., Hyldig-Nielsen, J. J., Jørgensen, P., and Marcker, K. A., 1981, The structure of a chromosomal leghaemoglobin gene from soybean, *Nature* **291**:677–679.

Jukes, T. H., 1980, Silent nucleotide substitutions and the molecular evolutionary clock, *Science* **210**:973–978.

Jung, A., Sippel, A. E., Grez, M., and Schutz, G., 1980, Exons encode functional and structural units of chicken lysozyme, *Proc. Natl. Acad. Sci. USA* **77**:5759–5763.

Kafatos, F. C., Efstratiadis, A., Forget, B. G., and Weissman, S. M., 1977, Molecular evolution of human and rabbit β-globin mRNAs, *Proc. Natl. Acad. Sci. USA* **74**:5618–5622.

Kedes, L. H., 1979, Histone genes and histone messengers, *Annu. Rev. Biochem.* **48:**837–870.

Kimura, M., 1981, Estimation of evolutionary distances between homologous nucleotide sequences, *Proc. Natl. Acad. Sci. USA* **78:**454–458.

Knöchel, W., Wittig, B., Wittig, S., John, M. E., Grundmann, U., Oberthür, W. Godovac, J., and Braunitzer, G., 1982, No evidence for "stress" α-globin genes in chicken, *Nature* **259:**710–712.

Konkel, D. A., Maizel, J. V., Jr., and Leder, P., 1979, The evolution and sequence comparison of two recently diverged mouse chromosomal β-globin genes, *Cell* **18:**865–873.

Kretschmer, P. J., Coon, H. C., Davis, A., Harrison, M., and Nienhuis, A. W., 1981, Hemoglobin switching in sheep, *J. Biol. Chem.* **256:**1975–1982.

Küpper, H., Keller, Kurz, C., Forss, S., Schaller, H., Franze, R., Strohmaier, K., Marquardt, O., Zaslavsky, V. G., and Hofschneider, P. H., 1981, Cloning of cDNA of major antigen of foot and mouth disease virus and expression in *E. coli*, *Nature* **289:**555–559.

Kurosky, A., Barnett, D. R., Lee, T.-H., Touchstone, B., Hay, R. E., Arnott, M. S., Bowman, B. H., and Fitch, W. M., 1980, Covalent structure of human haptoglobin: A serine protease homolog, *Proc. Natl. Acad. Sci. USA* **77:**3388–3392.

Lacy, E., and Maniatis, T., 1980, The nucleotide-sequence of a rabbit beta-globin pseudogene, *Cell* **21:**545–553.

Lawn, R. M., Efstratiadis, A., O'Connell, C., and Maniatis, T., 1980, The nucleotide-sequence of the human beta-globin gene, *Cell* **21:**647–651.

Lawn, R. M., Adelman, J., Dull, T. J., Gross, M., Goeddel, D., and Ullrich, A., 1981, DNA sequence of two closely linked human leukocyte interferon genes, *Science* **212:**1159–1162.

Lazawska, J., Jacq, C., and Slonimski, P. P., 1980, Sequence of introns and flanking exons in wild-type and box 3 mutations of cytochrome b reveals an interlaced splicing protein coded by an intron, *Cell* **22:**333–348.

Leder, P., Hansen, J. N., Konkel, D., Leder, A., Nishioka, Y., and Talkington, C., 1980, Mouse globin system: A functional and evolutionary analysis, *Science* **209:**1336–1342.

Leder, A., Swan, D., Ruddle, F., D'Eustachio, P., and Leder, P., 1981, Dispersion of α-like globin genes of the mouse to three different chromosomes, *Nature* **293:**196–200.

Lerner, M. R., Boyle, J. A., Mount, S. M., Wolin, S. L., and Steitz, J. A., 1980, Are snRNPs involved in splicing? *Nature* **283:**220–224.

Lewin, B., 1980a, Alternatives for splicing: Recognizing the ends of introns, *Cell* **22:**324–326.

Lewin, B., 1980b, Alternatives for splicing: An intron-coded protein, *Cell* **22:**645–646.

Li, W.-H., Gojobori, T., Nei, M., 1981, Pseudogenes as a paradigm of neutral evolution, *Nature* **292:**237–239.

Liebhaber, S. A., Goosens, M., Poon, R., and Kan, Y. W., 1980, Cloning and complete nucleotide sequence of the human 5′ α globin gene, *Proc. Natl. Acad. Sci. USA* **77:**7054–7058.

Loeb, L. A., Weymouth, L. A., Kunkel, T. A., Gopinathan, K. P., Beckman, R. A., and Duhe, D. K., 1979, On the fidelity of DNA replication, *Cold Spring Harbor Symp. Quant. Biol.* **43:**921–927.

Lomedico, P., Rosenthal, N., Efstratiadis, A., Gilbert, W., Kolodner, R., and Tizard, R., 1979, The structure and evolution of the two nonallelic rat preproinsulin genes, *Cell* **18:**545–558.

Marie, J., Simon, M.-P., Dreyfus, J.-C., and Kahn, A., 1981, One gene, but two messenger RNAs encode liver L and red cell L′ pyruvate kinase subunits, *Nature* **292:**70–72.

Marquardt, H., Todaro, G. J., Henderson, L. E., and Oroszlan, S., 1981, Purification and primary structure of a polypeptide with multiplication-stimulating activity from rat liver cell cultures. Homology with human insulin-like growth factor II, *J. Biol. Chem.* **256**:6859–6865.

Martin, S. L., Zimmer, E. A., Kan, Y. W., and Wilson, A. C., 1980, Silent δ-globin gene in Old World monkeys, *Proc. Natl. Acad. Sci. USA* **77**:3563–3566.

Maxam, A. M., and Gilbert, W., 1980, Sequencing end-labelled DNA with base-specific chemical cleavages, in: *Methods in Enzymology*, Volume 65, Academic Press, New York, pp. 499–560.

Michelson, A. M., and Orkin, S. H., 1980, The 3′ untranslated regions of the duplicated human α-globin genes are unexpectedly divergent, *Cell* **22**:371–377.

Miyata, T., and Yasunaga, T., 1981, Rapidly evolving mouse α-globin-related pseudo gene and its evolutionary history, *Proc. Natl. Acad. Sci. USA* **78**:450–453.

Murray, V., and Holliday, R., 1979, Mechanism for RNA splicing of gene transcripts, *FEBS Lett.* **106**:5–7.

Nishioka, Y., and Leder, P., 1979, The complete sequence of a chromosomal mouse α-globin gene reveals elements conserved throughout vertebrate evolution, *Cell* **18**:875–882.

Nishioka, Y., Leder, A., and Leder, P., 1980, Unusual α-globin-like gene that has cleanly lost both globin intervening sequences, *Proc. Natl. Acad. Sci. USA* **77**:2806–2809.

Ohno, S., 1970, *Evolution by Gene Duplication*, Springer-Verlag, New York.

Okayama, H., and Berg, P., 1982, High efficiency cloning of full-length cDNA, *Mol. Cell. Biol.* **2**:161–170.

Orkin, S. H., Goff, S. C., and Hechtman, R. L., 1981, Mutation in an intervening splice junction in man, *Proc. Natl. Acad. Sci. USA* **78**:5041–5045.

Partington, G. A., and Baralle, F. E., 1981, Isolation of a *Xenopus laevis* α-globin gene, *J. Mol. Biol.* **145**:463–469.

Patient, R. K., Elkington, J. A., Kay, R. M., and Williams, J. G., 1980, Internal organization of the major adult α- and β-globin genes of *X. laevis*, *Cell* **21**:565–573.

Perler, F., Efstratiadis, A., Lomedico, P., Gilbert, W., Kolodner, R., and Dodgson, J., 1980, The evolution of genes: The chicken preproinsulin gene, *Cell* **20**:555–566.

Popp, R. A., Lalley, P. A., and Whitney, J. B., 1981, Mouse α-globin genes and α-like pseudogenes are not syntenic, *Fed. Proc.* **40**:763.

Proudfoot, N., 1980, Pseudogenes, *Nature* **286**:840–841.

Proudfoot, N. J., and Baralle, F. E., 1979, Molecular cloning of human ϵ-globin gene, *Proc. Natl. Acad. Sci. USA* **76**:5435–5439.

Proudfoot, N. J., and Brownlee, G. G., 1974, Sequence at the 3′ end of globin mRNA shows homology with immunoglobin light chain mRNA, *Nature* **252**:359–362.

Proudfoot, N., and Brownlee, G. G., 1976, 3′ non-coding region sequences in eukaryotic messenger-RNA, *Nature* **263**:211–214.

Proudfoot, N. J., and Maniatis, T., 1980, The structure of a human alpha-globin pseudogene and its relationship to alpha-globin gene duplication, *Cell* **21**:537–544.

Proudfoot, N. J., Shander, M. H. M., Manley, J. L., Gefter, M. L., and Maniatis, T., 1980, Structure and *in vitro* transcription of human globin genes, *Science* **209**:1329–1336.

Richards R. I., and Wells, J. R. E., 1980, Chicken globin genes. Nucleotide sequence of cDNA clones coding for the α-globin expressed during hemolytic anemia, *J. Biol. Chem.* **255**:9306–9311.

Richards, R. I., Shine, J., Ullrich, A., Wells, J. R. E., and Goodman, H. M., 1979, Molecular cloning and sequence analysis of adult chicken β-globin cDNA, *Nucl. Acids Res.* **7**:1137–1146.

Rogers, J., and Wall, R., 1980, A mechanism for RNA splicing, *Proc. Natl. Acad. Sci. USA* **77**:1877–1879.

Sakano, H., Rogers, J. H., Hveppi, K., Brack, C., Traunecker, A., Maki, R., Wall, R., and Tonegawa, S., 1979, Domains and the hinge region of an immunoglobulin heavy-chain are encoded in separate DNA segments, *Nature* **277**:627–633.

Salser, W. A., Cummings, I., Liu, A., Strommer, J., Padayatty, J., and Clarke, P., 1979, Analysis of chicken globin cDNA clones: Discovery of a novel chicken α globin gene induced by stress in young chickens, in: *Cellular and Molecular Regulation of Hemoglobin Switching* (G. Stamatoyannopoulos and A. Nienhuis, eds.), Grune and Stratton, New York, pp. 621–643.

Sanger, F., 1981, Determination of nucleotide sequences in DNA, *Bioscience Reports* **1**:3–18.

Sanger, F., and Coulson, A. R., 1975, A rapid method for determining sequences in DNA by primed synthesis with DNA polymerase, *J. Mol. Biol.* **94**:441–448.

Sanger, F., and Coulson, A. R., 1978, The use of thin acrylamide gels for DNA sequencing, *FEBS Lett.* **87**:107–110.

Sanger, F., Nicklen, S., and Coulson, A. R., 1977, DNA sequencing with chain-terminating inhibitors, *Proc. Natl. Acad. Sci. USA* **74**:5463–5467.

Scarpulla, R. C., Agne, K. M., and Wu, R., 1981, Isolation and structure of a rat cytochrome c gene, *J. Biol. Chem.* **256**:6480–6486.

Shafritz, D. A., 1977, Messenger RNA and its translation, in: *Molecular Mechanisms of Protein Biosynthesis* (H. Weissbach and S. Pestka, eds.), Academic Press, New York, pp. 555–601.

Sharp, P. A., 1981, Speculations on RNA splicing, *Cell* **23**:643–646.

Shatkin, A. J., 1976, Capping of eukaryotic mRNAs (review), *Cell* **9**:645–653.

Slightom, J. L., Blechl, A. E., and Smithies, O., 1980, Human fetal $^{G}\gamma$- and $^{A}\gamma$-globin genes: Complete nucleotide sequences suggest that DNA can be exchanged between these duplicated genes, *Cell* **21**:627–638.

Smithies, O., and Poulik, M. D., 1972, Initiation of protein-synthesis at an unusual position in an immunoglobulin gene, *Science* **175**:187–189.

Solnick, D., 1981, An adenovirus mutant defective in splicing RNA from early region 1A, *Nature* **291**:508–510.

Springgate, C. F., Battilla, N., and Loeb, L. A., 1973, Infidelity of DNA synthesis by reverse transcriptase, *Biochem. Biophys. Res. Commun.* **52**:401–407.

Spritz, R. A., DeRiel, J. K., Forget, B., and Weissman, S., 1980, Complete nucleotide-sequence of the human delta-globin gene, *Cell* **21**:639–646.

Stein, J. P., Catterall, J. F., Kristo, P., Means, A. R., and O'Malley, B. W., 1980, Ovomucoid intervening sequences specify functional domains and generate protein polymorphism, *Cell* **21**:681–687.

Stewart, P. R., and Letham, D. S. (eds.), 1973, *The Ribonucleic Acids*, Springer-Verlag, New York.

Tonegawa, S., Maxam, A. M., Tizard, R., Bernard, O., and Gilbert, W., 1978, Sequence of a mouse germ-line gene for a variable region of an immunoglobulin light chain, *Proc. Natl. Acad. Sci. USA* **75**:1485–1489.

Ullrich, A., Dull, T. J., Gray, A., Brosius, J., and Sures, I., 1980, Genetic variation in human insulin genes, *Science* **209**:612–615.

van Ooyen, A., van den Berg, J., Mantei, N., and Weissmann, C., 1979, Comparison of total sequence of a cloned rabbit β-globin gene and its flanking regions with a homologous mouse sequence, *Science* **206**:337–344.

Vanin, E. F., Goldberg, G. I., Tucker, P. W., and Smithies, O., 1980, A mouse α-globin-related pseudogene lacking intervening sequences, *Nature* **286:**222–226.

Vogeli, G., Anvedimento, E. V., Sullivan, M., Maizel, J. V., Lozano, G., Adams, S. L., Pastan, I., and DeCrombruggle, B., 1980, Isolation and characterization of genomic DNA coding for $\alpha 2$ type I collagen, *Nucl. Acids Res.* **8:**1823–1837.

Wallis, M., 1980, Growth hormone: Deletions in the protein and introns in the gene, *Nature* **284:**512.

Wasylyk, B., Kédinger, C., Corden, J., Brison, O., and Chambon, P., 1980, Specific *in vitro* initiation of transcription on conalbumin and ovalbumin genes and comparison with adenovirus-2 early and late genes, *Nature* **285:**367–372.

Weatherall, D. J., and Clegg, J. B., 1979, Recent developments in the molecular genetics of human hemoglobin, *Cell* **16:**467–479.

Weissman, S. M., 1979, Current approaches to analysis of the nucleotide sequence of DNA, *Anal. Biochem.* **98:**243–253.

Wickens, M. P., Buell, G. N., and Schimke, R. T., 1978, Synthesis of double-stranded DNA complementary to lysozyme, ovomucoid, and ovalbumin mRNAs, *J. Biol. Chem.* **253:**2483–2495.

Williams, J. G., Kay, R. M., and Patient, R. K., 1980, The nucleotide sequence of the major β-globin mRNA from *Xenopus laevis*, *Nucl. Acids. Res.* **8:**4247–4258.

Wilson, J. T., Forget, B. G., Wilson, L. B., and Weissman, S. M., 1977, Human globin messenger RNA: Importance of cloning for structural analysis, *Science* **196:**200–202.

Wilson, J. T., Wilson, L. B., Reddy, V. B., Cavallesco, C., Ghosh, P. K., deRiel, J. K., Forget, B. G., and Weissman, S. M., 1980, Nucleotide sequence of the coding portion of human α globin messenger RNA, *J. Biol. Chem.* **255:**2807–2815.

Wu, R. (ed.), 1979, Recombinant DNA, in: *Methods in Enzymology*, Volume 68, Academic Press, New York.

Yang, V. W., Lerner, M. R., Steitz, J. A., and Flint, S. J., 1981, A small nuclear ribonucleoprotein is required for splicing of adenoviral early RNA sequences, *Proc. Natl. Acad. Sci. USA* **78:**1371–1375.

Young, R. A., Hagenbüchle, O., and Schibler, U., 1981, A single mouse α-amylase gene specifies two different tissue-specific mRNAs, *Cell* **23:**451–458.

Zimmer, E. A., Martin, S. L., Beverley, S. M., Kan, Y. W., and Wilson, A. C., 1980, Rapid duplication and loss of genes coding for the α chains of hemoglobin, *Proc. Natl. Acad. Sci. USA* **77:**2158–2162.

Zuckerkandl, E., and Pauling, L., 1965, Evolutionary divergence and convergence in proteins, in: *Evolving Genes and Proteins* (V. Bryson and H. J. Vogel, eds.), Academic Press, New York, p. 77.

Index

Aardvark-paenungulate relationship, α-crystalline A sequences in, 97–100
Active-site histidines, properties of, 59
Adenovirus, DNA-binding protein of, 205
ADP-ribosylation, 197
ALIGN program, 201–202
Allowed amino acids (*see also* Amino Acid sequences)
  mean numbers of per site, 250
  in RFAC model, 248–251
  maximum parsimony in, 139
  most parsimonious tree for, 144
"Alu" family DNAs, 348–349 (*see also* DNA)
Amino acid difference, vs. evolutionary time for cytochrome *c*, 262–263
Amino acid replacements, low-molecular-weight nucleotide substrates in, 59–61
Amino acid sequences (*see also* α-Crystallin A sequences)
  analysis of, 122–129
  in biologic tree of nucleases, 50–52
  of carbohydrate attachment sites in pancreatic ribonucleases, 65–68
  evolutionary trees from, 44–48
  genealogic evidence from, 116–148
  maximum parsimony method in, 123–129
  in most parsimonious tree of ribonucleases, 48–50
  polypeptide chains in, 122
Amino acid sequence analysis
  calmodulin and, 180
  species list for, 123–129
  species phylogeny and, 145–148
Amino-terminal halves, of bovine histones H2A and H4, 228
Amount of allowed variation
  cytochrome *c* and, 266–268
  and "efficiency" of evaluation, 269–273
  in macromolecular sequence divergence, 257–258
  in simulated macromolecular evolution, 259–266
Amphibia, amino acid sequence analysis in, 127
Anagenesis, in molecular evolution, 184
Ancestral nodes, in clock model of protein evolution, 152–158
Annelida, amino acid sequence analysis in, 128
Anthropoidea, 135
  ancestor nodes for, 161
  clock model and, 155
  monophyletic origin of, 143
Antilocapridae, ribonuclease sequences for, 46–47
Antilopinae, ribonuclease sequences for, 46–47
Archonta, 6, 8, 15
  tarsal features of, 23
  validity of, 16
"Archontans," Insecivora and, 21–25
Arthropoda, amino acid sequence analysis in, 128
Artiodactyla, 25–28, 46, 100, 131, 135, 138, 145, 186
  amino acid sequence analysis for, 124–125
  evolutionary rates of pancreatic ribonuclease for, 55
Augmented parsimony distance, 255
Australopithecus, Homo and, 163

Aves
amino acid sequence analysis in, 126
Mammalia and, 131, 158
Squamata and, 136
Aves-Mammalia ancestral node, clock date for, 158

Bats, as only true flying mammals, 22
Biologic tree, of ribonucleases, 50–52 (*see also* Most parsimonious tree)
Bohr effect, 175
Bovidae
evolutionary rates of pancreatic ribonuclease for, 54–55
ribonuclease sequence for, 46–47
Bovinae, ribonuclease sequence for, 46–47
Bovine nonhistone chromosomal proteins, 210, 229
Bovine nucleosome core histones
alignment of, 226–228
evolutionary tree of, 229–230
H2A, 228
H3, 219–220
H4, 228
Bovine pancreatic ribonuclease
main-chain conformation of, 56, 58
three-dimensional structure of, 56–58
Bovine seminal ribonuclease
amino acid sequence of, 44
double-stranded RNA and, 61
Bullae, auditory, 11

Calmodulin
gene phylogeny of, 181
natural selection and, 180
Calmodulin evolution, 179–183
Calmodulin family, nucleotide replacements in different members of, 182
Caprinae, ribonuclease sequence for, 46–47
Capybara, ribonuclease sequence for, 48
Carbohydrate attachment sites in pancreatic ribonuclease, amino acid sequence and, 65–68
Carbonic anhydrase sequences
evolution rates for, 166
most parsimonious tree and, 140, 142
Carnivora, 138, 145
amino acid sequence analysis for, 124
clock dates within, 157
Ferungulata and, 19
phylogenetic position of, 16

Catarrhini-Platyrrhini split, 155
Caviomorpha, pancreatic ribonuclease evolutionary rates for, 55
cDNA clones, full-length, 360
Cecal digestion, carbohydrate/ribonuclease in, 67
Ceratomorpha, 144
Cercopithecoidea, 145
Cervidae
evolutionary rates of pancreatic ribonuclease in, 54–55
ribonuclease sequence for, 46–47
Cetacea, 25–28, 135, 138, 145
amino acid sequence analysis for, 125
ribonuclease sequence for, 45
Charged residues per molecule, for various mammals, 61–63
Chiroptera, 6, 28, 138, 145
amino acid sequence analysis for, 126
Dermoptera and, 22
phylogenetic position of, 103
Choriovitelline placenta, in marsupials, 11
Chromatin, structures of, 193–194
Chromatin fiber, 194–195
Chromosomal change, satellite DNAs and, 347–348
Chromosomal proteins, 203–230 (*see also* Proteins)
computer methods for, 194–203
conservative, 183
defined, 193
evolution in, 193–230
mutation acceptance rates for, 224–225
nonhistone, 207–211
protein relationships in, 199–200
Chromosome II, human non-α globin gene cluster form of, 337
Chromosomes, eukaryote, *see* Eukaryote chromosomes
Clock model of protein evolution, 148–161
ancestral nodes in, 152–158
clock dates vs. paleontological dates in, 152–158
evolution of, 150–152
nonuniform rates in, 159–161
results obtained by, 152–161
Cloning, 358–362
Codons
base replacement frequencies at first position within, 286
fixed mutation distribution among and within, 286–291

Codons (*cont'd*)
random fixation of, 243–276 (*see also* Random fixation of allowed codons)
Coelenterata, amino acid sequence analysis in, 128
Computer simulation model, of macromolecular sequence evolution, 243–276
Conservative chromosomal proteins, 183
Constrained stochastic theory, 282–284
Core histone alignment, evolutionary tree from, 227–230
Core histone superfamily, 225–230
Cosmids, 358
Creodonta, Carnivora and, 19
Crocodilia, Aves and, 131
Crossopterygii, amino acid sequence analysis in, 127
Crystallin(s)
αA chain sequence analysis in, 88–91
amino acid sequence analysis in, 83–84
defined, 78
properties of, 79
sequence homology of, 80–81
variation between species, 82–84
of vertebrate eye lens, 78–82
α-Crystallin
isolation of, 8–87
phylogenetic reconstruction in, 91–93
structural analysis of, 84–93
α-Crystallin A chains
amino acid sequence reconstruction for aardvark, 92
electrophoresis of, 87–88
most parsimonious trees in, 98–99, 105
variable positions in, 94–95
α-Crystallin A sequences (*see also* α-Crystallin A chains)
aardvark-paenungulate relationship in, 97–100
in mammalian phylogeny, 93–97, 103–104
maximum parsimony in, 139
metatherian-eutherian divergence in, 97
phylogenetic inferences from, 93–106
phylogenetic trees of, 93–100
in vertebrate classes or subclasses, 104–106
α-Crystallin B chains, evolutionary change rates and, 106–108
α-Crystallin evolution
changes in charge avoidance in, 108–109
α-Crystalline evolution (*cont'd*)
clock dates and, 158–159
"covarions" in, 109
directional trend among substitutions in, 109–111
molecular aspects of, 106–111
parallel and back substitutions in, 108
unequal distribution of substitutions in, 110–111
and variable rates of change, 109–111
Cuis, ribonuclease sequence for, 48
Cyclic AMP, ubiquitin and, 217
Cytochrome *c*
amino acid difference vs. evolutionary time for, 262–263
amount of allowed variation and, 266–268
clock dates and, 159
evolution of, 178–179, 263
evolution rates for, 166
gene phylogenies and, 136
Homo phyletic line and, 162–163
most parsimonious tree, sequences and, 137
percent amino acid difference and, 263
in phylogenetic tree of vertebrates, insects, and plants, 252
selectional constraints on, 268
in simulated macromolecular evolution model, 261–264
starting DNA sequence in, 247

Darwinian theory, synchronized evolutionary tempos and, 167
Deinotheroidea, 26
Dermoptera, 6
Chiroptera and, 22
Dinocerata, 26
DISP program, 200
Divergence, nonuniform molecular, *see* Nonuniform molecular divergence
DNA
"Alu" family of, 348–349
digested and hybridized, 333–334
double-stranded complementary, 359
ethidium bromide-stained agarose gel of, 322
genomic, *see* Genomic DNA
^{125}I hybridization with, 323
hydroxylapatite fractionation and, 321
middle-order reiterated, 329–330
mRNA copying into, 343
primate, 334–335

DNA (*cont'd*)
protamines and, 206–207
in nucleosome core, 195–196
radiolabeled human male-specific, 331
rare sequences of, 321
ribosomal genes and, 331–336
satellite, 347–348
single-copy, 336–344
Y-chromosome, 327–329
DNA-binding proteins, 205–206
DNA mutation, in computer macromolecular sequence evolution, 245
DNA polymorphism, genomic DNA and, 344–346
DNA reassociation or melting characteristics, 320
DNA replication error/unit time, 245
DNA restriction and modification enzymes, 321–322
DNA sequencing, 362–364
in computer simulation of evolution, 244
DNA sequencing gel, schematic representation of, 363
Double-stranded RNA, pancreatic nuclease activity on, 61–63
DPG (diphosphoglycerate)-binding sites, 175–177
Drosophila melanogaster, 218–220, 222, 346
"libraries" of, 349

Early eutherians (*see also* Eutheria; Eutherians)
edentates as offshoots of, 101
phylogenies of, 7–8
Early Tertiary "condylarths," 16
Early Tertiary mammals, discoveries of, 7–8
Echinodermata, amino acid sequence analysis in, 128
Edentata, 6–7, 97 (*see also* Edentates)
amino acid sequence analysis for, 126
origin times for, 28
relationships in, 16
ribonuclease sequence for, 48
specializations in, 16–19
stapes in, 17
Edentates
as early eutherian offshoots, 101
Insectivora and, 16
Edentates (*cont'd*)
isolation in South America, 16–17, 35
pholidotans and, 9
as primitive eutherian group, 17
Recent orders and, 18
Elasmobranchii, amino acid sequence analysis in, 127
Elephant shrews (*see also* Macroscelideans)
lagomorphs and, 9
Rodentia and, 21
Embrithopoda, 26
Erinaceomorphs, 21, 24
Ernotheria, 8
Eukaryote chromatin, structure of, 194–195
Eukaryote chromosomes, structure of, 193–194
Eukaryotic gene coding, processing and expression of, 365
Eukaryotic genes
expression and organization of, 364–382
gene duplication and, 373–377
intervening gene sequences and, 367–373
introns and, 370–372
pseudogenes and, 377–382
Eurotamandua, Edentata and, 35
Eutheria (*see also* Eutherians)
amino acid sequence analysis for, 122–129
ancestor nodes for, 161
distinctive features of, 11–16
grandorders in, 15
monophyletic, 135
pancreatic ribonuclease evolutionary rates for, 55
putative ancestral stock of, 8
subclass and ordinal mammalian relationships in, 138
validity of, 10
Xenarthra and, 30
Eutherian history, 7–8
Eutherian monophyly, 10–12, 135
Eutherian orders
divergence times for, 27–30
operational taxonomic units in, 145
Eutherian phylogeny, 30–33
amino acid sequence evidence on, 115–187
anatomical and fossil evidence on, 3–35
comparative anatomy and paleontology factors in, 15–16

Eutherian phylogeny (*cont'd*)
higher level, 12–27
polytomy in, 30
Eutherian relationships, subordinal, 141–144
Eutherians (*see also* Eutheria)
auditory bullae in, 11
divergence times for, 32
epipubic elements in, 12
of Late Cretaceous, 31
operational taxonomic units in, 145, 165
primitive carotid arterial system of, 9
radition time frame for, 31
subordinal relationships in, 141–144
trophoblast in, 10
Eutherian tarsus, studies of, 9
Eutherian taxa, origin times for, 28
Evolution
acceleration-deceleration patterns in, 149, 162
macromolecular, *see* Macromolecular evolution
simulated vs. real results in, 258–269
Evolutionary "efficiency," amount of allowed variation and, 269–273
Evolutionary estimates, from protein and nucleic acid sequence data vs. nonrandom REH theory, 296–299
Evolutionary interest, 291–296
Evolutionary problems, new approaches to, 319–350
Evolutionary tree (*see also* Most parsimonious tree; Ribonuclease tree; Tree)
for amino acid sequences of mammalian pancreatic ribonuclease, 44–48
building strategy for, 384
most parsimonious, *see* Most parsimonious tree
nucleotide sequences in building of, 382–397
protein alignments and, 200–202
Exons, nucleotide sequences of, 371–372, 376
Eye lens (*see also* α-Crystallin)
accommodation in, 77
annular pad or *ringwulst* of, 75
characteristic proteins of, 78–80
comparative anatomy and evolution of, 75–78
crystallins of, 78–82
lens capsule in, 76–71
Eye lens (*cont'd*)
shape of, 76
yellow pigment in, 77
Eye lens proteins [*see also* Crystallin; Protein(s); Vertebrate lens proteins]
intraspecies variation in, 82
in molecular biologic studies, 75–111
variation in isoelectric points among, 85–86

Ferae, 6, 15
Ferungulata, 8
Carnivora and, 19, 25
Fibrinopeptides A and B
evolution rates for, 166
lowest NR length tree for, 144
phylogenies of, 253, 256
in simulated macromolecular evolution, 259–261
Fish protamines
alignment of, 208
evolutionary tree of, 209
Fixation constant, in simulated macromolecular evolution, 259–266
Fixation rate, macromolecular sequence divergence and, 257–258
Fixation rate alteration, in RFAC simulation, 266–268
Fixations
nonrandomicities in, 254
role of in macromolecular evolution, 273–274
Fixed mutations, apparent and true distribution with codons, 287–291
"Fossil tooth taxa," 10
Fungi, amino acid sequence analysis for, 129

Gamma locus, in globin phylogeny, 185–186
Genealogic evidence
from amino acid sequences, 116–148
interclass relationships and, 136–138
from other gene phylogenies, 136–145
Gene duplication, 118, 134, 164–165
Gene evolution theories, two classes of, 28
Gene expression, 118
lowest NR and, 134, 164–165
molecular evolution and, 357
Genomic DNA, 319–350
comparative studies of, 327–349
complexity of, 323

Genomic DNA (*cont'd*)
DNA polymorphisms and, 344–346
human vs. gorilla, 340
"libraries" of, 325–326, 350
molecular systematics and, 343
restriction mapping studies of, 323–325
ribosomal genes and, 331–336
study techniques in, 320–326
transportable elements in, 346–347
Genomic DNA cloning, cDNA cloning and, 361–362
Gene organization, molecular evolution and, 357–398
Gene phylogeny
amino acid changing/silent substitution ratio in, 392
fitting of into species phylogeny, 117–120
genealogic evidence for, 136–145
orthologous and paralogous sequences in, 118
Genes, eukaryotic, 364–382
Gene sequences
and genetic mechanisms in evolution, 341
most parsimonious solution for set of, 311
single vs. multiple pairs of, 310
Gene structure
constraints in, 283–284
molecular evolution and, 357–398
Genetic evolution mechanisms, gene sequences and, 341
Gene VP1, in REH theory, 292–293 (*see also* VP1 genes)
Giraffidae, ribonuclease sequence for, 46–47
Glires concept, 20–21
Globin evolution rate, 166–178
Globin functional regions, 393–397
δ-Globin gene, origin and distribution of, 380
Globin gene family, pseudogenes and, 379
Globin gene phylogeny
γ-loci and, 185–186
genealogic history and, 168
new cladistic findings in, 185–187
nucleotide substitutions and, 390–391
parsimonious, 169
Globin gene relationships, 386–389
Globin nucleotide sequence data, in phylogenetic tree building, 385
Globin sequences, most parsimonious tree and, 169–171
Glycosylation, of pancreatic ribonucleases, 64–71
Gnathostomata, 167–169
Gorilla genomic DNA, homologous cloned human DNA and, 340
Gorilla "library," 338
Gorilla Y chromosome, vs. human, 329
Grandorders, in Eutheria, 15

Haploid cell, genome of, 319
Haptoglobins, as duplicated genes, 378
α-Hemoglobin, of Papionini monkey tribe, 160
α-Hemoglobin chains
insertions and deletions in descent of, 174
percent amino acid difference vs. evolutionary time for, 265–266
in simulated macromolecular evolution, 265
β-Hemoglobin
globin phylogeny and, 185–186
lowest NR length tree for, 136, 139, 141, 144
mouse vs. rabbit mRNA sequences in relation to, 296–298
nucleic acid sequence for, in humans, 300, 303–304
in RNAs of rabbit or mouse, 304–305
β-Hemoglobin chains
insertions and deletions in descent of, 174
percentage amino acid difference vs. evolutionary time for, 266–267
RFAC curves and, 268
Hemoglobin genes
distribution and linkage of, 374
nucleotide sequences of, 375
vs. tumor virus genes VP1 and VP2, 307
Hemoglobin nucleotide tree, 384–386
Higher eutherian phylogeny, 12–27 (*see also* Eutheria)
anatomical and fossil evidence in, 3–35
higher resolution in, 4
molecular studies in, 3–35
Hippomorpa, 144
Hippopotamus, stomach fermentation in, 68–70

Hippotraginae, ribonuclease sequence for, 46–47
Histone evolution, duplications in, 227
Histone H1, 197–198, 212
evolutionary tree of, 214
histone H5 and, 212–214
Histone H2A, 212–217
evolutionary tree of, 215
variants of, 216
Histone H2B, 217–219
evolutionary tree of, 220
variants of, 219
Histone H3, 219–221
evolutionary tree of, 222
variants of, 221
Histone H4, 221–224
evolutionary tree of, 224
matrix of differences for, 223
variants of, 223
Histone H5, 212
evolutionary tree of, 214
Histone superfamily, 225–230
alignments in, 225–227
Hominoidea, genetic relationships within, 144–145
Homo
Anthropoidea ancestor of, 161
Australopithecus and *Ramapithecus* relationships with, 163
Homo-Pan-Gorilla branch, 145, 160
Homo sapiens
ribonuclease sequence for, 48
sequential proteins and, 184
Human β-hemoglobin mRNAs, vs. rabbit, 303–304
Hyracoidea, 25–28, 126
Hystricomorpha, 20
Hystricomorphy rodents
cecal digestion in, 70–71
in most parsimonious tree of ribonucleases, 49–50

Inhibitor proteins, pancreatic ribonuclease reactions with, 64 (*see also* Proteins)
Insectivora, 6, 15, 145
amino acid sequence analysis for, 126
"Archontans" and, 21–25
primates and, 24
Insects, phylogenetic tree of, 252
Intervening gene sequences, in eukaryotic gene expression and organization, 367–373
Introns, eukaryotic genes and, 370–372

"Jumping genes," 346

Lagomorpha, 8, 16, 20–21, 28, 135–136, 138, 140
amino acid sequence analysis of, 124
Primates and, 141, 145
Lagomorphs, 6
elephant shrews (macroscelideans) and, 9, 21
Lamprey myoglobin, tentative tryptic
Lens α-crystallin, gene phylogenies for, alignment of, 171
Late Cretaceous fossils, new evidence of, 8
Lemuriformes, 135
136 (*see also* α-Crystallin; Eye lens protein)
Lens proteins
comparative study of, 78–84
intraspecies variation in, 82
Leptictimorpha, 20
Lorsiformes, 120, 135
Lowest nucleotide replacement length tree
eutherian region of, 164–165
for β-hemoglobin, 136
GD + GE + cytochrome *c* in, 134–136
gene phylogeny and, 132–133
within Hominoidea, 144–145
inferences from, 130–136
Low-molecular-weight nucleotide substrates, enzymic activity on, 58–61

Macromolecular evolution
models of, 252–256
random fixation of neutral mutations in, 273–274
simulated, *see* Simulated macromolecular evolution
simulation model of, *see* RFAC
Macromolecular sequence divergence
rate, fixation, and amount of allowed variation in, 257–258
in real and simulated evolution, 256

Macromolecular sequences
computer simulation of evolution in, by random fixation of allowed codons, 243–276
evolution of, 243–276
Macroscelidea, 8, 20–21, 28
Magnorder Ernotheria, 8
Mammalia, amino acid sequence analysis for, 122–129
Mammalian classification, history of, 5–10
Mammalian fetal membranes, surveys of, 9
Mammalian pancreatic ribonucleases
amino acid sequences of, 44
evolution of, 43–71
Mammalian phylogeny
αA sequences of, 103–104
paleontology role in, 9
Mammalian relationships, subclass and ordinal, 138–141
Mammalian species, ribonuclease sequences for, 46–47
Marsupialia, ribonuclease sequences for, 48
MATTOP program, 201–202
Maximum parsimony method, in amino acid sequences, 123–129 (*see also* Most parsimonious tree)
Megachiroptera, 22
Menotyphyla, 6
Mesozoic animals, discovery of, 7
Metatherian-eutherian divergence, in α-crystallin A sequence, 97
Microchiroptera, 22
Molecular clock hypothesis, evaluation of, 148–152 (*see also* Clock model of protein evolution)
Molecular divergence, nonuniform, *see* Nonuniform molecular divergence
Molecular evolution (*see also* Evolution; Macromolecular evolution)
anagenesis and, 184
gene structure, organization, and expression in, 357–398
Molecular phylogenies, substitution patterns in, 15
Molecular-size proportions, variation in among vertebrate water-soluble lens proteins, 84
Mollusca, amino acid sequence analysis in, 128
Monophyletic groups, evidence for and against, 16
Monotremata, amino acid sequence analysis in, 126
Most parsimonious tree
in α-crystallin A chains, 98–99, 105
of amino acid sequences in mammalian pancreatic ribonucleases, 48–50
of carbonic anhydrase isozymes, 142
of carbonic anhydrase sequences, 140
cytochrome *c* sequences and, 137
eutherian region of, 164–165
of globin sequences, 169–171, 388
in tree construction strategy, 117
Mouse genes, sequence homology between, 342
mRNA, 285
copying of into DNA, 343
sequences for mouse vs. rabbit β-hemoglobin, 297
Muridae, evolutionary rates of pancreatic ribonuclease for, 55
Mutation acceptance rates, for chromosomal proteins, 203, 224–225
Mutation probability matrix, 283
Myoglobin evolution, functional groups in earlier and later stages of, 178
Myoglobin genealogy, Proboscidea-Artiodactyla relationship and, 139
Myoglobin-hemoglobin ancestor, NR lengths for hemoglobin positions in descent of, 176
Myoglobin molecule, slow evolution rate at sites in, 175
Myoglobin sequences, time scales in, 132–133
Myomorpha, 20, 55
Myomorph rodents, digestive system and diet in, 70
Myrmecophagidae, stapes in, 17

Natural selection
calmodulin evolution and, 180
central role of, 167–183
in conservation of perfected adaptations, 175
synchronized evolutionary tempos and, 167–168
NHCP, *see* Nonhistone chromosomal proteins
Non-α-globin gene cluster, structure of, 337

Non-α-globin gene cluster, analysis of, 339
Nonhistone chromosomal proteins
  bovine, 210
  evolutionary tree of, 211
  high-mobility group, 207–211
Nonuniform molecular divergence, 281–312
  accurate genetic distance estimates in, 306–309
  extension to noncoding regions, 305–306
  parameter estimation in, 291–296
NR, *see* Nucleotide replacement lengths
Nuclear magnetic resonance, in measurement of active-site histidines, 58–59
Nuclear protein A24, 212–217
  formation and structure of, 197–198
  histone H2A replacement by, 216
Nucleic acid sequence coding, for proteins, 383
Nucleosome core, 195–199, 229
Nucleosome structure, 194–199
Nucleotide codons, amino acid changing/silent substitution ratios for, 394
Nucleotide replacement lengths, 116, 130–136
  gene duplication + gene expression and, 162–165
  for hemoglobin residue positions, 176
Nucleotide replacement rates, for different globin gene categories, 396
Nucleotide sequencing
  in evolutionary tree building, 382–397
  lowest nucleotide replacement length in, 116, 130–136
Nucleotide substitutions
  distribution and type of, 390
  most parsimonious tree and, 387
  number of in pancreatic ribonuclease, 57
  pattern of, 389–390

Observational classes, 282
Odoicoileinae, ribonuclease sequences for, 46–47
Operational taxonomic units in eutherian orders, 145, 165
"Orders of Mammals, The" (Gregory), 5–6
Orthologous sequences, in gene phylogeny, 118
OTUs, *see* Operational taxonomic units

Paenungulata, 26, 138
Paenungulates, ungulates and, 100–101
PAM matrix, 201–203
Pancreatic ribonucleases
  activity in low-molecular-weight nucleotide substrates, 58–61
  amino acid sequences of, 44
  average number of substrates in, 57
  characteristics of, 57
  covalently attached carbohydrate function in relation to, 64–71
  double-stranded RNA and, 61–63
  evolutionary sites of, 54–55
  evolutionary significance of, 57
  glycosylation of, 64–71
  interaction with inhibitor proteins, 64
  relative affinity for ribonuclease inhibitor from human placenta, 64
  specific activity in RNA, 58–61
Pantodonta, 26
Paralogous sequences, in gene phylogeny, 118
Parameter estimation, in gene sequence, 291–294
Parsimonious globin gene phylogeny, 169
Pecora
  Camelidae separation from 186
  classification of based on feeding ecology and stomach structure, 70
  evolutionary rates of pancreatic ribonuclease for, 54–55
  monophyletic, 143
  ribonuclease sequence for, 46–47
Perissodactyla, 25–28, 100, 131, 135, 138, 140, 145
  amino acid sequence analysis for, 125
  ribonuclease sequence for, 47
  suborders of, 144
Pholidota, 16–19, 138
  α-crystallin A analysis of, 102
  amino acid sequence analysis for, 124
  origin times for, 28
  relationships in, 18–19, 102
Pholidotans, edentates and, 9
Phylogenetic inference, levels of, 120–122
Phylogenetic relationships, for Recent and selected fossil groups, 13–16
Phylogeny
  gene vs. species, 117–120
  teleostan, 9
Pinnipeds, monophyletic nature of, 101–102

Placental development, in eutherians and marsupials, 11
Plantae, amino acid sequence analysis for, 128–129
Plants, phylogenetic tree of, 252
Platyrrhini, 145
Polytomy, evidence of, 30
Ponginae, 145
Primate-lagomorph relationship, 145–147
Primate phylogeny
  α-crystallin A analysis of, 103
  amino acid sequence evidence in, 123, 165–187
Primates
  Archonta and, 6
  insectivorans and, 24
  Lagomorpha and, 145–147
  monophyletic nature of, 120
  origin times for, 28
  Perissodactyla and Artiodactyla in relation to, 131
Proboscidea, 25–28, 138, 145
  amino acid sequence analysis for, 126
  myoglobin genealogy and, 139
Prokaryota, amino acid sequence analysis for, 129
Prokaryote DNA-binding proteins, 203–205
Prosimii, 135
Protamines
  fish, 208–209
  nuclear DNA and, 206–207
Protein alignments, evolutionary trees and, 200–202
Protein evolution
  allowed variation in, 246
  amino acid sequence analysis in, 83–84
  clock model of, 148–161
  rate variations in, 161–167
  tempo and mode of, 161–183
Protein families and superfamilies, 202–203
Protein relationships, in chromosomal proteins, 199–200
Proteins
  amino acid sequences of, 357
  chromosomal, *see* Chromosomal proteins
  as "clocks," 155
  conservative chromosomal, 183
  prokaryote DNA-binding, 203–205
  sequence of in RFAC simulation model, 247
  "structural free degree" of, 248
  viral nucleic acid-binding, 205
Psammechinus miliaris, 214, 219–220, 222, 227
Pseudogenes, 377–382
Pyrotheria, 26

Rabbit α- and β-hemoglobin
  divergence in, 298–299, 305
  mRNAs and, 305
Ramapithecus, *Homo* and, 163
Random evolutionary hit model, 245, 255
  calculations for, 254
  hemoglobin genes vs. VP1 and VP2 virus genes in, 307–308
  nucleotide replacements estimated by, 272
  VP1 genes in, 294–296
  yeast ISO-1 vs. cytochrome *c* ISO-2 genes, 300
Random fixation of allowed codons model, *see* RFAC model
Random point mutation, in RFAC model, 248
Recent fossils, phylogenetic relationships for major groups in, 13–16
Recent Lagomorpha, 8
Recent taxa, higher level patterns among, 9
Recombinant DNA, 115
  cloning in, 358–362
  eukaryotic gene expression and, 364
  new methods of using, 320
  structural genes and, 338
  technology for, 358–364
Recombinant cDNA plasmid, steps in production of, 359
REH, *see* Random evolutionary hit model
RELATE program, 202
Reptilia
  α-hemoglobin sequences in, 131
  amino acid sequence analysis in, 127
  restriction endonucleases, human DNA and, 322
Restriction enzyme, Y chromosome DNA and, 328
Reverse transcriptase, 361
RFAC (random fixation of allowed codons) model, 247–254
  Allowed Amino Acids in, 247–251
  amount of allowed variation vs. "efficiency" of evolution in, 269–273
  branch length times in, 253
  comparison with other models, 254–256

RFAC (random fixation of allowed codons) model (*cont.*)
in computer simulation of macromolecular evolution, 243–276
cytochrome *c* in, 261–264
extreme case of, 276
fibrinopeptides in, 259–261
fixation in, 251
fixations vs. temporal branch length in, 253
input data in, 247
modification of, 276
mutation in, 248
observed variation in, 249
phylogenies in, 251–254
predictive value of, 275
real evolution and, 274
and real macromolecular evolution fluctuations, 275
as simulation model, 247–254
synonymous amino acids and, 248–250
RFAC simulation, doubling technique and, 264
RFAC technique, *see* Simulated macromolecular evolution
Ribonuclease
biological tree and, 50–52
most parsimonious tree and, 48–50
pancreatic, *see* Pancreatic ribonuclease
three-dimensional structure of, 56–58
Ribonuclease absorption, carbohydrate as protection against, 67
Ribonuclease amino acids, prediction of from α-helices and β-sheets, 59
Ribonuclease evolution
classical taxonomy and, 53
evolutionary rates and, 48–50, 52–56
Ribonuclease trees
vs. classical taxonomy, 53
distribution of from five placental mammalian orders, 32
most parsimonious, 48–50
in ribonuclease evolution, 48–52
Ribosomal genes, as tool for systematic comparisons, 335
Ribosomal RNA gene family, map of, 332
RNA
double-stranded, 348
enzymic activity on, 58–61
ribosomal, 331–332
three types of, 331–32
transcription and processing of, 364–367
RNA sequencing, 363–364
RNA splice points, exons and, 371
Rodentia, 6, 16, 20–21, 28, 145
amino acid sequence analysis for, 124
evolutionary pancreatic ribonuclease rates for, 55
monophyletic, 143
ribonuclease sequence for, 47
single-protein genealogies and, 141
Ruminant digestion, carbohydrate/ribonuclease function in, 67
Ruminantia
evolutionary rates of pancreatic ribonuclease in, 54–55
monophyletic, 143

Satellite DNAs, chromosomal organization and, 347–348
Scandentia, 28, 135–136, 140
amino acid sequence analysis for, 124
Sciuromorpha, 20
SEARCH program, 202
Selective constraints (*see also* Gene structure)
base composition and, 284–285
base replacement probabilities in, 285–286
magnitude of, 284–291
Sequence evolution, acceleration-deceleration pattern in, 162 (*see also* Evolution; Molecular evolution; Simulated macromolecular evolution)
Silent substitutions
amino acid changing and, 392–394
as molecular clock, 395
rate constancy in, 397
Simulated macromolecular evolution, 243–276
amount of allowed variation in, 259–266
fixation constant during, 259–266
RFAC simulation model in, 247–254
Single-copy DNA, 336–344 (*see also* DNA)
Sirena, 25–28, 138
amino acid sequence analysis in, 126
Squamata, 13
Species phylogeny, fitting gene phylogeny into, 117–120
S-peptide-bovine S protein complexes, enzymic activity in, 60
Sperm whale myoglobin, lamprey myoglobin and, 171

Stapes, in Edentata, 17–18
Starting DNA sequence, lengths of, 248
Starting protein sequence, in RFAC simulation model, 247
Stochastic theory, constrained, 282–284
Stomach fermentation, in artiodactyl species, 68–70
Stomach forms, radiation of, 45
Stomach structure, Pecora classification based on, 70
Strepsirhini, 135, 145
Strongylocentrotus purpuratus, 214–215, 219–220
Structural free degree, of proteins, 248
Synapomorphic amino acid substitutions, 136
Syncchocystis, DNA binding protein sequence for, 204
Synchronized evolutionary tempos, 167–168
Synonymous amino acids, in RFAC model, 248–250
Synonymous codons, 254, 270
Synopsis Methodica Animalium Quadrapedum et Serpentini Generis (Ray), 5

Tarsiformes, 135
Teleostei, amino acid sequence analysis in, 127
Tetrahymena thermophila, 222–223
Theory, falsifiable, 281
Theria
  "Of eutherian-metatherian grade," 27
  evolutionary pancreatic ribonuclease rates for, 55
Therictoidea, 6
Thermoplasma acidophilum, DNA-binding protein and, 204–205
Thymopoietin II, ubiquitin and, 218
Thymosin polypeptide β, 216
T lymphocytes, ubiquitin and, 217
Transposons, 346
Tree
  evolutionary, *see* Evolutionary tree
  lowest NR length, 116–117, 130–136
  most parsimonious, *see* Most parsimonious tree
Tree construction strategy, 116–122 (*see also* Evolutionary tree)
  rationale of, 116–117
Tubulidentata, 25–28, 126
Tupaioidea, 145

Ubiquitin, 216–218
Unguiculata, 8
Ungulata, 15
Ungulates
  condylarths and, 25
  divergence dates for, 29
  paenungulates and, 100–101
  radiation of, 25–27

Variable codons, density of fixed mutations in, 286–288
Vertebrate α-crystallin A sequences, most parsimonious trees for, 104–107
Vertebrate globins, phylogeny of, 168–175
Vertebrate lens proteins, variation in isoelectric points among, 85–86
Vertebrate-mollusc globin ancestor, descent from to vertebrate myoglobins, 172–173
Vertebrates
  amino acid sequence analysis of, 122–129
  phylogenetic tree of, 252
Villous chorioallantoic placenta, 11
Viral nucleic acid-binding proteins, 205–206
VP1 genes
  of BK and polyoma viruses, 292–293
  BK vs. SV 40 viruses, 301
  hemoglobin genes and, 307
  nonrandom REH theory and, 294–296, 299
  substitution rates within codon in, 293
  and SV40 vs. polyoma viruses, 301
VP2 capsid protein, gene coding for, 300
VP2 genes
  BK virus vs. polyoma, 302–303
  BK vs. SV40 viruses and, 302
  hemoglobin genes and, 307

Whalebone whalers, pancreatic ribonuclease levels in, 61
Whales, amino acid sequence analysis for, 125 (*see also* Cetecea)

Xenarthra, Eutheria and, 30
Xenopus laevis, 373–374
Xenopus tropicalis, 373–374

Yeast ISO-1 vs. ISO-2 cytochrome genes, in nonrandom REH theory, 300

"Zalambdodont insectivores," 20